D1583612

Fundamentals of Heat Engines

Wiley-ASME Press Series

Corrosion and Materials in Hydrocarbon Production: A Compendium of Operational and Engineering Aspects
Bijan Kermani, Don Harrop

Design and Analysis of Centrifugal Compressors
Rene Van den Braembussche

Case Studies in Fluid Mechanics with Sensitivities to Governing Variables
M. Kemal Atesmen

The Monte Carlo Ray-Trace Method in Radiation Heat Transfer and Applied Optics
J. Robert Mahan

Dynamics of Particles and Rigid Bodies: A Self-Learning Approach
Mohammed F. Daqaq

Primer on Engineering Standards, Expanded Textbook Edition
Maan H. Jawad, Owen R. Greulich

Engineering Optimization: Applications, Methods, and Analysis
R. Russell Rhinehart

Compact Heat Exchangers: Analysis, Design and Optimization Using FEM and CFD Approach
C. Ranganayakulu, Kankanhalli N. Seetharamu

Robust Adaptive Control for Fractional-Order Systems with Disturbance and Saturation
Mou Chen, Shuyi Shao, Peng Shi

Robot Manipulator Redundancy Resolution
Yunong Zhang, Long Jin

Stress in ASME Pressure Vessels, Boilers, and Nuclear Components
Maan H. Jawad

Combined Cooling, Heating, and Power Systems: Modeling, Optimization, and Operation
Yang Shi, Mingxi Liu, Fang Fang

Applications of Mathematical Heat Transfer and Fluid Flow Models in Engineering and Medicine
Abram S. Dorfman

Bioprocessing Piping and Equipment Design: A Companion Guide for the ASME BPE Standard
William M. (Bill) Huitt

Nonlinear Regression Modeling for Engineering Applications: Modeling, Model Validation, and Enabling Design of Experiments
R. Russell Rhinehart

Geothermal Heat Pump and Heat Engine Systems: Theory and Practice
Andrew D. Chiasson

Fundamentals of Mechanical Vibrations
Liang-Wu Cai

Introduction to Dynamics and Control in Mechanical Engineering Systems
Cho W. S. To

Fundamentals of Heat Engines

Reciprocating and Gas Turbine Internal Combustion Engines

Jamil Ghojel (PhD)

This Work is a co-publication between John Wiley & Sons Ltd and ASME Press

This edition first published 2020
© 2020 John Wiley & Sons Ltd

This Work is a co-publication between John Wiley & Sons Ltd and ASME Press

All rights reserved. No part of this publication may be reproduced, stored in a retrieval system, or transmitted, in any form or by any means, electronic, mechanical, photocopying, recording or otherwise, except as permitted by law. Advice on how to obtain permission to reuse material from this title is available at http://www.wiley.com/go/permissions.

The right of Jamil Ghojel to be identified as the author of this work has been asserted in accordance with law.

Registered Offices
John Wiley & Sons, Inc., 111 River Street, Hoboken, NJ 07030, USA
John Wiley & Sons Ltd, The Atrium, Southern Gate, Chichester, West Sussex, PO19 8SQ, UK

Editorial Office
The Atrium, Southern Gate, Chichester, West Sussex, PO19 8SQ, UK

For details of our global editorial offices, customer services, and more information about Wiley products visit us at www.wiley.com.

Wiley also publishes its books in a variety of electronic formats and by print-on-demand. Some content that appears in standard print versions of this book may not be available in other formats.

Limit of Liability/Disclaimer of Warranty
In view of ongoing research, equipment modifications, changes in governmental regulations, and the constant flow of information relating to the use of experimental reagents, equipment, and devices, the reader is urged to review and evaluate the information provided in the package insert or instructions for each chemical, piece of equipment, reagent, or device for, among other things, any changes in the instructions or indication of usage and for added warnings and precautions. While the publisher and authors have used their best efforts in preparing this work, they make no representations or warranties with respect to the accuracy or completeness of the contents of this work and specifically disclaim all warranties, including without limitation any implied warranties of merchantability or fitness for a particular purpose. No warranty may be created or extended by sales representatives, written sales materials or promotional statements for this work. The fact that an organization, website, or product is referred to in this work as a citation and/or potential source of further information does not mean that the publisher and authors endorse the information or services the organization, website, or product may provide or recommendations it may make. This work is sold with the understanding that the publisher is not engaged in rendering professional services. The advice and strategies contained herein may not be suitable for your situation. You should consult with a specialist where appropriate. Further, readers should be aware that websites listed in this work may have changed or disappeared between when this work was written and when it is read. Neither the publisher nor authors shall be liable for any loss of profit or any other commercial damages, including but not limited to special, incidental, consequential, or other damages.

Library of Congress Cataloging-in-Publication Data

Names: Ghojel, Jamil, author.
Title: Fundamentals of heat engines: reciprocating and gas turbine internal combustion engines / Jamil Ghojel.
Description: First edition. | Hoboken, NJ, USA : John Wiley & Sons, Inc.,
    2020. | Series: Wiley-ASME press series | Includes bibliographical
    references and index.
Identifiers: LCCN 2019047568 (print) | LCCN 2019047569 (ebook) | ISBN
    9781119548768 (hardback) | ISBN 9781119548782 (adobe pdf) | ISBN
    9781119548799 (epub)
Subjects: LCSH: Heat-engines.
Classification: LCC TJ255 .G45 2020 (print) | LCC TJ255 (ebook) | DDC
    621.402/5 – dc23
LC record available at https://lccn.loc.gov/2019047568
LC ebook record available at https://lccn.loc.gov/2019047569

Cover Design: Wiley
Cover Images: Turbine Blades © serts/Getty Images, Rad sports car silhouette © Arand/Getty Images

Set in 9.5/12.5pt STIXTwoText by SPi Global, Chennai, India

Printed and bound by CPI Group (UK) Ltd, Croydon, CR0 4YY

10 9 8 7 6 5 4 3 2 1

Contents

Series Preface

The Wiley-ASME Press Series in Mechanical Engineering brings together two established leaders in mechanical engineering publishing to deliver high-quality, peer-reviewed books covering topics of current interest to engineers and researchers worldwide.

The series publishes across the breadth of mechanical engineering, comprising research, design and development, and manufacturing. It includes monographs, references and course texts.

Prospective topics include emerging and advanced technologies in Engineering Design; Computer-Aided Design; Energy Conversion & Resources; Heat Transfer; Manufacturing & Processing; Systems & Devices; Renewable Energy; Robotics; and Biotechnology.

Preface

The reciprocating piston engine and the gas turbine engine are two of the most vital and widely used internal combustion heat engines ever invented. Piston engines are still dominant in the areas of land and marine transportation, mining, and agricultural industries. They also play a significant role in light aircraft and stand-by power-generation applications. Power that can be generated by piston engines ranges from a fraction of a kilowatt to more than $80\,MW$, with thermal efficiencies approaching 50%. Gas turbines are dominant in civil and military aviation and play a major role in base, midrange, and peak load electric power generation ranging from small stand-by units up to $300\,MW$ per engine with thermal efficiencies approaching 40% at the upper range and $500\,MW$ in combined cycle configurations with thermal efficiencies approaching 60%. Gas turbines are also ideal as power plants operating in conjunction with large renewable power plants to eliminate intermittency.

Demand for power and mobility in its different forms will continue to increase in the twenty-first century as hundreds of millions of people in the developing world become more affluent, and the cheapest and most efficient means of satisfying this demand will continue to be the heat engine. As a consequence, the heat engine will most likely remain an active area of research and development and engineering education for the foreseeable future. Traditionally, the piston engine has been an ideal tool for teaching mechanical engineering, as it features fundamental principles of the engineering sciences such as thermodynamics, engineering mechanics, fluid mechanics, chemistry (more specifically, thermochemistry), etc. In this book, gas turbine engine theory, which is based on the same engineering principles, is combined with piston engine theory to form a single comprehensive tool for teaching mechanical, aerospace, and automotive engineering in entry- and advanced-level undergraduate courses and entry-level energy-related postgraduate courses. Practicing engineers in industry may also find some of the material in the book beneficial.

The book comprises 3 parts, 15 chapters, and 4 appendices. The first chapter in Part I is a review of some principles of engineering science, and the second chapter covers a wide range of thermochemistry topics. The contribution of engineering science to heat engine theory is fundamental and is manifested over the entire energy-conversion chain, as this figure shows.

| Heat | ⇒ | Mechanical Work | ⇒ | Rotary Power |

↑ ↑ ↑

Thermochemistry | Fluid Mechanics | Engineering Mechanics
| Thermodynamics |
| Thermochemistry |

Part II covers theoretical aspects of the reciprocating piston engine starting with simple air-standard cycles, followed by theoretical cycles of forced induction engines and ending with more realistic cycles that can be used to predict engine performance as a first approximation. Part III on gas turbines also covers cycles with gradually increasing complexity, ending with realistic engine design-point and off-design calculation methods.

Representative problems are given at the end of each chapter, and a detailed example of piston-engine design-point calculations is given in Appendix C. Also, case studies of design-point calculations of gas turbine engines are provided in Chapters 12 and 13.

The book can be adopted for mechanical, aerospace, and automotive engineering courses at different levels using selected material from different chapters at the discretion of instructors.

Jamil Ghojel

Glossary

Symbols

A Area, air, Helmholtz function

a Acceleration, speed of sound, correlation coefficient

B Bulk modulus, correlation coefficient, bypass ratio

C Gas velocity, molar specific heat

c Mass specific heat, speed of sound

D Diameter, degree of reaction in reaction turbines

E Total energy, utilization factor in reaction turbines, modulus of elasticity

F Force, thrust, fuel

f Specific thrust

G Gibbs free energy

g Gravitational acceleration

H Enthalpy, heating value of fuel

h Specific enthalpy, blade height

I Moment of inertia

i Number of cylinders

j Number of strokes

K Degrees Kelvin, equilibrium constant, force, mole ratio of hydrogen to carbon monoxide

L Length

l Length, blade length

M Quantity in moles, Mach number, moment of force

m Mass

$\dot{m}$ Mass flow rate

N Rotational speed in revolution per minute, force

n Polytropic index (exponent), number of moles

p Pressure, cylinder gas pressure

Q Heat transfer, force

q Specific heat transfer

$\dot{Q}$ Rate of heat transfer

R Radius, gas constant, crank radius

$\bar{R}$ Universal gas constant

r Pressure ratio
S Entropy, stroke
s Specific entropy
T Absolute temperature, torque, fundamental dimension of time
t Time, temperature
U Internal energy, blade speed
u Specific internal energy
V Volume, velocity, relative velocity
v Specific volume, piston speed
W Work
$\dot{W}$ Power
w Specific work, blade row width, rate of heat release
x Distance, mass fraction, number of carbon atoms in a fuel, cumulative heat release
$\dot{x}$ Linear velocity
$\ddot{x}$ Linear acceleration
y Number of hydrogen atoms in a fuel
z Number of oxygen atoms in a fuel, height above datum

Greek Symbols

α Angle, pressure ratio in constant-volume combustion, angular acceleration
β Angle, volume ratio in constant-pressure combustion
γ Ratio of specific heats, V-angle (engine crank)
Δ Symbol for difference
δ Expansion ratio in an engine cylinder
ε Compression ratio (volume ratio)
ϵ Heat-exchanger effectiveness
η Efficiency
θ Angle, crank angle
$\dot{\theta}$ Angular velocity
$\ddot{\theta}$ Angular acceleration
κ Compressibility
λ Relative air-fuel ratio
μ Dynamic viscosity, coefficient of molecular change
ν Kinematic viscosity
Π Non-dimensional group
ρ Density, volume ratio during heat rejection at constant volume (generalized air-standard cycle)
σ Stress, rounding-off coefficient in piston engine cycles
τ Ratio of crank radius to connecting rod length
ϕ Flow coefficient, crank angle (Wiebe function), equivalence ratio
φ Angle (Wiebe function), heat utilization coefficient
ψ Loading coefficient, coefficient of molar change
ω Angular velocity, degree of cooling

Subscripts

a	Air, actual, total volume
b	Brake
C	Carbon mass fraction in liquid or solid fuel
c	Compressor, clearance (volume), crank
com	Compressor (volume ratio)
cp	Crank pin
cr	Critical
ct	Compressor turbine
cw	Crank web
e	exit
f	Fuel, frictional, formation
g	Gas, gravimetric
H	Hydrogen mass fraction in liquid or solid fuel
h	Higher
i	Inlet, intake, indicated, species, inertia
l	Liquid, lower
m	Mean
N	Nitrogen mole fraction in gaseous fuel
n	Nozzle
O	Mass fraction of oxygen in liquid or solid fuel
P	Product of combustion
p	Piston, propulsive
pc	Compressor polytropic efficiency
pp	Piston pin
pt	Turbine polytropic efficiency, power turbine
R	Reactants (air plus fuel)
r	Rod (connecting rod)
S	Sulfur mass fraction in liquid or solid fuel
s	Isentropic, stoichiometric, swept (volume)
t	Turbine, total (stagnation) condition
w	Whirl (velocity)

Superscripts

g	Gravimetric
0	Reference state (pressure)
v	Volumetric

Abbreviations

A/F	Air-fuel ratio
AFT	Adiabatic flame temperature
BDC	Bottom dead centre
ca	Crank angle
CI	Compression ignition
F/A	Fuel-air ratio
bmep	Brake mean effective pressure
GT	Gas turbine
bsfc	Brake specific fuel consumption
imep	Indicated mean effective pressure
ICE	Internal combustion engine
isfc	Indicated specific fuel consumption
mep	Mean effective pressure
NI	Natural-induction (engine)
Re	Reynolds number
rpm	Revolutions per minute
SI	Spark ignition
TDC	Top dead centre
TET	Turbine entry temperature

About the Companion Website

The companion website for this book is at

www.wiley.com/go/JamilGhojel_Fundamentals of Heat Engines

The website includes:

- Solution manual for instructors
- PPTs

Scan this QR code to visit the companion website.

Part I

Fundamentals of Engineering Science

Introduction I: Role of Engineering Science

For the last 200 years or so, humans have been living in the epoch of power in which the heat engine has been the dominant device for converting heat to work and power. The development of the heat engine was for most of that time slow and chaotic and carried out mainly by poorly qualified practitioners who had no knowledge of basic theories of energy and energy conversion to mechanical work. In the field of engineering mechanics, drawings of early steam engines depict various, at times strange, inefficient mechanisms to convert steam power to mechanical power, such as the walking beam and sun and planet gear systems. The piston-crank mechanism was first used in a steam engine in 1802 by Oliver Evans (Sandfort 1964) despite a design being proposed as early as 1589 for converting the rotary motion of an animal-driven machine to reciprocating motion in a pump. The first internal combustion engine (ICE) to be made available commercially was Lenoir's gas engine in 1860. This engine was also the first to employ a piston-crank mechanism to convert reciprocating motion of the piston to rotary motion, which has become, despite its shortcomings, a fixed feature and highly efficient mechanism in modern reciprocating engines. However, engine designers were never fully satisfied with this mechanism due to the need to balance numerous parasitic forces generated during operation and were constantly looking for alternative ways of obtaining direct rotary motion. This is said to have been one of the stimuli to develop steam and gas turbines in which a fluid, flowing through blades, causes the shaft to rotate, thus eliminating the need for a crankshaft. The results are smoother operation, lower levels of vibration, and low-cost support structures. All of these developments occurred over a very long period of time with advances in the science of engineering mechanics (more specifically, engineering dynamics), together with other engineering science branches such as fluid mechanics and thermodynamics.

Examples of the principles of fluid mechanics of relevance to the topics of Parts I and II in the book include the momentum equation used to calculate thrust in aircraft gas turbine engines, Bernoulli's equation to calculate flow in the induction manifold of piston engines, and dimensional analysis to determine the characteristics of turbomachinery for gas turbines.

The great scientific breakthroughs in the development of heat-engine theory came with the development of the science of thermodynamics, starting with the pioneering work of Nicolas Sadi Carnot (1796–1832) and followed by the monumental contributions of Rudolf Clausius (1822–1888) and William Thomson (Lord Kelvin, 1824–1907). Ever since, knowledge of thermodynamics has become essential to improving existing heat engine designs

Fundamentals of Heat Engines: Reciprocating and Gas Turbine Internal Combustion Engines, First Edition. Jamil Ghojel.
© 2020 John Wiley & Sons Ltd. This Work is a co-publication between John Wiley & Sons Ltd and ASME Press.
Companion website: www.wiley.com/go/JamilGhojel_Fundamentals of Heat Engines

and developing new types of engine processes for superior economy and reduced emissions. At the same time, the heat engine, particularly the reciprocating ICE, has become an ideal tool for teaching mechanical and automotive engineering, as it features, in addition to thermodynamics, fundamental principles of engineering mechanics and fluid mechanics as discussed earlier.

A chapter on thermochemistry (Chapter 2) is included in Part I, dealing with fuel properties and the chemistry of combustion reactions and the effect of control of the combustion temperature through control of air-fuel ratios in order to preserve the mechanical integrity of engine components. Extensive numerical data on gas properties and adiabatic flame temperature calculations are included, which can be used for preliminary design-point calculations of practical piston and gas turbine engine cycles.

1

Review of Basic Principles

1.1 Engineering Mechanics

Mechanics deals with the response of bodies to the action of forces in general, and *dynamics* is a branch of mechanics that studies bodies in motion. The principles of dynamics can be used, for example, to solve practical problems in aerospace, mechanical, and automotive engineering. These principles are basic to the analysis and design of land, sea, and air transportation vehicles and machinery of all types (pumps, compressors, and reciprocating and gas-turbine internal combustion engines). A review of some principles relevant to heat engines is presented here.

1.1.1 Definitions

Particle. A conceptual body of matter that has mass but negligible size and shape. Any finite physical body (car, plane, rocket, ship, etc.) can be regarded as a particle and its motion modelled by the motion of its centre of mass, provided the body is not rotating. The motion of a particle can be fully described by its location at any instant in time.

Rigid body. An assembly of a large number of particles that remain at a constant distance from each other at all times irrespective of the loads applied. To fully describe the motion of a rigid body, knowledge of both the location and orientation of the body at any instant is required. Gas turbine shafts are rigid bodies that are rotating at high speeds. The reciprocating piston-crank mechanism in piston engines is a complex system comprising rotating crank shaft and sliding piston connected through a rigid rod describing complex irregular motion.

Kinematics. Study of motion without reference to the forces causing the motion and allowing the determination of displacement, velocity, and acceleration of the body.

Kinetics. Study of the relationship between motion and the forces causing the motion, based on Newton's three laws of motion.

1.1.2 Newton's Laws of Motion

According to the *first law*, the momentum of a body keeps it moving in a straight line at a constant speed unless a force is applied to change its direction or speed.

Fundamentals of Heat Engines: Reciprocating and Gas Turbine Internal Combustion Engines, First Edition. Jamil Ghojel.
© 2020 John Wiley & Sons Ltd. This Work is a co-publication between John Wiley & Sons Ltd and ASME Press.
Companion website: www.wiley.com/go/JamilGhojel_Fundamentals of Heat Engines

The second law defines the force that can change the momentum of the body as a vector quantity whose magnitude is the product of mass and acceleration:

$$F = ma \text{ } kg.m/s^2 \text{ or } Newton \text{ } (N) \tag{1.1}$$

Another form of this law that is particularly pertinent to gas turbine practice states that force is equal to the rate of change of momentum or mass flow rate $\dot{m}$ multiplied by velocity change dv (the letter v will be used for velocity exclusively in the mechanics section of this chapter):

$$F = \frac{d(mv)}{dt} = \dot{m}(dv) \text{ } N \tag{1.2}$$

For an aircraft engine, the air flow into the engine diffuser is equal to the forward flight speed v_1, and engine exhaust gases accelerate to velocity v_2 in the engine nozzle. For a mass flow rate $\dot{m}$ of the gases, the thrust is therefore $F = \dot{m}(v_2 - v_1) \text{ } N$.

In heat engines, it is often necessary to use vector algebra to resolve the acting forces to determine the forces of interest that can produce work. For example, the pressure force of the combusting gases in the piston engine, which is the source of cycle work, does not act directly on the crank, as a result of which parasitic forces are generated, causing undesirable phenomena such as piston slap. Resolving the forces at the piston pin determines the force transmitted through the connecting rod to the crank, generating a torque. In a gas turbine, the gas force generated during flow through the blades has a component acting parallel to the turbine axis that causes bearings overload and needs to be balanced to prevent axial displacement of the rotor.

The *third law* simply states that 'for every force there is an equal and opposite reaction force'. In an aircraft jet engine, the change in momentum of a large flow rate of gases between the inlet and outlet of the engine generates a backward force known as *thrust*, which has an equal reaction that propels the aircraft forward.

1.1.3 Rectilinear Work and Energy

A force F does work on a particle when the particle undergoes displacement in the direction of the force:

Work = Force × Displacement (in $N.m$ or *Joule*)

If the force is variable and moving along a straight line,

$$W_{1-2} = \int_{s_1}^{s_2} F \text{ } ds$$

Newton' Second Law for a particle can be written as

$$F = ma = m\left(\frac{dv}{dt}\right) ; \text{hence, the equation for } W_{1-2} \text{ can be written as}$$

$$W_{1-2} = \int_{s_1}^{s_2} m\left(\frac{dv}{dt}\right) ds$$

For an incremental change in distance, $ds = vdt$; hence

$$W_{1-2} = \int_{s_1}^{s_2} m\left(\frac{dv}{dt}\right) vdt = m\int_{C_1}^{C_2} \frac{1}{2} d(v^2)$$

Finally,

$$W_{1-2} = \frac{m}{2}\left(v_2^2 - v_1^2\right) = KE_2 - KE_1 \tag{1.3}$$

The work done by a force is equal to the change in kinetic energy. This equation is the simplest form of the conservation of energy equation.

1.1.4 Circular Motion

Rotary motion is the most convenient means for transferring mechanical power in almost all driving and driven machinery. This is particularly so in heat engine practice where thermal energy is converted to mechanical work, which is then transferred via rotating shaft to a driven machinery (electrical generator, propeller, wheels of a vehicle, pump, etc.). Consider the non-uniform circular motion shown in Figure 1.1, in which particle P at angular position θ has linear tangential velocity v and angular velocity ω.

The components $\dot{x}$ and $\dot{y}$ of velocity v $(=\omega r)$ in the x and y directions are:

$$\dot{x} = -v\sin\theta = -\omega r\sin\theta \tag{1.4a}$$

$$\dot{y} = v\cos\theta = \omega r\cos\theta \tag{1.4b}$$

The accelerations in the same directions are

$$\ddot{x} = \frac{d\dot{x}}{dt} = -r\left(\omega\cos\theta\frac{d\theta}{dt} + \dot{\omega}\sin\theta\right)$$

$$\ddot{x} = \ddot{x}_1 + \ddot{x}_2 = -r\omega^2\cos\theta - r\dot{\omega}\sin\theta \tag{1.5}$$

where $\ddot{x}_1$ and $\ddot{x}_2$ are the first- and second-order acceleration components in the x direction (Figure 1.1b,c).

$$\ddot{y} = \frac{d\dot{y}}{dt} = r\left(-\omega\sin\theta\frac{d\theta}{dt} + \dot{\omega}\cos\theta\right)$$

$$\ddot{y} = \ddot{y}_1 + \ddot{y}_2 = -r\omega^2\sin\theta + r\dot{\omega}\cos\theta \tag{1.6}$$

$\ddot{y}_1$ and $\ddot{y}_2$ are the first-and second-order acceleration components in the y direction.

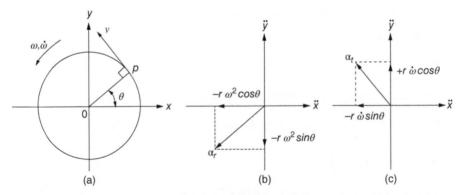

(a)   (b)   (c)

Figure 1.1 Non-uniform circular motion in Cartesian coordinates: (a) initial position and velocity; (b) first-order components of resultant acceleration; (c) second-order components of resultant acceleration.

The first-order components of the resultant acceleration in the radial direction towards 0 is

$$a_r = \dot{v}_1 = \sqrt{\ddot{x}_1^2 + \ddot{y}_1^2} = r\omega^2$$

Since $\omega = v/r$,

$$a_r = \frac{v^2}{r} \qquad (1.7a)$$

Radial acceleration a_r is directed opposite to OP in Figure 1.1b

The second-order components of the resultant acceleration in the tangential direction is

$$a_t = \dot{v}_2 = \sqrt{\ddot{x}_2^2 + \ddot{y}_2^2} = r\dot{\omega}$$

Since the angular acceleration $\alpha = \dot{\omega}$,

$$a_t = r\alpha \qquad (1.7b)$$

Tangential acceleration a_t is directed perpendicular to OP in Figure 1.1c.

The resultant acceleration is

$$a = \sqrt{a_r^2 + a_t^2} = \sqrt{(r\omega^2)^2 + (r\dot{\omega})^2}$$

$$a = \sqrt{\left(\frac{v^2}{r}\right)^2 + (r\alpha)^2} \qquad (1.8)$$

1.1.4.1 Uniform Circular Motion of a Particle

In the uniform circular motion, $r = $ const., $\dot{r} = \ddot{r} = 0$, $\omega = \dot{\theta} = $ const., and $\dot{\omega} = \ddot{\theta} = 0$.

Equations 1.4a, 1.7b, and 1.8 for velocity and acceleration become:

$$v = r\dot{\theta} = r\omega \qquad (1.9)$$

$$a = r\dot{\theta}^2 = r\omega^2 = \frac{v^2}{r} \qquad (1.10)$$

These equations apply to any point on the outer surface of a machinery shaft rotating at constant angular velocity, such as reciprocating and gas turbines engines.

1.1.5 Rotating Rigid-Body Kinetics

The motion of a particle can be fully described by its location at any instant. For a rigid body, on the other hand, knowledge of both the location and orientation of the body at any instant is required for full description of its motion.

The motion of the body about a fixed axis can be determined from the motion of a line in a plane of motion that is perpendicular to the axis of rotation (Figure 1.2). The angular position, displacement, velocity, and acceleration are, respectively, θ, $d\theta$,

$$\omega = d\theta/dt \quad \text{and} \quad \alpha = d\omega/dt = d^2\theta/dt^2$$

The tangential and radial components of the acceleration at P and the resultant acceleration are, respectively,

$$a_t = \alpha r, a_r = \omega^2 r, a = \sqrt{a_t^2 + a_r^2}$$

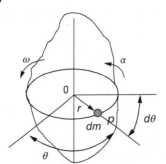

Figure 1.2   Rigid-body rotational motion.

Referring to Figure 1.2, the force required to accelerate mass dm at P is $dF = a_t dm$ and the moment required to accelerate the same mass is $dM = r\,a_t\,dm$.

The resultant moment needed to accelerate the total mass of the rotating rigid body is

$$M = \int dM = \int r\,a_t\,dm = \int r^2 \alpha\,dm$$

For a constant angular acceleration,

$$M = \alpha \int r^2\,dm = I\alpha \tag{1.11}$$

where $I = \int r^2\,dm$ is the moment of inertia of the whole mass of the rigid body rotating about an axis passing through 0. Equation (1.11) indicates that if the body has rotational motion and is being acted upon by moment M, its moment of inertia I is a measure of the resistance of the body to angular acceleration α. In linear motion, the mass m is a measure of the resistance of the body to linear acceleration a when acted upon by force F.

In planar kinetics, the axis chosen for analysis passes through the centre of mass G of the body and is always perpendicular to the plane of motion. The moment of inertia about this axis is I_G. The moment of inertia about an axis that is parallel to the axis passing through the centre of mass is determined using the parallel axis theorem

$$I = I_G + md^2 \tag{1.12}$$

where d is the perpendicular distance between the parallel axes.

For a rigid body of complex shape, the moment of inertia can be defined in terms of the mass m and radius of gyration k such that $I = mk^2$, from which $k = \sqrt{I/m}$. If I is in units of kg. m^2, k will be in metres. The radius of gyration k can be regarded as the distance from the axis to a point in the plane of motion where the total mass must be concentrated to produce the same moment of inertia as does the actual distributed mass of the body, i.e.

$$k = \sqrt{\frac{1}{m} \int r^2\,dm} \tag{1.13}$$

1.1.6   Moment, Couple, and Torque

The moment of force F about a point 0 is the product of the force and the perpendicular distance L of its line of action from 0 (Figure 1.3a):

$$M = F \cos \theta\, L \tag{1.14}$$

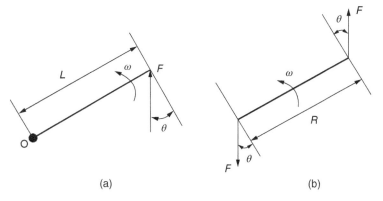

(a)   (b)

Figure 1.3 Definitions of moment, couple, and torque.

A *couple* is a pair of planar forces that are equal in magnitude, opposite in direction, and parallel to each other (Figure 1.3b). Since the resultant force is zero, the couple can only generate rotational motion. The moment of the couple is given by

$$M = F \cos \theta R \tag{1.15}$$

Torque is also a moment and is given by Eq. (1.14), but is used mainly to describe a moment tending to turn or twist a shaft of reciprocating and gas turbine engines, motors, and other rotating machinery. In machinery such as engines, force F will be applied to the arm L at a right angle ($\theta = 0$). In these applications, the power is often expressed in terms of the torque (see Eqs. 1.21 and 1.22 in Section 1.1.9).

1.1.7 Accelerated and Decelerated Shafts

Consider a shaft carrying a gas turbine rotor or piston engine flywheel with the moments and torques acting as shown in Figure 1.4. A heat engine is usually started by means of an external driver such as starting motor by accelerating the driving shaft from rest to the required speed. The driving torque required to accelerate the shaft T_d is balanced by the inertia couple $M_i = I\alpha$ (α is angular acceleration) and resistance couple M_R, which is mainly due to friction in the bearings, as shown in Figure 1.4a. The governing equation is

$$T_b = M_i + M_r = I\alpha + M_r \tag{1.16}$$

Figure 1.4 Kinetics of rotating shaft: (a) accelerating shaft; (b) decelerating shaft.

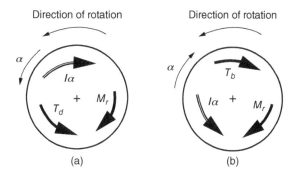

Direction of rotation

Direction of rotation

(a)   (b)

To stop an engine, a braking torque T_b is applied, which is assisted by the resistance moment M_r to decelerate the shaft from the rated speed to rest, as shown in Figure 1.4b The governing equation is

$$T_d + M_r = M_i = I\alpha \tag{1.17}$$

The engine can be brought to rest without applying a braking torque by cutting off the fuel supply and allowing the resistance couple to decelerate the shaft to rest. Note that when the shaft is decelerating, the angular acceleration vector is counter to the direction of rotation of the shaft.

1.1.8 Angular Momentum (Moment of Momentum)

The angular momentum of body about an axis is the moment of its linear momentum about the axis. Figure 1.5 shows a body rotating with angular velocity ω about an axis passing through 0 (perpendicular to the plane of the page):

Linear momentum of particle of mass $dm = dm\,\omega l$
Moment of momentum of particle about $0 = dm\,\omega l^2$
Total momentum H_0 of the body about 0 for constant angular velocity

$$H_0 = \int dm\,\omega l^2 = I_0\omega \tag{1.18}$$

If G is the centre of gravity of the body,

$$I_0 = I_G + mh^2 \tag{1.19}$$

and the angular momentum of the body can be written as

$$H_0 = I_G\omega + mh^2\omega = I_G\omega + mhv \tag{1.20}$$

The angular momentum of a rigid body about any axis remains constant, unless an external torque about the same axis is applied. This is known as the *law of conservation of angular momentum*.

1.1.9 Rotational Work, Power, and Kinetic Energy

If a rigid body rotates through incremental angle $d\theta$ under the action of constant torque T, the incremental rotational work will be

$$dW = Td\theta$$

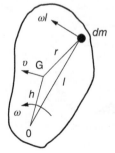

Figure 1.5 Angular momentum of a rigid body.

Table 1.1 Equations of motion for linear and rotational motions.

Linear	Rotational
$s = v_i t + at^2/2$	$\theta = \theta_i t + \alpha t^2/2$
$v_f = v_i + at$	$\theta_f = \theta_i + \alpha t$
$s = (v_i + V_f)t/2$	$\theta = (\omega_i + \omega_f)t/2$
$v_f^2 = v_i^2 2as$	$\omega_f^2 = \omega_i^2 2\alpha\theta$

And the rotational power is

$$\dot{W}_r = T\frac{d\theta}{dr} = T\omega \tag{1.21}$$

Power produced by heat engines is always rotational; hence, subscript *r* will be dropped henceforward. If the angular velocity of the engine shaft is expressed in terms of rotational speed *N* in revolutions per minute (rpm), Eq. (1.21) can be rewritten as

$$\dot{W} = T\omega = \frac{\pi}{30}NT \ W \tag{1.22}$$

The SI unit of power is the *watt* (*W*), but the old unit of *horsepower* (*HP*) is still widely used, where $1 \ HP \equiv 0.746 \ kW$.

The kinetic energy of a particle of mass *dm* (Figure 1.5) is $1/2dm(\omega l)^2$, and the total kinetic energy (*KE*) for the whole rigid body having a constant angular velocity is

$$KE = \frac{1}{2}\omega^2 \int l^2 dm = \frac{1}{2}\omega^2 I_0 \tag{1.23}$$

Making use of Eq. (1.19), we can write

$$KE = \frac{1}{2}\omega^2 I_G + \frac{1}{2}\omega^2 mh^2 = \frac{1}{2}\omega^2 I_G + \frac{1}{2}v^2 m \tag{1.24}$$

Table 1.1 summarises the equations of motion of uniformly accelerating bodies in linear and rotational motion. The following notation is used in the equations:

s, *v*, and *a*: linear displacement, velocity, and acceleration
θ, ω, and α: angular displacement, velocity, and acceleration.
Subscripts *i* and *f* denote *initial* and *final*, respectively.

1.2 Fluid Mechanics

Fluid mechanics deals with the behaviour of a fluid – liquid, gas, or vapour – in quiescent state and in a state of motion. Fluids are substances that cannot preserve a shape of their own. In heat engine processes, the fluids used are predominantly in gas form and include air at various degrees of compression and products of combustion at elevated pressures and temperatures. Understanding the principles of fluid mechanics will help students to better handle the processes in the reciprocating and gas turbine engines.

1.2.1 Fluid Properties

1.2.1.1 Mass and Weight
Mass is a measure of inertia and quantity of the body of matter (fluid), m (kg).

Weight is the force with which a body of the fluid is attracted towards the earth by gravity:

$$w = mg \ N$$

Density is the amount of mass per unit volume:

$$\rho = \frac{m}{V} \ kg/m^3$$

Specific weight is the weight of a unit volume of a substance:

$$\gamma = \frac{w}{V} = \rho g \ N/m^3$$

Specific gravity is

$$sg = \frac{\gamma_f}{\gamma_w @ 4°C} = \frac{\rho_f}{\rho_w @ 4°C}$$

where subscripts f and w are for *fluid* and *water*, respectively.

$$\gamma_w @ 4°C = 9.81 \ kN/m^3$$
$$\rho_w @ 4°C = 1000 \ kg/m^3$$

1.2.1.2 Pressure
Pressure is the force exerted by a fluid on a unit area of its surroundings:

$$p = \frac{F}{A} \ N/m^2 \text{ or } Pa$$

Pressure acts perpendicular to the walls of the container surrounding the fluid. A column of fluid of height h m having a cross sectional area of A m^2 and density ρ kg/m^3 will exert a pressure of

$$p = \frac{hA\rho g}{A} = h\rho g = \gamma h \ kPa$$

1.2.1.3 Compressibility
Compressibility is the change in volume of a substance when subjected to a change in pressure exerted on it. The usual parameter used to measure compressibility of liquids is the bulk modulus of elasticity E:

$$E = \frac{-\Delta p}{(\Delta V)/V} \ N/m^2$$

The compressibility of a gas at constant temperature is defined as

$$\kappa = -\frac{1}{v}\left(\frac{\partial v}{\partial p}\right)_T$$

For a perfect gas:

$$\kappa = \frac{1}{p} \ m^2/N$$

1.2.1.4 Viscosity

Generally, the shearing stress τ developed in a moving fluid between a stationary surface and a moving fluid body is proportional to the velocity gradient $\Delta v/\Delta y$, and the constant of proportionality is the dynamic viscosity μ:

$$\tau = \mu \frac{\Delta v}{\Delta y}$$

Fluids such as water, oil, gasoline, alcohol, kerosene, benzene, and glycerine behave in accordance with this equation and are known as *Newtonian* fluids. Fluids that behave otherwise (viscosity changes with stress) are known as *non-Newtonian* fluids.

The previous equation can be rewritten in terms of the viscosity as

$$\mu = \frac{\tau}{\Delta v/\Delta y} = \tau \left(\frac{\Delta y}{\Delta v} \right)$$

The units of μ can be developed as follows:

$$\left(\frac{N}{m^2} \right) \left(\frac{m}{m/s} \right) \to \left(\frac{N}{m^2} \right) s \to Pa.s \to \left(\frac{kg.m}{s^2} \right) \left(\frac{1}{m^2} \right) s \to \frac{kg}{m.s}$$

The ratio of dynamic viscosity to density of the fluid is the kinematic viscosity v:

$$v = \frac{\mu}{\rho} \; m^2/s$$

Viscosity of liquids decreases with increasing temperature, and that of gases increases with increasing temperature.

1.2.2 Fluid Flow

If a fluid body with cross-sectional area A is flowing at velocity C, its volumetric flow rate Q is given by

$$Q = AC \; m^3/s \tag{1.25}$$

And the mass flow rate $\dot{m}$ is given by

$$\dot{m} = Q\rho \; kg/s \tag{1.26}$$

Consider now the flow of this fluid through the control volume shown in Figure 1.6. The mass flow equations at inlet 1 and exit 2 are given by

$$\dot{m}_1 = \rho_1 A_1 C_1 \text{ and } \dot{m}_2 = \rho_2 A_2 C_2$$

The continuity equation or equation of conservation of mass for this flow is obtained by equating the mass flow rates at sections 1 and 2, $\dot{m}_1 = \dot{m}_2$, or

$$\rho_1 A_1 C_1 = \rho_2 A_2 C_2 \tag{1.27}$$

The total energy (in units of $N. m$) for an element of fluid of mass m at sections 1 and 2 of the control volume shown in Figure 1.6 is given by

$$E_1 = \frac{mp_1}{\rho} + \frac{mC_1^2}{2} + mgz_1 \text{ and } E_2 = \frac{mp_2}{\rho} + \frac{mC_2^2}{2} + mgz_2$$

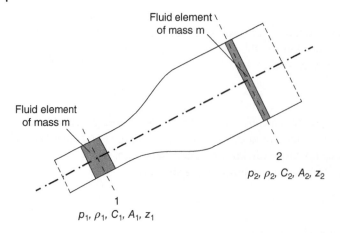

Figure 1.6 Fluid flow through a control volume.

where

mp/ρ : flow energy required to move fluid element m against pressure p
$mC^2/2$: kinetic energy of element m travelling at velocity C
mgz : potential energy of the element due to its elevation z relative to a reference level

If there is no energy addition, storage, or loss between sections 1 and 2, the energy will be conserved, and $E_1 = E_2$:

$$\frac{p_1}{\rho} + \frac{C_1^2}{2} + gz_1 = \frac{p_2}{\rho} + \frac{C_2^2}{2} + gz_2 \tag{1.28}$$

or, in terms of pressure heads (in metres, for example)

$$\frac{p_1}{\rho g} + \frac{C_1^2}{2g} + z_1 = \frac{p_2}{\rho g} + \frac{C_2^2}{2g} + z_2 \tag{1.29}$$

If both sides of Eq. (1.28) are multiplied by ρ, it can be rewritten in terms of fluid pressure as

$$p + \frac{1}{2}\rho C^2 + \rho gz = \text{const.} \tag{1.30}$$

Equation (1.30) is known as Bernoulli's equation. If it is rewritten in differential form, it gives Euler's equation:

$$dp = C\rho dC = 0 \tag{1.31}$$

1.2.2.1 General Energy Equation

If there is energy addition, storage, or loss between sections 1 and 2 in Figure 1.6, the energy equation can be written as

$$\left(\frac{p_1}{\rho g} + \frac{C_1^2}{2g} + z_1\right) + \sum f_l = \left(\frac{p_2}{\rho g} + \frac{C_2^2}{2g} + z_2\right) \tag{1.32}$$

where $\sum f_l$ is the algebraic sum of all losses and gains between points 1 and 2. These could include mechanical energy gained from a booster pump, mechanical energy lost by running a fluid motor or turbine, and energy lost due to friction in the control volume.

1.2.3 Acoustic Velocity (Speed of Sound)

The acoustic velocity of a fluid is the speed of sound in the fluid under isentropic conditions and is given by

$$a = \sqrt{\left(\frac{dp}{d\rho}\right)_{isen.}} = \sqrt{\gamma R T} \tag{1.33}$$

For a gas of molecular mass μ_f, $R = \overline{R}/\mu_f$, where $\overline{R}$ is the universal gas constant (=8314.4 $J/kmole\ K$).

Rewriting Eq. (1.33) in terms of the universal gas constant $\overline{R}$,

$$a = \sqrt{\gamma \frac{\overline{R}}{\mu_f} T} \tag{1.34}$$

At a given temperature and ratio of specific heats, the acoustic velocity can be written as

$$a \propto \sqrt{\frac{\overline{R}}{\mu_f}}$$

The Mach number M_a (in honour of Ernst Mach) is defined as

$$M_a = \frac{C}{a} \tag{1.35}$$

The Mach number is used to indicate speed of flow or forward speed of aircraft and rockets and also to indicate different flow regimes:

Mach number	Flow regime
$M_a < 1$	Subsonic flow
$M_a = 1$	Sonic flow
$1 < M_a < 5$	Supersonic flow
$M_a > 5$	Hypersonic flow

1.2.4 Similitude and Dimensional Analysis

Many problems in fluid mechanics can be solved analytically; however, in a large number of cases, problems can only be solved by experimentation. Similitude and dimensional analysis make it possible to use measurements obtained in a laboratory under specific conditions to describe the behaviour of other similar systems without the need for further experimentation.

1.2.4.1 Dimensional Analysis

Dimensional analysis is based on representing physical quantities with a combination of fundamental dimensions, noting that units of two sides of an equation must be consistent.

In fluid mechanics, as in other branches of engineering sciences, the fundamental dimensions are mass (M), length (L), and time (T). Temperature, if applicable, can be assigned a fundamental dimension such as (θ). These fundamental dimensions can be used to provide qualitative descriptions of physical quantities: for example, velocity can be described as

Table 1.2 Symbols, units, and dimensions of common physical quantities.

Quantity	Symbol	Units	Dimensions
Length	l	m	L
Time	t	s	T
Mass	m	kg	M
Force	F	N	MLT^{-2}
Temperature	T	K	θ
Velocity	C or V	m/s	LT^{-1}
Volume	m^3	m^3	L^3
Acceleration	a	m/s^2	LT^{-2}
Angular velocity	ω	rad	T^{-1}
Area	m^2	m^2	L^2
Volume flow rate	$\dot{Q}$	m^3/s	$L^3 T^{-1}$
Mass flow rate	$\dot{m}$	kg/s	MT^{-1}
Pressure	p	N/m^2	$ML^{-1}T^{-2}$
Density	ρ	kg/m^3	ML^{-3}
Specific weight	γ	N/m^3	$ML^{-2}T^{-2}$
Dynamic viscosity	μ	$N.s/m^2$	$ML^{-1}T^{-1}$
Kinematic viscosity	ν	m^2/s	$L^2 T^{-1}$
Work	W	J	$ML^2 T^{-2}$
Power	$\dot{W}$	W	$ML^2 T^{-3}$
Surface tension	σ	N/m	MT^{-2}
Bulk modulus	B	N/m^2	$ML^{-1}T^{-2}$
Momentum	G	$kg.m/s$	MLT^{-1}
Torque, moment of force	T, M	$N.m$	$ML^2 T^{-2}$

LT^{-1}, density as MT^{-3}, and so on. Table 1.2 lists the symbols, units, and dimensions of common physical quantities. For effective application of dimensional analysis, it is essential to state which independent variables are relevant to the problem.

In cases where temperature is a basic physical quantity and it is preferable to avoid using an extra fundamental dimension such as θ, the gas constant is usually lumped together with the temperature, and the combined variable $RT = p/\rho$ (from the equation of state) will have the dimensions

$$\frac{ML^{-1}T^{-2}}{ML^{-3}} = L^2 T^{-2}$$

1.2.4.2 Buckingham Pi (π) Theorem
The theory states that 'A relationship between m different variables can be reduced to a relationship between $m - n$ dimensionless groups in terms of the n fundamental units'. If

$$u_1 = f(u_2, u_3, u_4, u_5 \ldots u_m),$$

or

$$f(u_1, u_2, u_3, u_4, u_5 \dots .u_m) = 0 \tag{1.36}$$

then

$$\Pi_1 = f(\Pi_2, \Pi_3, \Pi_4, \dots .\Pi_{m-n}),$$

or

$$f(\Pi_1, \Pi_2, \Pi_3, \Pi_4, \dots .\Pi_{m-n}) = 0 \tag{1.37}$$

Consider the steady flow of an incompressible fluid through a long, smooth, horizontal pipe. The pressure drop per unit length Δp caused by friction can be written as a general mathematical function:

$$\Delta p = f(D, \rho, \mu, C)$$

where

D: pipe diameter, m
ρ: fluid density, kg/m^3
μ: dynamic viscosity of the fluid, $kg/m.s$
C: average fluid velocity, m/s

Experimental solution of this problem would require changing one variable while keeping the other three constant and plotting the results on four graphs. The downside of this approach to solving this problem is that the plots are valid only for a specific fluid and pipe, and the obtained results are difficult to fit to a general functional relationship.

According to the Buckingham theorem, for five variables ($m = 5$) and three fundamental units M, L, and T ($n = 3$), there will be $m - n = 2$ nondimensional Π groups. The dimensions of the independent variables are

Independent variable	Δp	D	C	μ	ρ
Dimensions	$ML^{-2}T^{-2}$	L	LT^{-1}	$ML^{-1}T^{-1}$	ML^{-3}

First, n repeating variables that are dimensionally the simplest are selected (three in this case) – D, ρ, and C – and form the first Π group as

$$\Pi_1 = \Delta p D^a C^b \rho^c,$$

or as

$$M^0 L^0 T^0 = (ML^{-2}T^{-2})(L)^a(LT^{-1})^b(ML^{-3})^c$$

For M: $0 = 1 + c$
For L: $0 = -2 + a + b - 3c$
For T: $0 = -2 - b$

Solving these simultaneous equations, we obtain $a = 1, b = -2, c = -1$ and

$$\Pi_1 = \Delta p D^1 C^{-2} \rho^{-1} = \frac{\Delta p D}{\rho C^2}$$

The second Π group is

$$\Pi_2 = \mu D^a C^b \rho^c$$

or

$$M^0 L^0 T^0 = (ML^{-1}T^{-1})(L)^a(LT^{-1})^b(ML^{-3})^c$$

For M: $0 = 1 + c$

For L: $0 = -1 + a + b - 3c$

For T: $0 = -1 - b$

Solving these simultaneous equations, we obtain $a = -1, b = -1, c = -1$ and

$$\Pi_2 = \mu D^{-1} C^{-1} \rho^{-1} = \frac{\mu}{DC\rho}$$

The final functional form can be written as

$$\Pi_1 = f(\Pi_2)$$

or

$$\frac{\Delta p D}{\rho C^2} = f\left(\frac{\mu}{DC\rho}\right)$$

The Reynolds number $Re = DC\rho/\mu$; hence, the functional relationship can be written as

$$\frac{\Delta p D}{\rho C^2} = \psi(Re)$$

Dimensional analysis will not provide the forms of the functions f and ψ. These can be obtained only from carefully set experiments.

This methodology is used in turbomachinery to develop functions for compressor and turbine performance characteristics, as will be discussed in Chapter 14. The function that was found reasonable for compressors is

$$f(D, N, \dot{m}, p_{1t}, p_{2t}, RT_{1t}, RT_{2t}) = 0 \tag{1.38}$$

where D is the impeller diameter, N is the rotational speed (usually rpm), $\dot{m}$ is the mass flow rate of the fluid (usually air), p_{1t} and T_{1t} are the total pressure and temperature at the compressor inlet, and p_{2t} and T_{2t} are the total pressure and temperature at the compressor outlet (Saravanamuttoo et al. 2001). According to the Buckingham theorem, the number of nondimensional Π groups that can be formed is $m - n = 7 - 3 = 4$. These can be shown to be

$$f\left(\frac{p_{2t}}{p_{1t}}, \frac{T_{2t}}{T_{1t}}, \frac{\dot{m}\sqrt{RT_{1t}}}{D^2 p_{1t}}, \frac{ND}{\sqrt{RT_{1t}}}\right) = 0 \tag{1.39a}$$

For a compressor with fixed size and specified gas, the terms R and D can be dropped and Eq. (1.39a) becomes

$$\psi\left(\frac{p_{2t}}{p_{1t}}, \frac{T_{2t}}{T_{1t}}, \frac{\dot{m}\sqrt{T_{1t}}}{p_{1t}}, \frac{N}{\sqrt{T_{1t}}}\right) = 0 \tag{1.39b}$$

The graphical form of this function is obtained by plotting one group, such as the compressor pressure ratio p_{2t}/p_{1t}, against the mass flow rate parameter $\dot{m}\sqrt{T_{1t}}/p_{1t}$ for a number of constant values of compressor speed parameter $N/\sqrt{T_{1t}}$. The physical interpretations of the mass flow rate and speed parameters will be explained in the following section.

1.3 Thermodynamics

Thermodynamics is the study of the interaction of heat and fluids in motion. It is closely related to work-producing or work-absorbing machines such as engines, refrigerators, and compressors and the working substances used in each machine. *Substances* are fluids that are capable of phase change (liquid to gas, gas to liquid, liquid to solid) and are expandable and compressible. The most widely used working substances are shown in the following table:

Machine	Working substance
Air compressor	Air
Reciprocating and gas turbine engines	Combustion products
Steam turbines	Water vapour (steam)
Refrigerators	Ammonia, CFCs, HCFCs

1.3.1 Work and Heat as Different Forms of Energy

Energy is the capacity a body or substance possesses that can result in the performance of mechanical work. The presence of energy can only be observed by its effects, which can appear in different forms such as work and heat. Other forms of energy include potential, kinetic, and internal. Energy cannot be created or destroyed, but it can be converted from one form to another. Mechanical *work* is the product of force and distance:

$$W = FS$$

If F and S are at right angles to each other, no work can be done.

Heat is a manifestation of the degree of agitation of atoms and molecules making up a body as its temperature changes. It is also the effect of one system on another as a result of temperature inequality. Alternately, it can be said that heat is interaction that is not work.

$$Q \propto \Delta T$$

A *system* is a region in space with boundaries across which matter, work, and heat can cross, as shown schematically in Figure 1.7. An example of a real system is a mass of gas or vapour contained in an engine cylinder, the boundary of which is comprised of the cylinder walls, cylinder head, and piston crown when the valves are closed.

Figure 1.7 Schematic diagram of a thermodynamic system.

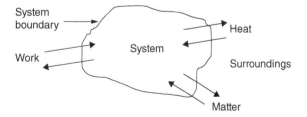

There are two types of systems:

Open system in which matter, heat, and work can pass through the system boundary
Closed system in which neither matter, heat, nor work pass through the system boundary

The *state* of the working fluid (matter) in a system is fully defined by two independent properties.

A *property* is a measurable characteristic of a system such as pressure, volume, temperature, energy, density, or a combination of the former such as internal energy, enthalpy, and entropy. Some measured properties were defined in Section 1.2, and temperature will be discussed in detail later, so we will define the derived (or combination) properties here.

Specific internal energy u is the intrinsic energy per unit mass of a fluid that is not in motion. Its value depends on its pressure and temperature unless the fluid is a perfect gas, in which case the specific internal energy will be dependent on temperature only. For a perfect gas, $u = c_v T$
and for a mass m of the fluid,

$$U = mc_v T$$

Enthalpy is defined as $H = U + pV$ kJ. Specific enthalpy is $h = u + pv$ kJ/kg.

Entropy can be defined from the equation $dQ = Tds$, where T is the absolute thermodynamic temperature and s is the property entropy in units of $J/mole. K$.

A *process* is a change in a system from one state to another. Hence, the change in the pressure and temperature of a mass of gas from (p_1, T_1) to (p_2, T_2) is a process. The state defined by (p_1, T_1) is the initial state of the system, and the state defined by (p_2, T_2) is the final state of the system. Processes can be reversible or irreversible. In a *reversible* process, both the system and the surroundings are returned to their original conditions after the process and the reverse process had been carried out. In an *irreversible* process, reversal cannot be effected without leaving some change in the system or surroundings. Any process with friction (dry friction or viscous friction) is irreversible because some energy of the system is expended in overcoming friction and dissipated in the surroundings.

1.3.2 Mixture of Gases

For a mixture of N gaseous components, the total mass and total moles are, respectively,

$$m_{mix} = m_1 + m_2 + m_3 + \ldots + m_N = \sum_{i=1}^{N} m_i$$

$$n_{mix} = n_1 + n_2 + n_3 + \ldots + n_N = \sum_{i=1}^{N} n_i$$

where
$\quad m_i$: mass of the ith component
$\quad n_i$: amount of substance in moles of the ith component
$\quad$If the molecular mass of ith component is μ_i, then

$$m_i = \mu_i n_i \tag{1.40}$$

The mass fraction (or mass concentration) is

$$c_i = \frac{m_i}{m_{mix}} = \frac{n_i \mu_i}{\sum n_i \mu_i} \tag{1.41}$$

Similarly, the mole fraction (or mole concentration) can be found as follows:

$$y_i = \frac{n_i}{n_{mix}} = \frac{m_i/\mu_i}{\sum m_i/\mu_i} = \frac{\dfrac{m_i}{\mu_i m_{mix}}}{\sum \dfrac{m_i}{\mu_i}} = \frac{\dfrac{1}{\mu_i}\left(\dfrac{m_i}{m_{mix}}\right)}{\sum \dfrac{1}{\mu_i}\left(\dfrac{m_i}{m_{mix}}\right)} = \frac{c_i/\mu_i}{\sum c_i/\mu_i} \tag{1.42}$$

The molar mass of the total mixture is

$$\mu_{mix} = \frac{m_{mix}}{n_{mix}} = \frac{\sum m_i}{n_{mix}} = \frac{\sum n_i \mu_i}{n_{mix}} = \sum y_i \mu_i, \text{ or}$$

$$\mu_{mix} = \frac{m_{mix}}{n_{mix}} = \frac{m_{mix}}{\sum \dfrac{m_i}{\mu_i}} = \frac{1}{\dfrac{1}{m_{mix}}\sum \dfrac{m_i}{\mu_i}} = \frac{1}{\sum \dfrac{1}{\mu_i}\left(\dfrac{m_i}{m_{mix}}\right)} = \frac{1}{\sum \dfrac{c_i}{\mu_i}} \tag{1.43}$$

From Eqs. (1.40), (1.41), and (1.42)

$$c_i = y_i \mu_i / \mu_{mix} \tag{1.44}$$

Example 1.1 A gas mixture has the following mass composition:

$CO_2 - 17.55\% \; O_2 - 4.26\%$

$N_2 - 76.33\% \; CO - 1.86\%$

Determine the molar composition of the mixture.

Solution

$$y_i = \frac{c_i/\mu_i}{\sum c_i/\mu_i}$$

Gas	% Mass fraction	Mass fraction, c_i	Molecular mass, μ_i	Mole fraction, c_i/μ_i	% Mole fraction
CO_2	17.55	0.175 5	44	0.003 99	$100 \times \dfrac{0.00399}{0.03324} = 12.0$
O_2	4.26	0.042 6	32	0.001 33	$100 \times \dfrac{0.00133}{0.03324} = 4.0$
N_2	76.33	0.763 3	28	0.027 26	$100 \times \dfrac{0.02726}{0.03324} = 82.0$
CO	1.86	0.018 6	28	0.000 66	$100 \times \dfrac{0.00066}{0.03324} = 2.0$
	100	$\sum c_i = 1.0$		$\sum c_i/\mu_i = 0.03324$	Total = 100

1.3.2.1 Dalton Model of Gas Mixtures

If a gas mixture of two components A and B is at pressure p and temperature T in a container with volume V, each gas in the mixture exists separately and independently at the temperature and volume of the mixture, and their respective pressures are p_A and p_B. For the mixture

$$pV = n\overline{R}T$$
$$n = n_A + n_B$$

For the components,

$$p_A V = n_A \overline{R}T$$
$$p_B V = n_B \overline{R}T$$

Since $n = n_A + n_B$,

$$\frac{pV}{\overline{R}T} = \frac{p_A V}{\overline{R}T} + \frac{p_B V}{\overline{R}T},$$

or

$$p = p_A + p_B \qquad (1.45)$$

p_A and p_B are known as the *partial pressures.*

$$p_A = n_A \frac{\overline{R}T}{V}$$

$$p_B = n_B \frac{\overline{R}T}{V}$$

But $\quad \dfrac{\overline{R}T}{V} = \dfrac{p}{n}$

Therefore,

$$p_A = \frac{n_A}{n} p = y_A p, \text{ and } p_{AB} = \frac{n_B}{n} p = y_B p$$

It can be shown that the internal energy and enthalpy of a mixture of two gases (A and B) can be written as

$$U = mu = m(c_A u_A + c_B u_B)$$
$$H = mh = m(c_A h_A + c_B h_B)$$

The gas constants for the *i*th component and gas mixture are, respectively,

$$R_i = \frac{\overline{R}}{\mu_i}, R_{mix} = \frac{\overline{R}}{\mu_{mix}}$$

Using Eq. (1.43), we obtain

$$R_{mix} = \frac{\overline{R}}{\dfrac{1}{\sum \dfrac{c_i}{\mu_i}}} = \overline{R} \sum \frac{c_i}{\mu_i} \qquad (1.46)$$

For the two-gas mixture

$$R_{mix} = \overline{R}\left(\frac{c_A}{\mu_A} + \frac{c_B}{\mu_B}\right) = \left(c_A \frac{\overline{R}}{\mu_A} + c_B \frac{\overline{R}}{\mu_B}\right) = c_A R_A + c_B R_B$$

1.3.3 Processes in Ideal Gas Systems

The state of a gas may be completely specified by combining three variable properties (pressure p, temperature T, and volume V) in an equation called the *ideal gas equation of state*:

$$
\left.
\begin{aligned}
pV &= mRT \\
pv &= RT \\
pV &= n\overline{R}T
\end{aligned}
\right\}
\tag{1.47}
$$

The terms in Eq. (1.47) are as follows:

Pressure p is in N/m^2 or Pa.
Volume V is in m^3.
Mass m is in kg.
Temperature T is in K.
Specific volume v is in m^3/kg.
Molar amount of gas n is in $kmole$ (1 kmole of any gaseous substance occupies a volume of 22.41 m^3 at the standard temperature $0°C$ and pressure 101.325 Pa).
Gas constant $R = \overline{R}/\mu$ in $J/kg. K$.
Molecular mass of any gas μ is in $kg/kmole$.
Universal gas constant $\overline{R} = 8314.4 \, J/kmole.K$

1.3.3.1 Adiabatic Processes

An *adiabatic* process is one where the energy of the system changes only by means of work transfer, and there is no heat crossing the boundary. The relationship between the state properties can be written as

$$
pV^\gamma = \text{const.}, \quad TV^{\gamma-1} = \text{const.}, \quad T/p^{(\gamma-1)/\gamma} = \text{const.}
\tag{1.48}
$$

The volume V can be replaced by the specific volume $v = V/m$, which yields the additional equation $p_1/p_2 = (v_2/v_1)^\gamma$. The exponent γ is the ratio of specific heat capacities. The *specific heat capacity* (the word *capacity* will be dropped in future references) is defined as the amount of heat energy required to raise the temperature of a unit quantity of matter by one degree Celsius (on a mass basis c in $J/kg. K$ and on a mole basis C in $J/kmole. K$). The specific heat at constant pressure is written as c_p in $J/kg. K$ or C_p in $J/kmole. K$ and at constant volume as c_v in $J/kg. K$ or C_v in $J/kmole. K$. Both c_p and c_v increase with temperature. Table A.1 in Appendix A shows the molar specific heats at constant pressure of some gases as a function of temperature. The specific heat at constant volume can be determined from the following equations, assuming the gases behave as ideal gases:

$$
c_p - c_v = R, C_p - C_v = \overline{R}
$$
$$
\gamma = c_p/c_v = C_p/C_v
$$
$$
c_p = \gamma R/(\gamma - 1), C_p = \gamma \overline{R}/(\gamma - 1)
$$

1.3.3.2 Heat-Only Process

A hot object tends to cool to the temperature of colder surroundings, and a cold object is warmed to the temperature of hotter surroundings. The phenomenon is caused by the heat-transfer process, in which the energy of a system changes while no work is done on or by the system (no work exchange with the surroundings). This process can occur with the volume remaining constant, and the change of energy in the system for an ideal gas is then

$$(\Delta Q)_{V=const} = mc_v \Delta T = \Delta U \tag{1.49}$$

ΔU is the internal energy of the system.

If the pressure remains constant during the process, the change of energy in the system is

$$(\Delta Q)_{p=const} = mc_p \Delta T \tag{1.50}$$

The source of the heat in both these cases could be external or internal. Examples of processes with internal heat sources are the spark ignition engine (constant-volume combustion of the fuel) and the gas turbine combustor (constant-pressure combustion of the fuel).

1.3.3.3 Isothermal Process

An isothermal process takes place at constant temperature, and the equation of state can be written as

$$PV = \text{const.} \tag{1.51}$$

1.3.3.4 Isochoric Process

An isochoric process is a constant-volume process for which the equation of state is reduced to

$$P/T = \text{const.} \tag{1.52}$$

1.3.3.5 Polytropic Process

A process is referred to as *polytropic* when it deviates from the adiabatic as a result of heat crossing the boundary in addition to work. The relationship is similar to the adiabatic with the adiabatic exponent γ replaced by a polytropic exponent n $(n < \gamma)$:

$$PV^n = \text{const.,} \ TV^{n-1} = \text{const.,} \ \ T/p^{(n-1)/n} = \text{const.} \tag{1.53}$$

The volume V can also be written in terms of specific volumes $v = V/m$, which yields the additional equation $p_1/p_2 = (v_2/v_1)^n$.

1.3.4 Cycles

When a fluid undergoes a series of processes and then returns to its initial state, the fluid executes a thermodynamic *cycle*. A cycle that consists only of reversible processes is a reversible cycle.

To illustrate the application of some of the processes discussed earlier when calculating cycles, let us consider the theoretical thermodynamic cycle shown in Figure 1.8, which is known as the Diesel cycle in honour of the German inventor Rudolf Diesel.

The cycle has two heat-only processes and two polytropic processes:

Process 2 – 3: Constant-pressure heat addition $Q_{in} = mc_p(T_3 - T_2)$.
Process 4 – 1: Constant-volume heat rejection $Q_{out} = mc_v(T_4 - T_1)$.
Process 1 – 2: Adiabatic compression $PV^\gamma = C$ (C is the constant in Eq. 1.48).

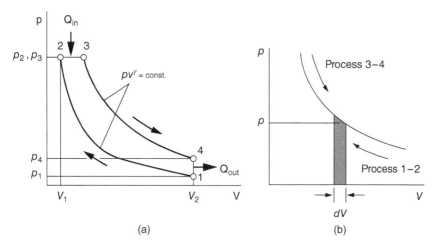

(a)   (b)

Figure 1.8 Application of process equations in theoretical cycles: (a) Diesel cycle; (b) calculation scheme for compression and expansion work.

The magnitude of the compression work done on the gas can be determined as shown in Figure 1.8b:

$$W_C = \int_{V_1}^{V_2} p \, dV$$

where

p: average pressure for the shaded area

dV: volume increment

Since $p = C/V^\gamma$,

$$W_C = C \int_{V_1}^{V_2} \frac{dV}{V^\gamma} = \frac{C}{-\gamma + 1} [V^{-\gamma+1}]_{V_1}^{V_2} = \frac{-C}{\gamma - 1} \left(\frac{1}{V_2^{\gamma-1}} - \frac{1}{V_1^{\gamma-1}} \right)$$

Now

$$C = p_1 V_1^\gamma = p_2 V_2^\gamma,$$

and hence,

$$W_C = \frac{1}{\gamma - 1} (p_1 V - p_2 V_2) \tag{1.54}$$

Since $pV = mRT$, the compression work can be rewritten as

$$W_C = \frac{mR}{\gamma - 1} (T_1 - T_2) \tag{1.55}$$

Process 3 – 4: Adiabatic expansion $PV^\gamma = C$. The magnitude of the expansion work done by the gas can be deduced in a similar way:

$$W_E = \frac{1}{\gamma - 1} (p_3 V_3 - p_4 V_4) \tag{1.56}$$

or

$$W_E = \frac{mR}{\gamma - 1} (T_3 - T_4) \tag{1.57}$$

If the sign convention given in Figure 1.9 is used, the compression work is negative work and the expansion work is positive work.

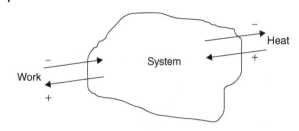

Figure 1.9 Sign convention for heat and work.

1.3.5 First Law of Thermodynamics

This law is a statement of the law of conservation of energy: when a system undergoes a thermodynamic cycle, the net heat exchange $\sum Q$ between the system and its surroundings plus the net work exchange $\sum W$ between the system and the surroundings is equal to zero. This implies that energy can neither be created nor destroyed in a system; it can only be converted from one form to another:

$$\sum Q + \sum W = 0$$

Considered here are applications of the first law in two engineering energy systems: non-flow system and steady-flow system.

1.3.5.1 Non-Flow Energy Equation

For a closed system (no flow of fluid) that does not execute a cycle, the energy equation is

$$\sum Q = \sum W + \Delta U \tag{1.58}$$

where

$\sum Q$: total heat transfer
$\sum W$: total work transfer
ΔU: internal energy change

The sign convention for work and heat is shown in Figure 1.9

1.3.5.2 Steady-Flow Energy Equation

Consider 1 *kg* of a working fluid element that may be a liquid, gas, or vapour, or any combination, and which is flowing steadily through a control volume (Figure 1.10). The fluid may possess energy in a number of forms between which conversions are possible:

1. Flow work or pressure work, given by pv.
2. Kinetic energy $C^2/2$, due to the movement of the fluid element with velocity C.
3. Internal (thermal) energy u, due to the energy of the fluid molecules.
4. Potential or gravimetric energy, due to the height z above some datum line and given by zg.
5. Chemical, electrical, or magnetic energies may also be added, but these are not involved in the overwhelming cases encountered in thermal power cycles.
6. Heat Q may enter or leave the control volume.
7. Mechanical energy W may be added or removed, with some of the added energy being used to pump the fluid into the control volume or expel it out again.
8. Accumulated (stored) energy in the control volume as a whole e_{cv}.

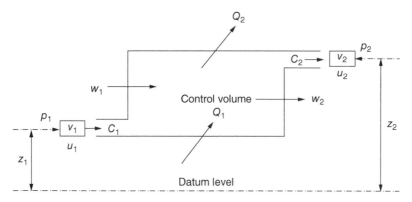

Figure 1.10 Steady-state, steady-flow control volume.

From the law of conservation of energy, the net energy input should be equal to the net energy output plus the energy that accumulates in the control volume. For properties per unit mass (specific properties denoted by lowercase letters) and consistent energy units (J) for all terms, we can write

$$\left(p_1v_1 + \frac{C_1^2}{2} + u_1 + gz_1 \right) + q_1 + w_1 = e_{cv} + \left(p_2v_2 + \frac{C_2^2}{2} + u_2 + gz_2 \right) + q_2 + w_2$$

Energy is not usually allowed to accumulate in the control volume of practical thermal power plants operating on thermodynamic cycles, and therefore the term e_{cv} will be henceforward ignored and the energy equation is reduced to

$$\left(p_1v_1 + \frac{C_1^2}{2} + u_1 + gz_1 \right) + q_1 + w_1 = \left(p_2v_2 + \frac{C_2^2}{2} + u_2 + gz_2 \right) + q_2 + w_2 \tag{1.59}$$

Since specific enthalpy $h = u + pv$, Eq. (1.59) can be rewritten as

$$\left(h_1 + \frac{C_1^2}{2} + gz_1 \right) + q_1 + w_1 = \left(h_2 + \frac{C_2^2}{2} + gz_2 \right) + q_2 + w_2 \tag{1.60}$$

For a fluid flowing steadily at the rate of $\dot{m}$ kg/s, the energy equation becomes

$$\dot{m}\left(h_1 + \frac{C_1^2}{2} + gz_1 \right) + \dot{Q}_1 + \dot{W}_1 = \dot{m}\left(h_2 + \frac{C_2^2}{2} + gz_2 \right) + \dot{Q}_2 + \dot{W}_2 \tag{1.61}$$

where

$$\dot{Q}_1 = \dot{m}q_1, \dot{W}_1 = \dot{m}w_1, \dot{Q}_2 = \dot{m}q_2, \dot{W}_2 = \dot{m}w_2$$

All terms in Eq. (1.61) have units of power (Nm/s, J/s, or W).

For a control volume with multiple flows into and out of the system, the general steady-flow energy equation can be written as

$$\sum \dot{m}\left(h_1 + \frac{C_1^2}{2} + gz_1 \right) + \dot{Q}_1 + \dot{W}_1 = \sum \dot{m}\left(h_2 + \frac{C_2^2}{2} + gz_2 \right) + \dot{Q}_2 + \dot{W}_2 \tag{1.62}$$

1.3.5.3 Stagnation Properties

Stagnation properties are those thermodynamic properties a flowing compressible fluid would possess if it were brought to rest adiabatically and reversibly, i.e. isentropically, and without heat and work transfer. The stagnation state is a convenient hypothetical state that simplifies many of the equations involving flow by taking account of the kinetic energy terms in the steady flow energy equation implicitly.

The stagnation enthalpy h_t is the enthalpy that a gas stream of enthalpy h and velocity C would possess when brought to rest adiabatically and without work transfer. The energy equation thus becomes

$$h_t - h = \frac{1}{2}(0 - C^2)$$

$$h_t = h + \frac{1}{2}C^2 \tag{1.63}$$

For a perfect gas, $h = c_p T$ and the corresponding stagnation temperature T_t is

$$T_t = T + \frac{C^2}{2c_p} \tag{1.64}$$

Applying the concept of stagnation properties to an adiabatic compression, the energy Eq. (1.60) becomes

$$W_c = (h_2 - h_1) + \frac{1}{2}(C_2^2 - C_1^2) = c_p(T_2 - T_1) + \frac{1}{2}(C_2^2 - C_1^2)$$

Rearranging, we get

$$W_c = \left(c_p T_2 + \frac{C_2^2}{2}\right) - \left(c_p T_1 + \frac{C_1^2}{2}\right) = c_p T_{2t} - c_p T_{1t} \tag{1.65}$$

Temperature-measuring devices such as thermometers and thermocouples in reality measure the stagnation temperature of the flow and not the static temperature. Thus, introduction of stagnation temperatures simplifies solving the energy equation by eliminating the kinetic energy term and the need to measure flow velocity.

The stagnation pressure p_t is defined as the pressure the gas stream would possess if the gas were brought to rest adiabatically and reversibly. Using Eqs. (1.48) and (1.64), p_t can be written as

$$p_t = p\left(\frac{T_t}{T}\right)^{\gamma/(\gamma-1)} = p\left(1 + \frac{C^2}{2c_p T}\right) \tag{1.66}$$

1.3.5.4 Isentropic Flow

Examples include flow in ducts, nozzles, and diffusers without heat transfer and work being done. Knowing $C = M_a a = M_a \sqrt{\gamma R T}$, where a is the speed of sound and M_a is the Mach number at the inlet, and rewriting Eq. (1.64) as

$$\frac{T_t}{T} - 1 = \frac{C^2}{2c_p T}$$

we obtain

$$\frac{T_t}{T} - 1 = \frac{M_a^2 \gamma R T}{2c_p T} = \frac{M_a^2 \gamma R}{2c_p}$$

Also,

$$\frac{R}{c_p} = \frac{\gamma - 1}{\gamma}$$

Combining the last two equations, we get

$$\frac{T_t}{T} = 1 + \frac{1}{2}(\gamma - 1)M_a^2 \qquad (1.67)$$

From Eqs. (1.48) and (1.67)

$$\frac{p_t}{p} = \left(\frac{T_t}{T}\right)^{\gamma/(\gamma-1)} = \left[1 + \frac{1}{2}(\gamma - 1)M_a^2\right]^{\gamma/(\gamma-1)} \qquad (1.68)$$

The pressure ratio for $M_a = 1$ at $\gamma = 1.4$ is equal to 1.893. This is the critical pressure ratio for air. To achieve supersonic flow, the stagnation pressure needs to be such that $p_t > 1.893p$.

1.3.5.5 Speed Parameter

It was shown in Section 1.2 that dimensional analysis and similitude can be used to derive functional representations of complex flow systems, such as compressors, using a reduced number of nondimensional groups of properties. Among the groups discussed were the velocity parameter $ND/\sqrt{T_{1t}}$ and mass flow parameter $\dot{m}\sqrt{RT_{1t}}/D^2 p_{1t}$ (Eq. 1.39a). To find a physical interpretation of these seemingly arbitrary combinations of physical properties, consider first the ratio $C/\sqrt{T_t}$:

$$\frac{C}{\sqrt{T_t}} = M_a\sqrt{\gamma R}\sqrt{\frac{T}{T_t}}$$

Combining this equation with Eq. (1.67) we obtain

$$\frac{C}{\sqrt{T_t}} = \frac{M_a\sqrt{\gamma R}}{\left[1 + \frac{1}{2}(\gamma - 1)M_a^2\right]^{1/2}} \qquad (1.69)$$

Now, for a given compressor blade design and impeller tip speed U, $C = f(U)$ and $U = f(ND)$; hence, from Eq. (1.69) for the compressor inlet conditions

$$\frac{ND}{\sqrt{T_{1t}}} \propto \frac{M_a\sqrt{\gamma R}}{\left[1 + \frac{1}{2}(\gamma - 1)M_a^2\right]^{1/2}}$$

The compressor speed parameter is a function of the flow Mach number at the inlet (flight Mach number for a turbojet engine) and thermodynamic properties of the fluid. For a given gas with known thermodynamic properties and a compressor of fixed size,

$$\frac{N}{\sqrt{T_{1t}}} \propto M_a \qquad (1.70)$$

Hence, the nondimensional speed parameter is directly proportional to the Mach number.

1.3.5.6 Mass Flow Parameter

The mass flow rate and density of a fluid are $\dot{m} = \rho A C$ and $\rho = p/RT$. Combining these equations with the equation for the Mach number, we obtain

$$\dot{m} = \frac{pA}{RT} M_a \sqrt{\gamma RT}$$

or, rearranging,

$$\frac{\dot{m}\sqrt{T}}{Ap} = M_a \sqrt{\frac{\gamma}{R}} \tag{1.71}$$

Now

$$\frac{\dot{m}\sqrt{T_t}}{Ap} = \frac{\dot{m}\sqrt{T}}{Ap} \sqrt{\frac{T_t}{T}}$$

Also

$$\frac{\dot{m}\sqrt{T_t}}{Ap} = \frac{\dot{m}\sqrt{T_t}}{Ap_t} \frac{p_t}{p},$$

hence

$$\frac{\dot{m}\sqrt{T_t}}{Ap_t} = \frac{\dot{m}\sqrt{T}}{Ap} \frac{p}{p_t} \sqrt{\frac{T_t}{T}}$$

Combining this equation with Eqs. (1.67), (1.68), and (1.71) we obtain

$$\frac{\dot{m}\sqrt{T_t}}{Ap_t} = M_a \sqrt{\frac{\gamma}{R}} \frac{\left[1 + \frac{1}{2}(\gamma - 1)M_a^2\right]^{1/2}}{\left[1 + \frac{1}{2}(\gamma - 1)M_a^2\right]^{\gamma/(\gamma-1)}}$$

Finally, for the compressor inlet conditions

$$\frac{\dot{m}\sqrt{T_{1t}}}{Ap_{1t}} = \frac{M_a \sqrt{\frac{\gamma}{R}}}{\left[1 + \frac{1}{2}(\gamma - 1)M_a^2\right]^{(\gamma+1)/2(\gamma-1)}} \tag{1.72}$$

Rearranging Eq. (1.72) yields

$$\frac{\dot{m}\sqrt{RT_{1t}}}{D^2 p_{1t}} = \frac{A M_a \sqrt{\gamma}}{D^2 \left[1 + \frac{1}{2}(\gamma - 1)M_a^2\right]^{(\gamma+1)/2(\gamma-1)}} \tag{1.73}$$

For a compressor of fixed size and constant fluid properties, the flow parameter is a function of the Mach number:

$$\frac{\dot{m}\sqrt{T_{1t}}}{p_{1t}} = f(M_a) \tag{1.74}$$

1.3.5.7 Applications of the Energy Equation

Simple nozzle or diffuser. A *nozzle* is a device that increases the velocity of a gas or vapour at the expense of pressure (Figure 1.11a). A *diffuser* is a device that increases the pressure of a gas or vapour at the expense of velocity (Figure 1.11b).

There is no work or heat transfer and negligible potential energy change in both systems. If the input kinetic energy is of considerable value (the gas is approaching the nozzle in Figure 1.11a at a high velocity C_1), the energy equation for the nozzle is

$$h_1 + \frac{C_1^2}{2} = h_2 + \frac{C_2^2}{2} \tag{1.75}$$

The kinetic energy of the fluid leaving the diffuser is usually ignored due to the low velocity at the exit, C_2 (Figure 1.11b). The energy equation is then reduced to

$$h_1 = h_2 + \frac{C_2^2}{2} \tag{1.76}$$

Reciprocating internal combustion engine. Figure 1.12 shows a schematic diagram of the engine as a single steady-flow system. Assuming negligible pressure work and potential and kinetic energies, energy Eq. (1.62) can be written as follows:

$$(\dot{m}h)_{fuel} + (\dot{m}h)_{air} + (\dot{m}h)_{coldcoolant} + \dot{Q}_{fuel} + \dot{W}_{fan}$$
$$= (\dot{m}h)_{exhaust} + (\dot{m}h)_{hcoolant} + \dot{Q}_{waste} + \dot{W}_{shaft} \tag{1.77}$$

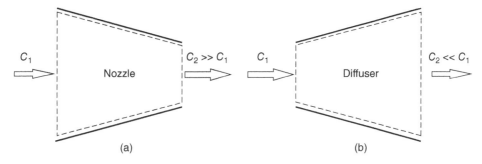

Figure 1.11   Schematic diagrams of a (a) nozzle; (b) diffuser.

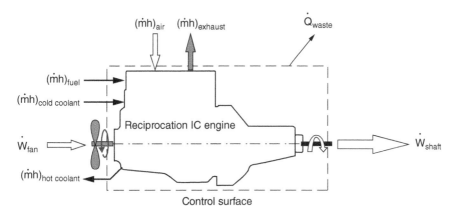

Figure 1.12   The reciprocating internal combustion engine as a steady-flow system.

$\dot{Q}_{waste}$ is the total heat wasted as a result of heat transfer from the hot surfaces of the engine to the surroundings.

The energy equation is often used for the combined combustion/expansion (power) stroke in reciprocating engines in order to assess the process of heat release by the burning fuel. As both inlet and exhaust valves are closed during this process, it is a non-flow closed system with no added mechanical work. Applying the energy Eq. (1.58) written in terms of specific values, we obtain

$$\sum q = \sum w + \Delta u \tag{1.78}$$

Both the volume and pressure change as the gases expand in the cylinder, producing work $w = \int pdv$. There is no work addition to the process, but there is a heat loss to the surroundings (unless it is assumed that the process is adiabatic); hence, the energy equation then becomes

$$q_{comb} = du + \int pdv + q_{loss} \tag{1.79}$$

This simple and convenient form of the energy equation equates the energy input as heat from the combustion of the fuel to the sum of the change of internal energy of the gases as their temperature changes, work done by the gases as they expand in the cylinder during the power stroke, and the heat loss to the surroundings.

Gas turbine. The expansion process is steady-state, steady-flow with heat exchange with the surroundings and negligible potential and kinetic energy changes (Figure 1.13).

From Eq. (1.61)

$$\dot{W}_{out} = \dot{m}(h_1 - h_2) + \dot{Q}_{in} - \dot{Q}_{out} \tag{1.80}$$

(work input = enthalpy change + net heat loss)

If the expansion process in the turbine is adiabatic, the output power is simply

$$\dot{W}_{out} = \dot{m}(h_1 - h_2) \tag{1.81}$$

Air compressor. The potential energy, input heat, inlet velocity, and output mechanical work can be ignored in the case of the compressor shown in Figure 1.14. Equation (1.61) is then reduced to

$$\dot{W}_{in} = \dot{m}(h_2 - h_1) + \dot{m}\frac{C_2^2}{2} + \dot{Q}_{out} \tag{1.82}$$

(work input = enthalpy change + kinetic energy of discharge + heat losses

Figure 1.13 Schematic diagram of a turbine.

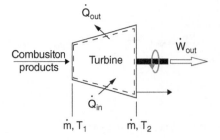

$\dot{Q}_{out}$

Combusiton products

Turbine

$\dot{W}_{out}$

$\dot{Q}_{in}$

$\dot{m}, T_1$ $\dot{m}, T_2$

Figure 1.14 Schematic diagram of air compressor.

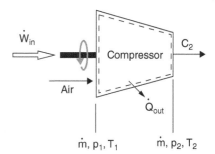

If the velocity of the gas is reduced at the exit from the compressor so that the kinetic energy of discharge is negligible and there is no appreciable heat loss, Eq. (1.82) is reduced to

$$\dot{W}_{in} = \dot{m}(h_2 - h_1) \tag{1.83}$$

For a perfect gas, Eq. (1.83) can be written as

$$\dot{W}_{in} = \dot{m}c_p(T_2 - T_1) \tag{1.84}$$

Heating gas at constant pressure or constant volume. When a gas is heated at constant volume without work or heat transfer, Eq. (1.58) is written as

$$q_1 = \Delta u = (u_2 - u_1) \tag{1.85a}$$

For a perfect gas,

$$q_1 = c_v(T_2 - T_1) \tag{1.85b}$$

When the gas is heated at constant pressure under steady flow conditions, Eq. (1.60) is reduced to

$$q_1 = (h_2 - h_1) + q_2 \tag{1.86}$$

For a perfect gas with negligible heat losses, $q_2 = 0$ and Eq. (1.86) becomes

$$q_1 = c_p(T_2 - T_1) \tag{1.87}$$

1.3.6 Second Law of Thermodynamics

The first law of thermodynamics states that energy cannot be created or destroyed but it can be converted from one form to another; and when heat is converted to work, the latter can never be greater than the former. However, it does not state how much of the heat energy, for example, can be converted to work and how efficiently. The second law, in its various statements, gives the answers to these questions. A clear statement of the second law (Rogers and Mayhew, 1992) that is relevant to the subject matter of this book and based on Planck's statement is as follows:

> "It is impossible to construct a system which will operate as a cycle, extract heat from a reservoir and do an equivalent amount of work on the surroundings."

It follows that part of the extracted heat must be rejected to another reservoir at a lower temperature. Two cases can be identified:

- Heat transfer will occur down a temperature gradient as heat from high-temperature source, such as combustion chamber in a gas turbine, is partly converted to mechanical

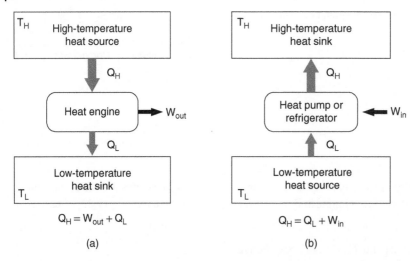

Figure 1.15 Schematic arrangements of a (a) heat engine; (b) heat pump or refrigerator.

work with the balance rejected to a low-temperature reservoir (sink) such as the atmosphere (Figure 1.15a). This system is known as a *heat engine*.

- Heat can be transferred from a low-temperature source, such as the cooling compartment in a refrigerator, up a temperature gradient to a high-temperature reservoir (sink), such as the kitchen, with the assistance of external mechanical work (Figure 1.15b). This system is known as a *heat pump*, *air conditioner*, or *refrigerator*.

The second law of thermodynamics is also stated as the *law of degradation of energy* whereby the quantity of energy is conserved, but its quality (the potential to produce useful work) is not. Every time energy changes form or is transferred from one system to another, its potential to produce useful work is reduced irreversibly forever. It is then said that energy has *degraded*.

This law is the reason we may face an energy and/or climate crisis. All the energy that we use ultimately ends up as waste heat transferred to the earth's atmosphere and then to space.

1.3.6.1 Entropy

Entropy is a thermodynamic property that is a measure of process irreversibility or energy degradation and is defined as

$$dS = dQ/T \text{ in } J/K, \text{ or } ds = dq/T \text{ in } J/kg.K \tag{1.88}$$

where
    dS: total entropy change
    ds: specific entropy change
    dQ: heat transferred reversibly
    T: absolute temperature at which heat is transferred

- If heat is added to a system, ds will be positive (entropy increases).
- If heat is removed from a system, ds will be negative (entropy decreases)
- If $ds = 0$ during a process, the process is isentropic. The frictionless adiabatic process is an isentropic process.

A reversible process occurs when both the system and the surroundings are returned to their original conditions after the process and reverse process have been carried out. Processes in nature are irreversible, however, because reversal always causes some change to occur in the system and/or surroundings. Factors causing irreversibility include:

- Friction
- Unrestricted expansion
- Heat transfer through a finite temperature difference
- Mixing of two different gases
- Chemical reactions

1.3.7 The Carnot Principle

Nicolas Sadi Carnot (1796–1832) was a French engineer who made significant contributions to the science of thermodynamics by recognising that heat engines must operate with cyclic processes. A cycle occurs when a thermodynamic system, having undergone a series of processes, arrives at a final state that is exactly the same as its initial state. In Carnot's own words (Sandfort, 1964): 'The thermal agency by which mechanical effect may be obtained is the transference of heat from one body to another at a lower temperature'. Carnot also investigated the problem of determining the maximum work that can be extracted from the transfer of heat from high to low temperature. He eventually came up with a definition of a perfect thermodynamic engine as follows: 'Whatever amount of mechanical effect it can derive from a certain thermal agency, if an equal amount be spent in working it backwards, an equal reverse thermal effect will be produced'. Such an engine has come to be known as the *reversible engine*, and the quotation as the *Carnot principle*. Furthermore, Carnot stated that the maximum limits of temperature between which any actual heat engine can work are the temperature of combustion of fuel and the temperature of the coldest body we can easily find and use in nature, usually the water in rivers and lakes. Figure 1.15a is the Carnot engine, and the Carnot cycle is shown in Figure 1.16 in $p - V$ and $T - s$ coordinates:

Process 1–2: Isothermal expansion ($pV =$ const.) with heat addition
Process 2–3: Reversible adiabatic expansion ($pV^\gamma =$ const)
Process 3–4: Isothermal compression ($pV =$ const) with heat rejection
Process 4–1: Reversible adiabatic compression ($pV^\gamma =$ const)

Thermal efficiency of this heat engine is

$$\eta = \frac{Q_H - Q_L}{Q_H} = 1 - \frac{Q_L}{Q_H} = 1 - \frac{T_L}{T_H} \tag{1.89}$$

Since Eq. (1.89) is obtained without reference to a specific working fluid, it can be surmised that all reversible cycles operated between the same temperatures will have the same thermal efficiency.

Based on accumulated experimental knowledge, scientists and engineers have come to the conclusion that it is impractical to build the Carnot engine, and it remains to date as the ideal cycle against which real heat engine cycles are measured. If such an engine were to operate between combustion temperature of iso-octane (gasoline) $T_H = 2300\,K$ and the standard ambient temperature $T_L = 298.15\,K$, the Carnot efficiency would be 78%. By comparison, the most efficient reciprocating internal combustion engines can hardly achieve 50%.

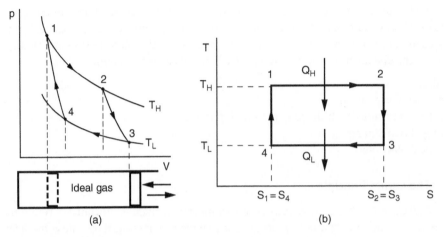

Figure 1.16 Ideal Carnot engine cycle in (a) p-V and (b) T-s coordinate systems.

1.3.8 Zeroth Law of Thermodynamics

If two systems are in thermal equilibrium with a third system, then they are in thermal equilibrium with each other and the three systems are said to be at the same temperature.

This law was added to the laws of thermodynamics early in the twentieth century because it was realised that the concept of equal-in-temperature is a prerequisite to a logical development of those laws. And to be logical, it was named the *zeroth law of thermodynamics*.

1.3.8.1 Thermodynamic Scale of Temperature

Temperature is a fundamental concept, not expressible in terms of other units or physical properties of the devices used to measure temperature, such as alcohol or mercury in glass thermometers or the electromotive force generated in a thermocouple. Physicists have established that temperature measures the kinetic energy of molecules, and the higher the molecular agitation, the higher the temperature and vice versa. The physicist William Thomson (Lord Kelvin) is credited with the establishment in 1848 of the absolute temperature scale (hence the symbol K for temperature) on the basis of Carnot's reversible cycle.

To show how it is possible to arrive at an absolute scale, a hypothetical experiment can be conducted to show that an absolute zero of temperature must exist, and then extend this line of reasoning to develop an 'energy' or 'thermodynamic' temperature scale.

It was shown earlier that the efficiency of the Carnot cycle is written as

$$\eta = 1 - \frac{Q_L}{Q_H} = 1 - \frac{T_L}{T_H}$$

This equation shows that T_L can never be zero or less than zero, for if T_L were equal to zero, the thermal efficiency of the heat engine would be 1, or 100%, which would be a violation of the second law. Also, if T_L were less than zero, the thermal efficiency would be greater than 100%, which would mean a reversal of the direction of Q_L, and heat would be drawn from the low-temperature source. This also would be a violation of the second law. So, theoretically, the absolute zero of temperature could never be reached.

If a Carnot engine is operated between a high-temperature source at the temperature of boiling water T_H and a low-temperature sink at the temperature of melting ice T_L, and the amount of heat received Q_{steam} and rejected Q_{ice} were measurable accurately, the ratio Q_{steam}/Q_{ice} would be equal to 1.3662. Since $Q_{steam}/Q_{ice} = T_{steam}/T_{ice}$,

$$\frac{T_{steam}}{T_{ice}} = 1.3662 \tag{1.90}$$

If the temperature range $T_{steam} - T_{ice}$ is arbitrarily divided into 100 equal divisions (call it $100\,°C$), a second equation can be obtained

$$T_{steam} - T_{ice} = 100 \tag{1.91}$$

The simultaneous solution of Eqs. (1.90) and (1.91) yields the temperature of boiling water and melting ice on a thermodynamic scale:

$$\left.\begin{array}{l} T_{steam} \approx 373 \\ T_{ice} \approx 273 \end{array}\right\} \text{degrees Kelvin}(K)$$

If the temperature range $T_{steam} - T_{ice}$ is divided into 180 equal divisions (call it degrees Fahrenheit, F), we get

$$T_{steam} - T_{ice} = 180 \tag{1.92}$$

$$\left.\begin{array}{l} T_{steam} \approx 672 \\ T_{ice} \approx 492 \end{array}\right\} \text{degrees Rankine}(R)$$

The determination of a thermodynamic scale using reversible Carnot engine as described here is practically impossible. To determine where the number 1.3662 comes from, we start with the equation of state for a perfect gas at constant volume and constant pressure, which yields

$$\left(\frac{v_1}{v_2}\right)_p = \left(\frac{p_1}{p_2}\right)_v = \frac{T_1}{T_2} \tag{1.93}$$

Next, constant-volume and constant-pressure thermometers are placed alternately in boiling water and melting ice, and the volume and pressure ratios of the gases in the devices (carbon dioxide CO_2, hydrogen H_2, and helium He) are measured at an initial pressure. If the procedure is repeated for different initial pressures and the results plotted versus pressure, straight lines are obtained, which converge at zero absolute pressure to a value of 1.3662, which must be the ratio of absolute temperature of boiling water and melting ice.

Currently, the international temperature scale (ITS-90) is the standard used to represent the thermodynamic (absolute) temperature scale as closely as possible with improved accuracy over gas thermometry.

1.3.9 Third Law of Thermodynamics

This law, based on empirical evidence, postulates that absolute entropy of a pure crystalline substance in complete internal equilibrium is zero at temperature zero degree absolute. The third law allows the determination of absolute entropies from thermal data.

The entropy change of a gas on molar basis is

$$d\bar{s} = C_p \frac{dT}{T} - \bar{R}\frac{dp}{p}$$

The entropy change for process 1–2 is then

$$s_2 - s_1 = \int_{T_1}^{T_2} C_p \frac{dT}{T} - \bar{R} ln \frac{p_2}{p_1} \tag{1.94}$$

If the specific heat C_p is assumed constant, Eq. (1.94) becomes

$$s_2 - s_1 = C_{p(avg)} ln \frac{T_2}{T_1} - \bar{R} ln \frac{p_2}{p_1} \tag{1.95}$$

More accurate results can be obtained if the variability of specific heat with temperature is accounted for. Taking the absolute zero as the reference temperature, s_T^0 can be defined as

$$s_T^0 = \int_0^T C_p \frac{dT}{T} \tag{1.96}$$

The values of s_T^0 are usually calculated for different temperatures and can be found in tabular form as $s^0 = f(T)$ in most thermodynamic reference books. Using Eq. (1.96), we can write

$$\int_{T_1}^{T_2} C_p \frac{dT}{T} = s_{T_2}^0 - s_{T_1}^0 \tag{1.97}$$

Substituting Eq. (1.97) in Eq. (1.94), we obtain

$$s_2 - s_1 = s_{T_2}^0 - s_{T_1}^0 - \bar{R} ln \frac{p_2}{p_1} \; kJ/kmole.K \tag{1.98}$$

As Eq. (1.98) shows, entropy changes with both temperature and pressure. When using Eq. (1.98) in chemical reactions, the pressure ratio in the last term is replaced by the mole concentration of each substance.

The $s_T^0 = f(T)$ data for some commonly used gases in heat engine practice such as CO_2, CO, H_2O, H_2, O_2, and N_2 are tabulated in Appendix A. They can be calculated to a very high degree of accuracy for the temperature range 100–6000 K by third-order logarithm functions of the form

$$s_T^0 = a + b \, ln\left(\frac{T}{1000}\right) + c \, ln\left(\frac{T}{1000}\right)^2 + d \, ln\left(\frac{T}{1000}\right)^3 kJ/kmole.K \tag{1.99}$$

The coefficients of correlation Eq. (1.99) for six gases are given in Table 1.3.

Table 1.3 Coefficients of Eq. (1.99) for the calculation of $s_T^0 = \int_0^T C_p dT/T$ of combustion products, $p_{ref} = 0.1$ MPa $T_{ref} = 0$ K.

	s_T^0, kJ/kmole.K					
Coefficient	CO_2	CO	H_2O	H_2	O_2	N_2
a	269.7726	234.9291	233.3053	166.0444	243.7299	228.5929
b	53.72045	33.10685	42.08792	30.81348	34.40598	32.65506
c	4.955586	1.661032	4.707912	2.02213	2.175759	1.661282
d	−0.61724	9.19E-02	0.667926	0.447018	0.19782	0.150999

Table 1.4 Coefficients of Eq. (1.100) for the calculation of $s_T^0 = \int_0^T C_p dT/T$ of air.

A	B_1	B_2	B_3	B_4	B_5	B_6	B_7	B_8
				B_n				
29.44	230.18	−1.61	−5.99	22.94	−24.56	12.98	−3.48	0.38

As for air, the following modified function (Rivkin 1987) can be used for the temperature range −50 to 1500 °C with the constant coefficients shown in Table 1.4:

$$s^0 = A \ln\left(\frac{T}{1000}\right) + \sum_{n=1}^{n=8} B_n\left(\frac{T}{1000}\right)^{n-1} kJ/kmole.K \tag{1.100}$$

Finally, if the second law of thermodynamics can be stated as the *law of degradation of energy*, then entropy can be regarded as the measure of the degree of energy degradation.

Problems

Engineering Mechanics

1.1 Determine the velocity required to keep satellite P in orbit at altitude S km above the surface of the earth. Force F_2 is the balancing force that keeps the satellite in circular orbit. The gravitational force acting on the satellite, treated as a particle, is

$$F_1 = mgR_E^2/R^2$$

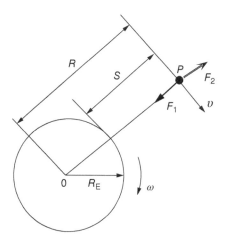

1.2 A crank-slider has a crank radius $R = 100$ *mm*, a connecting rod length $L = 200$ *mm*, and a crank angular speed $\omega = 100$ *rpm* counterclockwise.
(a) Write the expressions for the displacement x and velocity $\dot{x}$ of piston P
(b) Determine the velocity when $\theta = 45^0$.

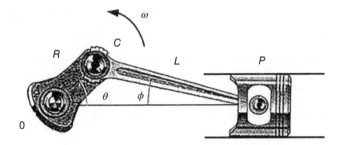

1.3   Determine the moment of inertia and radius of gyration about an axis through the centre of a composite body with a central inner disk of mass 4 *kg* and radius 300 *mm* and a ring of mass 5 *kg* and radius 400 *mm*. The masses of the four spokes shown can be ignored.

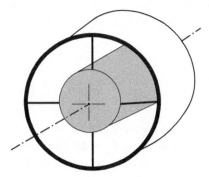

1.4   A constant torque of 150 *Nm* applied to a turbine rotor is sufficient to overcome the constant bearing friction torque and give it a speed of 75 *rpm* from rest after 9 revolutions. When the 150 *Nm* torque is removed, the turbine rotor turns for a further 23 revolutions before stopping. Determine the moment of inertia of the rotor and the bearing friction torque.

1.5   A motor is required to reach an operating speed of 500 *rpm* in a time of 30 seconds from rest at a constant acceleration. Determine
(a) The required constant acceleration
(b) The number of revolutions the motor turns to reach this operating speed.
   The motor is then turned off and comes to rest under a constant deceleration of 0.1 *rad/s²* due to bearing friction. How long and how many revolutions does it take for the motor to come to reset?

Fluid Mechanics

1.6   The following nondimensional parameters are suggested for graphical representation of compressor characteristics:

$$\text{Mass parameter } \frac{\dot{m}\sqrt{RT}}{D^2 p}, \text{ speed parameter } \frac{N}{\sqrt{T}}$$

where $\dot{m}$ is the mass flow rate, R is the gas constant, T is the gas temperature, D is the impeller diameter, and p is the gas pressure. Verify whether these parameters are truly nondimensional. If they are not, suggest a corrected form of the parameters.

1.7 Dimensional analysis has identified the following function for the pressure loss in a pipe:

$$\Pi_1 = \psi(\Pi_2)$$

where

$$\Pi_1 = \frac{\Delta p D}{\rho C^2}, \quad \Pi_2 = Re = \frac{D C \rho}{\mu}$$

(a) Check if these two groups are nondimensional, and suggest remedies if they are not.

(b) Determine the actual correlation for flow of water in a 1500 mm long pipe with 12.5 mm diameter using experiments conducted at 20 °C with the following results:

C, m/s	0.375	0.594	0.887	1.780	3.392	5.157	7.114	8.757
Δp, Pa	300	747	1 480	5 075	15 753	32 606	57 456	82 832

The water is at 20 °C, the density $\rho = 998.2$ kg/m^3, and the dynamic viscosity is

$$\mu = 1.005 \times 10^{-3} \ N.s/m^2$$

1.8 Air is drawn into engine cylinder as the piston moves away from top dead centre (TDC) during the induction stroke, as shown in the following figure. Determine the pressure drop in the induction manifold for these conditions:

(a) Ambient pressure $p_a = 0.1$ MPa.

(b) Flow velocity at the inlet C_a is negligible, and at the valve throat $C = 56$ m/s.

(c) Loss head at the valve restriction is given by $h_{loss} = \beta C^2/2g$, where the loss factor $\beta = 7$.

(d) Density of air $\rho_a = 1.21$ kg/m^3.

(Due to the small pressure drop during the induction process, air can be treated as incompressible fluid.)

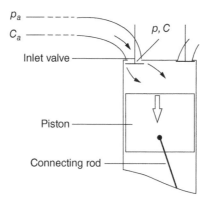

1.9 In a frictionless cylinder and piston device, the piston is forced against a gas by a spring, which exerts a force directly proportional to the gas volume. In addition, the atmospheric pressure of $101.3\,kPa$ acts on the outer face of the piston.

(a) Considering the gas as a system, calculate the work for the process from an initial state of $0.2\,MPa$, $0.1\,m^3$ to final volume of $0.3\,m^3$.

(b) If the spring is taken as a system, find the work for the same process.

(c) Explain the difference in work terms for the two systems in (a) and (b).

Thermodynamics

1.10 A gas mixture has the following molar (volume) composition:

CO_2	12.0%
O_2	4.0%
N_2	82.0%
CO	2.0%

Determine the mixture mass composition.

1.11 The gravimetric analysis of a gaseous mixture shows the following components:

H_2	18.0%
H_2O	4.0%
O_2	25.0%
N_2	21.0%
CO	3.0%
CO_2	13.0%
CH_4	16.0%

(a) What is the volumetric composition of the mixture?

(b) Calculate the specific gas constant and molecular mass of the mixture.

1.12 In a non-flow process, there is a heat-transfer loss of $1055\,kJ$ and an internal energy increase of $210\,kJ$. Determine the work transfer, and state whether the process is an expansion or compression.

1.13 During the power stroke of an engine, the heat transferred to the cooling system was found to be $150\,kJ/kg$ of working substance. The internal energy also decreased by $400\,kJ/kg$ of working substance. Determine the work done, and state whether it is work done on the engine or by the engine.

1.14 The ventilation system servicing a lecture theatre seating 300 students suddenly fails. Assuming the average heat transfer rate from a person not actively working is about $400\,kJ\,h^{-1}$,

(a) What is the change in internal energy of the air 15 minutes after shutdown?

(b) If all the students and the lecture theatre are taken to be the system, how much does the system's internal energy change?

(c) How do you account for the increase in air temperature?

1.15 The collecting panels of a small solar boiler receive radiation energy at the rate of $3400\,kJ/m^2$ each hour during daylight. For a small rural power plant with $10\,kW$ output, determine the required collector area if the electrical output is only 6% of the incident radiation.

1.16 In a steady-flow open system, a fluid flows at the rate of $4\,kg/s$. It enters the system at a pressure of $600\,kPa$, a velocity of $220\,m/s$, internal energy $2200\,kJ/kg$, and specific volume $0.42\,m^3/kg$. It leaves the system at a pressure of $150\,kPa$, velocity of $145\,m/s$, internal energy $1650\,kJ/kg$, and specific volume $1.5\,m^3/kg$. During its passage through the system, the substance has a loss by heat transfer of $40\,kJ/kg$ to the surroundings. Determine the power of the system, stating whether it is from or to the system. Neglect any change of gravitational potential energy.

1.17 Air enters a gas turbine system with a velocity of $105\,m/s$ and has a specific volume of $0.8\,m^3/kg$. The inlet area of the turbine is $0.05\,m^2$. At exit, the air has a velocity of $135\,m/s$ and specific volume of $1.5\,m^3/kg$. In its passage through the turbine system, the specific enthalpy of the air is reduced by $145\,kJ/kg$ and the air has a heat loss of $27\,kJ/kg$. Determine
(a) The mass flow rate of the air through the turbine system
(b) The exit area of the turbine system in m^2
(c) The power developed by the turbine system in kilowatts

1.18 A $1.0\,m^3$ closed vessel is filled with air at pressure $p = 7\,bar$ and temperature $T = 288\,K$. The vessel is heated until the air temperature reaches $373\,K$. Determine
(a) The mass of air in the vessel
(b) The air pressure at the end of the heating process
(c) The change of internal energy, enthalpy, and entropy of air as a result of heating

1.19 A cylinder with a moving piston is filled with $2.5\,kg$ of air. Initially, the pressure and temperature in the cylinder are, respectively, $0.8\,MPa$ and $25\,°C$. If the air is heated to $100\,°C$ at constant pressure, determine
(a) The amount of heat input
(b) Work done
(c) Change of internal energy, enthalpy, and entropy of the air
(d) Specific volume of the air at the start and end of the process.

1.20 $0.1\,kg$ of air in the cylinder of a compression ignition (CI) engine undergoes polytropic compression during which the volume decrease 16 times and the pressure increases 45 times. If the initial temperature is $320\,K$, determine
(a) The polytropic index of the compression process
(b) Temperature of the air at the end of compression
(c) Compression work
(d) Change of internal energy, enthalpy ant entropy
(e) Heat added to the process

1.21 1.0 kg of air undergoes a Carnot cycle in which the heat source is at $900\,K$ and the heat sink at $300\,K$. if the minimum and maximum pressures in the cycle are 0.1 and 6.0 MPa, respectively, determine

(a) Properties at the characteristic points

(b) Work done

(c) Thermodynamic efficiency

1.22 The following figure depicts a thermodynamic model for a reciprocating internal combustion engine with the control volume shown by the dashed line. The heat losses by heat transfer from the hot engine surfaces and the enthalpy of fuel can be assumed negligible.

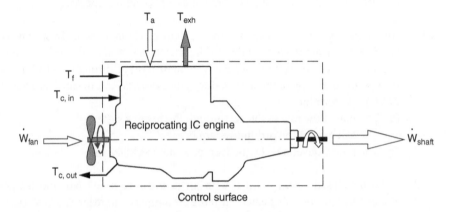

(a) Write the general energy equation for the system.

(b) Determine the required rate of heat input if the engine is producing 120 kW of power, and the flow rate of inlet air and water flow rate in the cooling system are 0.15 kg/s and 4.4 kg/s, respectively. Take the exhaust (assumed air) temperature $T_{exh} = 507\,°C$, and the temperatures of the coolant $T_{c,in} = 80\,°C$ and $T_{c,out} = 84\,°C$. The inlet air and the fuel are at 295 K.

(c) Determine the fuel consumption (in *litres/h*) if the lower heating value of the fuel is 43 500 kJ/kg and the fuel density is 850 kg/m^3.

2

Thermodynamics of Reactive Mixtures

When studying theoretical cycles, the source of the heat input into a particular cycle is immaterial. In fuel-air and practical cycles, however, the heat source is usually a hydrocarbon fuel that reacts with an oxidant, producing heat and various gaseous products. Thermodynamics of reactive mixtures, or *thermochemistry*, is essentially the basic chemistry behind the reaction of hydrocarbon fuels with the oxygen of the air and broadly referred to as *combustion*. Combustion is a rapid exothermic chemical reaction between a fuel and the oxygen of the air, resulting in liberation of heat accompanied by bright flame.

2.1 Fuels

Fuels used in internal combustion engines, such as piston engines and gas turbines, are mainly complex hydrocarbon compounds and originate from fossil fuels, and are predominantly composed of chemical combinations of carbon and hydrogen in gaseous or liquid forms. Since the main composition is carbon and hydrogen, they are referred to as *hydrocarbon fuels*. Gaseous hydrocarbon fuels can be naturally occurring or petroleum based. Liquid hydrocarbon fuels are products of refining raw petroleum or biomass and are normally tailored to the specific requirements of different engine types. Fuels, some of which are given in Table 2.1, typically consist of

- Hydrogen
- Carbon
- Oxygen and sulfur (small amounts)

2.2 Stoichiometry

Stoichiometry is the relationship between quantities of substances taking part in a reaction or forming a compound. The origin of the word *stoichiometry* is from the Greek word *stoikhion* meaning 'element'. For the combustion process to occur, an oxidant is required. Atmospheric air is the most common source of the oxidant used in combustion, namely oxygen. The compositions of air by mass and volume fractions are shown in Table 2.2.

Fundamentals of Heat Engines: Reciprocating and Gas Turbine Internal Combustion Engines, First Edition. Jamil Ghojel.
© 2020 John Wiley & Sons Ltd. This Work is a co-publication between John Wiley & Sons Ltd and ASME Press.
Companion website: www.wiley.com/go/JamilGhojel_Fundamentals of Heat Engines

Table 2.1 Typical carbon, hydrogen, and sulfur mass contents of common fuels.

	Gasoline (%)	Diesel (%)	Light fuel oil (%)	Heavy fuel oil (%)
Carbon	85.5	86.3	86.2	86.1
Hydrogen	14.4	12.8	12.4	11.8
Sulfur	0.1	0.9	1.4	2.1

Table 2.2 Composition of air.

	By volume (%)	By mass (%)
Oxygen O_2	21	23
Nitrogen N_2	79	77

A mixture of fuel and air is stoichiometric if there is theoretically just enough oxygen (air) to burn all the combustible elements of the fuel completely. The products of complete combustion are carbon dioxide CO_2, water vapour H_2O and inert nitrogen N_2 (it will be shown later that under specific conditions, nitrogen can react with oxygen).

The strength of an air-fuel mixture is defined in terms of the relative air-fuel ratio λ as

$$\lambda = \frac{actual\ air - fuel\ ratio}{stoichiometric\ air - fuel\ ratio}$$

If we denote the actual air-fuel ratio by $(A/F)_a$ and the stoichiometric air-fuel ratio by $(A/F)_s$, we obtain

$$\lambda = \frac{(A/F)_a}{(A/F)_s} \tag{2.1}$$

Another term defining the strength of an air-fuel mixture is the equivalence ratio ϕ, which is defined as

$$\phi = \frac{actual\ fuel - air\ ratio}{stoichiometric\ fuel - air\ ratio}$$

or

$$\phi = \frac{(F/A)_a}{(F/A)_s} \tag{2.2}$$

Hence,

$$\phi = 1/\lambda$$

Air-fuel ratios can be expressed on mass (gravimetric), molar, or volume bases as follows:

On a mass basis:
  kg of air per *kg* of fuel
  kg of air per *kmole* of fuel
On a molar basis:
  kmole of air per *kmole* of fuel
  kmole of air per *kg* of fuel

On a volume basis:

m^3 of air per m^3 of fuel

m^3 of air per kg of fuel

The relative air-fuel ratio can be equal to, less than, or greater than 1, depending on the mixture strength:

Stoichiometric mixture: $\lambda = 1$

Lean mixture (excess air): $\lambda > 1$

Rich mixture (excess fuel/insufficient air): $\lambda < 1$

2.3 Chemical Reactions

2.3.1 Fuels Having a Single Chemical Formula

Fuels in the chemical reaction of combustion can be divided into four categories:

1. Gaseous fuels that exist naturally or produced in chemical processes, examples of which are hydrogen (H_2), methane (CH_4), propane (C_3H_8), and butane (C_4H_{10}).
2. Combustible (chemically active) compounds that do not occur naturally but result from combustion reactions, including compounds such as carbon monoxide CO and hydroxyl OH.
3. Liquid fuels produced in chemical processes, examples of which are pentane (C_5H_{12}), hexane (C_6H_{14}), isooctane (C_8H_{18}), methanol (CH_4O), and ethanol (C_2H_6O).
4. Surrogate fuels used to model actual hydrocarbon fuels used in engines to simplify combustion calculations. For example, isooctane (C_8H_{18}) is often used to model gasoline and dodecane ($C_{12}H_{26}$) to model diesel fuel. Other formulas suggested to model gasoline and diesel fuels are $C_xH_{1.87x}$ and $C_xH_{1.8x}$, respectively (Haywood, 1988). The hypothetical fuel $C_{11}H_{21.34}$ can be used to model kerosene, which is the basic gas turbine fuel. Dodecene ($C_{12}H_{24}$) will be used in this book as a surrogate gas turbine fuel. $C_{8.5}H_{17}$, $C_{12}H_{22}$, and $C_{11}H_{21}$ could be used to model aviation gas turbine fuels (Martel et al., 2000). For biomass-based fuels, the formulas $C_{13.4}H_{26.7}O_2$, $C_{21}H_{39.5}O_2$, and $C_{18.9}H_{3.6}O_2$ are suggested for coconut, rapeseed, and linseed oils, respectively (van Basshuysen and Schafer, 2007).

In the combustion equations that follow, liquid fuels are assumed to be in gaseous form when the elements react with oxygen of the air. Generally, for a gaseous fuel containing x atoms of carbon, y atoms of hydrogen, and z atoms of oxygen, the stoichiometric (chemically correct) reaction with oxygen can be written as

$$C_xH_yO_z + \left(x + \frac{y}{4} - \frac{z}{2}\right)O_2 \rightarrow xCO_2 + \frac{y}{2}H_2O$$

Taking into account the adopted composition of air, this reaction can be rewritten as

$$C_xH_yO_z + \left(x + \frac{y}{4} - \frac{z}{2}\right)(O_2 + 3.7619N_2) \rightarrow xCO_2 + \frac{y}{2}H_2O + 3.7619\left(x + \frac{y}{4} - \frac{z}{2}\right)N_2$$

(2.3)

The stoichiometric air-fuel ratio on volume basis for this reaction is

$$\left(\frac{A}{F}\right)_s^v = \left(x + \frac{y}{4} - \frac{z}{2}\right)(O_2 + 3.7619N_2), \text{ or}$$

$$\left(\frac{A}{F}\right)_s^v = \frac{1}{0.21}\left(x + \frac{y}{4} - \frac{z}{2}\right) kmole\ air/kmole\ fuel \tag{2.4}$$

$$\left(\frac{A}{F}\right)_s^v = \frac{\frac{1}{0.21}\left(x + \frac{y}{4} - \frac{z}{2}\right)}{(12x + y + 16z)} kmole\ air/kg\ fuel \tag{2.5}$$

The letters *s* and *v* stand for *stoichiometric* and *volumetric*, respectively.

And on a mass basis:

$$\left(\frac{A}{F}\right)_s^g = \frac{(32 + 3.7619 \times 28)}{(12x + y + 16z)}\left(x + \frac{y}{4} - \frac{z}{2}\right) kg\ air/kg\ fuel \tag{2.6}$$

For a reaction with excess air:

$$C_xH_yO_z + \lambda\left(x + \frac{y}{4} - \frac{z}{2}\right)(O_2 + 3.7619N_2)$$

$$= xCO_2 + \frac{y}{2}H_2O + (\lambda - 1)\left(x + \frac{y}{4} - \frac{z}{2}\right)O_2 + 3.7619\lambda\left(x + \frac{y}{4} - \frac{z}{2}\right)N_2 \tag{2.7}$$

where $(O_2 + 3.7619N_2)$ or $(O_2 + 0.79/0.21)$ is the amount of air corresponding to 1 *kmole* of oxygen in the atmospheric air.

The amount of reactants are

$$M_R = 1 + \lambda\left(x + \frac{y}{4} - \frac{z}{2}\right)(O_2 + 3.7619N_2)\ kmole/kmole\ fuel$$

We can also write M_R in terms of the fuel composition or stoichiometric air-fuel ratio as follows:

$$M_R = 1 + \frac{\lambda}{0.21}\left(x + \frac{y}{4} - \frac{z}{2}\right)\ kmole/kmole\ fuel, \text{ or}$$

$$M_R = 1 + \lambda\left(\frac{A}{F}\right)_s^v\ kmole/kmole\ fuel \tag{2.8}$$

The amounts of products of combustion

$$M_{CO_2} = x\ kmole/kmole\ fuel \tag{2.9}$$

$$M_{CO_2} = \frac{y}{2}\ kmole/kmole\ fuel \tag{2.10}$$

$$M_{O_2} = \left(x + \frac{y}{4} - \frac{z}{2}\right)(\lambda - 1), \text{ or}$$

$$M_{O_2} = 0.21(\lambda - 1)\left(\frac{A}{F}\right)_s^v\ kmole/kmole\ fuel \tag{2.11}$$

$$M_{N_2} = 3.7619\lambda\left(x + \frac{y}{4} - \frac{z}{2}\right), \text{ or}$$

$$M_{N_2} = 0.79\lambda\left(\frac{A}{F}\right)_s^v\ kmole/kmole\ fuel \tag{2.12}$$

The total amount of products of combustion are

$$M_P = x + \frac{y}{2} + 0.21(\lambda - 1)\left(\frac{A}{F}\right)_s^v + 0.79\lambda\left(\frac{A}{F}\right)_s^v = x + \frac{y}{2} + (\lambda - 0.21)\left(\frac{A}{F}\right)_s^v \tag{2.13}$$

Equations (2.8) and (2.13) show that the volumes of the reactants and products are not equal, and

$$\Delta M = M_P - M_R = \left(x + \frac{y}{2}\right) + \lambda\left(\frac{A}{F}\right)_s^v - 0.21\left(\frac{A}{F}\right)_s^v - \left[1 + \lambda\left(\frac{A}{F}\right)_s^v\right]$$

$$\Delta M = \left(x + \frac{y}{2} - 1\right) - 0.21\left(\frac{A}{F}\right)_s^v \tag{2.14}$$

The molar change ΔM is dependent on the elemental composition of the fuel, and it could be positive, equal to zero, or negative, according to the following conditions.

The coefficient of molar change is defined as

$$\mu_o = \frac{M_P}{M_R} = \frac{M_R + \Delta M}{M_R} = 1 + \frac{\Delta M}{M_R} \tag{2.15}$$

For hydrocarbon fuels having the chemical formula $C_xH_yO_z$ reacting with stoichiometric or excess air $(\lambda \geq 1)$

$$\mu_o = 1 + \frac{\Delta M}{M_R} = 1 + \frac{\frac{y}{4} + \frac{z}{2} - 1}{1 + \lambda\left(\frac{A}{F}\right)_s^v} \tag{2.16}$$

2.3.2 Multi-Component Gaseous Fuels

These are hydrocarbon fuels occurring naturally or as distillation products. Examples of the former are natural gas and shale gas and of the latter petroleum gases and coal gas. Natural gas is formed from decomposing animal and plant matter deep underground. Shale gas is also found underground, trapped within the porous structure of shale sediments. Natural gas and shale gas exist as gaseous mixtures consisting mainly of methane. Typical compositions of the two fuels are given in Table 2.3. Examples of gaseous fuels resulting from distillation processes are petroleum gases and coal gas. Petroleum gases, which are obtained

Table 2.3 Volume composition of some gaseous fuels.

Element	Natural gas	Coal gas	Shale gas (USA)
Methane CH_4	92.6	17	94.3
Ethylene C_2H_4	—	5	—
Ethane C_2H_6	3.6	—	2.7
Propane C_3H_8	0.8	—	0.6
Butane C_4H_{10}	0.3	—	0.2
Pentane C_5H_{12}	—	—	0.2
Nitrogen N_2	2.6	6	1.5
Carbon dioxide CO_2	0.1	4.5	0.5
Carbon monoxide CO	—	19	—
Hydrogen H_2	—	48	—
Oxygen O_2	—	0.5	—

from natural gas processing and oil refining, are mixtures of butane and propane and are commercialised as liquefied petroleum gas (LPG), which is used for heating and cooking and as a fuel in internal combustion engines. Coal gas is obtained by the coal gasification process and can be used as a fuel in combined-cycle power plants or as feedstock to produce synthetic natural gas (syngas). Syngas in its turn can be converted into gasoline and diesel fuels for use in internal combustion engines. A typical composition of coal gas in shown in Table 2.3.

The molar (volumetric) composition of a multi-component gaseous fuel can be written as

$$\sum_{i=1}^{n} (C_x H_y O_z)_i + V_N = 1 \ kmole \tag{2.17}$$

where

$(C_x H_y O_z)_i$ is the volumetric fraction of the ith hydrocarbon fuel component and V_N is the volumetric fraction of nitrogen in the fuel.

For the combustion of a single constituent in pure oxygen in mole terms:

$$C_x H_y O_z + \left(x + \frac{y}{4} - \frac{z}{2}\right) O_2 \rightarrow x CO_2 + \frac{y}{2} H_2 O$$

Note that the amount of oxygen in the reactants is reduced by the amount of oxygen in the fuel itself. The stoichiometric air-fuel ratio for this fuel with n components on volume basis is

$$\left(\frac{A}{F}\right)_s^v = \frac{1}{0.21} \sum_{i=1}^{n} \left(x + \frac{y}{4} - \frac{z}{2}\right) (C_x H_y O_z)_i \ \ kmole \ air/kmole \ fuel \tag{2.18}$$

The letters s and v stand for *stoichiometric* and *volumetric*, respectively.

The total amount of the reactants per *kmole* of fuel for stoichiometric and non-stoichiometric mixtures are, respectively,

$$M_R = 1 + \left(\frac{A}{F}\right)_s^v \ kmole/kmole \ fuel; \ M_R = 1 + \lambda \left(\frac{A}{F}\right)_s^v \ kmole/kmole \ fuel \tag{2.19}$$

Upon complete combustion of the gaseous fuel with excess air, the products of complete combustion are:

$$M_{CO_2} = \sum_{i=1}^{n} x(C_x H_y O_z)_i \ \ kmole/kmole \ fuel \tag{2.20}$$

$$M_{H_2O} = \sum_{i=1}^{n} \frac{y}{2}(C_x H_y O_z)_i \ \ kmole/kmole \ fuel \tag{2.21}$$

$$M_{O_2} = 0.21(\lambda - 1) \left(\frac{A}{F}\right)_s^v \ \ kmole/kmole \ fuel \tag{2.22}$$

$$M_{N_2} = 0.79\lambda \left(\frac{A}{F}\right)_s^v + V_N \ \ kmole/kmole \ fuel \tag{2.23}$$

The quantities in *kmole* in Eqs. (2.20–2.23) can be replaced by m^3 if need be.

The total amount of combustion products is:

$$M_P = \sum_{i=1}^{n} \left(x + \frac{y}{2}\right)(C_x H_y O_z)_i + \lambda \left(\frac{A}{F}\right)_s^v - 0.21 \left(\frac{A}{F}\right)_s^v + V_{N_2} \tag{2.24}$$

From Eq. (2.17)

$$V_N = 1 - \sum_{i=1}^{n} (C_x H_y O_z)_i$$

From Eq. (2.18)

$$0.21 \left(\frac{A}{F}\right)_s^v = \sum_{i=1}^{n} \left(x + \frac{y}{4} - \frac{z}{2}\right) (C_x H_y O_z)_i$$

Hence, Eq. (2.24) can be written as

$$M_P = 1 + \lambda \left(\frac{A}{F}\right)_s^v + \sum_{i=1}^{n} \left(\frac{y}{4} + \frac{z}{2} - 1\right) (C_x H_y O_z)_i \qquad (2.25)$$

The molar change resulting from the reaction is

$$\Delta M = M_P - M_R = \sum_{i=1}^{n} \left(\frac{y}{4} + \frac{z}{2} - 1\right) (C_x H_y O_z)_i \qquad (2.26)$$

The molar change for a gaseous fuels depends on the elemental composition of the fuel:

If $\left(\frac{y}{4} + \frac{z}{2} - 1\right) > 1$, the molar change is positive.

If $\left(\frac{y}{4} + \frac{z}{2} - 1\right) = 0$, the molar change is zero.

If $\left(\frac{y}{4} + \frac{z}{2} - 1\right) < 1$, the molar change is negative.

Example 2.1   Determine the stoichiometric air-fuel ratio for a gaseous fuel with three components: 90% CH_4, 7% C_2H_6, 3% C_3H_8.
   The air-fuel ratio on a volume basis is

$$\left(\frac{A}{F}\right)_s^v = \frac{1}{0.21} \sum_{i=1}^{n} \left(x + \frac{y}{4} - \frac{z}{2}\right) (C_x H_y O_z)_i$$

$$= \frac{1}{0.21} [(1 + 4/4)(0.9) + (2 + 6/4)(0.07) + (3 + 8/4)(0.03)]$$

$$\left(\frac{A}{F}\right)_s^v = 10.45 \, kmole \, air/kmole \, fuel$$

On mass (gravimetric) basis:

$$\left(\frac{A}{F}\right)_s^g = 10.45 \times \frac{molecular \, mass \, of \, air \, (\mu_a)}{molecular \, mass \, of \, fuel \, (\mu_f)}$$

The molecular mass of air, taking into account the volume of concentrations of oxygen and nitrogen, is

$$\mu_a = 32(0.21) + 28(0.79) = 28.84 \, kg/kmole \, fuel$$

If argon and carbon dioxide in the air are also accounted for, $\mu_a = 28.97$. The molecular mass of the fuel considering the volume concentrations of the three components is

$$\mu_f = 16(0.9) + 30(0.07) + 44(0.03) = 17.82 \, kg/kmole \, fuel$$

Finally,

$$\left(\frac{A}{F}\right)_s^g = \frac{10.45(28.84)}{17.82} = 16.91 \ kg \ air/kg \ fuel$$

2.3.3 Fuels with Known Mass Concentration of the Constituent Elements

This category includes liquid and solid fuels with known mass concentrations of the constituent elements (carbon, hydrogen, sulfur, and oxygen). Since solid fuels are not used in reciprocating and gas turbine engine, only liquid fuels will be considered. The mass concentrations of carbon, hydrogen, sulfur, and oxygen, based on the gravimetric analysis of the fuel, can be used to analyse the combustion reactions of the fuel. The same approach can also be applied to solid fuels. Sulfur and oxygen contents are taken into account mainly in fuels with high concentrations of these two elements, such as heavy diesel fuels. Sulfur content in regular gasolines and diesel fuels is currently limited by legislation in most industrialised countries to less than $10 \ mg/kg$. Table 2.4 shows the average mass composition of liquid fuels used in reciprocating and gas turbine engines.

Hydrocarbon fuels react with oxygen only when they are in the gaseous phase; hence, liquid fuels must be brought to their gaseous phase before burning them.

Consider $1 \ kg$ of solid or liquid fuel containing carbon, hydrogen, sulfur, and oxygen. If x_c, x_H, x_S, and x_O are, respectively, the mass fractions of carbon, hydrogen, sulfur, and oxygen in $1 \ kg$ of fuel, we can write the composition as

$$x_c + x_H + x_O + x_S = 1 \ kg \ fuel \tag{2.27}$$

For complete combustion, each combustible element of the fuel can react with oxygen, as shown in Table 2.5.

2.3.3.1 Air-Fuel Ratios

From the equations in Table 2.5, the stoichiometric air-fuel ratio on mass and volume bases, taking into account the mass and volume oxygen fractions in the air and the presence of oxygen in the fuel, can be written as follows:

$$\left(\frac{A}{F}\right)_s^g = \frac{1}{0.23}\left(\frac{8}{3}x_c + 8x_H + x_S - x_O\right) kg \ air/kg \ fuel \tag{2.28}$$

$$\left(\frac{A}{F}\right)_s^v = \frac{1}{0.21}\left(\frac{1}{12}x_c + \frac{1}{4}x_H + \frac{1}{32}x_S - \frac{1}{32}x_O\right) kmole \ air/kg \ fuel \tag{2.29}$$

Table 2.4 Typical carbon, hydrogen, oxygen, and sulfur mass contents of common liquid fuels.

Element	Gasoline%	Kerosene%	Diesel%	Light fuel oil %	Heavy fuel oil %
Carbon	87.47	86.08	86.6	86.2	86.1
Hydrogen	12.14	13.92	13.4	12.4	11.8
Oxygen	0.13	—	—	—	—
Sulfur	—	—	—	1.4	2.1

Table 2.5 Stoichiometric reactions of the elements of a liquid or solid fuel.

$C + O_2 \rightarrow CO_2$	$12\ kg\ C + 32\ kg\ O_2 \rightarrow 44\ kg\ CO_2$	$12\ kg\ C + 1\ kmole\ O_2 \rightarrow 1\ kmole\ CO_2$
	$1\ kg\ C + \dfrac{8}{3}\ kg\ O_2 \rightarrow \dfrac{11}{3}\ kg\ CO_2$	$1\ kg\ C + \dfrac{1}{12}\ kmole\ O_2 \rightarrow \dfrac{1}{12}\ kmole\ CO_2$
	$x_c\ kg\ C + \dfrac{8}{3} x_c\ kg\ O_2 \rightarrow \dfrac{11}{3} x_c\ kg\ CO_2$	$x_c\ kg\ C + \dfrac{1}{12} x_c\ kmole\ O_2 \rightarrow \dfrac{1}{12} x_c\ kmole\ CO_2$
$2H_2 + O_2 \rightarrow 2H_2O$	$4\ kg\ H_2 + 32\ kg\ O_2 \rightarrow 36\ kg\ H_2O$	$4\ kg\ H_2 + 1\ kmole\ O_2 \rightarrow 2\ kmole\ H_2O$
	$1\ kg\ H_2 + 8\ kg\ O_2 \rightarrow 9\ kg\ H_2O$	$1\ kg\ H_2 + \dfrac{1}{4}\ kmole\ O_2 \rightarrow \dfrac{1}{2}\ kmole\ H_2O$
	$x_H\ kg\ H_2 + 8x_H\ kg\ O_2 \rightarrow 9x_H\ kg\ H_2O$	$x_H\ kg\ H_2 + \dfrac{1}{4} x_H\ kmole\ O_2 \rightarrow \dfrac{1}{2} x_H\ kmole\ H_2O$
$S + O_2 \rightarrow SO_2$	$32\ kg\ S + 32\ kg\ O_2 \rightarrow 64\ kg\ SO_2$	$32\ kg\ S + 1\ kmole\ O_2 \rightarrow 1\ kmole\ SO_2$
	$1\ kg\ S + 1\ kg\ O_2 \rightarrow 2\ kg\ SO_2$	$1\ kg\ S + \dfrac{1}{32}\ kmole\ O_2 \rightarrow \dfrac{1}{32}\ kmole\ SO_2$
	$x_S\ kg\ S + x_S\ kg\ O_2 \rightarrow 2\ x_S\ kg\ SO_2$	$x_S\ kg\ S + \dfrac{1}{32} x_S\ kmole\ O_2 \rightarrow \dfrac{1}{32} x_S kmole\ SO_2$

The mass amount of the reactants (air-fuel mixture) G_1 per kg of fuel with excess air is

$$G_R = 1 + \lambda \left(\frac{A}{F}\right)_s^g \ kg/kg\ fuel \tag{2.30}$$

The mole amount of the air-fuel mixture M_R per kg of fuel with excess air is

$$M_R = \frac{1}{\mu_f} + \lambda \left(\frac{A}{F}\right)_s^v \ kmole/kg\ fuel \tag{2.31}$$

where μ_f is the molecular mass of the fuel.

2.3.3.2 Products of Complete Combustion ($\lambda \geq 1$)

In addition to the products shown in Table 2.5, the combustion products will include excess oxygen and all the nitrogen in the air.

The products of complete combustion with excess air on mass basis are

$$G_{CO_2} = \frac{11}{3} x_c \ kg/kg\ fuel$$

$$G_{H_2O} = 9x_H \ kg/kg\ fuel$$

$$G_{SO_2} = 2 x_S \ kg/kg\ fuel$$

$$G_{O_2} = 0.23(\lambda - 1) \left(\frac{A}{F}\right)_s^g \ kg/kg\ fuel$$

$$G_{N_2} = 0.77\lambda \left(\frac{A}{F}\right)_s^g \ kg/kg\ fuel$$

The total mass amount of the combustion products are

$$G_p = G_{CO_2} + G_{H_2O} + G_{SO_2} + G_{O_2} + G_{N_2}$$

$$G_p = \frac{11}{3} x_c + 9x_{H_2} + 2 x_S + (\lambda - 0.23) \left(\frac{A}{F}\right)_s^g \ kg/kg\ fuel \tag{2.32}$$

The products of complete combustion with excess air on a molar (volume) basis are

$$M_{CO_2} = \frac{x_c}{12} kmole/kg \ fuel$$

$$M_{H_2O} = \frac{x_H}{2} kmole/kg \ fuel$$

$$M_{SO_2} = \frac{x_S}{32} kmole/kg \ fuel$$

$$M_{O_2} = 0.21(\lambda - 1) \left(\frac{A}{F}\right)_s^v kmole/kg \ fuel$$

$$M_{N_2} = 0.79\lambda \left(\frac{A}{F}\right)_s^v kmole/kg \ fuel$$

The total molar amount of the combustion products is

$$M_p = M_{CO_2} + M_{H_2O} + M_{SO_2} + M_{O_2} + M_{N_2} \tag{2.33a}$$

$$M_p = \frac{x_c}{12} + \frac{x_H}{2} + \frac{x_S}{32} + (\lambda - 0.21)\left(\frac{A}{F}\right)_s^v \tag{2.33b}$$

The molar change resulting from the combustion process is

$$\Delta M = M_P - M_R = \frac{x_H}{4} + \frac{x_O}{32} - \frac{1}{\mu_f}$$

The coefficient of molar change is

$$\mu_o = 1 + \frac{\Delta M}{M_R} = 1 + \frac{\frac{x_H}{4} + \frac{x_O}{32} - \frac{1}{\mu_f}}{\frac{1}{\mu_f} + \lambda \left(\frac{A}{F}\right)_s^v} \tag{2.34}$$

2.3.3.3 Products of Incomplete Combustion ($\lambda < 1$)

Combustion in compression ignition (CI) engines always takes place with excess air with a minimum values of the relative air-fuel ratio around $\lambda = 1.3 - 1.4$. In spark ignition (SI) engines, λ could be less than 1 when operating close to maximum output power and when idling (Figure 2.1). Under these conditions, lack of oxygen leads to the formation of carbon monoxide in addition to carbon dioxide, and not all the hydrogen in the fuel will react with oxygen, causing hydrogen to form in the products. In the following discussion relating to

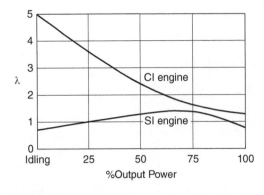

Figure 2.1 Relative air-fuel ratio as a function of power output.

combustion in SI engines, when $\lambda < 1$, sulfur content in the fuel is ignored, reducing the composition to

$$x_c + x_H + x_O = 1 \ kg \ fuel$$

It is found that the mole ratio of hydrogen to carbon monoxide in the combustion products of SI engine fuels is almost constant, i.e.

$$K = \frac{M_{H_2}}{M_{CO}} \approx const.$$

For $x_H/x_c = 0.17 - 0.19$, $K = 0.45 - 0.5$; and for $x_H/x_c = 0.13$, $K = 0.3$ (Arkhangelsky et al., 1971). It can be shown that the products of incomplete combustion of liquid gasoline are

$$M_{CO_2} = \frac{x_c}{12} - 0.42 \left(\frac{1-\lambda}{1+K} \right) \left(\frac{A}{F} \right)_s^v \ kmole/kg \ fuel$$

$$M_{CO} = 0.42 \left(\frac{1-\lambda}{1+K} \right) \left(\frac{A}{F} \right)_s^v \ kmole/kg \ fuel$$

$$M_{H_2O} = \frac{x_H}{2} - 0.42K \left(\frac{1-\lambda}{1+K} \right) \left(\frac{A}{F} \right)_s^v \ kmole/kg \ fuel$$

$$M_{H_2} = 0.42K \left(\frac{1-\lambda}{1+K} \right) \left(\frac{A}{F} \right)_s^v \ kmole/kg \ fuel$$

$$M_{N_2} = 0.79\lambda \left(\frac{A}{F} \right)_s^v \ kmole/kg \ fuel$$

The total amount of the products of incomplete combustion is

$$M_p = M_{CO_2} + M_{CO} + M_{H_2O} + M_{H_2} + M_{N_2} \tag{2.35a}$$

$$M_p = \frac{x_c}{12} + \frac{x_H}{2} + 0.79\lambda \left(\frac{A}{F} \right)_s^v \tag{2.35b}$$

As before,

$$M_R = \frac{1}{\mu_f} + \lambda \left(\frac{A}{F} \right)_s^v \ kmole/kg \ fuel$$

Hence,

$$\Delta M = M_p - M_R = \frac{x_c}{12} + \frac{x_H}{2} - 0.21\lambda \left(\frac{A}{F} \right)_s^v - \frac{1}{\mu_f}$$

and

$$\mu_o = 1 + \frac{\Delta M}{M_R} = 1 + \frac{\frac{x_c}{12} + \frac{x_H}{2} - 0.21\lambda \left(\frac{A}{F} \right)_s^v - \frac{1}{\mu_f}}{\frac{1}{\mu_f} + \lambda \left(\frac{A}{F} \right)_s^v} \tag{2.36}$$

Example 2.2 Consider gasoline having the composition $x_c = 0.85$, $x_H = 0.15$ and molecular mass $\mu_f = 120$, $\lambda = 0.9$:

$$\left(\frac{A}{F} \right)_s^v = \frac{1}{0.21} \left(\frac{0.85}{12} + \frac{0.15}{4} \right) = 0.516 \ kmole/kg \ fuel$$

$$\mu_o = 1 + \frac{\frac{0.85}{12} + \frac{0.15}{2} - 0.21(0.85)(0.516) - \frac{1}{120}}{\frac{1}{120} + (0.85)(0.516)} = 1 + 0.084 = 1.084$$

The volume of the products is greater than the volume of the reactants by 8.4%.

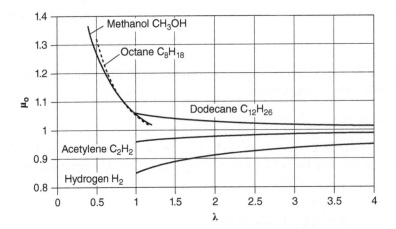

Figure 2.2 Coefficient of molar change versus relative air-fuel ratio for some liquid and gaseous fuels.

Figure 2.2 shows plots of the coefficients of the molar change versus relative air-fuel ratio for a number of liquid and gaseous fuels. Octane and dodecane are liquid fuels often used to model gasoline and light diesel fuels, respectively. Methanol is a biomass-based liquid fuel that has a very high octane number and can, in theory, be used as an alternative fuel to gasoline in high-compression-ratio engines. Ethanol (C_2H_5OH), which is also a liquid fuel derived from biomass, is already in use as an additive to boost the octane number of gasoline. The curve of the coefficient of molar change for ethanol is not shown in Figure 2.2 as it is almost identical to the curve for octane. The coefficient of molar change for the liquid fuels is always greater than 1 but decreases as λ increases. Note the sharp decrease in the case of octane and methanol compared with dodecane. The coefficient of molar change for the gaseous fuels acetylene and hydrogen is always less than 1 but increases as λ increases.

2.4 Thermodynamic Properties of the Combustion Products

Combinations of the combustion products CO_2, CO, H_2, O_2, and N_2 resulting from the reaction of liquid or gaseous hydrocarbon fuels form a gaseous mixture with specific species concentration. The combustion products are usually treated as ideal gases, and the gas mixture laws can be applied.

The amount of substance (number of moles) is related to the mass m and molecular mass μ of the substances as follows:

$$n = m/\mu$$

The number of moles n of a gas mixture of i components is

$$n = n_1 + n_2 + n_3 + \ldots = \sum_1^i n_i$$

The total mass m of the mixture is

$$m = m_1 + m_2 + m_3 + \ldots = \sum_1^i m_i$$

The average molecular mass μ_{av} of the mixture is

$$\mu_{av} = \frac{n_1\mu_1 + n_2\mu_2 + n_3\mu_3 + \ldots}{n_1 + n_2 + n_3 + \ldots} = \frac{\sum_1^i n_i\mu_i}{\sum_1^i n_i}$$

The average mass m_{av} of the mixture is

$$m_{av} = n_1\mu_1 + n_2\mu_2 + n_3\mu_3 + \ldots = \sum_1^i n_i\mu_i$$

If the universal gas constant is $\overline{R}$ ($8.3143\ kJ/kmole\ K$), the average gas constant of the mixture is then

$$R_{av} = \frac{\overline{R}}{\mu_{av}} = \sum_1^i \left(\frac{m_i}{m}\right) R_i \text{ where } m_i/m \text{ is the mass fraction of a constituent species.}$$

The average molar heat capacity at constant pressure is

$$C_{pav} = \frac{\sum_1^i n_i C_{pi}}{\sum_1^i n_i} J/mole, kJ/kmole, \text{ or } MJ/kmole \tag{2.37}$$

The average molar heat capacity at constant volume can be determined from the following equation:

$$C_{vav} = C_{pav} - \overline{R}$$

The ratio of specific heats is

$$\gamma = \frac{C_{pav}}{C_{vav}}$$

Other properties such as internal energy, enthalpy, and entropy can be determined using equations similar to Eq. (2.37).

The molar thermodynamic properties of the products of combustion, mostly based on the JANAF tables, can be found in tabular form in most books on thermodynamics. The molar specific heats at constant pressure for CO_2, CO, H_2, O_2, N_2, and *air* as functions of ($T/1000$) based on the JANAF tables are presented as tenth-order polynomials in the following compact form:

$$C_{p(gas)} = \sum_{n=0}^{10} a_n \left(\frac{T}{1000}\right)^n kJ/kmole\ K \tag{2.38}$$

The coefficients for Eq. (2.38) are given in Table 2.6. Specific heats versus temperature for the gases in Table 2.6 are tabulated in Appendix A.

The specific heat of air at constant pressure is calculated from Eq. (2.39) on the basis of the specific heats of oxygen and nitrogen:

$$C_{p(air)} = 0.21 C_{p(O_2)} + 0.79 C_{p(N_2)} \tag{2.39}$$

The results thus obtained correlate closely with data by Lemmon et al. (2000) in the temperature range $T = 300 - 2000\ K$, Jones and Dugan (1996) in the temperature range $T = 300 - 3000\ K$, and Rivkin (1987) in the temperature range $T = 223 - 1773\ K$.

Table 2.6 Coefficients a_n in Eq. (2.38) for specific heats at constant pressure of air and combustion products (enthalpy reference temperature $T_r = 298.15\ K$).

Coefficients	$C_{p(gas)}$ kJ/kmole. K						$C_{p(air)}$ kJ/kmole. K
	CO_2	CO	H_2O	H_2	O_2	N_2	Air
a_0	$1.850545E+01$	$3.302900E+01$	$3.316928E+01$	$2.513306E+01$	$3.025975E+01$	$3.275234E+01$	$3.222890E+01$
a_1	$8.416804E+01$	$-3.138893E+01$	$-3.820820E+00$	$2.644710E+01$	$-1.952694E+01$	$-2.735711E+01$	$-2.571277E+01$
a_2	$-8.923812E+01$	$8.587979E+01$	$1.933991E+01$	$-6.929132E+01$	$8.311367E+01$	$6.957493E+01$	$7.241806E+01$
a_3	$6.465750E+01$	$-9.886403E+01$	$-4.975866E+00$	$9.449683E+01$	$-1.173565E+02$	$-7.381263E+01$	$-8.295684E+01$
a_4	$-3.347177E+01$	$6.603562E+01$	$-6.353983E+00$	$-7.179134E+01$	$9.043058E+01$	$4.550827E+01$	$5.494196E+01$
a_5	$1.239974E+01$	$-2.809166E+01$	$5.700697E+00$	$3.353642E+01$	$-4.274806E+01$	$-1.793481E+01$	$-2.314560E+01$
a_6	$-3.230182E+00$	$7.853733E+00$	$-2.195554E+00$	$-1.005971E+01$	$1.295578E+01$	$4.665608E+00$	$6.406544E+00$
a_7	$5.736699E-01$	$-1.438759E+00$	$4.825460E-01$	$1.947187E+00$	$-2.529855E+00$	$-7.989831E-01$	$-1.162466E+00$
a_8	$-6.582423E-02$	$1.663162E-01$	$-6.273150E-02$	$-2.353271E-01$	$3.080278E-01$	$8.673568E-02$	$1.332070E-01$
a_9	$4.386358E-03$	$-1.100489E-02$	$4.507030E-03$	$1.615523E-02$	$-2.128097E-02$	$-5.413669E-03$	$-8.745802E-03$
a_{10}	$-1.286906E-04$	$3.176718E-04$	$-1.383728E-04$	$-4.810416E-04$	$6.371577E-04$	$1.480348E-04$	$2.507506E-04$

2.5 First Law Analysis of Reacting Mixtures

Two combustion systems will be analysed:

1. Closed, non-flow system with chemical reactions applicable to the combustion process in the cylinder of a piston engine
2. Steady-state, steady-flow system with chemical reactions applicable to the combustion process in the combustion chamber of a gas turbine

2.5.1 Non-Flow Process with Chemical Reactions

From the first law for a non-flow process, say in a closed cylinder, with heat and work transfer,

$$Q - W = E_2 - E_1 \tag{2.40}$$

W is the work done on or by the system, and Q is the heat added to or rejected by the system, causing the state of the system to change from E_1 to E_2 during the process. If changes in kinetic and potential energies are ignored, the terms on right-hand side in Eq. (2.40) can be replaced by internal energies:

$$Q - W = U_2 - U_1 \tag{2.41}$$

If Eq. (2.41) is applied to a non-flow process with chemical reactions, it can be rewritten as

$$Q - W = U_{P2} - U_{R1} \tag{2.42}$$

U_{P2} is the internal energy of the products of combustion at state 2, and U_{R1} is the internal energy of the reactants taking part in the combustion process at state 1. The change of state is now accompanied by changes in the chemical and thermodynamic properties of the substances involved. Thermodynamic properties of some combustion species are often tabulated relative to a reference temperature T_0 ($T_0 = 298.15\ K$). Hence, it is preferable to rewrite Eq. (2.42) as follows:

$$Q - W = U_{P2} - U_{R1} = (U_{P2} - U_{P0}) + (U_{P0} - U_{R0}) + (U_{R0} - U_{R1})$$

or

$$Q - W = U_{P2} - U_{R1} = (U_{R0} - U_{R1}) + (U_{P0} - U_{R0}) + (U_{P2} - U_{P0}) \tag{2.43}$$

Reading the internal energy terms in the brackets on the right-hand side of Eq. (2.43) from back to front, the path from reactants to products during the combustion process will be as follows:

1. Reactants at state 1 (p_1, T_1, U_{R1}) are brought to a reference state 0 (p_0, T_0, U_{R0}) (no chemical reactions).
2. Chemical reactions convert the reactants from state 0 (p_0, T_0, U_{R0}) to products of combustion at state 0 (p_0, T_0, U_{P0}).
3. The products of combustion at state 0 (p_0, T_0, U_{P0}) are brought to the final state 2 (p_2, T_2, U_{P2}).

Ignoring work transfer during the combustion process ($W = 0$) and considering that

$$(U_{R1} - U_{R0}) = \sum_R n_{Ri}(U_{R1} - U_{R0})_i = \sum_R n_{Ri}C_v(T_2 - T_0)_i$$

$$(U_{P2} - U_{P0}) = \sum_P n_{Pi}(U_{P2} - U_{P0})_i = \sum_R n_{Pi}C_v(T_2 - T_0)_i$$

$$\Delta U_0 = (U_{P0} - U_{R0})$$

Eq. (2.43) can be written as

$$Q = U_{P2} - U_{R1} = \sum_P n_{Pi}(U_{P2} - U_{P0})_i - \sum_R n_{Ri}(U_{R1} - U_{R0})_i + \Delta U_0 \tag{2.44}$$

where n_{Ri}, n_{Pi} are the molar concentrations of the ith components of the reactants and products, respectively, and ΔU_0 is the internal energy of reaction of the fuel at constant volume and temperature T_0. The internal energy of reaction ΔU_0 can be determined in the laboratory by measuring heat rejected during constant-volume combustion of a liquid or solid fuel with no work done and negligible kinetic energy change in a *bomb calorimeter*. The reference pressure and temperature for the combustion process are $p_o = 0.1\ MPa$, $T_o = 298.15\ K$. If the products and reactants are both at the same temperature, Eq. (2.44) becomes

$$Q = U_{P2} - U_{R1} = \Delta U_0 \tag{2.45}$$

Combustion is an exothermic process, and the sign of ΔU_0 could be positive or negative depending on the sign convention used. In this book, heat added to a system and work done by a system are both positive. Conversely, heat rejected by a system and work done on a system are both negative. Hence

$$Q = -\Delta U_0 \tag{2.46}$$

The non-flow combustion process $1 - 2$ is shown on a U-T diagram in Figure 2.3. T_2 is the combustion temperature when the process takes place without heat and work transfer. The combustion temperature reduces to $T_{2'}$ when the process is accompanied by heat and work transfer. The U-T diagram is based on the fact that the gases in the reactants and products are assumed ideal and their properties (internal energy in this case) are functions

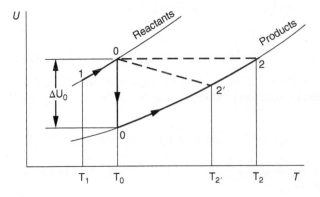

Figure 2.3 U-T diagram of the non-flow combustion process.

of temperature only, and they can be represented by two curves separated by the value of the internal energy of reaction ΔU_0.

The heat released Q as a result of the combustion of any fuel at any temperature in a non-flow process can be determined from Eq. (2.44) using tabulated data on internal energies or specific heats at constant volume of the species involved and the internal energy of reaction of the specific fuel. The internal energies of air and combustion products of hydrocarbon fuels based on the JANAF tables can be represented to a high degree of accuracy by tenth-order polynomials of the following form:

$$U_T = \sum_{n=0}^{10} a_n \left(\frac{T}{1000}\right)^n MJ/kmole \tag{2.47}$$

The coefficients of the polynomials are given in Table 2.7. The internal energy is equal to zero at the reference temperature $T_0 = 0\,K$, and the correlations are valid for the temperature range 0–6000 K. Internal energies of the gases are tabulated in Table A.2.

2.5.2 Steady-State System with Chemical Reactions

2.5.2.1 Enthalpy of Formation
We first define the enthalpy of formation of a substance. Figure 2.4 shows the steady-state, steady-flow combustion process of carbon C with oxygen without work transfer at constant temperature and pressure. In this reaction, $1\,kmole$ of carbon C reacts with $1\,kmole$ of oxygen O_2 forming $1\,kmole$ of carbon dioxide CO_2 and releasing $Q\,kJ$ of heat. Essentially, substance CO_2 is formed from the constituent elements C and O_2 that occur naturally, accompanied by heat released Q, which is known as the *enthalpy of formation*. Although Q can be measured in the laboratory, the usual practice is to determine it from observed spectroscopic data of the substance. The enthalpy of formation is defined as the change in enthalpy when a compound is formed from its constituent elements in their natural forms and in a standard state of pressure and temperature (0.1 MPa, 298.15 K). The value of 0 is assigned to the enthalpy of all the elements in their natural forms at 0.1 MPa, 298.15 K. The symbol H_f^0 will be used throughout this book for the enthalpy of formation. The superscript 0 denotes properties at pressure 0.1 MPa and reference temperature 298.15 K.

The first law of thermodynamics of the process in Figure 2.4 can be written as

$$Q = H_P - H_R \tag{2.48}$$

Subscripts R and P stand for reactants and products, respectively. For a reaction involving multiple reactants and products, Eq. (2.48) can be rewritten as

$$Q = H_P - H_R = \sum_P n_{Pi} H_P - \sum_R n_{Ri} H_P \tag{2.49}$$

$$Q = H_P - H_R = 1 \times (H_f^0)_{CO_2} - 1 \times (H_f^0)_C + 1 \times (H_f^0)_{O_2} = (H_f^0)_{CO_2} - 1 \times (0) - 1 \times (0)$$

$$Q = H_P = (H_f^0)_{CO_2} = -393{,}522\,kJ/kg$$

Data for H_f^0 for selected hydrocarbon substances are given in Table A.9.

Table 2.7 Coefficients a_n in Eq. (2.47) for internal energies of air and combustion products (enthalpy reference temperature $T_r = 298.15\ K$).

| Coefficients | U_T, MJ/kmole | | | | | | | U_T, MJ/kg |
	CO_2	CO	H_2O	H_2	O_2	N_2	Air	Air
a_0	6.99398$E-$02	$-$1.56296$E-$02	$-$5.08529$E-$03	2.39396$E-$02	2.50505$E-$03	$-$1.82911$E-$02	$-$1.39239$E-$02	$-$4.80633$E-$04
a_1	1.67712$E+$01	2.16939$E+$01	2.54112$E+$01	1.85390$E+$01	2.12816$E+$01	2.16685$E+$01	2.15873$E+$01	7.45160$E-$01
a_2	2.18406$E+$01	$-$6.00327$E+$00	$-$3.65829$E+$00	6.20120$E+$00	$-$5.72126$E+$00	$-$5.29138$E+$00	$-$5.38165$E+$00	$-$1.85766$E-$01
a_3	9.43870$E-$01	1.28888$E+$01	9.50305$E+$00	$-$9.67249$E+$00	1.81274$E+$01	1.02997$E+$01	1.19435$E+$01	4.12272$E-$01
a_4	$-$1.05289$E+$01	$-$9.79156$E+$00	$-$4.47658$E+$00	9.25364$E+$00	$-$1.74914$E+$01	$-$6.89577$E+$00	$-$9.12086$E+$00	$-$3.14838$E-$01
a_5	7.68722$E+$00	4.29032$E+$00	8.98662$E-$01	$-$4.90817$E+$00	9.39927$E+$00	2.59118$E+$00	4.02088$E+$00	1.38794$E-$01
a_6	$-$2.91753$E+$00	$-$1.18942$E+$00	$-$1.48044$E-$03	1.58163$E+$00	$-$3.10545$E+$00	$-$5.92934$E-$01	$-$1.12056$E+$00	$-$3.86801$E-$02
a_7	6.60984$E-$01	2.12086$E-$01	$-$3.77546$E-$02	$-$3.19246$E-$01	6.45588$E-$01	8.22414$E-$02	2.00544$E-$01	6.92248$E-$03
a_8	$-$8.98160$E-$02	$-$2.36157$E-$02	7.96302$E-$03	3.95654$E-$02	$-$8.24470$E-$02	$-$6.41353$E-$03	$-$2.23805$E-$02	$-$7.72542$E-$04
a_9	6.76349$E-$03	1.49574$E-$03	$-$7.21841$E-$04	$-$2.75839$E-$03	5.91546$E-$03	2.23929$E-$04	1.41915$E-$03	4.89869$E-$05
a_{10}	$-$2.17228$E-$04	$-$4.11632$E-$05	2.54719$E-$05	8.29234$E-$05	$-$1.82660$E-$04	$-$9.01231$E-$07	$-$3.90705$E-$05	$-$1.34865$E-$06

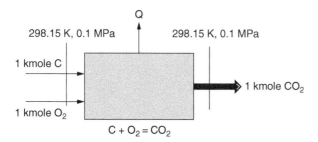

Figure 2.4 Steady-state, steady-flow combustion process without change of state.

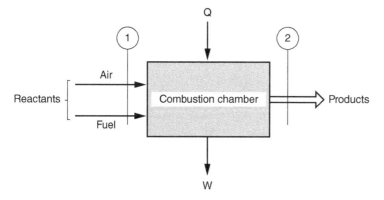

Figure 2.5 Schematic diagram of the steady-flow system with chemical reactions.

2.5.2.2 Enthalpy of Reaction

The steady-state steady flow process discussed can be developed to cover a process in which the products and reactants are at two different states, as shown in Figure 2.5:

$$Q - W = H_{P2} - H_{R1} \tag{2.50}$$

Following the same methodology described for the non-flow combustion process, an H-T diagram of the steady-state, steady-flow combustion process can be constructed a shown in Figure 2.6 and Eq. (2.50) rewritten as

$$Q - W = H_{P2} - H_{R1} = (H_{R0} - H_{R1}) + \Delta H_0 + (H_{P2} - H_{P0}) \tag{2.51}$$

Equation (2.50) can also be written as

$$Q - W = H_{P2} - H_{R1} = \sum_P n_{pi} H_P - \sum_R n_{Ri} H_R \tag{2.52}$$

The enthalpy of a substance at any temperature T above or below the reference temperature $T_0 = 298.15\ K$ with no change in pressure is given by

$$H_T = H_f^0 + \Delta H_T \tag{2.53}$$

ΔH_T is the enthalpy of a substance at any temperature relative to a reference temperature $T_0 = 298.15\ K$ and pressure $p_0 = 0.1\ MPa$ at which the enthalpy is taken equal to zero.

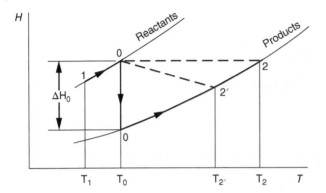

Figure 2.6 H-T diagram of the steady-state, steady-flow combustion process.

Substituting H_T from Eq. (2.53) for the enthalpies in Eq. (2.52), we obtain

$$Q - W = \sum_P n_{pi}(H_f^0 + \Delta H_T)_i - \sum_R n_{Ri}(H_f^0 + \Delta H_T)_i \tag{2.54}$$

For a steady-flow combustion process, $W = 0$, and Eq. (2.54) can be solved using thermodynamic data for the relative enthalpies ΔH_T and enthalpies of formation H_f^0 of the species in the reactants and products. Rearranging Eq. (2.54) we get

$$Q = \left(\sum_P n_{pi}(\Delta H_T)_i - \sum_R n_{Ri}(\Delta H_T)_i \right) + \left(\sum_P n_{pi}(H_f^0)_i - \sum_R n_{Ri}(H_f^0)_i \right) \tag{2.55}$$

If the reaction takes place at a constant pressure 0.1 *MPa* and reference temperature 398.15 *K*, the first two terms on the right-hand side of Eq. (2.55) equate to zero, and it is reduced to

$$Q = \left(\sum_P n_{pi}(H_f^0)_i - \sum_R n_{Ri}(H_f^0)_i \right) \tag{2.56}$$

If we replace Q with $-\Delta H_0$, then

$$-\Delta H_0 = \sum_P n_{pi}(H_f^0)_i - \sum_R n_{Ri}(H_f^0)_i \tag{2.57}$$

The negative sign indicates an exothermic reaction (heat lost by the system). ΔH_0 is called the *standard enthalpy of reaction* (or *heat of combustion*) of the fuel at constant pressure p_0 and reference temperature T_0. Since the change in enthalpy is independent of the path by which the end state is reached, the reaction of fuel C_xH_y can be viewed in two stages: a breaking up of the reactant into elements followed by a recombination of the elements with oxygen to form the products. Therefore, the enthalpy of reaction of the fuel can be found from the enthalpies of formation of the products formed from the twin-path reaction of the elements:

$$C_xH_y \diagup \begin{matrix} xC + xO_2 = xCO_2 \\ \\ \dfrac{y}{2} H_2 + \dfrac{y}{4} O_2 = \dfrac{y}{2} H_2O \end{matrix}$$

ΔH_0 of a gaseous fuel can be determined in the laboratory by measuring heat rejected during steady-flow combustion with no work done and negligible kinetic energy change using a device such as Boy's Calorimeter. The reference pressure and temperature are $p_o = 0.1\ MPa$, $T_o = 298.15\ K$.

Example 2.3 Determine the enthalpy of reaction of methane with oxygen:

$$CH_4 + 2O_2 = CO_2 + 2H_2O$$

$$\Delta H_0 = \sum_P n_{pi}(H_f^0)_i - \sum_R n_{Ri}(H_f^0)_i = 1 \times (H_f^0)_{CO_2} + 2 \times (H_f^0)_{H_2O} - 1 \times (H_f^0)_{CH_4} - 2 \times (H_f^0)_{O_2}$$

Using data from Table A.9,

$$\Delta H_{0(CH_4)} = 1 \times (H_f^0)_{CO_2} + 2 \times (H_f^0)_{H_2O} - 1 \times (H_f^0)_{CH_4} - 2 \times (0)$$

$$\Delta H_{0(\Delta CH_4)} = 1 \times (-393{,}520) + 2(-241{,}820) - 1 \times (-74{,}400)$$
$$= -803160\ kJ/kmole\ (-50{,}197\ kJ/kg)$$

The relationship between ΔU_0 and ΔH_0 can be determined as follows:

$$\Delta H_0 = H_{P0} - H_{R0} = (U_{P0} + p_{P0}V_{P0}) - (U_{R0} + p_{R0}V_{R0})$$
$$= (U_{P0} - U_{R0}) + (p_{P0}V_{P0} - p_{R0}V_{R0})$$

Using the equation of state $pV = n\bar{R}T$ and noting that combustion takes place at the reference temperature T_0, we obtain

$$\Delta H_0 = \Delta U_0 + \bar{R}T_0(n_P - n_R) \tag{2.58}$$

If $n_P \approx n_R$, then

$$\Delta H_0 \approx \Delta U_0 \tag{2.59}$$

In engineering practice, it is common to use other terms when referring to energy transfer in combustion processes: the *calorific value* and *heating value* of the fuel. These two terms are synonymous and are defined as the amount of heat released when a unit mass of fuel is burnt completely in a calorimeter under specified conditions. Unlike ΔU_0 and ΔH_0, which are negative values according to the adopted sign convection, the heating value is always quoted as a positive value. Additionally, the heating value depends on the phase of the water in the combustion products. When the water in the products is in liquid form, the heating value is known as the *higher* heating value (H_h); and when the water is in vapour form, it is known as the *lower* heating value (H_l).

For gaseous fuels in a steady-flow (constant-pressure) process:

$$H_h = H_l + (mh_{fg})_{water}$$

For solid or liquid fuels in a constant-volume process:

$$H_h = H_l + (mu_{fg})_{water}$$

where m is the mass of the water in the products per unit mass of fuel, and h_{fg} and u_{fg} are, respectively, the enthalpy and internal energy of vaporisation of water at the reference temperature $T_0 = 298.15\ K$.

Table 2.8 Composition and heating values of some liquid, gaseous, and solid fuels.

Fuels	Formula	x_c	x_H	x_O	x_S	Published data H_h	H_l	Eq. (2.60) H_h	Eq. (2.61) H_l
SI engine fuels	—	—	—	—	—	MJ/kg	MJ/kg	MJ/kg	MJ/kg
Gasoline									
Regular unleaded (l)		87.47	12.14	0.13		44.17	41.52	44.98	42.24
Premium unleaded (l)		88.4	11.28	0.24		43.44	40.98	44.21	41.66
Premium plus (l)		87.63	10.94	1.33		42.69	40.31	43.40	40.93
Octane (l)	C_8H_{18}	84.21	15.79					48.47	44.90
Alcohols									
Methanol (l)	CH_3OH	37.5	12.5	50		23.85	21.11	23.00	20.18
Ethanol (l)	C_2H_5OH	52.17	13.04	34.78		30.59	27.72	30.34	27.39
CI engine fuels									
Diesel fuels									
Light diesel (l)	$C_{12.3}H_{22.2}$	86.92	13.08			46.1	43.20	45.99	43.04
Heavy diesel (l)	$C_{14.6}H_{24.8}$	87.6	12.4			45.5	42.80	45.37	42.57
Vegetable oils									
Coconut oil (l)		73.3	12.2	14.5			35.30	38.67	35.92
Pape seed oil (l)		77.2	12	10.8			37.20	40.15	37.44
Linseed oil (l)		77.5	11.6	10			37.00	39.84	37.22
Gas turbine engines									
Kerosene (l)		86.3	13.6		0.100	46.2	43.25	46.45	43.37
Aviation fuel (l)		85	15					47.75	44.36
Miscellaneous fuels									
Gaseous fuels									
Methane (g)	CH_4	75	25			55.51	50.02	56.91	51.26
Propane (g)	C_3H_8	81.82	18.18			50.35	46.36	50.67	46.55
Butane (g)	C_4H_{10}	82.76	17.24			49.72	46.05	49.80	45.91
Solid fuels									
Anthracite (s)		88.2	2.7	1.7	1.20	33.5	32.45	33.34	32.73
Medium-rank coal (s)		81.8	4.9	4.4	1.90			33.70	32.60
Low-rank coal (s)		75	4.6	10.9	2.10			30.33	29.29
Coke (s)		90	0.4	1.9		30.75	30.50	30.91	30.82
Dry wood		50.3	6.2	43.08		19.9		20	18.8
Wood (40% moisture)		30.18	3.72	25	85	10.9		12.12	10.27

Heating values can be found in thermodynamic and internal combustion (IC) engine texts for various gaseous, liquid, and solid fuels in tabular form. Heating values can also be approximated mathematically if the mass composition of a fuel is known, using empirical equations such as Mendeleev's equation (Arkhangelsky et al., 1971):

$$H_h = 34.013x_c + 125.6x_H - 10.9(x_O - x_S) - 2.512(9x_H + x_{H_2O}) \, MJ/kg \tag{2.60}$$

$$H_l = 34.013x_c + 125.6x_H - 10.9(x_O - x_S) \, MJ/kg \tag{2.61}$$

where, x_c, x_H, x_O, x_S are the mass fractions of carbon, hydrogen, oxygen, and sulfur, respectively, in 1 kg of fuel, and x_{H_2O} is the moisture content, if any, in the fuel.

Table 2.8 lists heating value data for some fuels from the literature and the corresponding values computed from Eqs. (2.60) and (2.61).

The molar enthalpy changes of the products of complete combustion relative to the reference temperature 25 °C, adapted from the JANAF tables, can be represented by a tenth-order polynomial of the form

$$\Delta H_T = \sum_{n=0}^{10} a_n \left(\frac{T}{1000} \right)^n kJ/kmole \tag{2.62}$$

The polynomial coefficients for CO_2, CO, H_2O, H_2, O_2, N_2, and air are shown in Table 2.9. Tabulated data can be found in Table A.3. The enthalpies of the gases are taken equal to 0 at reference temperature 298.15 K and pressure 0.1 MPa. The polynomial are valid for the temperature range $T = 298.15$–6000 K.

2.6 Adiabatic Flame Temperature

2.6.1 Steady-State Process

Consider the H-T diagram shown in Figure 2.7 for combustion process in which the reactants are at three different initial states: state 1 ($T_1 < T_0$), state 0 (at which the temperature is the reference value T_0), and state 5 ($T_5 > T_0$). At T_1, T_0, and T_5, two distinct processes can be identified:

- Steady-flow isothermal combustion processes $1 - 3$, $0 - 0$, and $5 - 7$, during which heat equal to the enthalpy of combustion is generated at constant temperature with no work transfer
- Steady-flow constant enthalpy (adiabatic) combustion processes $1 - 2$, $0 - 4$, and $5 - 6$, during which the energy released is used to raise the temperature of the products from T_1 to T_2, T_0 to T_4, and T_5 to T_6, respectively, without work transfer

At the reference temperature T_0, process $0 - 4$ yields the maximum temperature increase for steady-state, steady-flow combustion process, which is known as the *adiabatic flame temperature* (AFT). This path-independent process can also be imagined taking place along the path $0 - 0 - 4$ shown by the arrows. Similarly, the adiabatic flame temperatures for reactant temperatures T_1 and T_5 can be found by following the paths $1 - 0 - 0 - 2$ and $5 - 0 - 0 - 6$. It is apparant that the adiabatic flame temperature increases with increasing temperature of the reactants. Generally, applying Eq. (2.54) to any constant enthalpy

Table 2.9 Coefficients a_n in Eq. (2.62) for air and combustion products ($T = 300{-}6000\,K$)

Coefficients	ΔH_T, MJ/kmole			
	CO_2	CO	H_2O	H_2
a_0	$-8.75035E+00$	$-8.79549E+00$	$-9.97897E+00$	$-8.65178E+00$
a_1	$1.98233E+01$	$3.09993E+01$	$3.40028E+01$	$2.86562E+01$
a_2	$3.81012E+01$	$-9.24622E+00$	$-4.60987E+00$	$2.22978E+00$
a_3	$-2.34710E+01$	$1.79305E+01$	$1.09847E+01$	$-5.13309E+00$
a_4	$1.02904E+01$	$-1.41821E+01$	$-5.75463E+00$	$6.18173E+00$
a_5	$-3.19713E+00$	$6.61689E+00$	$1.56920E+00$	$-3.59589E+00$
a_6	$6.94534E-01$	$-1.96865E+00$	$-2.24252E-01$	$1.21806E+00$
a_7	$-1.02394E-01$	$3.77856E-01$	$9.34966E-03$	$-2.54050E-01$
a_8	$9.69029E-03$	$-4.53299E-02$	$1.82025E-03$	$3.22641E-02$
a_9	$-5.26987E-04$	$3.09266E-03$	$-2.71534E-04$	$-2.29429E-03$
a_{10}	$1.24152E-05$	$-9.16141E-05$	$1.12781E-05$	$7.01428E-05$

Coefficients	ΔH_T, MJ/kmole			ΔH_T, MJ/kg
	O_2	N_2	Air	Air
a_0	$-8.09901E+00$	$-8.96855E+00$	$-8.78736E+00$	$-3.03326E-01$
a_1	$2.54662E+01$	$3.21683E+01$	$3.07553E+01$	$1.06162E+00$
a_2	$5.20982E+00$	$-1.16998E+01$	$-8.08606E+00$	$-2.79118E-01$
a_3	$3.26310E+00$	$1.96777E+01$	$1.60992E+01$	$5.55720E-01$
a_4	$-5.66051E+00$	$-1.47654E+01$	$-1.27274E+01$	$-4.39331E-01$
a_5	$3.51907E+00$	$6.65939E+00$	$5.93146E+00$	$2.04745E-01$
a_6	$-1.22789E+00$	$-1.93166E+00$	$-1.76104E+00$	$-6.07883E-02$
a_7	$2.60641E-01$	$3.63304E-01$	$3.36977E-01$	$1.16319E-02$
a_8	$-3.34816E-02$	$-4.28545E-02$	$-4.02771E-02$	$-1.39030E-03$
a_9	$2.39941E-03$	$2.88203E-03$	$2.73714E-03$	$9.44817E-05$
a_{10}	$-7.37573E-05$	$-8.43160E-05$	$-8.07633E-05$	$-2.78783E-06$

process without work transfer yields

$$\sum_P n_{pi}(H_f^0 + \Delta H_T)_i - \sum_R n_{Ri}(H_f^0 + \Delta H_T)_i = 0 \tag{2.63}$$

or

$$\sum_P n_{pi}(\Delta H_T)_i + \Delta H_0 - \sum_R n_{Ri}(\Delta H_T)_i = 0 \tag{2.64}$$

If the enthalpy of combustion ΔH_0 of a hydrocarbon fuel is unknown, Eq. (2.63) can be used together with the values of enthalpies of formation and relative enthalpies. If the enthalpy of combustion for the fuel is known, Eq. (2.64) is easier to handle.

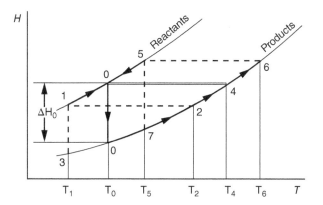

Figure 2.7 H-T diagram of the steady-state combustion process for different initial conditions.

Consider the general equation of hydrocarbon fuel $C_xH_yO_z$ for stoichiometric and lean-mixture reactions:

$$C_xH_yO_z + \lambda\left(x + \frac{y}{4} - \frac{z}{2}\right)(O_2 + 3.7619N_2)$$

$$= xCO_2 + \frac{y}{2}H_2O + (\lambda - 1)\left(x + \frac{y}{4} - \frac{z}{2}\right)O_2 + 3.7619\lambda\left(x + \frac{y}{4} - \frac{z}{2}\right)N_2$$

Equation (2.63) for the reaction with the reactants initially at temperature T_1 (Figure 2.7) can now be written as

$$x(H_f^0 + \Delta H_{T_2})_{CO_2} + \frac{y}{2}\left(H_f^0 + \Delta H_{T_2}\right)_{H_2O} + (\lambda - 1)\left(x + \frac{y}{4} - \frac{z}{2}\right)\left(H_f^0 + \Delta H_{T_2}\right)_{O_2}$$

$$+ 3.7619\lambda\left(x + \frac{y}{4} - \frac{z}{2}\right)\left(H_f^0 + \Delta H_{T_2}\right)_{N_2}$$

$$= \left[(H_f^0)_{Fuel} + (H_{T_1} - H_{T_0})_{Fuel}\right] + \lambda\left(x + \frac{y}{4} - \frac{z}{2}\right)\left(H_f^0 + \Delta H_{T_1}\right)_{O_2}$$

$$+ 3.7619\lambda\left(x + \frac{y}{4} - \frac{z}{2}\right)\left(H_f^0 + \Delta H_{T_1}\right)_{N_2} \tag{2.65}$$

The enthalpies of formation of oxygen and nitrogen are taken as 0 in this equation.

Equation (2.64) for the same reaction of a fuel with known enthalpy of combustion ΔH_0 can be written as

$$x(\Delta H_{T_2})_{CO_2} + \frac{y}{2}(\Delta H_{T_2})_{H_2O} + (\lambda - 1)\left(x + \frac{y}{4} - \frac{z}{2}\right)(\Delta H_{T_2})_{O_2}$$

$$+ 3.7619\lambda\left(x + \frac{y}{4} - \frac{z}{2}\right)(\Delta H_{T_2})_{N_2} + \Delta H_0$$

$$= (H_{T_1} - H_{T_0})_{Fuel} + \lambda\left(x + \frac{y}{4} - \frac{z}{2}\right)\left[(\Delta H_{T_1})_{O_2} + 3.7619(\Delta H_{T_1})_{N_2}\right] \tag{2.66}$$

The adiabatic flame temperature can be found from either equation by iteration, by successively assuming a temperature and looking up the properties until the right-hand side becomes equal to the left-hand side of the equation.

For the case of incomplete combustion $\lambda < 1$ (excess fuel or insufficient air), the reaction equation depends on the products of incomplete combustion. If we assume that carbon monoxide (CO) is the only product of incomplete combustion, the following reaction can be used:

$$C_xH_yO_z + \lambda\left(x + \frac{y}{4} - \frac{z}{2}\right)(O_2 + 3.7619N_2) = aCO_2 + bCO + dH_2O + fN_2$$

The concentration of the products can be determined for any fuel with a known formula by balancing the elements on both sides of the reaction equation and the adiabatic flame temperature determined as described previously.

If hydrogen molecules H_2 are also present in the products, it is convenient to write the reaction in terms of reactants (Eq. (2.31)) and products (Eq. 2.35a) in *kmole* per *kg* fuel thus:

$$\frac{1}{\mu_f} + \lambda\left(\frac{A}{F}\right)_s^v = M_{CO_2} + M_{CO} + M_{H_2O} + M_{H_2} + M_{N_2} \tag{2.67}$$

This method can also be used in stoichiometric and lean-mixture reactions ($\lambda \geq 1$), for which Eq. (2.67) can be rewritten as

$$\frac{1}{\mu_f} + \lambda\left(\frac{A}{F}\right)_s^v = M_{CO_2} + M_{H_2O} + M_{O_2} + M_{N_2} \tag{2.68}$$

The combustion products in this case are as in Eq. (2.33a) without sulfur dioxide and, if the mixture is stoichiometric ($\lambda = 1$), $M_{O_2} = 0$. The term $1/\mu_f$ is multiplied by the enthalpy of formation of the fuel and the term $\lambda(A/F)_s^v$ by the enthalpy of air in Eqs. (2.67) and (2.68).

Combustion Eq. (2.63) applied to reaction (2.67) can be written as

$$M_{CO_2}(H_f^0 + \Delta H_{T_2})_{CO_2} + M_{CO}(H_f^0 + \Delta H_{T_2})_{CO} + M_{H_2O}(H_f^0 + \Delta H_{T_2})_{H_2O}$$
$$+ M_{H_2}(H_f^0 + \Delta H_{T_2})_{H_2} + M_{N_2}(H_f^0 + \Delta H_{T_2})_{N_2}$$
$$= [(H_f^0)_{Fuel}/\mu_f + (H_{T_1} - H_{T_0})_{Fuel}/\mu_f] + \lambda(A/F)_s^v(\Delta H_{T_1})_{air} \tag{2.69}$$

The units of the concentrations are in *kmole/kg*, relative enthalpies ΔH_{T_2} in *kJ/kmole*, and enthalpies of formation in *kJ/kmole*. Fuel is usually delivered to the combustion chamber at the ambient temperature; hence, the term $(H_{T_1} - H_{T_0})_{Fuel}/\mu_f$ can be ignored.

Example 2.4   Calculate the adiabatic flame temperature of the steady-state, steady-flow combustion of liquid octane (C_8H_{18}) as a function of relative air-fuel ratio $0.6 \leq \lambda \leq 4.0$ for the following cases when the reactants and products are at a pressure of 0.1 *MPa*:

Case 1: Initial temperature of the fuel and air 298.15 K
Case 2: Initial temperature of the fuel and air 278 K
Case 3: Initial temperature of the fuel 298.15 K and of air 400 K.

Assume the water in the products is in vapour phase, and take the enthalpy of formation and specific heat at constant pressure for octane as -250100 and 254.6 *kJ/kmole. K*, respectively.

Solution
Gravimetric composition of fuel C_8H_{18}:

$$x_c = 12 \times 8/(12 \times 8 + 1 \times 18) = 96/114 = 0.842$$
$$x_{H_2} = 1 \times 18/(12 \times 8 + 1 \times 18) = 18/114 = 0.158$$

Table 2.10 Adiabatic flame temperature of octane (C_8H_{18}) vs. relative air-fuel ratio and relative fuel-air ratio (equivalence ratio).

ϕ	λ	AFT (K)		
		Case 1	Case 2	Case 3
1.67	0.6	1830	1815	1895
1.43	0.7	2020	2005	2100
1.25	0.8	2170	2158	2245
1.11	0.9	2295	2275	2355
1.05	0.95	2355	2330	2410
1.00	1	2400	2378	2460
0.91	1.1	2275	2262	2335
0.83	1.2	2120	2105	2198
0.71	1.4	1915	1895	1985
0.63	1.6	1755	1735	1820
0.56	1.8	1615	1600	1690
0.50	2	1505	1490	1585
0.33	3	1155	1130	1235
0.25	4	962	930	1050
0.20	5	840	820	930

Combustion Eq. (2.63) is applied to reaction (2.67) for $\lambda < 1$ and to reaction (2.68) for $\lambda \geq 1$, taking into account that $M_{O_2} = 0$ for $\lambda = 1$. The results of the calculations are shown in Table 2.10 and Figure 2.8. It can be concluded that

- Maximum AFT in each case is obtained when the mixture is stoichiometric ($\lambda = 1$).
- AFT decreases, relative to the maximum value in each case, when the air-fuel mixture is made rich ($\lambda < 1$) or lean ($\lambda > 1$).
- Reducing the temperature of the air below the standard temperature causes the AFT to decrease over the entire range of air-fuel ratios.
- Increasing the temperature of the air above the standard temperature causes the AFT to increase over the entire range of air-fuel ratios.
- For a given temperature of the reactants, making the mixture lean (increasing λ) allows one to reduce the combustion temperature if engine-design considerations dictate low-temperature operation, such as industrial gas turbines.

2.6.2 Constant-Volume Combustion Process

The adiabatic constant-volume combustion process with no work and heat transfer can be rewritten as

$$U_{P2} - U_{R1} = \sum_P n_{Pi}(U_{P2} - U_{P0})_i - \sum_R n_{Ri}(U_{R1} - U_{R0})_i + \Delta U_0 = 0$$

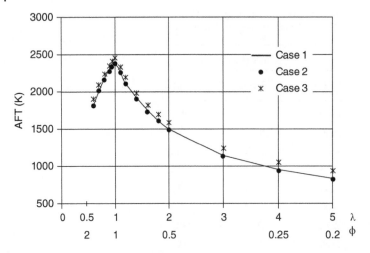

Figure 2.8 Adiabatic flame temperature of octane (C_8H_{18}) as a function of λ and ϕ in a steady-state, steady flow combustion process.

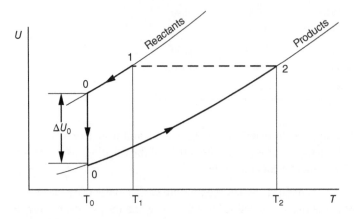

Figure 2.9 Constant-volume process in a U-T diagram.

or

$$\sum_P n_{Pi}(U_{P2} - U_{P0})_i = \sum_R n_{Ri}(U_{R1} - U_{R0})_i - \Delta U_0 \tag{2.70}$$

The process is shown schematically on the U-T diagram in Figure 2.9. The constant internal energy process $1-2$ can be replaced by process $1-0-0-2$ since properties are path-independent.

As an example, consider the constant-volume combustion process $1-2$ in the Otto cycle shown in Figure 2.10. The products (assumed pure air at this stage) are initially at state 0 (p_0, T_0, U_0), which could be the reference state or any other state. The reactants at state 0 (p_0, T_0, U_0) are compressed adiabatically to state 1 (p_1, T_1, U_1). At the end of the constant-volume combustion process $1-2$ (Figures 2.9 and 2.10), the products at point 2 will be at state (p_2, T_2, U_2). Knowing the fuel used and temperature at point 1, Eq. (2.70) can be applied to any air-fuel mixture and product composition to estimate the final

Figure 2.10 Otto cycle with inlet conditions at the reference point T_0 and p_0.

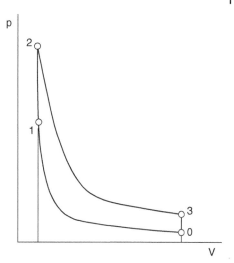

temperature of the products at point 2 by iteration. Internal energies $U_T = f(T/1000)$ for CO_2, CO, H_2O, H_2, N_2, and air can be found from Eq. (2.47) and Table 2.7. The internal energy of combustion ΔU_0 is the heating value (higher value if water in the products is liquid; lower value if water in the products is gaseous) of the fuel with a negative sign to indicate heat transfer from the system. Values of ΔU_0 for selected fuels are given in Table 2.8. For the reaction

$$\frac{1}{\mu_f} + \lambda \left(\frac{A}{F}\right)_s^{\nu} = M_{CO_2} + M_{CO} + M_{H_2O} + M_{H_2} + M_{N_2}$$

Eq. (2.70) is rewritten as

$$M_{CO_2}(\Delta U_{T_2})_{CO_2} + M_{CO}(\Delta U_{T_2})_{CO} + M_{H_2O}(\Delta U_{T_2})_{H_2O} + M_{H_2}(\Delta U_{T_2})_{H_2}$$
$$+ M_{N_2}(\Delta U_{T_2})_{N_2} + \Delta U_0 = (U_{T1} - U_{T0})_{fuel}/\mu_f + \lambda(A/F)_s^{\nu}(\Delta U_{T_1})_{air} \qquad (2.71)$$

2.7 Entropy Change in Reacting Mixtures

2.7.1 Absolute Entropy

It was shown in Chapter 1 that entropy change for an ideal substance between points 1 and 2 can be written as

$$s_2 - s_1 = s_{T_2}^0 - s_{T_1}^0 - \overline{R} \ln \left(\frac{p_2}{p_1}\right) \ kJ/kmole.K \qquad (2.72)$$

The absolute entropy of the substance at temperature T and pressure p, derived from Eq. (2.72), can be written as

$$s(T,p) = s^0(T) - \overline{R} \ln \left(\frac{p}{p_0}\right) \ kJ/kmole.K \qquad (2.73)$$

For component i in a multi-species chemical reaction, Eq. (2.73) can be written as

$$s_i(T,p_i) = s_i^0(T) - \overline{R} \ln \left(\frac{y_i p_{mix}}{p_0}\right) \ kJ/kmole.K \qquad (2.74)$$

and the total entropy for a mixture of ideal gases in the reaction is

$$S_{mix} = \sum_i n_i s_i(T, p_i) \, kJ/K \tag{2.75}$$

where p_i, y_i, and n_i are the partial pressure, mole fraction, and number of moles of component i in the mixture respectively, and p_{mix} is the total pressure of the mixture.

2.8 Second Law Analysis of Reacting Mixtures

Any working system must satisfy the first law as well as the second law. Up to now, chemical reactions have been examined based on the first law only. To examine compliance with the second law, consider the following entropy balance for any isolated system including chemically reacting systems

$$(S_1 - S_2) + S_{gen} = \Delta S_{sys} \tag{2.76}$$

where

$S_1 - S_2$: net entropy transfer in the system

S_{gen} : entropy generation within the system

ΔS_{sys} : entropy change of the system

Taking heat transfer to the system to be positive, the entropy balance for an isolated combustion system with heat transfer Q at system boundary temperature T can be written as

$$\sum \frac{Q}{T} + S_{gen} = (S_P - S_R) \tag{2.77}$$

Subscripts P and R denote products and reactants, respectively. For an adiabatic process, $Q = 0$, and Eq. (2.77) is reduced to

$$S_{gen} = (S_P - S_R) = \Delta S_{isol} \tag{2.78}$$

For an isolated adiabatic system (such as the combustion system), the second law of thermodynamics requires that

$$\Delta S_{isol} \geq 0 \tag{2.79}$$

$\Delta S_{isol} = 0$ for a reversible systems and $\Delta S_{isol} > 0$ for irrevsrsible systems.

Example 2.5   Propane gas is burned in a steady-state, steady-flow adiabatic process with 200% theoretical air ($\lambda = 2$). If both the propane and air enter the system at 0.1 MPa and 25 °C, determine the entropy change as a result of combustion.

General reaction of combustion of propane in air:

$$C_3H_8 + 5\lambda(O_2 + 3.7619N_2) \rightarrow 3CO_2 + 4H_2O + 5(\lambda - 1)O_2 + 18.8\lambda N_2$$

If $\lambda = 2$, the reaction is

$$C_3H_8 + 10(O_2 + 3.7619N_2) \rightarrow 3CO_2 + 4H_2O + 5O_2 + 37.619N_2$$

Table 2.11 Entropy change data for combustion of methane.

	n_i	y_i	$s_i^0(T)$	$-\bar{R}\,ln(y_i p_{mix}/p_0)$	$n_i s_i(T,p_i)$
			Reactants		
C_3H_8	1	1	270.30	0	270.3
O_2	10	0.21	205.15	12.98	2 181.22
N_2	37.619	0.79	191.61	1.96	7 281.86
				$S_R = \sum_R n_i s_i(T,p_i)\,kJ/K$	9733.39
			Products		
CO_2	3	0.060	316.02	23.33	1 018.04
H_2O	4	0.081	270.52	20.94	1 165.84
O_2	5	0.101	272.96	19.08	1 451.48
N_2	37.61905	0.758	256.08	2.30	9 657.84
				$S_P = \sum_P n_i s_i(T,p_i)\,kJ/K$	13 364.38

The adiabatic flame temperature is determined to be $1620\,K$ by applying Eq. (2.69) to this reaction.

The absolute entropies for individual components i of the reactants and products are determined from the following equation:

$$s_i(T,p_i) = s_i^0(T) - \bar{R}\,ln\left(\frac{y_i p_{mix}}{p_0}\right)\,kJ/kmole.K$$

and the total absolute entropies of the reactants and products are found from

$$S_{mix} = \sum_i n_i s_i(T,p_i)\,kJ/K$$

The results of the calculations are detailed in the Table 2.11. Note that the fuel and air (reactants) are treated as two separate streams entering the combustion chamber.

The change of entropy for the chemical reaction of propane combustion is

$$\Delta S_{isol} = \sum_P n_i\,s_i(T,p_i) - \sum_R n_i s_i(T,p_i) = 13364.38 - 9733.39 = 3630.99\,kJ/K$$

Since $\Delta S_{isol} \gg 0$, it can be concluded that the steady state-steady-flow adiabatic combustion process of propane gas is highly irreversible despite being complete.

2.9 Chemical and Phase Equilibrium

The combustion processes analysed so far, both complete and incomplete, were based on the assumption that the final products of combustion were a direct result of the chemical reaction of the hydrocarbon fuel elements and oxygen of the air. In the case of complete stoichiometric combustion, the products were carbon dioxide (CO_2), water (H_2O, vapour or liquid form), and nitrogen (N_2). If the reaction occurred with excess oxygen, any amount in excess of the stoichiometric was included in the products in the form of free oxygen molecules (O_2). The products in the case of incomplete combustion were assumed to include CO, or CO and H_2 in addition to CO_2, H_2O, and N_2. In all these cases, the combustion products were assumed to form instantaneously following the start of the process, and their composition remained unchanged in frozen equilibrium until the final state was reached.

In reality, combustion product formation is never instantaneous, and fluid equilibrium can exist before the process is completed and final chemical equilibrium is reached. During the fluid or interim equilibrium state, multiple products of incomplete combustion can form, particularly at high temperatures, even when there is sufficient time and oxygen for the reaction to occur.

A criterion for chemical equilibrium during irreversible process, applicable to reacting ideal gas mixtures, and its use in the analysis of reacting systems is described next.

2.9.1 Gibbs and Helmholtz Functions and Equilibrium

Consider the following composite properties, which are combinations of other basic thermodynamic properties such as internal energy, pressure, volume, temperature, and entropy:

Enthalpy: $H = U + pV$

Helmholtz function: $A = U - TS$

Gibbs function: $G = H - TS$

Enthalpy arises in connection with the application of the first law to flow processes. Helmholtz and Gibbs functions are used in processes involving chemical reactions and change of phase. Combining the energy equation (first law) and the principle of increasing entropy (second law) for a reacting system yields

$$\delta Q = dU + pdV$$

$$dS \geq \frac{\delta Q}{T}$$

$$dU + pdV \leq TdS, \text{ or}$$

$$dU + pdV - TdS \leq 0 \tag{2.80}$$

The enthalpy and Gibbs functions in differential form are

$$dH = dU + pdV + Vdp$$

$$dG = dH - Tds - sdT$$

Combining these two functions, noting that $dT = 0$ and $dp = 0$, the differential Gibbs function at constant temperature and pressure can be written as

$$dG_{(p,T)} = dU + pdV - Tds \tag{2.81}$$

From Eqs. (2.80) and (2.81):

$$dG_{(T,p)} \leq 0 \tag{2.82}$$

The differential Helmholtz function at constant temperature ($dT = 0$) and volume ($dV = 0$) is

$$dA = dU - TdS \tag{2.83}$$

Figure 2.11 Criteria for chemical equilibrium.

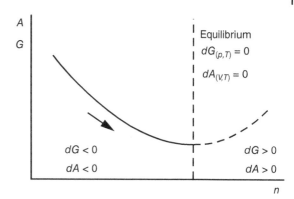

Combining Eqs. (2.80) and (2.83) yields

$$dA_{(V,T)} \leq 0 \tag{2.84}$$

Equation (2.82) states that a chemical reaction at constant pressure and temperature proceeds in the direction of decreasing Gibbs function until chemical equilibrium is reached at $dG_{(T,p)} = 0$. Similarly, Eq. (2.84) states that a chemical reaction at constant volume and temperature proceeds in the direction of decreasing Helmholtz function until chemical equilibrium is reached at $dA_{(V,T)} = 0$. The Helmholtz and Gibbs functions for any system in complete equilibrium must be a minimum with respect to the concentration of any substance in the reactants or products, as shown schematically in Figure 2.11. An increase of the Gibbs or Helmholtz functions as shown in the region to the right of the vertical dashed line is a violation of the second law of thermodynamics.

2.9.2 Equilibrium Constant

For an equilibrium chemical reaction involving four compounds A, B, C, and D, the reaction can be written as

$$aA + bB \rightleftarrows cC + dD \tag{2.85}$$

where a, b, c, and d are stoichiometric coefficients. The arrows indicate that the reaction can be forward or backward in direction. We will limit the discussion for the time being to the forward reaction only.

It can be shown (Borganakke and Sonntag, 2009; Moran and Shapiro, 2008) that for the reaction in Eq. (2.85),

$$\ln \left[\frac{y_C^c y_D^d}{y_A^a y_B^b} \left(\frac{P}{P_0} \right)^{(c+d-a-b)} \right] = \frac{-\Delta G^0}{\overline{R}T} \tag{2.86}$$

where y_A, y_B, y_C, y_D are the molar concentrations of gases A, B, C, D in the mixture, p is the total mixture pressure, p_0 is the reference pressure (0.1 MPa or bar), and ΔG^0 is the change in the Gibbs function for reaction (2.85):

$$\Delta G^0 = c(g_C^0) + d(g_D^0) - a(g_A^0) - b(g_B^0) \tag{2.87}$$

(g_A^0), (g_B^0), (g_C^0), and (g_D^0) are the Gibbs functions of formation of substances A, B, C, and D at temperature T and standard pressure $p_0 = 0.1$ *MPa*. Values of these functions for various substances are tabulated in thermodynamic references such as the JANAF tables.

By definition, the equilibrium constant K_p is equal to the term in square brackets in Eq. (2.86):

$$K_p = \frac{y_C^c y_D^d}{y_A^a y_B^b} \left(\frac{P}{P_0} \right)^{(c+d-a-b)} \tag{2.88}$$

Hence, Eq. (2.86) can be rewritten as

$$ln(K_p) = \frac{-\Delta G^0}{\overline{R}T} \tag{2.89}$$

2.9.2.1 Equilibrium Constant of Formation

Equilibrium constants are applicable not only to reaction processes but also to any chemical compound that is formed from basic constituents. The individual constant in this case is known as the *equilibrium constant of formation* K_f and is defined in a similar way to the enthalpies of formation, with basic elements being given a zero value. Thermochemical property tables such as the JANAF tables include values of $logK_f$ as a function of temperature for most known compounds. The equilibrium constant K_p for any reaction can be found from the equilibrium constant of formation of the species in the reaction using the following equation:

$$logK_p = \sum_P n_p logK_{f,i} - \sum_R n_R logK_{f,j} \tag{2.90}$$

where

$logK_{f,i}$: ith component in the product
$logK_{f,j}$: jth component in the reactants

As an example, we will calculate the equilibrium constant for the following carbon dioxide reaction at 2000 K:

$$CO + \frac{1}{2}O_2 \rightleftarrows CO_2$$

$$logK_p = 1 \times logK_{fCO_2} - \left(1 \times logK_{fCO} + \frac{1}{2}logK_{fO_2} \right)$$

From JANAF property tables, $log\,K_{fCO} = 7.47$, $log\,K_{fO_2} = 0$, and $log\,K_{fCO_2} = 10.351$; hence,

$$logK_p = 10.351 - 7.47 = 2.881, \text{ from which } K_p = 760.32.$$

Values of K_p for some common combustion reactions were calculated using Eq. (2.90) and tabulated as shown in Table 2.12, which is an abridged version of more comprehensive data for 17 reactions (see Table A.5).

Due to the large magnitude of K_p at temperatures below 2000 K, it is common to come across some references that tabulate the logarithmic values of K_p. For example, in Borganakke and Sonntag (2009), the natural logarithmic values are tabulated; and in Moran and Shapiro (2008), the logarithmic values to base 10 are tabulated. Furthermore, in both these references, the values of K_p are based on reactions with twice the coefficients and in

Table 2.12 Equilibrium constant $K_p = p_C^c p_D^d / p_A^a p_B^b$ for the reaction $aA + bB \rightleftarrows cC + dD$; reference pressure = 1 atm.

T	$H \rightleftarrows \frac{1}{2}H_2$	$O \rightleftarrows \frac{1}{2}O_2$	$N \rightleftarrows \frac{1}{2}N_2$	$H_2 + \frac{1}{2}O_2 \rightleftarrows H_2O$
600	2.17E + 16	3.75E + 18	1.21E + 38	4.30E + 18
1000	4.43E + 08	6.41E + 09	3.37E + 21	1.15E + 10
1400	2.07E + 05	1.06E + 06	2.55E + 14	2.22E + 06
1800	2.81E + 03	8.28E + 03	2.74E + 10	1.86E + 04
2200	1.78E + 02	3.72E + 02	8.04E + 07	8.75E + 02
2600	2.61E + 01	4.33E + 01	1.41E + 06	1.05E + 02
3000	6.35E + 00	8.89E + 00	7.21E + 04	2.20E + 01
3400	2.15E + 00	2.65E + 00	7.38E + 03	6.67E + 00
3800	9.08E − 01	1.02E + 00	1.22E + 03	2.59E + 00
4200	4.52E − 01	4.68E − 01	2.82E + 02	1.20E + 00
4600	2.54E − 01	2.47E − 01	8.40E + 01	6.35E − 01
5000	1.56E − 01	1.44E − 01	3.03E + 01	3.72E − 01
5400	1.03E − 01	9.08E − 02	1.27E + 01	2.34E − 01
5800	7.23E − 02	6.10E − 02	5.94E + 00	1.57E − 01

the opposite direction from the reactions shown in Table 2.12. The equilibrium constants K_p in Table 2.12 can be converted to the values tabulated in the two aforementioned references thus: $K_{p1} = 2\,ln\,(1/K_p)$ and $K_{p2} = 2\,log\,(1/K_p)$, respectively.

By definition,

$$y_A = \frac{n_A}{N}, y_B = \frac{n_B}{N}, y_C = \frac{n_C}{N}, y_D = \frac{n_D}{N}$$

where N is the total number of moles of the mixture including any inert components n_{inert} present: $N = n_A + n_B + n_C + n_D + n_{inert}$

Equation (2.88) can now be rewritten as

$$K_p = \frac{n_C^c n_D^d}{n_A^a n_B^b} \left(\frac{p/p_0}{N} \right)^{(c+d-a-b)} \tag{2.91}$$

If the *atm* (1 *atm* = 0.101325 *MPa*) is used as the unit for pressure, then $p_0 = 1$ *atm*, and Eq. (2.91) becomes

$$K_p = \frac{n_C^c n_D^d}{n_A^a n_B^b} \left(\frac{p}{N} \right)^{(c+d-a-b)} = \frac{\left(\frac{n_C}{N} p \right)^c \left(\frac{n_D}{N} p \right)^d}{\left(\frac{n_A}{N} p \right)^a \left(\frac{n_B}{N} p \right)^b} = \frac{p_C^c p_D^d}{p_A^a p_B^b} \tag{2.92}$$

where p_A, p_B, p_C, and p_D are the partial pressures of the components in reaction (2.83).

A similar formulation can be used if p_0 and p are in *bar* (1 *bar* = 100*kPa* = 0.1 *MPa*). In these cases Eq. (2.86) is rewritten as,

$$ln \left[\frac{p_C^c p_D^d}{p_A^a p_B^b} (p)^{(c+d-a-b)} \right] = \frac{-\Delta G^0}{\overline{R}T}$$

Equation (2.92), written for the reactions involved in a dissociating mixture of known temperature and pressure, forms a set of simultaneous equations that can be solved to find the equilibrium composition of the mixture.

2.9.3 Dissociation and Equilibrium Composition

As mentioned previously, the combustion products in real processes are never in frozen equilibrium, and products of incomplete combustion can form, particularly at high temperatures, even when there is sufficient time and oxygen for the reaction to occur. These additional products are the result of *dissociation*, which is defined as the separation of molecules in the products of combustion into simpler compounds or even atoms. For dissociation to occur, a high-temperature environment is required, which is provided in most cases when hydrocarbon fuels are burned. The consequences of dissociation include the production of harmful chemical pollutants that are emitted into the atmosphere and the decrease of the adiabatic flame temperature (combustion temperature). The reason for the decrease in combustion temperature is that dissociation of some chemical compounds is accompanied by energy absorption, i.e. the reaction is endothermic. The decrease in combustion temperature will have an adverse effect on the thermal efficiency of the heat engine. An example of an endothermic reaction is the dissociation of carbon dioxide in an environment with high temperature:

$$CO_2 \rightarrow CO + \frac{1}{2}O_2$$

Not all dissociation reactions are endothermic, however. The following reaction is accompanied by heat generation (exothermic reaction):

$$CO + \frac{1}{2}O_2 \rightarrow CO_2$$

The dissociation reaction of carbon dioxide is usually written as

$$CO_2 \rightleftarrows CO + \frac{1}{2}O_2$$

which means the reaction can go forward or backward depending on the conditions of the reaction environment. A state of chemical equilibrium between the reactants and products at any particular temperature and pressure exists when the two forward and backward reactions proceed at the same rate, i.e. the number of CO_2 molecules dissociated and formed become equal.

Other possible reactions involving dissociation during combustion of hydrocarbon fuels are

$$H_2 + \frac{1}{2}O_2 \rightleftarrows H_2O$$

$$OH + \frac{1}{2}H_2 \rightleftarrows H_2O$$

$$H_2O + CO \rightleftarrows H_2 + CO_2$$

$$OH \rightleftarrows \frac{1}{2}O_2 + \frac{1}{2}H_2$$

$$O_2 \rightleftarrows O + O$$

$$H_2 \rightleftarrows 2H$$

$$N_2 \rightleftarrows 2N$$

$$C + 2H_2 \rightleftarrows CH_4$$

For the reaction

$$H_2 + \frac{1}{2}O_2 \rightleftarrows H_2O$$

the forward equilibrium constant from Eq. (2.92) for $p_0 = 1$ bar is

$$K_{pf} = \frac{p_{H_2O}}{p_{H_2}(p_{O_2})^{1/2}}$$

And the backward equilibrium constant is

$$K_{pb} = \frac{p_{H_2}(p_{O_2})^{1/2}}{p_{H_2O}}$$

Hence

$$K_{pf} = 1/K_{pb}$$

The second subscript will be dropped at this stage of the analysis as only the forward values are usually tabulated in thermodynamic references. The equilibrium constant K_p is not dimensionless, and its value depends on the units of the pressure used and the stoichiometric coefficients of the reaction. Also, in addition to the units of the pressure, the form of the reaction equation must be specified when quoting values of K_p. For example, the reaction forming water can be written as

$$2H_2 + O_2 \rightleftarrows 2H_2O$$

for which

$$K_{pf} = \frac{(p_{H_2O})^2}{(p_{H_2})^2 p_{O_2}}$$

The equilibrium constant can be written in terms of species concentrations in moles per unit volume by incorporating the equation of state $pV = N\overline{R}T$ into Eq. (2.91) as follows:

$$K_p = \frac{n_C^c n_D^d}{n_A^a n_B^b}\left(\frac{1}{p_0}\frac{\overline{R}T}{V}\right)^{(c+d-a-b)} = \left[\frac{(n_C^c/V)(n_D^d/V)}{(n_A^a/V)(n_B^b/V)}\right]\left(\frac{\overline{R}T}{p_0}\right)^{(c+d-a-b)} = K_c\left(\frac{\overline{R}T}{p_0}\right)^{(c+d-a-b)}$$

K_c is the concentration equilibrium constant and is given by

$$K_p = K_c\left(\frac{\overline{R}T}{p_0}\right)^{(a+b-c-d)} \tag{2.93}$$

The equilibrium constant K_p is a measure of the degree of dissociation in the reaction. A high value indicates that the mixture contains a large proportion of un-dissociated components, i.e. more complete reaction.

2.10 Multi-Species Equilibrium Composition of Combustion Products

2.10.1 Frozen Composition

These reactions have already been discussed in detail and will be presented here as a basis for comparison with equilibrium composition of combustion products. *Frozen* composition means the absence of dissociation in the products, and the number of species in the

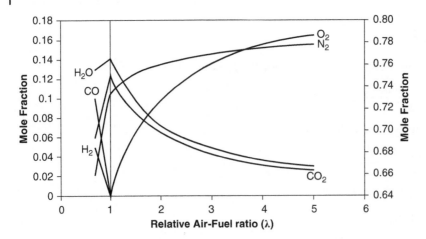

Figure 2.12 Frozen composition of the combustion products of octane (C_8H_{18}).

products depends on the air-fuel ratio and whether only CO is included in the products of rich mixtures or both CO and H_2. Three reactions were used for this case:

Lean mixture: $\lambda > 1$

$$C_xH_yO_z + \lambda \left(x + \frac{y}{4} - \frac{z}{2}\right)(O_2 + 3.7619N_2)$$
$$= xCO_2 + \frac{y}{2}H_2O + (\lambda - 1)\left(x + \frac{y}{4} - \frac{z}{2}\right)O_2 + 3.7619\lambda\left(x + \frac{y}{4} - \frac{z}{2}\right)N_2$$

Stoichiometric mixture: $\lambda = 1$

$$C_xH_yO_z + \left(x + \frac{y}{4} - \frac{z}{2}\right)(O_2 + 3.7619N_2) \rightarrow xCO_2 + \frac{y}{2}H_2O + 3.7619\left(x + \frac{y}{4} - \frac{z}{2}\right)N_2$$

Rich mixture: $\lambda < 1$

$$\frac{1}{\mu_f} + \lambda\left(\frac{A}{F}\right)_s^v = M_{CO_2} + M_{CO} + M_{H_2O} + M_{H_2} + M_{N_2}$$

The last reaction is based on the gravimetric (mass) composition of the fuel, and the products are given by Eqs. (2.35a).

Figure 2.12 shows the frozen concentration as a function of relative air fuel ratio λ for the combustion of octane (C_8H_{18}) in air. Note that M_{CO} and M_{H_2} tend to zero at $\lambda = 1$.

2.10.2 Equilibrium Composition

The equilibrium composition in this case accounts for the presence of dissociation of one species or more depending on the number of components in the combustion products. Knowledge of the equilibrium constants enables the combustion equation to be established and the adiabatic combustion temperature to be computed. For the following discussion, a computer program was developed to investigate the effect of dissociation on the adiabatic flame temperature, product composition, and the ratio of specific heats of the products for the combustion of liquid and gaseous hydrocarbon fuels in air within a wide range of air-fuel ratios (Ghojel, 1994).

2.10.2.1 Six Species in the Products

To explain the methodology for the reaction with the formation of equilibrium products, consider the following general reaction of a generic hydrocarbon fuel $(C_xH_yO_z)$ with air in which some of the CO_2 and H_2O molecules in the products dissociate:

$$C_xH_yO_z + \lambda m_s(O_2 + 3.7619N_2) \rightarrow n_1CO_2 + n_2H_2O + n_3CO + n_4H_2 + n_5O_2 + n_6N_2$$

where $m_s = x + y/4 - z/2$ (amount of oxygen used in the stoichiometric reaction of the fuel).

This reaction contains six unknowns $(n_1, n_2, n_3, n_4, n_5, n_6)$ that need to be determined. Balancing carbon, oxygen, hydrogen, and nitrogen atoms yields four equations:

$$x = n_1 + n_3 \tag{i}$$

$$2\lambda m_s + z = 2n_1 + n_2 + n_3 + 2n_5 \tag{ii}$$

$$y = 2n_2 + 2n_4 \tag{iii}$$

$$\lambda m_s(3.7619) = n_6 \tag{iv}$$

The other two equations required, together with the equilibrium constants for the reactions, are shown in Table 2.13. If the total pressure of the mixture is p and the total number of moles of products is $N = n_1 + n_2 + n_3 + n_4 + n_5 + n_6$, then

$$p_{CO_2} = \frac{n_1}{N}p, \quad p_{H_2O} = \frac{n_2}{N}p, \quad p_{CO} = \frac{n_3}{N}p, \quad p_{O_2} = \frac{n_4}{N}p, \quad p_{H_2} = \frac{n_5}{N}p, \quad p_{N_2} = \frac{n_6}{N}p$$

Since pressure appears in these equations, the total pressure of the system must be specified. Also, since the equilibrium constants are temperature dependent, a combustion temperature must be assumed to find the equilibrium constants from property tables to complete the set of simultaneous equations needed to find the six unknowns. As an example of the calculations, Figure 2.13 shows the equilibrium concentration as a function of relative air fuel ratio λ for the combustion of octane (C_8H_{18}) in air with six species in the products.

2.10.2.2 Eleven Species in the Products

The reaction for octane in this case can be written as follows:

$$C_xH_yO_z + \lambda m_s(O_2 + 3.7619N_2)$$
$$\rightarrow n_1CO_2 + n_2CO + n_3H_2O + n_4H_2 + n_5OH + n_6H + n_7N_2 + n_8NO$$
$$+ n_9N + n_{10}O + n_{11}O_2$$

Table 2.13 Equilibrium constants for dissociation reactions in products with six species.

Reaction	Formula	Reaction equilibrium constant	
1	$CO + \frac{1}{2}O_2 \rightleftarrows CO_2$	$K_{pCO_2} = \dfrac{p_{CO_2}}{p_{CO}(p_{O_2})^{1/2}}$	$K_{pCO_2} = \dfrac{n_1}{n_3\left(\dfrac{n_4}{N}\right)^{1/2}} \dfrac{1}{p^{1/2}}$
2	$H_2 + \frac{1}{2}O_2 \rightleftarrows H_2O$	$K_{pH_2O} = \dfrac{p_{H_2O}}{p_{H_2}(p_{O_2})^{1/2}}$	$K_{pH_2O} = \dfrac{n_2}{n_5\left(\dfrac{n_4}{N}\right)^{1/2}} \dfrac{1}{p^{1/2}}$

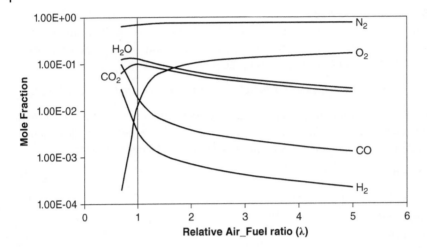

Figure 2.13 Equilibrium composition for the combustion of octane (C_8H_{18}) (six species in the products, $T_f = 298\ K$, $p = 30\ atm$, $T_{mix} = 2750\ K$).

This reaction contains 11 unknowns ($n_1, ..., n_{11}$) that need to be determined. Balancing carbon, oxygen, hydrogen, and nitrogen atoms yields four equations:

$$x = n_1 + n_2 \qquad\qquad (i)$$
$$2\lambda m_s + z = 2n_1 + n_2 + n_3 + n_5 + n_8 + n_{10} + 2n_{11} \qquad\qquad (ii)$$
$$y = 2n_3 + 2n_4 + n_5 + n_6 \qquad\qquad (iii)$$
$$2\lambda m_s(3.7619) = 2n_7 + n_8 + n_9 \qquad\qquad (iv)$$

With a given mixture pressure, the partial pressures can be determined as before. The seven additional equations required for the determination of the composition and the equilibrium products are shown in Table 2.14. In the previous case with six species, only two dissociation reactions were taken into account to determine the composition. In the current case, 7 dissociation reactions need to be considered to solve a set of 11 simultaneous equations. Figure 2.14 shows the equilibrium concentration as a function of relative air fuel ratio λ for the combustion of octane (C_8H_{18}) in air with 11 species in the products.

2.10.2.3 Eighteen Species in the Products

The methodology for calculating the equilibrium composition of the products of combustion of hydrocarbon fuels can be applied to any number of species, reaching hundreds in some studies. We will conclude this topic by outlining a case of 18 species in the products as per the following reaction:

$$C_xH_yO_z + \lambda m_s(O_2 + 3.7619N_2)$$
$$\rightarrow n_1CO_2 + n_2CO + n_3C + n_4CH_4 + n_5H_2O + n_6H_2 + n_7OH + n_8H + n_9NO_2$$
$$+ n_{10}NO + n_{11}N_2 + n_{12}N + n_{13}NH_3 + n_{14}O_3 + n_{15}O_2 + n_{16}O + n_{17}HNO_3$$
$$+ n_{18}HCN$$

Table 2.14 Equilibrium constants for dissociation reactions in products with 11 species

Reaction	Formula	Reaction equilibrium constant	
1	$H \rightleftarrows \frac{1}{2}H_2$	$K_{p1} = \dfrac{(p_{H_2})^{1/2}}{p_H}$	$K_{p1} = \dfrac{\left(\frac{n_4}{N}\right)^{1/2}}{\left(\frac{n_6}{N}\right)} \dfrac{1}{p^{1/2}}$
2	$O \rightleftarrows \frac{1}{2}O_2$	$K_{p2} = \dfrac{(p_{O_2})^{1/2}}{p_O}$	$K_{p2} = \dfrac{\left(\frac{n_{11}}{N}\right)^{1/2}}{\left(\frac{n_{10}}{N}\right)} \dfrac{1}{p^{1/2}}$
3	$N \rightleftarrows \frac{1}{2}N_2$	$K_{p3} = \dfrac{(p_{N_2})^{1/2}}{p_N}$	$K_{p3} = \dfrac{\left(\frac{n_7}{N}\right)^{1/2}}{\left(\frac{n_9}{N}\right)} \dfrac{1}{p^{1/2}}$
4	$H_2 + \frac{1}{2}O_2 \rightleftarrows H_2O$	$K_{p4} = \dfrac{p_{H_2O}}{p_{H_2}(p_{O_2})^{1/2}}$	$K_{p4} = \dfrac{n_3}{n_4\left(\frac{n_{11}}{N}\right)^{1/2}} \dfrac{1}{p^{1/2}}$
5	$OH + \frac{1}{2}H_2 \rightleftarrows H_2O$	$K_{p5} = \dfrac{p_{H_2O}}{p_{OH}(p_{H_2})^{1/2}}$	$K_{p5} = \dfrac{n_3}{n_5\left(\frac{n_4}{N}\right)^{1/2}} \dfrac{1}{p^{1/2}}$
6	$\frac{1}{2}N_2 + \frac{1}{2}O_2 \rightleftarrows NO$	$K_{p6} = \dfrac{p_{NO}}{(p_{N_2})^{1/2}(p_{O_2})^{1/2}}$	$K_{p6} = \dfrac{n_8}{(n_7)^{1/2}(n_{11})^{1/2}}$
7	$H_2O + CO \rightleftarrows H_2 + CO_2$	$K_{p7} = \dfrac{p_{H_2}p_{CO_2}}{p_{H_2O}p_{CO}}$	$K_{p7} = \dfrac{n_4 n_1}{n_3 n_2}$

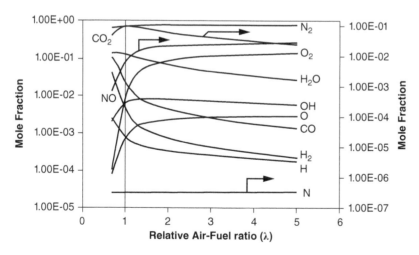

Figure 2.14 Equilibrium composition for the combustion of octane (C_8H_{18}) (11 species in the products, $T_f = 298\ K$, $p = 30\ atm$, $T_{mix} = 2750\ K$).

This reaction contains 18 unknowns (n_1, ..., n_{18}) that need to be determined. Balancing carbon, oxygen, hydrogen, and nitrogen atoms yields four equations:

$$x = n_1 + n_2 + n_3 + n_4 \tag{i}$$

$$2\lambda m_s + z = 2n_1 + n_2 + n_5 + n_7 + 2n_9 + n_{10} + 3n_{14} + 2n_{15} + n_{16} + 3n_{17} \tag{ii}$$

$$y = 4n_4 + 2n_5 + 2n_6 + n_7 + n_8 + 3n_{13} + n_{17} + n_{18} \tag{iii}$$

$$2\lambda m_s(3.7619) = n_9 + n_{10} + 2n_{11} + n_{12} + n_{13} + n_{17} + n_{18} \tag{iv}$$

The additional equations required for the determination of the composition and the equilibrium constants for the reaction are shown in Table 2.15. The pressure appears in these equations also; hence, the total pressure of the system must be specified. Also,

Table 2.15 Equilibrium constants for dissociation reactions in products with 18 species.

Reaction	Formula	Reaction equilibrium constant	
1	$H \rightleftarrows \frac{1}{2}H_2$	$K_{p1} = \dfrac{(p_{H_2})^{1/2}}{p_H}$	$K_{p1} = \dfrac{(n_6)^{1/2}}{(n_8)}\left(\dfrac{N}{p}\right)^{1/2}$
2	$O \rightleftarrows \frac{1}{2}O_2$	$K_{p2} = \dfrac{(p_{O_2})^{1/2}}{p_O}$	$K_{p2} = \dfrac{(n_{15})^{1/2}}{(n_{16})}\left(\dfrac{N}{p}\right)^{1/2}$
3	$O_3 \rightleftarrows \frac{3}{2}O_2$	$K_{p3} = \dfrac{(p_{O_2})^{3/2}}{p_{O_3}}$	$K_{p3} = \dfrac{(n_{15})^{3/2}}{(n_{14})}\left(\dfrac{p}{N}\right)$
4	$H_2 + \frac{1}{2}O_2 \rightleftarrows H_2O$	$K_{p4} = \dfrac{p_{H_2O}}{p_{H_2}(p_{O_2})^{1/2}}$	$K_{p4} = \dfrac{n_5}{n_6(n_{15})^{1/2}}\left(\dfrac{N}{p}\right)^{1/2}$
5	$\frac{1}{2}O_2 + \frac{1}{2}H_2 \rightleftarrows OH$	$K_{p5} = \dfrac{p_{OH}}{(p_{O_2})^{1/2}(p_{H_2})^{1/2}}$	$K_{p5} = \dfrac{n_7}{(n_{15})^{1/2}(n_6)^{1/2}}$
6	$C + O_2 \rightleftarrows CO_2$	$K_{p6} = \dfrac{p_{CO_2}}{p_C p_{O_2}}$	$K_{p6} = \dfrac{n_1}{n_3 n_{15}}\left(\dfrac{N}{p}\right)$
7	$C + \frac{1}{2}O_2 \rightleftarrows CO$	$K_{p7} = \dfrac{p_{CO}}{p_C(p_{O_2})^{1/2}}$	$K_{p7} = \dfrac{n_2}{n_3(n_{15})^{1/2}}\left(\dfrac{N}{p}\right)^{1/2}$
8	$C + 2H_2 \rightleftarrows CH_4$	$K_{p8} = \dfrac{p_{CH_4}}{p_C(p_{H_2})^2}$	$K_{p8} = \dfrac{n_4}{n_3(n_6)^2}\left(\dfrac{N}{p}\right)^2$
9	$N \rightleftarrows \frac{1}{2}N_2$	$K_{p9} = \dfrac{(p_{N_2})^{1/2}}{p_N}$	$K_{p9} = \dfrac{(n_{11})^{1/2}}{(n_{12})}\left(\dfrac{N}{p}\right)^{1/2}$
10	$\frac{1}{2}N_2 + \frac{1}{2}O_2 \rightleftarrows NO$	$K_{p10} = \dfrac{p_{NO}}{(p_{N_2})^{1/2}(p_{O_2})^{1/2}}$	$K_{p10} = \dfrac{n_{10}}{(n_{11})^{1/2}(n_{15})^{1/2}}$
11	$NO + \frac{1}{2}O_2 \rightleftarrows NO_2$	$K_{p11} = \dfrac{p_{NO_2}}{p_{NO}(p_{O_2})^{1/2}}$	$K_{p11} = \dfrac{n_9}{n_{10}(n_{15})^{1/2}}\left(\dfrac{N}{p}\right)^{1/2}$
12	$\frac{1}{2}N_2 + \frac{3}{2}H_2 \rightleftarrows NH_3$	$K_{p12} = \dfrac{p_{NH_3}}{(p_{N_2})^{1/2}(p_{H_2})^{3/2}}$	$K_{p12} = \dfrac{n_{13}}{(n_{11})^{1/2}(n_6)^{3/2}}\left(\dfrac{N}{p}\right)$
13	$3NO_2 + H_2O \rightleftarrows 2HNO_3 + NO$	$K_{p13} = \dfrac{(p_{HNO_3})^2 p_{NO}}{(p_{NO_2})^3 p_{H_2O}}$	$K_{p13} = \dfrac{(n_{17})^2 n_{10}}{(n_9)^3 n_5}\left(\dfrac{N}{p}\right)$
14	$C + \frac{1}{2}N_2 + \frac{1}{2}H_2 \rightleftarrows HCN$	$K_{p14} = \dfrac{p_{HCN}}{p_C(p_{N_2})^{1/2}(p_{H_2})^{1/2}}$	$K_{p14} = \dfrac{n_{18}}{n_3(n_{11})^{1/2}(n_6)^{1/2}}\left(\dfrac{N}{p}\right)$
15	$H_2O + CO \rightleftarrows H_2 + CO_2$	$K_{p15} = \dfrac{p_{H_2} p_{CO_2}}{p_{H_2O} p_{CO}}$	$K_{p15} = \dfrac{n_6 n_1}{n_5 n_2}$

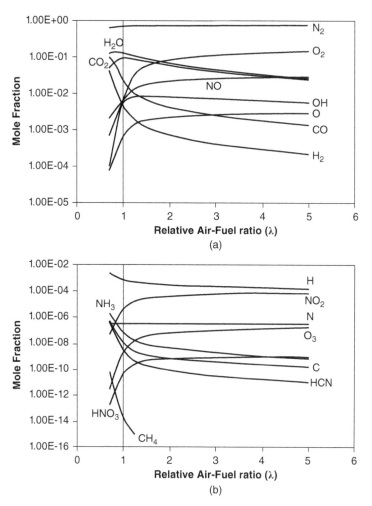

Figure 2.15 Equilibrium composition for the combustion of octane (C_8H_{18}) in air (18 species in the products, $T_f = 298\ K$, $p = 30\ atm$, $T_{mix} = 2750\ K$).

since the equilibrium constants are temperature dependent, a combustion temperature must be assumed to find the equilibrium constants from property tables to complete the set of simultaneous equations needed to find the 18 unknowns. Figure 2.15 shows the equilibrium concentration as a function of relative air-fuel ratio λ for the combustion of octane (C_8H_{18}) in air with 18 species in the products.

The use of equilibrium constants is a powerful tool for solving complex combustion processes involving multiple reactions occurring in the reacting mixture simultaneously. By writing the reaction equations for all possible reactions and combining them with the mass balance equations, the concentration of all the species in the products can be determined by solving the resulting set of simultaneous equations. The adiabatic flame temperature for the reaction with dissociation at a given pressure can be found by substituting the concentrations of the species in the products into Eq. (2.64) or (2.70) and iterating for AFT until the equation balances.

Irrespective of the number of dissociation reactions involved and number of species in the products, the general methodology for solving for the unknowns is as follows:

1. Assume a final combustion temperature – a knowledge of typical adiabatic flame temperatures for common hydrocarbon fuels can help in reducing the number of iterations.
2. Determine the values of the equilibrium constants.
3. Solve the simultaneous equations to determine the number of moles of species.
4. If the energy equation balances with the calculated species concentrations, the assumed temperature will be equal to the AFT.
5. If the energy equation does not balance, another temperature is assumed, and the calculations are repeated from step 2 until the desired agreement is reached.

Despite the fact that dissociation is accompanied by energy absorption, the enthalpy of combustion (heating value, calorific value) is unaffected since this value is determined at a temperature close to the ambient temperature. Dissociation also has no effect on the combustion efficiency. However, the thermal efficiency of a power plant could be affected adversely if the temperature of the reaction is high enough for significant dissociation activity. This is due to the heat being transferred to the working fluid at a lower temperature as a result of dissociation. Dissociation has an insignificant effect on the thermal efficiencies of industrial gas turbines owing to the relatively low maximum combustion temperatures that are necessary due to metallurgical limitations. Dissociation is important, however, in modern aviation jet engines, piston engines, and rocket engines, as it tends to lower the adiabatic combustion temperature; this is evident from Table 2.16 and Figure 2.16, which show the adiabatic flame temperature as a function of λ for the constant-pressure combustion of octane (C_8H_{18}) in air. The fuel is delivered to the reactor at 298 K, where it is mixed with air at pressure $p_a = 30$ bar and temperature $T_a = 700$ K (typical state of air at the end of a compression process in piston engines). The adiabatic flame temperature is calculated for: (i) frozen products with 3 species $(\lambda = 1)$, 4 species $(\lambda > 1)$, and 6 species $(\lambda < 1)$; (ii) equilibrium components with 6, 11, and 18 species in the combustion products for all values of λ.

Figures 2.17 and 2.18 show the effect of the initial mixture pressure and temperature on the adiabatic flame temperature. For a given initial mixture and fuel temperatures and air-fuel ratio, the AFT increases with pressure as shown in Figure 2.17. The increase is relatively steep up to about 10 atm, tapering off afterwards to increase at a moderate rate. Increasing the initial mixture temperature from 298 to 500 K causes the AFT to increase over the entire range of pressure rise. The higher the initial mixture temperature, the greater the rate at which AFT increases with pressure.

The effect of initial mixture temperature on AFT at constant mixture pressure and fuel temperature is shown in Figure 2.18. The AFT increases linearly with increasing initial mixture temperature for fixed mixture pressure, fuel temperatures, and air-fuel ratio. The rate of AFT rise increases as the mixture becomes leaner.

In conclusion:

- The AFT increases when the mixture temperature and/or pressure in the constant-pressure reactor is increased (Figures 2.17 and 2.18).
- Accounting for dissociation in the combustion products causes the AFT to decrease significantly in the region of stoichiometric combustion compared with the frozen species in the products.

Table 2.16 Adiabatic flame temperature for frozen and equilibrium compositions of octane combustion ($T_f = 298\ K$, $T_{mix} = 700\ K$, $p_a = 30\ bar$).

			AFT (K)		
ϕ	λ	Frozen	Equilibrium: 6 species	Equilibrium: 11 species	Equilibrium: 18 species
1.667	0.6	2105	2160	2088	2085
1.429	0.7	2295	2326	2272	2270
1.250	0.8	2446	2455	2420	2415
1.176	0.85	2511	2509	2487	2476
1.111	0.9	2569	2554	2524	2520
1.053	0.95	2623	2573	2548	2546
1.000	1	2671	2565	2539	2536
0.909	1.1	2530	2480	2460	2456
0.833	1.2	2408	2380	2360	2356
0.769	1.3	2302	2284	2266	2262
0.714	1.4	2208	2192	2178	2174
0.667	1.5	2125	2112	2100	2096
0.500	2	1819	1805	1807	1803
0.333	3	1490	1480	1480	1476
0.250	4	1310	1301	1300	1299
0.200	5	1195	1191	1190	1190

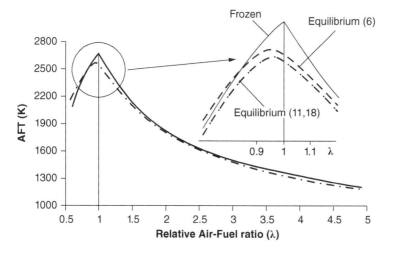

Figure 2.16 Effect of dissociation on the adiabatic flame temperature of isooctane combustion with frozen and equilibrium species in the products ($T_f = 298\ K$, $T_{mix} = 700\ K$, $p_a = 30\ bar$).

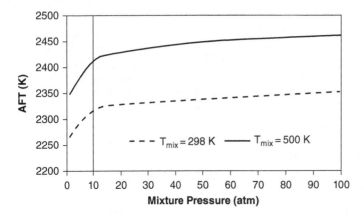

Figure 2.17 Effect of mixture pressure on the AFT of liquid octane (C_8H_{18}, $\lambda = 1.0$, $T_f = 298\ K$).

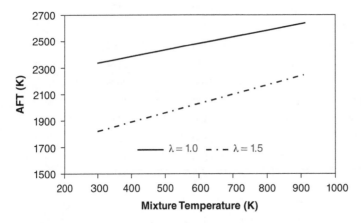

Figure 2.18 Effect of initial mixture temperature on AFT of liquid octane (C_8H_{18}, $p_{mix} = 30\ atm$, $T_f = 298\ K$).

- Increasing the number of equilibrium reactions from 6 to 11 causes the AFT to decrease appreciably.
- Increasing the number of species further has no noticeable effect on the AFT.
- At values of $\lambda > 2$, the temperatures for all reactions become almost equal.
- The location of the maximum temperature for equilibrium reactions shifts slightly away from the stoichiometric air-fuel ratio towards richer mixtures ($\lambda \approx 0.95$).

Problems

2.1 The gravimetric analysis of jet fuel (kerosene) shows that 1 *kg* of fuel contains 0.863 *kg* of carbon, 0.136 *kg* of hydrogen, and 0.001 *kg* of sulfur. If the fuel is burned with 300% excess air, determine
   (a) The stoichiometric air-fuel ratio on both mass and volume bases
   (b) The dry composition of the products on both mass and volume bases

2.2 Ethylene (C_2H_4) reacts in a steady-flow calorimeter with 100% theoretical air (stoichiometric mixture) at 25 °C and 1 *bar*. Determine the enthalpy of formation of the fuel if the heat transfer to the calorimeter during the combustion (assumed complete) is 1408 *MJ/kmole* of fuel.

2.3 Gasoline is burned in an internal combustion engine with 15% excess air at 25 °C. Assuming the fuel can be treated as equivalent to liquid octane (C_8H_{18}) and the combustion is complete, determine
(a) The stoichiometric air-fuel ratios on mass and volume bases
(b) The enthalpy of reaction of the fuel for both wet and dry products.

2.4 An unknown hydrocarbon fuel C_xH_y reacts with dry air. A volumetric analysis of the dry products gives the following results: 9.57% CO_2, 6.39% O_2, 84.04% N_2. Determine
(a) The composition of the fuel
(b) The reaction equation for the process
(c) The actual air-fuel ratio
(d) The relative air-fuel ratio
(e) The percent of excess air
(f) The enthalpy of reaction of the fuel for dry products

2.5 The following table shows the concentration of the constituent components of consumer natural gas and their enthalpies of formation:

Component	% Volume	ΔH_f kJ/kmole
CH_4	91.108	−74 898
C_2H_6	5.073	−84 724
C_3H_8	0.895	−103 916
C_4H_{10}	0.0942	−124 817
C_5H_{12}	0.0139	−146 538
C_6H_{14}	0.0039	−167 305
C_7H_{16}	0.0148	−187 946
CO_2	1.789	−393 522
N_2	1.009	0

Determine
(a) The stoichiometric air-fuel ratio on both volumetric and gravimetric bases
(b) The wet volumetric composition of the combustion products if 10% excess air is provided
(c) The enthalpy of reaction for both wet and dry products
(d) The adiabatic flame temperature if the fuel is burned with 10% excess air in a steady-flow process, the combustion is complete, and water is in vapour phase.

2.6 Liquid octane (C_8H_{18}) is burned with 400% excess air at 25 °C in a steady-flow process. Determine the adiabatic flame temperature.

2.7 Determine the adiabatic flame temperature for liquid heptane in Problem (2.6) if the mixture is stoichiometric.

2.8 Hydrogen [H_2 (g)] enters a rocket engine at 100 K and reacts with the stoichiometric amount of O_2 (g) at 100 K. The reaction occurs in a steady-flow process, and combustion is complete. The standardised specific enthalpies for the entering H_2 (g) and O_2 (g) are -5300 and $-5780\,kJ/kmole$, respectively. Assuming ideal gas behaviour for the reactants and products, determine the adiabatic flame temperature for the reaction.

2.9 Automotive fuel for SI engines has the gravimetric composition $x_C = 0.855$ and $x_H = 0.145$ and molecular mass $\mu_m = 114$. For a relative air-fuel ratio $\lambda = 1.15$, determine
(a) The stoichiometric air/fuel ratio on both mass and mole bases
(b) The quantity of reactants
(c) The mass and molar composition of the combustion products
(d) The theoretical coefficient of molar change

2.10 Repeat Problem (2.9) for $\lambda = 0.85$. Assume the products CO_2, CO, H_2O, H_2, N_2, and take $K = M_{H_2}/M_{CO} = 0.3$.

2.11 A stoichiometric mixture of methane (CH_4) and air at 25 $°C$ and 0.1 MPa burns in an adiabatic steady-flow system. The products of combustion include, in addition to the products of complete combustion, CO, O_2, and H_2. Determine the adiabatic flame temperature.

2.12 In an engine test, the dry-volumetric analysis of the combustion products of an unknown fuel C_xH_y showed the following quantities: 0.0527 $kmole$ CO_2, 0.1338 $kmole$ O_2, and 0.8135 $kmole$ N_2. Assuming the fuel is completely burnt,
(a) Estimate the carbon to hydrogen mass ratio of the fuel.
(b) Determine the air-fuel ratio by mass and the percentage of excess air.
(c) If the fuel is pentane (C_5H_{12}), calculate the enthalpy of combustion (in $kJ/kmole$ and kJ/kg) for the stoichiometric reaction given the enthalpies of formation of the fuel, reactants, and products shown in the following table. The water in the products is in vapour form.

Substance	ΔH_f, $kJ/kmole$
C_5H_{12}	$-146\,538$
CO_2	$-393\,520$
H_2O (g)	$-241\,820$
O_2	0
N_2	0

2.13 In the reaction of octane (C_8H_{18}) with 10% excess air ($\lambda = 1.1$), six species are formed in the products: CO_2, H_2O, CO, H_2, O_2, N_2. If the reactants are at a pressure 10 *bar* and temperature 500 K, calculate the composition of the products, assuming *constant pressure* and adiabatic combustion temperature of 1600 K. Assume that only carbon dioxide and water dissociate.

2.14 In the reaction of octane (C_8H_{18}) with 10% excess air ($\lambda = 1.1$), six species are formed in the products: CO_2, H_2O, CO, H_2, O_2, N_2. If the reactants are initially at a pressure of 10 *bar* and temperature of 500 K and ignited to burn at *constant volume*, calculate the conditions at the end of combustion. Assume that the combustion is adiabatic and only carbon dioxide and water dissociate during the reaction.

Part II

Reciprocating Internal Combustion Engines

Introduction II: History and Classification of Reciprocating Internal Combustion Engines

The reciprocating internal combustion engine can be classified under the broad term *heat engine* as an air-breathing, internal combustion piston engine (Chart II.1). The types of engines and modes of air induction discussed in this book are highlighted in the chart.

Precise technical terms are used in the text for the two piston engine types to be considered. The engine in which combustion is initiated by means of a spark plug will be referred to as a *spark-ignition engine* (SI engine). This engine is generally known as a *gasoline engine* (or motor) in the United States and Spain; *petrol engine* in the UK and Australia, *benzinmotor* in Germany, and *motore benzina* in Italy. The engine in which combustion is initiated by highly compressed air in the cylinder without an external aid will be referred to as a *compression–ignition engine* (CI engine). The term used almost universally for the CI engine is *diesel engine*. In this book, theoretical cycles get comprehensive coverage, and one of the cycles discussed in detail is the *Diesel cycle,* named in honour of Rudolf Diesel; in order to avoid confusion between the Diesel cycle and a diesel engine, the term *compression-ignition engine* will be used throughout the book.

The first commercially produced internal combustion piston engine was designed by the Belgian inventor Étienne Lenoir in 1860. The two-stroke engine used illuminating gas for fuel and developed little power; it had a thermal efficiency of about 4%. Despite these shortcomings, several hundred units were sold.

The principle of operation of the four-stroke engine was first described in a paper published in 1862 by the French engineer Alphonse Beau de Rochas, as follows (Sandfort 1964):

1. Suction of the charge into the cylinder during the entire outstroke of the piston
2. Compression of the charge during the following instroke
3. Ignition at dead centre and expansion during the next outstroke (power stroke)
4. Forced exit of the burned gases out of the cylinder during the following instroke, thus completing the cycle

The credit for building the first actual engine operating on such a cycle (known as the Otto cycle) goes to the German inventor Nikolaus August Otto (1832–1891). The first four-stroke engine was designed and built by Otto in 1876. It was a single-acting, single-cylinder horizontal engine with an output of 3 *hp* at 180 *rpm*. Engines were built under license in several countries; at the time of his death in 1891, more than 45 000 Otto cycle engines were in operation, some with multi-cylinder designs and ratings up to 100 *hp*. During the period from 1867 to 1908, numerous improvements were made on the design so that a thermal

Fundamentals of Heat Engines: Reciprocating and Gas Turbine Internal Combustion Engines, First Edition. Jamil Ghojel.
© 2020 John Wiley & Sons Ltd. This Work is a co-publication between John Wiley & Sons Ltd and ASME Press.
Companion website: www.wiley.com/go/JamilGhojel_Fundamentals of Heat Engines

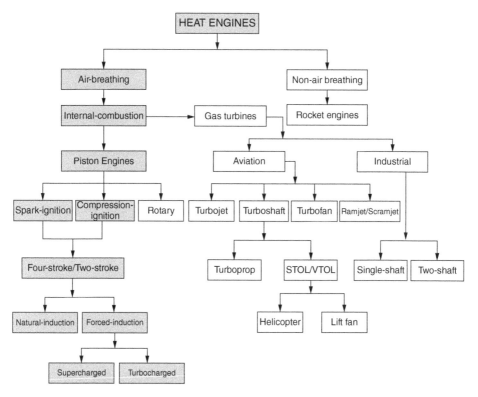

Chart II.1 Piston engine types covered in Part II are highlighted.

efficiency of more than 30% was achieved by 1908. Remarkably, few modern piston engines operate at efficiencies higher than that at present.

The credit for inventing the compression-ignition engine goes to the German engineer Rudolf Diesel (1858–1913). Diesel published a paper in 1893 entitled 'The Theory and Construction of a Rational Heat Motor' in which he described the operating principle of the engine, which can be summarised as follows:

1. Air is compressed adiabatically to one-sixteenth of its original volume to raise the temperature to about $800\,K$.
2. Fuel is injected into the engine and is ignited automatically without the need of an electric spark.
3. Work done by the high expansion ratio results in thermal efficiencies that are superior to the Otto cycle.

Diesel's engine, which was put on display in Munich in 1898, used high-pressure air to inject and atomise the liquid fuel (break it up into minute drops), which burned at an almost-constant pressure. With continuous improvements over the years, the compression-ignition engine has become the heat engine of choice in heavy transport, mining, industrial, and agricultural machinery, stationary and mobile generator plants; and with the introduction of the common rail fuel injection system, it has become a competitor to the SI engine in the passenger car market.

Remarkably, almost 143 years after Otto's first SI engine and 121 years after Diesel's CI engine, the reciprocating internal combustion engine is still dominant in the areas mentioned. This is the result of intensive research and development in most industrialised countries ever since its invention. Modern engines are a marvel of reliability and dependable performance, with easy starting characteristics in all climates and geographic conditions. There is an air of anticipating change in the passenger car market, with plans in some developed countries to phase out the sale of cars powered by IC engines and replace them with electric cars in the coming decades. However, it is highly unlikely that all 193 countries in the world will be able to make or can afford the switch to electric vehicles any time soon. Furthermore, any challenger to the dominance of IC engines in the other areas discussed will have to be formidable indeed.

3

Ideal Cycles for Natural-Induction Reciprocating Engines

An engine that draws air directly from the surroundings at ambient temperature and pressure is known as natural-induction engine or naturally aspirated engine. In this book, the term *natural-induction engine* will be predominantly used. As for cycles, the terms *ideal cycle* and *air-standard cycle* will be used interchangeably. In this chapter, four air-standard cycles will be examined to establish the theoretical potential of the reciprocating internal combustion piston engine (piston engine henceforward).

A generalised air-standard cycle is first analysed, and equations for the thermal efficiency and cycle specific work are deduced. The other three cycles are then obtained as special cases of the generalised cycle. These ideal cycles, widely known as air-standard cycles, are usually identified as the theoretical basis of current actual reciprocating piston engines:

- Constant-volume combustion cycle (Otto cycle)
- Constant-pressure combustion cycle (Diesel cycle)
- Dual-combustion cycle (a hybrid of the previous two cycles, known also as the mixed-combustion cycle or pressure-limited cycle)

The cycles are referred to as air-standard cycles because the working fluid they operate on is assumed to be pure air. For the remainder of the analysis, the word *combustion* will be dropped from the names of the cycles for the sake of brevity.

3.1 Generalised Cycle

The generalised cycle of the reciprocating piston engine is based on the combination of idealised reciprocating piston motion in a cylinder and the schematic thermodynamic representation of the heat engine shown in Figure 3.1.

The cycle ignores gas exchange processes (the engine has no valves) and is based on the following assumptions:

1. A constant amount of gas (air) is trapped in the cylinder and is referred to as the *charge*.
2. The specific heats of the charge remain unchanged during the entire cycle.
3. Heat transfer to the cycle (heat input q_{in}) is effected externally from a heat source at a high temperature (T_h) and heat transfer from the cycle (heat rejection q_{out}) is to a heat sink at low temperature (T_l).

Fundamentals of Heat Engines: Reciprocating and Gas Turbine Internal Combustion Engines, First Edition. Jamil Ghojel.
© 2020 John Wiley & Sons Ltd. This Work is a co-publication between John Wiley & Sons Ltd and ASME Press.
Companion website: www.wiley.com/go/JamilGhojel_Fundamentals of Heat Engines

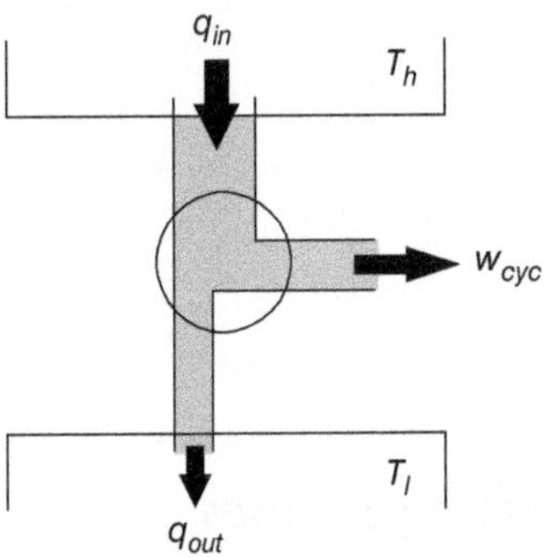

Figure 3.1 First law representation of the heat engine.

4. The heat-transfer processes to and from the cycle are partially at constant volume and partially at constant pressure.
5. There is no heat exchange with the surroundings during the compression and expansion processes (adiabatic processes).
6. Work W_{cyc} is produced by the cycle during the expansion process.

The processes in the generalised cycle, plotted in p-V and T-s coordinates, are shown in Figure 3.2:

Process 1–2: Adiabatic and reversible compression accompanied by a pressure increase from p_1 to p_2, a temperature increase from T_1 to T_2, no change in entropy, and a volume decrease from V_1 to V_2
Process 2–3: Heat addition at a constant volume accompanied by a pressure increase from p_2 to p_3, a temperature increase from T_2 to T_3, and an entropy increase from s_2 to s_3
Process 3–4: Heat addition at constant pressure accompanied by a gas expansion from V_3 to V_4, a temperature increase from T_3 to T_4, and an entropy increase from s_3 to s_4

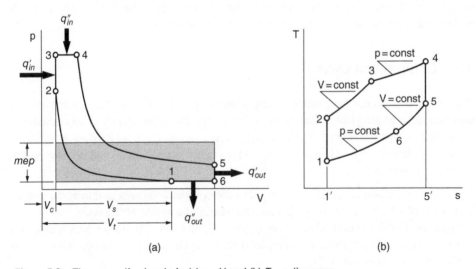

(a)   (b)

Figure 3.2 The generalised cycle in (a) $p - V$ and (b) $T - s$ diagrams.

Process 4–5: Adiabatic and reversible expansion accompanied by a pressure decrease from p_4 to p_5, a temperature decrease from T_4 to T_5, no change in entropy, and a volume increase from V_4 to V_5

Process 5–6: Heat rejection at constant volume accompanied by a pressure decrease from p_5 to p_6, a temperature decrease from T_5 to T_6, and an entropy decrease from s_5 to s_6

Process 6–1: Heat rejection at constant pressure accompanied by a volume decrease from V_6 to V_1, a temperature decrease from T_6 to T_1, and an entropy decrease from s_6 to s_1

The heat input, heat rejected, and heat equivalent of the cycle work per unit mass of charge q_{in}, q_{out}, and w_{cyc} are shown in the T-s diagram of the generalised cycle in Figure 3.3 by the shaded areas. The heat input is represented by the area $(1' - 1 - 2 - 3 - 4 - 5 - 5')$ (Figure 3.3a), the heat rejected by the area $(1' - 1 - 6 - 5 - 5')$ (Figure 3.3b), and the work output by the area enclosed by the cycle $(1 - 2 - 3 - 4 - 5 - 6 - 1)$ (Figure 3.3c).

From Figure 3.2:

$$q_{in} = q'_{in} + q''_{in}$$

$$q_{in} = c_v(T_3 - T_2) + c_p(T_4 - T_3)\,J/kg \tag{3.1}$$

c_v, c_p are the specific heats of the charge at constant volume and constant pressure, respectively, in $J/kg.K$.

Similarly, we can write the relationships for the heat rejected as

$$q_{out} = q'_{out} + q''_{out}$$

$$q_{out} = c_v(T_5 - T_6) + c_p(T_6 - T_1)\,J/kg \tag{3.2}$$

Let us define the following:

Ratio of specific heats $\gamma = c_p/c_v$
Compression ratio $\varepsilon = V_1/V_2$
Pressure ratio at constant volume $\alpha = p_3/p_2$
Expansion ratio at constant pressure $\beta = V_4/V_3$
Expansion ratio $\delta = V_5/V_4$
Volume ratio at constant pressure $\rho = V_6/V_1$

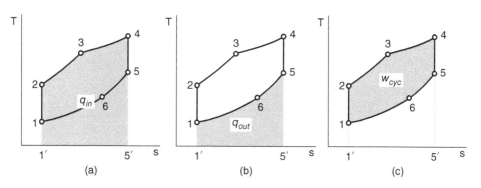

Figure 3.3 $T - s$ diagram of the generalised cycle (a) specific heat input; (b) heat rejected; (c) work output.

Combining these ratios with the ideal gas equations, we get the volume, pressure, and temperature at the characteristic points of the cycle in terms of the properties at point 1:

$$V_2 = \frac{V_1}{\varepsilon}$$

$$p_2 = p_1 \left(\frac{V_1}{V_2}\right)^\gamma = p_1 \varepsilon^\gamma$$

$$T_2 = T_1 \left(\frac{V_1}{V_2}\right)^{\gamma-1} = T_1 \varepsilon^{\gamma-1}$$

$$V_3 = V_2 = \frac{V_1}{\varepsilon}$$

$$p_3 = \alpha\, p_2$$

$$T_3 = T_2 \left(\frac{p_3}{p_2}\right) = T_2 \alpha = T_1 \alpha \varepsilon^{\gamma-1}$$

$$V_4 = \beta\, V_3 = \beta \frac{V_1}{\varepsilon}$$

$$p_4 = p_3$$

$$T_4 = T_3 \left(\frac{V_4}{V_3}\right) = T_3 \beta = T_1 \alpha \beta \varepsilon^{\gamma-1}$$

$$V_5 = V_6 = \rho V_1$$

$$p_5 = p_4 \left(\frac{V_4}{V_5}\right)^\gamma = p_4 \left(\frac{1}{\delta}\right)^\gamma$$

$$T_5 = T_4 \left(\frac{V_4}{V_5}\right)^{\gamma-1} = T_4 \left(\frac{1}{\delta^{\gamma-1}}\right) = T_1 \alpha \beta \left(\frac{\varepsilon}{\delta}\right)^{\gamma-1}$$

Since

$$\frac{\varepsilon}{\delta} = \frac{V_1}{V_2}\frac{V_4}{V_5} = \frac{V_4}{V_2} \bigg/ \frac{V_5}{V_1} = \frac{\beta}{\rho}$$

we obtain

$$T_5 = T_1 \alpha \beta \left(\frac{\beta}{\rho}\right)^{\gamma-1}$$

Finally,

$$V_6 = \rho\, V_1$$

$$p_6 = p_1$$

$$T_6 = T_1 \left(\frac{V_6}{V_1}\right) = \rho\, T_1$$

The equation for the heat input can now be written as

$$q_{in} = c_v T_1 (\alpha \varepsilon^{\gamma-1} - \varepsilon^{\gamma-1}) + c_p T_1 (\alpha \beta \varepsilon^{\gamma-1} - \alpha \varepsilon^{\gamma-1})$$

$$q_{in} = c_v T_1 \varepsilon^{\gamma-1} [(\alpha - 1) + \alpha \gamma (\beta - 1)] \tag{3.3}$$

and the rejected heat as

$$q_{out} = c_v T_1 \rho \left[\alpha \left(\frac{\beta}{\rho} \right)^\gamma - 1 \right] + c_p T_1 (\rho - 1)$$

$$q_{out} = c_v T_1 \left\{ \rho \left[\alpha \left(\frac{\beta}{\rho} \right)^\gamma - 1 \right] + \gamma(\rho - 1) \right\} \tag{3.4}$$

The thermal efficiency of the cycle is

$$\eta_t = \frac{W_{cyc}}{q_{in}}$$

$$\eta_t = \frac{q_{in} - q_{out}}{q_{in}} = 1 - \frac{q_{out}}{q_{in}}$$

Combining q_{in} and q_{out} in the previous equation with Eqs. (3.3) and (3.4) yields the thermal efficiency of the generalised air-standard cycle:

$$\eta_t = 1 - \frac{\rho \left[\alpha \left(\frac{\beta}{\rho} \right)^\gamma - 1 \right] + \gamma(\rho - 1)}{\varepsilon^{\gamma-1}[(\alpha - 1) + \alpha\gamma(\beta - 1)]} \tag{3.5}$$

The concept of *mean effective pressure* (*mep*) is very useful when comparing the specific performance of engines of various sizes and speeds. The work done by any closed thermodynamic cycle W_{cyc} plotted in $p - V$ coordinate system is equal to the area enclosed by the processes making up the cycle, which can be replaced by the rectangular shaded area shown in Figure 3.2a. The base of the rectangle is the difference between the maximum and minimum volumes, and the height is the *mep*. The *mep* is a constant arbitrary pressure that is assumed to act on the piston as it executes the different thermodynamic processes of the cycle.

$$mep = \frac{W_{cyc}}{(V_{max} - V_{min})} = \frac{W_{cyc}}{(V_6 - V_2)} = \eta_t \frac{q_{in}}{(V_6 - V_2)} \tag{3.6}$$

$$V_6 - V_2 = V_2 \left(\frac{V_6}{V_2} - 1 \right) = V_2 \left(\frac{V_6}{V_1} \frac{V_1}{V_2} - 1 \right)$$

$$V_6 - V_2 = V_2(\rho\varepsilon - 1) = \frac{V_1}{\varepsilon}(\varepsilon\rho - 1)$$

$$mep = \frac{\varepsilon}{V_1(\varepsilon\rho - 1)} \eta_t c_v T_1 \varepsilon^{\gamma-1}[(\alpha - 1) + \alpha\gamma(\beta - 1)]$$

Since

$$c_v = \frac{R}{\gamma - 1}, p_1 = \frac{RT_1}{V_1}$$

we can write

$$\frac{c_v T_1}{V_1} = \frac{p_1}{\gamma - 1}$$

Finally,

$$mep = \frac{p_1}{\gamma - 1} \frac{\varepsilon^\gamma}{(\varepsilon\rho - 1)} \eta_t[(\alpha - 1) + \alpha\gamma(\beta - 1)] \tag{3.7}$$

3.2 Constant-Volume Cycle (Otto Cycle)

Spark ignition engines are based on the constant volume cycle, more commonly known as the *Otto cycle*. Heat addition and heat rejection in this cycle occur at constant volume, as shown in Figure 3.4. The Otto cycle is a special case of the generalised cycle, and equations for thermal efficiency and *mep* can be obtained by substituting $\beta = 1$ and $\rho = 1$ in the corresponding generalised Eqs. (3.5) and (3.7):

$$\eta_t = 1 - \frac{1}{\varepsilon^{\gamma-1}} \tag{3.8}$$

$$mep = \frac{p_1}{(\gamma - 1)} \left(\frac{\varepsilon^\gamma}{\varepsilon - 1}\right) \eta_t[(\alpha - 1)] \tag{3.9}$$

For a given gas, γ is constant, and the thermal efficiency is a function of the compression ratio ε only (Figure 3.5). For a given value of γ, η_t increases continuously, with the rate of increase tapering off at the high range of ε. For a given compression ratio, the higher the value of γ, the higher the thermal efficiency. The values of γ used in Figure 3.5 are those at 300 K.

The function for the *mep* (Eq. 3.7) is more complex with multiple variables (p_1, ε, γ, η_t, and α). However, for a given gas, it can be reduced to a function of only two variables: ε and α. From Figure 3.4a and the definition of the compression ratio,

$$\varepsilon = \frac{V_1}{V_2} = \frac{V_1}{V_c} = \frac{V_1}{V_1 - V_s} = \frac{1}{1 - \frac{V_s}{V_1}}$$

where V_s is the swept volume. Hence,

$$V_1 = \left(\frac{\varepsilon}{\varepsilon - 1}\right) V_s$$

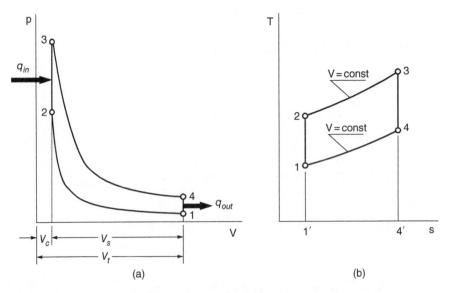

(a)   (b)

Figure 3.4 The Otto cycle in $p - V$ (a) and $T - s$ (b) coordinate systems.

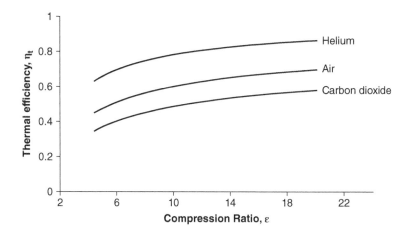

Figure 3.5 Thermal efficiency of the Otto cycle as a function of compression ratio ε and ratio of specific heats γ ($\gamma_{He} = 1.667$, $\gamma_{air} = 1.4$, $\gamma_{CO2} = 1.289$).

For 1 kg of charge:

$$p_1 = \frac{RT_1}{V_1} = \frac{RT_1}{V_s}\left(\frac{\varepsilon - 1}{\varepsilon}\right)$$

This equation indicates that for given temperature and swept volume, the pressure at the start of compression varies with the compression ratio. Practically, p_1 remains undistinguishable from the ambient pressure up to $\varepsilon = 4$ (Arkhangelsky et al., 1971). Assuming $p_1 = 0.1\ MPa$ at $\varepsilon = 4$:

$$\frac{RT_1}{V_s} = \left(\frac{4}{4-1}\right)p_1 = \frac{4}{3} \times 0.1 = \frac{0.4}{3}$$

At values of $\varepsilon > 4$, p_1 varies as follows:

$$p_1 = \frac{0.4}{3}\left(\frac{\varepsilon - 1}{\varepsilon}\right)$$

The corresponding volume of the charge at the start of compression for a given temperature is

$$V_1 = \frac{RT_1}{p_1}$$

The equation for the *mep* can now be written as

$$mep = \frac{0.4}{3}\frac{\varepsilon^{\gamma-1}}{(\gamma - 1)}\left(1 - \frac{1}{\varepsilon^{\gamma-1}}\right)(\alpha - 1) \tag{3.10}$$

For a given γ, *mep* can be plotted as a function of ε at constant values of the ratio of pressure rise at constant volume α (Figure 3.6), or as a function of α at constant values of ε (Figure 3.7). The *mep* increases steadily with increasing compression ratio ε and α. The rate of increase of *mep* is greater with increasing ε at higher values of α. Similarly, the rate of increase of *mep* with α is higher at the higher values of ε and γ. The effect of the ratio of specific heats of the charge can be seen in Figure 3.7 by comparing the continuous and dashed lines.

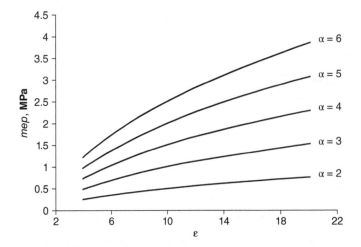

Figure 3.6 Mean effective pressure of the Otto cycle as a function of compression ratio (ε) at various ratios of pressure rise at constant volume (α) ($\gamma = 1.4$).

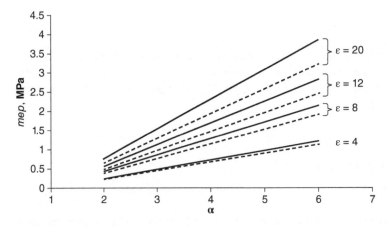

Figure 3.7 Mean effective pressure of the Otto cycle as a function of ratio of pressure rise at constant volume α at various compression ratios ε ($\gamma = 1.4$ solid lines, $\gamma = 1.3$ dashed lines).

3.3 Constant Pressure (Diesel) Cycle

Slow-speed modern compression ignition engines are loosely based on the constant pressure cycle (Diesel cycle) shown in Figure 3.8. This cycle is also a special case of the generalised cycle in which heat addition is wholly at constant pressure and heat rejection is at constant volume.

The equations of thermal efficiency and the *mep* for the Diesel cycle shown next are obtained by substituting $\alpha = 1$ and $\rho = 1$ in the corresponding generalised cycle Eqs. (3.5) and (3.7):

$$\eta_t = 1 - \left(\frac{1}{\varepsilon^{\gamma-1}}\right)\frac{(\beta^{\gamma} - 1)}{\gamma(\beta - 1)} \tag{3.11}$$

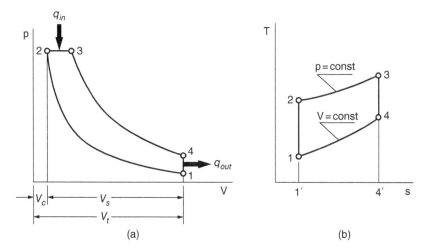

Figure 3.8 The Diesel cycle in $p - V$ (a) and $T - s$ (b) coordinate systems.

$$mep = \frac{p_1}{(\gamma - 1)} \left(\frac{\varepsilon^{\gamma}}{\varepsilon - 1} \right) \eta_t \gamma (\beta - 1) \tag{3.12}$$

The thermal efficiency of the diesel cycle is a function of three variables γ, ε, and β. If β is fixed, η_t can be represented as a function of ε and γ as in Figure 3.9a. Alternatively, if γ is fixed, η_t can be represented as a function of ε and β as shown in Figure 3.9b. The thermal efficiency increases as ε and γ increase and β decreases. Figure 3.9b shows that, for a given ε, the thermal efficiency decreases with increasing β. As the value of β approaches 1, the thermal efficiency of the Diesel cycle approaches that of the Otto cycle. The values of η_t for the Diesel cycle at $\beta = 1.1$ is slightly lower than the Otto cycle. At $\beta = 1.01$, the two cycles are almost indistinguishable.

If we replace p_1 by

$$p_1 = \frac{0.4}{3} \left(\frac{\varepsilon - 1}{\varepsilon} \right)$$

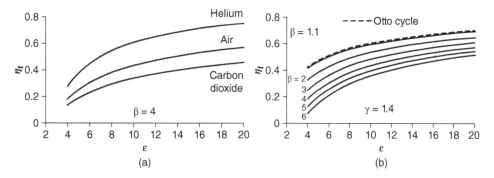

Figure 3.9 Thermal efficiency of the Diesel cycle: (a) as a function ε and γ at $\beta = 4$ ($\gamma_{He} = 1.667$, $\gamma_{air} = 1.4$, $\gamma_{CO2} = 1.289$); (b) as a function of ε and β at $\gamma = 1.4$.

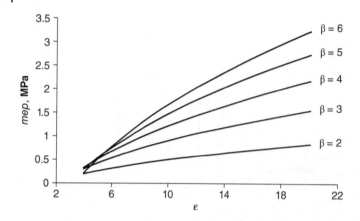

Figure 3.10 Mean effective pressure of the Diesel cycle as a function of compression ratio ε and ratio of volume change at constant pressure β.

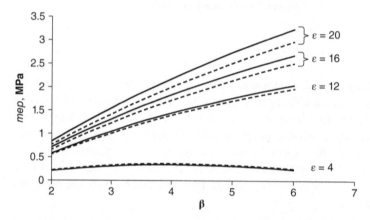

Figure 3.11 Mean effective pressure of the Diesel cycle as a function of β, ε, and γ ($\gamma = 1.4$ solid lines, $\gamma = 1.3$ dashed lines).

and substitute for η_t from Eq. (3.11) into Eq. (3.12), we obtain

$$mep = \frac{0.4}{3}\left(\frac{\gamma}{\gamma-1}\right)\left[1-\left(\frac{1}{\varepsilon^{\gamma-1}}\right)\frac{(\beta^\gamma-1)}{\gamma(\beta-1)}\right](\beta-1)\varepsilon^{\gamma-1} \qquad (3.13)$$

Figure 3.10 shows the *mep* as a function of ε and β at constant $\gamma = 1$. For a given value of β, *mep* increases with ε and, unlike the thermal efficiency, it also increases with β. The latter is clearly evident in Figure 3.11, which also shows the effect of γ. At a low compression ratio ($\varepsilon = 4$), *mep* exhibits a maximum at $\beta = 4$.

3.4 Dual Cycle (Pressure-Limited Cycle)

Modern high-speed compression ignition engines (popularly referred to as diesel engines) are based on this cycle, in which heat addition to the cycle is partially at constant volume and partially at constant pressure. Heat rejection is at constant volume only, i.e. $\rho = 1$ in the generalised cycle. The cycle is shown in Figure 3.12.

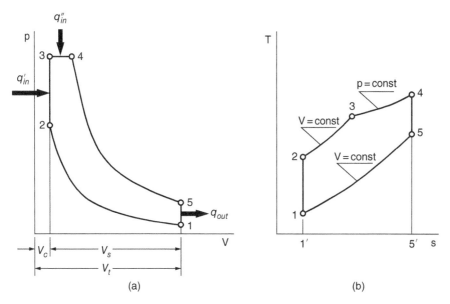

Figure 3.12 The dual cycle in $p - V$ (a) and $T - s$ (b) coordinate systems.

Substituting $\rho = 1$ in Eqs. (3.5) and (3.7), we obtain respectively the equations of the thermal efficiency and *mep* of the dual cycle:

$$\eta_t = 1 - \frac{1}{\varepsilon^{\gamma-1}} \frac{\alpha\beta^\gamma - 1}{[(\alpha - 1) + \alpha\gamma(\beta - 1)]} \tag{3.14}$$

$$mep = \frac{p_1}{(\gamma - 1)(\varepsilon - 1)} \varepsilon^\gamma \eta_t[(\alpha - 1) + \alpha\gamma(\beta - 1)] \tag{3.15}$$

These functions are much more complex than the ones for the Otto and Diesel cycles. To develop the methodology for the analysis of this cycle, we start with the combined heat input at constant volume and constant pressure (Eq. 3.3):

$$q_{in} = c_v T_1 \varepsilon^{\gamma-1}[(\alpha - 1) + \alpha\gamma(\beta - 1)]$$

Let

$$K = [(\alpha - 1) + \alpha\gamma(\beta - 1)] \tag{3.16}$$

The three characteristic functions of the cycle become

$$q_{in} = c_v T_1 \varepsilon^{\gamma-1} K$$

$$\eta_t = 1 - \left(\frac{1}{\varepsilon^{\gamma-1}}\right) \frac{\alpha\beta^\gamma - 1}{K}$$

$$mep = \frac{p_1}{(\gamma - 1)} \left(\frac{\varepsilon^\gamma}{\varepsilon - 1}\right) \eta_t K$$

Since

$$c_v = \frac{R}{(\gamma - 1)},$$

$$\frac{q_{in}}{T_1} = \frac{R}{(\gamma - 1)} \varepsilon^{\gamma-1} K$$

The last equation can be written as

$$K = \frac{q_{in}}{T_1} \frac{(\gamma - 1)}{R} \left(\frac{1}{\varepsilon^{\gamma-1}} \right) \tag{3.17}$$

Substituting for K in the efficiency equation yields

$$\eta_t = 1 - \frac{RT_1}{q_{in}(\gamma - 1)}(\alpha\beta^{\gamma} - 1) \tag{3.18}$$

Substituting for K in the *mep* equation and remembering that

$$p_1 = \frac{0.4}{3} \left(\frac{\varepsilon - 1}{\varepsilon} \right),$$

the final equation for *mep* can be written as

$$mep = \frac{0.4}{3} \frac{1}{(\gamma - 1)} \eta_t \varepsilon^{\gamma-1} K \tag{3.19}$$

The thermal efficiency and *mep* for the dual combustion cycle are complex functions with several variables that need to be reduced to a manageable number. The methodology presented here to establish a clear graphical representation of the performance parameters of the dual cycle (η_t and *mep*) is broadly based on the analysis used in Arkhangelsky et al. (1971):

1. Fix the total heat input q_{in}.
2. Assume constant compression ratio ε, ratio of specific heats γ, gas constant R, and temperature at the start of compression.
3. Calculate the value of K from Eq. (3.17).
4. For given values of α, calculate the corresponding values of β from Eq. (3.16).
5. Use Eq. (3.18) to calculate the thermal efficiency as a function of α at constant ε and varying q_{in}. The values of β can be plotted on the same coordinate system as shown in Figure 3.13. Each curve in this figure corresponds to a constant value of the input heat q_{in}.
6. Use Eq. (3.19) to calculate the *mep* as a function of α at constant ε and varying q_{in}. The values of β can be plotted on the same coordinate system as shown in Figure 3.14.

Inspection of Figures 3.13 and 3.14 shows the following:

- The larger the input heat, the larger the range of variation of α and β.
- For a given input heat and compression ratio, both the thermal efficiency and *mep* increase with α (i.e. with the fraction of heat supplied at constant volume).
- The maximum efficiency for each value of the input heat is reached when $\beta = 1$ (i.e. all the heat is supplied at constant volume as in the Otto cycle). This maximum remains constant independent of the heat input.
- The maximum *mep* for each value of the input heat is reached when $\beta = 1$ (i.e. all the heat is supplied at constant volume as in the Otto cycle). This maximum, unlike thermal efficiency, increases steadily with the heat input.
- The minimum efficiency for each value of the input heat is reached when $\alpha = 1$ (i.e. all the heat is supplied at constant pressure as in the Diesel cycle). This minimum varies with the heat input and increases as the input heat decreases.

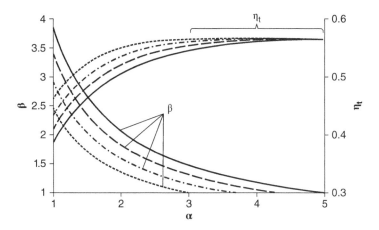

Figure 3.13 Thermal efficiency of the dual cycle and β as functions of α at given values of γ and ε.

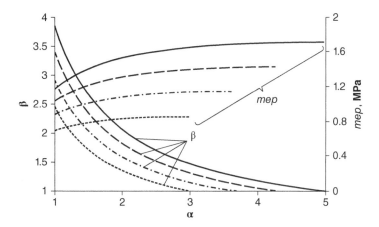

Figure 3.14 Mean effective pressure of the dual cycle and β as functions of α at given values of γ and ε.

- The minimum *mep* for each value of the input heat is reached when $\alpha = 1$ (i.e. all the heat is supplied at constant pressure as in the Diesel cycle). This minimum varies with the heat input and increases with the input heat.

In real engines, the heat input into the cycle will be through the delivery of a certain quantity of fuel. The parameters α and β represent the fractions of fuel delivered at constant volume and constant pressure, respectively, and their values can determine how much power the engine will produce and how efficiently. The effect of α and β on the thermal efficiency and *mep* can be gauged from Figure 3.15. If β is kept constant at 1.5 and the input heat is increased by increasing α, the thermal efficiency increases slightly while the *mep* increases several fold (Figure 3.15a). Similarly, if α is kept constant at 1.5 and the input heat is increased by increasing β, the thermal efficiency decreases significantly while the *mep*

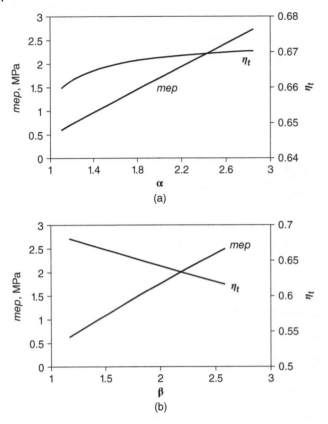

Figure 3.15 Thermal efficiency and mean effective pressure of the dual cycle at $\varepsilon = 18$: (a) $\beta = 1.5$; (b) $\alpha = 1.5$.

increases several fold (Figure 3.15b). The way to increase the *mep* (power) without loss of efficiency is to fix β and increase α.

The effect of the compression ratio ε on the thermal efficiency and *mep* of the dual cycle for a given heat input can be evaluated from the carpet plots shown in Figures 3.16 and 3.17. Figure 3.16 shows the thermal efficiency versus α at constant ε and constant β lines. Figure 3.17 shows the *mep*. If β is kept constant, both the thermal efficiency and *mep* increase with the compression ratio while α decreases. The rates of increase of η_t and *mep* decrease as ε increases, as indicated by the narrowing of the gaps between the constant ε lines.

The other cycle parameters that are of importance are the pressures and temperatures at different points of the cycle. In real engines, these parameters need to be selected carefully to insure component integrity and safety of operation. Figure 3.18 shows the effect of the compression ratio on the compression pressure p_2 and maximum cycle pressure p_3. The maximum pressure increases almost linearly with the compression ratio, and its values at all compression ratios are lower when α is kept constant while β is allowed to change.

These changes can be seen clearly from the p-V diagrams of the cycles plotted to scale at three values of the compression ratio with the heat input fixed (Figure 3.19). One way

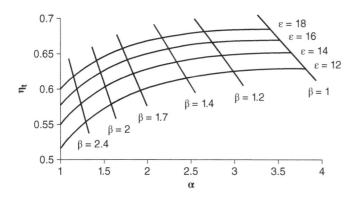

Figure 3.16 Carpet plot for the effect of α, β, and ε on the thermal efficiency of the dual cycle ($q_{in} = 1625$ kJ/kg).

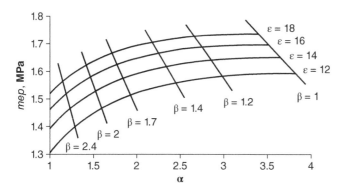

Figure 3.17 Carpet plot for the effect of α, β, and ε on the mean effective pressure of the dual cycle ($q_{in} = 1625$ kJ/kg).

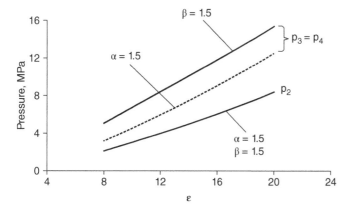

Figure 3.18 Effect of compression ratio on cycle pressures p_2 and p_3 ($q_{in} = 1500$ kJ/kg, solid lines $\beta = 1.5$, broken lines $\alpha = 1.5$).

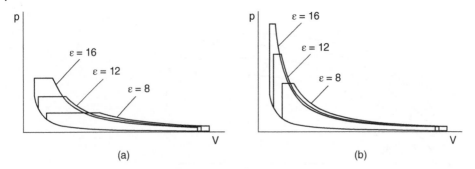

Figure 3.19 $p - V$ diagrams for the dual cycle at three compression rations (a) $\alpha = const$; $\beta = var$; (b) $\alpha = var$, $\beta = const$.

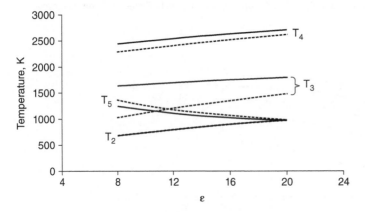

Figure 3.20 Effect of the compression ratio on the temperatures of four points of the dual combustion cycle ($q_{in} = 1500$ kJ/kg, solid lines $\beta = 1.5$, broken lines $\alpha = 1.5$).

to keep the maximum cycle pressure at an acceptable level at high compression ratios is to increase the fraction of fuel delivered at constant pressure.

The effect of the compression ratio on the temperatures at various points of the cycle (see Figure 3.12b) can be seen in Figure 3.20. All temperatures increase with the compression ratio with the exception of T_5. T_2 is independent of α and β. T_4 and T_5 tend to be higher when β is kept constant as ε increases. T_3 is much more sensitive to the change in ε when α is kept constant.

3.5 Cycle Comparison

Comparison of the three theoretical cycles of the reciprocating internal combustion engine in terms of the thermal efficiency is shown in Figure 3.21 and in terms of the *mep* in Figure 3.22. The performance parameters of the Otto and Diesel cycles represent the upper and lower limits achievable for a given input heat, whereas the dual cycle is somewhere in between. The thermal efficiency of the Otto cycle is, as we saw before, a function of the compression ratio only and, therefore, remains constant as the heat input increases.

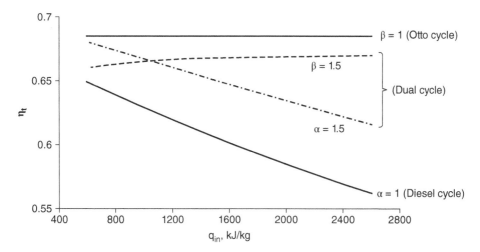

Figure 3.21 Comparison of the thermal efficiency for the Otto, Diesel, and dual cycles.

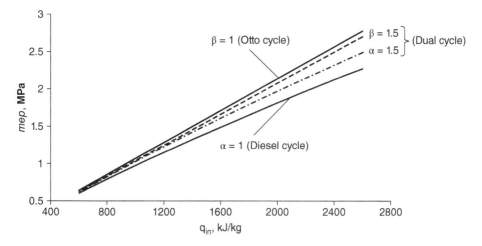

Figure 3.22 Comparison of the mean effective pressures of the Otto, Diesel, and dual cycles.

In the Diesel cycle, the thermal efficiency η_t decreases continuously with the increase in heat input. The values of the performance parameters of the dual combustion cycle depend on the fractions of heat supplied at constant volume and constant pressure, i.e. on α and β. In Figure 3.21, two efficiency curves are drawn, representing two distinct methods of regulating the heat input: in the first method, $\beta = 1.5$ and α increases with heat input; and in the second, $\alpha = 1.5$ and β increases with heat input. At the lower range of heat input, the second method yields higher η_t. As the value of β or α decreases and approaches 1, η_t increases with the first method approaching that of the Otto cycle and decreases with the second method approaching that of the Diesel cycle.

The same trends are also noticed for the *mep*, as shown in Figure 3.22. The only difference is that *mep* increases continuously with the heat input for all cycles. The differences in performance of the three cycles can also be gauged from the comparison of $p - V$ and $T - s$

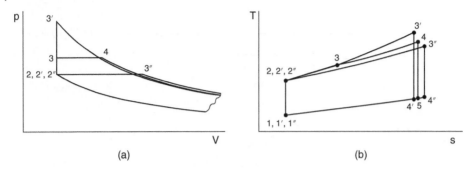

(a)   (b)

Figure 3.23 $p-V$ and $T-s$ diagrams of three cycles of reciprocating engines ($q_{in} = 600$ kJ/kg, $\varepsilon = 16$). Dual cycle: 1-2-3-4-5, $\alpha = 1.3$, $\beta = 1.3437$; Otto cycle: 1'-2'-3'-4', $\beta = 1$; Diesel cycle: 1''-2''-3''-4'', $\alpha = 1$.

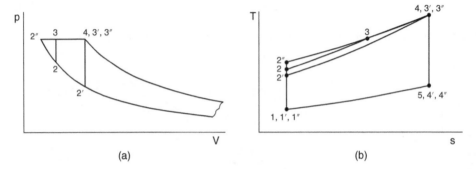

(a)   (b)

Figure 3.24 $p-V$ and $T-s$ diagrams of three cycles of reciprocating engines ($p_4 = 4.8$ MPa). Dual cycle: 1-2-3-4-5, $q_{in} = 585$ kJ/kg, $\varepsilon = 12$, $\alpha = 1.496$, $\beta = 1.247$; Otto cycle: 1'-2'- 3'-4', $q_{in} = 550$ kJ/kg, $\varepsilon = 9.5$, $\alpha = 2.041$, $\beta = 1$; Diesel cycle: 1''-2''-3''-4'', $q_{in} = 600$ kJ/kg, $\varepsilon = 16$, $\beta = 1.661$, $\alpha = 1$.

diagrams (Figures 3.23 and 3.24). Figure 3.23, in which the compression ratio and heat input are fixed, shows that the dual cycle falls somewhere between the Otto and Diesel cycles in terms of the maximum cycle pressure and temperature and the amount of heat rejected (represented by the area under process 1–5). From the $T-s$ diagram, it can be concluded that under the given conditions, the Otto cycle is the most efficient and the Diesel cycle least efficient.

In Figure 3.24, the maximum pressure and temperature of the three cycles are kept constant and equal while varying the compression ratio and heat input. Again, the performance of mixed cycle falls in between the other two cycles in term of the magnitude of the work produced (represented by the area enclosed by the cycle when plotted in $T-s$ coordinate system). In this comparison, the Diesel cycle is the most efficient and the Otto cycle the least efficient.

Problems

3.1 Consider the following data for the ideal piston-engine cycle with constant-volume heat addition (Otto cycle): $p_1 = 0.1$ MPa, $T_1 = 293$ K, $\varepsilon = 4.5$, $\alpha = 3.5$, $\gamma = 1.4$. Assuming constant specific heat, determine

(a) Properties at the characteristic points of the cycle
(b) Amounts of heat supplied and rejected
(c) Thermal efficiency
(d) Net cycle work

3.2 The following information provided is for an ideal piston-engine cycle with constant-pressure heat addition (Diesel cycle): $p_1 = 0.1\ MPa$, $T_1 = 293\ K$, $\varepsilon = 12.7$, $\gamma = 1.4$, $c_{pa} = 1.0117 kJ/kg.\ K$. Determine
(a) Properties at the characteristic points of the cycle
(b) Amounts of heat supplied and rejected
(c) Thermal efficiency
(d) Net cycle work

3.3 The hypothetical cycle $1-2-3-4$ has two constant-pressure and two constant-volume processes. If the specific heats at constant pressure and constant volume are assumed constant, and $\alpha = p_2/p_1$, $\rho = v_3/v_2$, show that the thermal efficiency is given by the following equation:

$$\eta_t = 1 - \frac{\beta(\alpha - 1 + \gamma) - \gamma}{\alpha - 1 + \gamma\alpha(\beta - 1)}$$

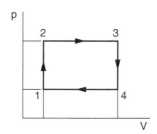

3.4 The initial pressure and temperature in an air-standard cycle with dual combustion process are $90\ kPa$ and $340\ K$, respectively. The total heat input into the cycle is $Q_{in} = 1090 kJ/kg$, and the compression ratio $\varepsilon = 10$. If the maximum cycle pressure is not to exceed $6.5\ MPa$, calculate the percentage of the total heat that must be provided at constant volume. Assume constant heat capacity throughout.

3.5 A small truck has a four-cylinder, four-litre compression ignition (CI) engine operating on the air-standard dual cycle shown. The heat q_{in} provided externally per cycle to each cylinder is from the combustion of $0.05\ g$ of light diesel fuel with lower heating value of $42\ 500\ kJ/kg$. The compression ratio of the engine is 15:1, and the cylinder bore is $100\ mm$. At the start of the compression stroke, conditions in the cylinder are $60\ °C$ and $100\ kPa$. It can be assumed that half of the heat input from combustion is added at constant volume and half at constant pressure. Calculate
(a) Temperature and pressure at each characteristic point of the cycle
(b) Net work of the cycle

(c) Thermal (indicated) efficiency of the cycle

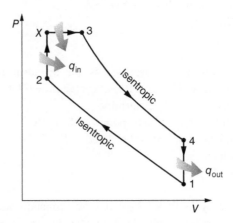

3.6 Consider the following cycle comprising these processes:
1 – 2 adiabatic compression
2 – 3 constant-volume combustion
3 – 4 adiabatic expansion
4 – 1 constant-pressure heat rejection

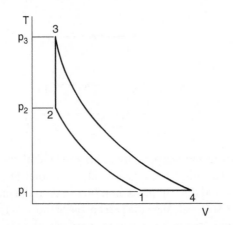

Sketch the T-s diagram, and show that the thermodynamic efficiency of the cycle is given by

$$\eta_t = 1 - \frac{\gamma(\alpha^{1/\gamma} - 1)}{\varepsilon^{\gamma-1}(\alpha - 1)}$$

where γ is the ratio of specific heats for air, $\varepsilon = V_1/V_2$, $\alpha = p_3/p_2$.

4

Ideal Cycles for Forced-Induction Reciprocating Engines

The equations for the mean effective pressure (*mep*) show that it can be boosted by increasing the pressure p_1 at the start of the compression process. For a given heat input, boosting the *mep* results in increased cycle work and torque. This can be done in real engines by employing a small compressor to raise the atmospheric pressure of the air before it is delivered to the engine. If the compressor is driven by the high-temperature and high-pressure exhaust gases (the former can be as high as $1000\,K$ and the latter significantly higher than the ambient pressure) the boosting system is known as *turbocharging*. If, on the other hand, the compressor is driven mechanically by the engine itself, the system is known as *supercharging*. In some cases, the two systems can be compounded to provide very high degrees of boosting.

4.1 Turbocharged Cycles

Turbocharging is defined here as a boosting process that utilises a freestanding turbine/compressor unit (known as a turbocharger). The turbine is connected to the exhaust manifold and provides the required power to run the compressor, which supplies air to the engine at temperature and pressure above ambient conditions.

4.1.1 Turbocharged Engine with Constant-Pressure Turbine

Figure 4.1 shows a schematic diagram of a turbocharged (TC) engine in which receiver R in the exhaust manifold provides constant pressure exhaust gases to turbine T. The high-temperature gases expand in the turbine, which is rigidly linked to compressor C, to provide the power required to raise the pressure and temperature of the air going into the compressor from ambient conditions (p_0, T_0) to engine inlet conditions (p_1, T_1).

The combined air-standard cycle comprising a dual cycle and a constant-pressure turbine cycle (Brayton cycle) is shown in Figure 4.2 plotted in $p - V$ and $T - s$ coordinate systems. The processes involved are as follows:

1) *Process $1 - c$*: Heat q'_{out} rejected from process $5 - 1$ of the dual cycle is supplied to the constant-pressure process $1 - c$ of the turbine.
2) *Process $c - b$*: Air (exhaust gases in real engines) expands isentropically in the turbine blades.

Fundamentals of Heat Engines: Reciprocating and Gas Turbine Internal Combustion Engines, First Edition. Jamil Ghojel.
© 2020 John Wiley & Sons Ltd. This Work is a co-publication between John Wiley & Sons Ltd and ASME Press.
Companion website: www.wiley.com/go/JamilGhojel_Fundamentals of Heat Engines

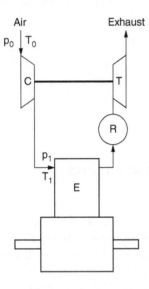

Air
$p_0 \mid T_0$

Exhaust

Figure 4.1 Schematic diagram of a turbocharged engine with constant-pressure turbine: E, engine; C, compressor; T, turbine; R, receiver.

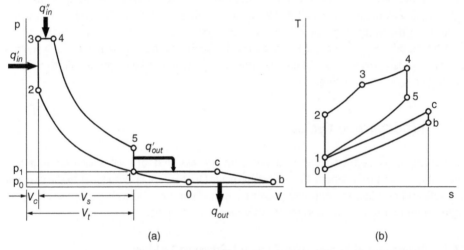

(a)   (b)

Figure 4.2 Combined dual cycle and constant-pressure turbine cycle in $p-V$ and $T-s$ coordinate systems.

3) *Process b – 0*: Heat q_{out} is rejected from the turbine cycle at constant pressure.
4) *Process 0 – 1*: Air is compressed isentropically in the compressor before it is supplied to the base cycle at pressure p_1 and temperature T_1, which are higher than the ambient pressure and temperature p_0 and T_0.
5) The dual cycle $(1-2-3-4-5)$ is executed thereafter.

4.1.1.1 Thermal Efficiency
The thermal efficiency of the dual cycle $1-2-3-4-5$ is

$$\eta_{t(dual)} = 1 - \frac{q'_{out}}{q_{in}} \tag{4.1}$$

where

$$q_{in} = q'_{in} + q''_{in}$$

Equation (3.14) written for the dual cycle is

$$\eta_{t(dual)} = 1 - \frac{1}{\varepsilon^{\gamma-1}} \frac{\alpha\beta^{\gamma} - 1}{[(\alpha - 1) + \alpha\gamma(\beta - 1)]} \tag{4.2}$$

From Eqs. (4.1) and (4.2):

$$q'_{out} = q_{in} \frac{1}{\varepsilon^{\gamma-1}} \frac{\alpha\beta^{\gamma} - 1}{[(\alpha - 1) + \alpha\gamma(\beta - 1)]}$$

From Figure 4.2a, the thermal efficiency of the Brayton (Joule) cycle $0-1-c-b$ can be written in terms of the pressure ratio $r_c = p_1/p_o$ as $\eta_{t(Brayton)} = 1 - 1/r_c^{(\gamma-1)/\gamma}$, or in terms of the compression ratio $\varepsilon_{com} = V_o/V_1$ as follows:

$$\eta_{t(Brayton)} = 1 - \frac{q_{out}}{q'_{out}} = 1 - \frac{1}{\varepsilon_{com}^{\gamma-1}}$$

Combining the last two equations yields

$$q_{out} = q'_{out} \frac{1}{\varepsilon_{com}^{\gamma-1}} = q_{in} \frac{1}{\varepsilon_{com}^{\gamma-1}} \frac{1}{\varepsilon^{\gamma-1}} \frac{\alpha\beta^{\gamma} - 1}{[(\alpha - 1) + \alpha\gamma(\beta - 1)]}$$

$$\frac{q_{out}}{q_{in}} = \frac{1}{\varepsilon_t^{\gamma-1}} \frac{\alpha\beta^{\gamma} - 1}{[(\alpha - 1) + \alpha\gamma(\beta - 1)]}$$

ε_t is the total (or overall) compression ratio for the combined cycle $\varepsilon_t = \varepsilon_{com}\varepsilon$. The thermal efficiency of the combined cycle is

$$\eta_t = 1 - \frac{q_{out}}{q_{in}}$$

or

$$\eta_t = 1 - \frac{1}{\varepsilon_t^{\gamma-1}} \frac{\alpha\beta^{\gamma} - 1}{[(\alpha - 1) + \alpha\gamma(\beta - 1)]} \tag{4.3}$$

This equation is the exact equation for the thermal efficiency of the natural-induction air-standard dual cycle with the compression ratio of the cycle replaced by the total compression ratio ε_t. The thermal efficiency characteristics of the boosted cycle with the same compression ratio as the natural-induction dual cycle will be identical to Figure (3.13).

The thermal efficiencies of the Otto and Diesel cycles boosted with constant-pressure turbine are, respectively,

$$\eta_t = 1 - \frac{1}{\varepsilon_t^{\gamma-1}} \tag{4.4}$$

$$\eta_t = 1 - \frac{1}{\varepsilon_t^{\gamma-1}} \frac{\beta^{\gamma} - 1}{[\gamma(\beta - 1)]} \tag{4.5}$$

4.1.1.2 Mean Effective Pressure

By definition, the *mep* is the work produced by the cycle per unit volume displaced by the piston in the engine cylinder:

$$mep = \frac{W_{net}}{(V_{max} - V_{min})} = \frac{W_{net}}{(V_1 - V_2)} = \frac{W_{net}}{V_1\left(1 - \dfrac{V_2}{V_1}\right)} = \frac{W_{net}}{V_1\left(1 - \dfrac{1}{\varepsilon}\right)}$$

As we have seen before in Chapter 3,

$$V_1 = \frac{c_v T_1 (\gamma - 1)}{p_1}$$

$$mep = \frac{W_{net}}{V_1\left(1 - \dfrac{1}{\varepsilon}\right)} = \frac{\varepsilon W_{net}}{V_1(\varepsilon - 1)} = \frac{\varepsilon p_1 W_{net}}{c_v T_1(\gamma - 1)(\varepsilon - 1)}$$

Since

$$\eta_t = \frac{q_{in} - q_{out}}{q_{in}} = \frac{W_{net}}{q_{in}},$$

$$W_{net} = \eta_t q_{in}$$

$$mep = \frac{\varepsilon p_1 \eta_t q_{in}}{c_v T_1(\gamma - 1)(\varepsilon - 1)}$$

The heat input into the cycle is

$$q_{in} = c_v T_1 \varepsilon^{\gamma - 1}[(\alpha - 1) + \alpha\gamma(\beta - 1)]$$

Hence,

$$mep = \frac{p_1 \varepsilon^{\gamma}}{(\gamma - 1)(\varepsilon - 1)} \eta_t[(\alpha - 1) + \alpha\gamma(\beta - 1)] \tag{4.6}$$

Substitution for η_t yields

$$mep = \frac{p_1 \varepsilon^{\gamma}}{(\gamma - 1)(\varepsilon - 1)} \left\{ [(\alpha - 1) + \alpha\gamma(\beta - 1)] - \frac{\alpha\beta^{\gamma} - 1}{\varepsilon_t^{\gamma - 1}} \right\} \tag{4.7}$$

The *mep* of the boosted Otto cycle is

$$mep = \frac{(\alpha - 1)p_1 \varepsilon^{\gamma}}{(\gamma - 1)(\varepsilon - 1)} \left(1 - \frac{1}{\varepsilon_t^{\gamma - 1}} \right) \tag{4.8}$$

and for the boosted Diesel cycle

$$mep = \frac{p_1 \varepsilon^{\gamma}}{(\gamma - 1)(\varepsilon - 1)} \left\{ [\gamma(\beta - 1)] - \frac{\beta^{\gamma} - 1}{\varepsilon_t^{\gamma - 1}} \right\} \tag{4.9}$$

The *mep* characteristics are identical in shape to that of the dual cycle but higher in magnitude, as shown in Figure 4.3 for three values of q_{in}.

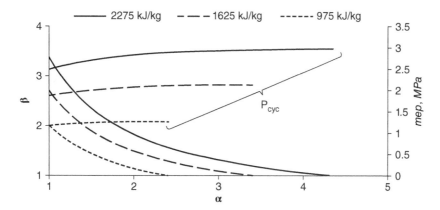

Figure 4.3 Mean effective pressure of the dual cycle with constant-pressure turbocharging ($\varepsilon_{comp} = 1.5$, $\varepsilon = 12$, $\varepsilon_{tot} = 18$, $p_0 = 0.1\ MPa$, $T_0 = 298\ K$, $R = 0.287\ kJ/kg.K$).

Figure 4.4 Schematic diagram of a turbocharged engine with variable-pressure turbine: E, engine; C, compressor; T, turbine.

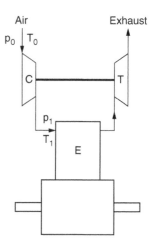

4.1.2 Turbocharged Engine with Variable-Pressure Turbine

The schematic diagram of a turbocharged engine with a variable-pressure turbine (impulse turbine) is shown in Figure 4.4. The exhaust gases enter the turbine without pressure modulation (no receiver) and expand in the turbine blades directly.

The air standard cycle, comprising a dual cycle and turbocharger with variable-pressure turbine, is shown in Figure 4.5 in $p-V$ and $T-s$ coordinate systems. The processes involved are:

1) *Process 5 – a*: Air from the dual cycle continues expanding isentropically in the turbine from point 5 to atmospheric pressure at point a, well beyond the point at which the heat is rejected by the basic dual cycle.
2) *Process a – 0*: Heat q_{out} is rejected from the turbine cycle at constant pressure.
3) *Process 0 – 1*: Air is compressed isentropically in the compressor before it is supplied to the base cycle at pressure p_1 and temperature T_1, which are higher than the ambient pressure and temperature p_0 and T_0.
4) The dual combustion cycle ($1 - 2 - 3 - 4 - 5$) is executed thereafter.

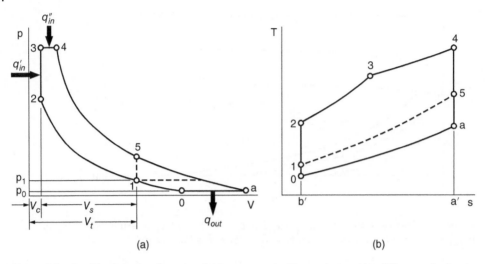

Figure 4.5 Combined dual cycle and variable-pressure turbine cycle in p-V and T-s coordinate systems.

4.1.2.1 Thermal Efficiency

As Figure 4.5a shows, this cycle is equivalent to the generalised cycle with heat rejection at constant pressure only and total compression ratio of $\varepsilon_t = V_0/V_2 = \varepsilon_{comp}\varepsilon$. Therefore, $T_5 = T_6$ in the generalised cycle (Figure 3.2), or

$$T_1\alpha\beta\left(\frac{\beta}{\rho}\right)^{\gamma-1} = \rho T_1$$

from which

$$\alpha\left(\frac{\beta}{\rho}\right)^{\gamma} = 1$$

and

$$\rho = \beta\alpha^{1/\gamma}$$

The thermal efficiency of the generalised cycle with compression ratio ε_t was deduced in Chapter 3 as

$$\eta_t = 1 - \frac{\rho\left[\alpha\left(\frac{\beta}{\rho}\right)^{\gamma} - 1\right] + \gamma(\rho - 1)}{\varepsilon_t^{\gamma-1}[(\alpha - 1) + \alpha\gamma(\beta - 1)]}$$

To determine the thermal efficiency of the turbocharged dual cycle with a variable-pressure turbine (cycle $0-1-2-3-4-a$), shown in in Figure 4.5, we substitute for $\alpha(\beta/\rho)^{\gamma}$ and ρ in the previous equation to get

$$\eta_t = 1 - \frac{\gamma[\beta\alpha^{1/\gamma} - 1]}{\varepsilon_t^{\gamma-1}[(\alpha - 1) + \alpha\gamma(\beta - 1)]} \tag{4.10}$$

The thermal efficiency of the Otto cycle boosted with variable-pressure turbine ($\beta = 1$) is

$$\eta_t = 1 - \frac{\gamma(\alpha^{1/\gamma} - 1)}{\varepsilon_t^{\gamma-1}(\alpha - 1)} \tag{4.11}$$

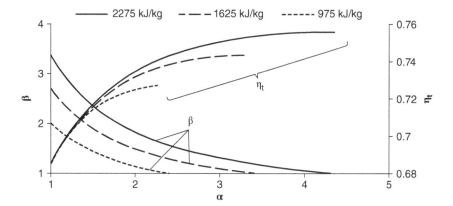

Figure 4.6 Thermal efficiency of the dual cycle with variable-pressure turbocharging ($\varepsilon_{comp} = 1.5$, $\varepsilon = 12$, $\varepsilon_{tot} = 18$, $p_0 = 0.1$ MPa, $T_0 = 298$ K, R = 0.287 kJ/kg.K).

The thermal efficiency of the Diesel cycle boosted with a variable-pressure turbine ($\alpha = 1$) is

$$\eta_t = 1 - \frac{1}{\varepsilon_t^{\gamma-1}} \tag{4.12}$$

Figure 4.6 shows the thermal efficiency of the dual cycle boosted with a variable-pressure turbine plotted versus α for three values of the heat input (q_{in}). It is apparent that the characteristics for this cycle are different from those of the natural-induction dual cycle both in shape and magnitude.

At $\rho = 1$, the cycle is reduced to a boosted Diesel cycle with the efficiency as a function of the total compression ratio only (Eq. 4.12). Hence the convergence of the thermal efficiencies for all input heat values when $\alpha = 1$. Also, for any given value of α, the thermal efficiency increases with the input heat q_{cyc}.

4.1.2.2 Mean Effective Pressure

$$mep = \frac{W_{net}}{V_1\left(1 - \frac{1}{\varepsilon}\right)} = \frac{\varepsilon W_{net}}{V_1(\varepsilon - 1)} = \frac{\varepsilon p_1 W_{net}}{c_v T_1(\gamma - 1)(\varepsilon - 1)} \text{ and } W_{net} = \eta_t q_{in}$$

Hence,

$$mep = \frac{\varepsilon p_1 \eta_t q_{in}}{c_v T_1(\gamma - 1)(\varepsilon - 1)}$$

Substituting for q_{in} from Eq. (3.3), we obtain

$$mep = \frac{p_1 \varepsilon^\gamma}{(\gamma - 1)(\varepsilon - 1)} \eta_t[(\alpha - 1) + \alpha\gamma(\beta - 1)] \tag{4.13}$$

Substituting for η_t from Eq. (4.10), we finally obtain

$$mep = \frac{p_1 \varepsilon^\gamma}{(\gamma - 1)(\varepsilon - 1)} \left\{ [(\alpha - 1) + \alpha\gamma(\beta - 1)] - \frac{\gamma[\beta\alpha^{1/\gamma} - 1]}{\varepsilon_t^{\gamma-1}} \right\} \tag{4.14}$$

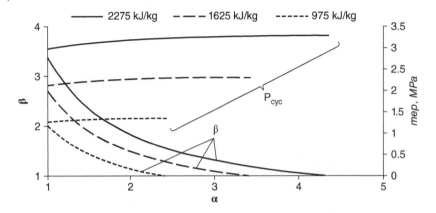

Figure 4.7 Mean effective pressure of the dual cycle with variable-pressure turbocharging ($\varepsilon_{comp} = 1.5$, $\varepsilon = 12$, $\varepsilon_t = 18$, $p_0 = 0.1\ MPa$, $T_0 = 298\ K$, $R = 0.287\ kJ/kg.K$).

Figure 4.7 shows the *mep* and β plotted versus α for three heat input values. These characteristics are similar in shape to the ones for the dual cycle but different in magnitude.

The *mep*s of the Otto and Diesel cycles boosted with a variable-pressure turbine are, respectively,

$$mep = \frac{p_1\varepsilon^\gamma}{(\gamma-1)(\varepsilon-1)}\left[(\alpha-1) - \frac{\gamma(\alpha^{1/\gamma}-1)}{\varepsilon_t^{\gamma-1}}\right] \tag{4.15}$$

$$mep = \frac{\gamma p_1\varepsilon^\gamma(\beta-1)}{(\gamma-1)(\varepsilon-1)}\left(1 - \frac{1}{\varepsilon_t^{\gamma-1}}\right) \tag{4.16}$$

4.2 Supercharged Cycles

The boosting system in this case is shown schematically in Figure 4.8. Compressor C is driven by shaft power directly taken through gearing from the engine crank shaft. The resultant increase in cycle work due to supercharging will not be as high as in the case of turbocharging because of the work required to run the compressor.

Figure 4.9 shows the supercharged dual cycle in $p - V$ and $T - s$ coordinate systems. Heat is rejected from the cycle at constant volume and lost to the surroundings (process $5 - 1$).

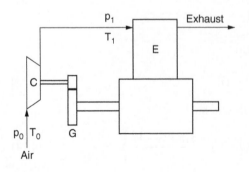

Figure 4.8 Schematic diagram of a supercharged engine: E, engine; G, gearing; C, compressor.

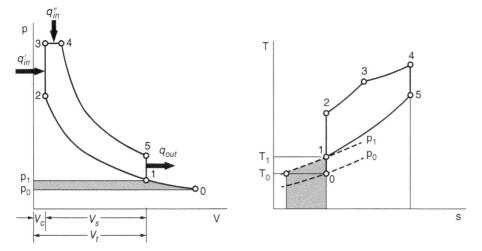

Figure 4.9 Supercharged dual combustion cycle in $p - V$ and $T - s$ coordinate systems.

The compressor shaft work W_{comp} required to raise the pressure from p_0 to p_1 and temperature from T_0 to T_1 (process $0-1$) is represented by the shaded areas.

4.2.1 Thermal Efficiency

The heat equivalent of the compressor shaft work per kg of air is

$$W_{com} = \frac{\gamma R}{\gamma - 1}(T_1 - T_0) \tag{4.17}$$

This equation shows that the heat equivalent of the work done in the reversible adiabatic compressor has the same value as the heat required to increase the temperature of the compressor at constant pressure from T_0 to T_1 (shaded area in the $T - s$ diagram).

Equation (4.17) can be rewritten in terms of the volume and pressure compression ratios as follows:

$$W_{com} = \frac{\gamma R T_0}{\gamma - 1}[(V_o/V_1)^{\gamma-1} - 1] = \frac{\gamma R T_0}{\gamma - 1}(\varepsilon_{com}^{\gamma-1} - 1) \tag{4.18}$$

or as

$$W_{com} = \frac{\gamma R T_0}{\gamma - 1}[(p_1/p_0)^{(\gamma-1)/\gamma} - 1] = \frac{\gamma R T_0}{\gamma - 1}[r_c^{(\gamma-1)/\gamma} - 1] \tag{4.19}$$

The net work W_{net} output from the boosted cycle is the work produced by the dual cycle W_{dual} (Eq. 3.3) minus the work required by the compressor to raise the pressure of air from p_a to p_1. The thermal efficiency of the boosted cycle is then

$$\eta_t = \frac{q_{in} - q_{out}}{q_{in}} = \frac{W_{net}}{q_{in}} = \frac{W_{dual} - W_{com}}{q_{in}}$$

$$\eta_t = \frac{W_{dual}}{q_{in}} - \frac{W_{com}}{q_{in}} = \eta_{t(dual)} - \frac{\frac{\gamma R T_0}{\gamma - 1}(\varepsilon_{com}^{\gamma-1} - 1)}{c_v T_1 \varepsilon^{\gamma-1}[(\alpha - 1) + \alpha\gamma(\beta - 1)]} \tag{4.20}$$

$\eta_{t(dual)}$ is the thermal efficiency of the dual cycle with compression ratio ε.

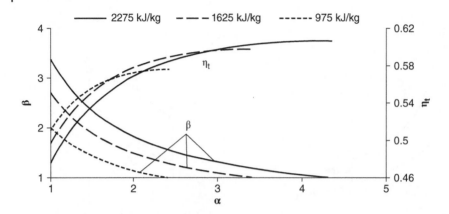

Figure 4.10 Thermal efficiency of the dual cycle with supercharging (ε_{comp} = 1.5, ε = 12, ε_{tot} = 18, p_0 = 0.1 MPa, T_0 = 298 K, R = 0.287 kJ/kg.K).

This equation shows that the thermal efficiency of the supercharged cycle for a given heat input is always less than the efficiency of the natural-induction dual cycle, and it decreases with an increasing degree of supercharging (compression ratio of the compressor).

Knowing $T_1 = T_0 \varepsilon_{comp}^{\gamma-1}$, $c_v = R/(\gamma - 1)$, and $\eta_{t(dual)}$ from Eq. (3.14), Eq. (4.20) can be rewritten as

$$\eta_t = 1 - \frac{1}{\varepsilon^{\gamma-1}} \frac{\alpha\beta^\gamma - 1}{[(\alpha - 1) + \alpha\gamma(\beta - 1)]} - \frac{\gamma(\varepsilon_{com}^{\gamma-1} - 1)}{\varepsilon_{com}^{\gamma-1}\varepsilon^{\gamma-1}[(\alpha - 1) + \alpha\gamma(\beta - 1)]}$$

$$\eta_t = 1 - \frac{1}{\varepsilon^{\gamma-1}} \frac{\alpha\beta^\gamma - 1}{[(\alpha - 1) + \alpha\gamma(\beta - 1)]} - \frac{\gamma(1 - 1/\varepsilon_{com}^{\gamma-1})}{\varepsilon^{\gamma-1}[(\alpha - 1) + \alpha\gamma(\beta - 1)]}$$

$$\eta_t = 1 - \frac{(\alpha\beta^\gamma - 1) + \gamma(1 - 1/\varepsilon_{com}^{\gamma-1})}{\varepsilon^{\gamma-1}[(\alpha - 1) + \alpha\gamma(\beta - 1)]} \tag{4.21}$$

Since $\varepsilon_{comp}^{\gamma-1} = (p_1/p_0)^{(\gamma-1)/\gamma}$, the thermal efficiency can also be written as

$$\eta_t = 1 - \frac{(\alpha\beta^\gamma - 1) + \gamma[1 - (p_0/p_1)^{(\gamma-1)/\gamma}]}{\varepsilon^{\gamma-1}[(\alpha - 1) + \alpha\gamma(\beta - 1)]} \tag{4.22}$$

The plots for the thermal efficiency of the supercharged dual cycle are shown in Figure 4.10. The graphs are quite different from those of the basic and turbocharged dual cycles discussed earlier. The maximum value at each input heat is reached when α is at maximum and β = 1, i.e. when the cycle approaches the supercharged Otto cycle. The thermal efficiency decreases with the heat input at the lower values of α and increases with the higher values of α.

The thermal efficiencies of the supercharged Otto and Diesel cycles are, respectively, ·

$$\eta_t = 1 - \frac{(\alpha - 1) + \gamma(1 - 1/\varepsilon_{com}^{\gamma-1})}{\varepsilon^{\gamma-1}[(\alpha - 1)]} \tag{4.23}$$

$$\eta_t = 1 - \frac{(\beta^\gamma - 1) + \gamma(1 - 1/\varepsilon_{com}^{\gamma-1})}{\gamma\varepsilon^{\gamma-1}(\beta - 1)} \tag{4.24}$$

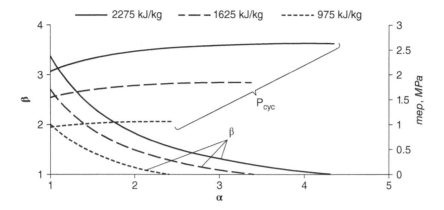

Figure 4.11 Mean effective pressure of the dual cycle with supercharging ($\varepsilon_{comp} = 1.5$, $\varepsilon = 12$, $\varepsilon_{tot} = 18$, $p_0 = 0.1$ MPa, $T_0 = 298$ K, $R = 0.287$ kJ/kg.K).

4.2.2 Mean Effective Pressure

$$mep = \frac{W_{net}}{V_1 - V_2} = \frac{W_{net}}{V_1\left(1 - \dfrac{1}{\varepsilon}\right)} = \frac{\varepsilon W_{net}}{V_1(\varepsilon - 1)}$$

Since $W_{net} = \eta_t q_{in}$ and $V_1 = c_v T_1(\gamma - 1)/p_1$,

$$mep = \frac{\varepsilon \eta_t q_{in} p_1}{(\varepsilon - 1)c_v T_1(\gamma - 1)}$$

Finally,

$$mep = \frac{p_1 \varepsilon^\gamma}{(\varepsilon - 1)(\gamma - 1)}\eta_t[(\alpha - 1) + \alpha\gamma(\beta - 1)] \tag{4.25}$$

The characteristics of the *mep* of the dual cycle with supercharging are shown in Figure 4.11. These are similar to the characteristics of the previous turbocharged cycles but lower in magnitude due to the fact that part of the cycle work is expended to run the compressor.

The *meps* of the supercharged Otto and Diesel cycles are, respectively,

$$mep = \frac{p_1 \varepsilon^\gamma}{(\varepsilon - 1)(\gamma - 1)}\eta_t[(\alpha - 1)] \tag{4.26}$$

$$mep = \frac{p_1 \varepsilon^\gamma}{(\varepsilon - 1)(\gamma - 1)}\eta_t[\gamma(\beta - 1)] \tag{4.27}$$

4.3 Forced Induction Cycles with Intercooling

The performance of boosted engines can potentially be further improved if the temperature of the compressed air at the inlet into the engine is cooled. Cooling the compressed air increases the density, thereby allowing the delivery of a greater air mass into the engine. Considered here are the turbocharged and supercharged cycles discussed earlier with added intercooling processes.

4.3.1 Cycle with Constant-Pressure Turbine and Intercooling

Figure 4.12 shows a schematic diagram of the engine with constant-pressure turbocharging, with the air at the end of the compression process in compressor C being cooled at constant pressure in a heat exchanger (HE) from T_1 to $T_{1'}$ before entering engine E.

Figure 4.13 shows the intercooled (shaded areas) and non-intercooled thermodynamic cycles plotted in $p - V$ and $T - s$ coordinate systems. The input heat and maximum cycle pressure are kept constant for both cycles. The intercooled cycle comprises the dual cycle $(1' - 2' - 3' - 4' - 5')$, constant pressure cooling process $(1 - 1')$ in a HE, and Brayton cycle $(0 - 1 - c' - b')$.

4.3.1.1 Cooling Process

The degree of cooling is defined as $\omega = V_1/V_{1'} = T_1/T_{1'}$ and the HE effectiveness as $\epsilon = (T_1 - T_{1'})/(T_1 - T_0)$. For the compression ratio in terms of volume ratios ε_{comp}, the isentropic compression process $(0 - 1)$ in the compressor yields

$$\varepsilon_{com}^{\gamma-1} = T_1/T_0$$

From the definition of the HE effectiveness,

$$T_1 - T_{1'} = \epsilon(T_1 - T_0)$$

$$T_{1'}\left(\frac{T_1}{T_{1'}} - 1\right) = \epsilon(T_0 \varepsilon_{com}^{\gamma-1} - T_0),$$

or

$$(\omega - 1) = \frac{T_0}{T_{1'}} \epsilon(\varepsilon_{com}^{\gamma-1} - 1)$$

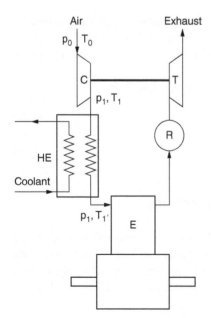

Air   Exhaust

p_0 | T_0

Figure 4.12 Schematic diagram of a turbocharged and intercooled engine: E, engine; C, compressor; T, turbine; HE, heat exchanger/intercooler.

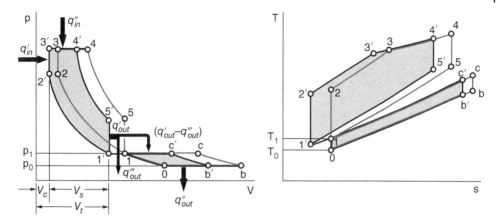

Figure 4.13 $p - V$ and $T - s$ diagrams for dual cycles with constant pressure turbocharging without intercooling $(0 - 1 - 2 - 3 - 4 - 5 - 1 - c - b)$ and with intercooling $(0 - 1 - 1' - 2' - 3' - 4' - 5' - 1' - 1 - c' - b')$.

From the definition of ω and the isentropic compression process,

$$\frac{T_0}{T_{1'}} = \frac{\omega}{\varepsilon_{com}^{\gamma-1}}$$

which finally leads to

$$\omega = \frac{1}{1 - \epsilon(1 - 1/\varepsilon_{com}^{\gamma-1})} \tag{4.28}$$

When ϵ tends to unity (perfect HE and $T_{1'} = T_0$), ω tends to $\varepsilon_{com}^{\gamma-1}$.

4.3.1.2 Thermal Efficiency

The thermal efficiency of the combined cycle is

$$\eta_t = 1 - \frac{q_{out}}{q_{in}}$$

where q_{in} and q_{out} are the heat input into and heat rejected from the turbocharged and intercooled cycle, respectively:

$$q_{in} = q'_{in} + q''_{in} = c_v T_{1'} \varepsilon^{\gamma-1}[(\alpha - 1) + \alpha\gamma(\beta - 1)]$$

$$q_{out} = q''_{out} + q'''_{out}$$

The output heat from the dual cycle q'_{out} at constant volume (process $5' - 1'$) is rejected at constant pressure during process $1' - 1 - c'$. Part of this heat q'''_{out} is dissipated in the HE (process $1' - 1$), and the remainder $(q'_{out} - q'''_{out})$ is supplied to the Brayton cycle as heat input (process $1 - c'$).

The heat rejected to the HE is

$$q'''_{out} = c_p(T_1 - T_{1'}) = c_p T_{1'}(T_1/T_{1'} - 1) = c_p T_{1'}(V_1/V_{1'} - 1) = c_p T_{1'}(\omega - 1)$$

The thermal efficiency of the Brayton cycle in the intercooled case $(0 - 1 - c' - b')$ is

$$\eta_{t(Brayton)} = 1 - \frac{q''_{out}}{q'_{out} - q'''_{out}} = 1 - \frac{1}{\varepsilon_{com}^{\gamma-1}}$$

from which

$$q''_{out} = \frac{q'_{out} - q'''_{out}}{\varepsilon_{com}^{\gamma-1}},$$

or

$$q''_{out} = \frac{c_v(T_{5'} - T_{1'}) - c_p T_{1'}(\omega - 1)}{\varepsilon_{com}^{\gamma-1}}$$

From the analysis of the basic dual cycle $1' - 2' - 3' - 4' - 5'$:

$$T_{5'} = T_{1'} \alpha\beta \left(\frac{\varepsilon}{\delta}\right)^{\gamma-1} \text{ and } \delta = \frac{\varepsilon}{\beta}$$

$$T_{5'} = T_{1'} \alpha\beta^\gamma$$

$$q''_{out} = \frac{c_v T_{1'}(\alpha\beta^\gamma - 1) - \gamma c_v T_{1'}(\omega - 1)}{\varepsilon_{com}^{\gamma-1}}$$

The total heat q_{out} rejected from the turbocharged and intercooled cycle is

$$q_{out} = \frac{c_v T_{1'}(\alpha\beta^\gamma - 1) - \gamma\, c_v T_{1'}(\omega - 1)}{\varepsilon_{com}^{\gamma-1}} + \gamma\, c_v T_{1'}(\omega - 1)$$

$$q_{out} = \frac{c_v T_{1'}}{\varepsilon_{com}^{\gamma-1}}\{(\alpha\beta^\gamma - 1) + \gamma[\varepsilon_{comp}^{\gamma-1}(\omega - 1) - (\omega - 1)]\}$$

The thermal efficiency of the combined cycle is

$$\eta_t = 1 - \frac{q_{out}}{q_{in}} = \frac{\dfrac{c_v T_{1'}}{\varepsilon_{com}^{\gamma-1}}[\alpha\beta^\gamma - 1 + \gamma(\omega - 1)(\varepsilon_{comp}^{\gamma-1} - 1)]}{c_v T_{1'} \varepsilon^{\gamma-1}[(\alpha - 1) + \alpha\gamma(\beta - 1)]}$$

$$\eta_t = 1 - \frac{1}{\varepsilon^{\gamma-1}\varepsilon_{com}^{\gamma-1}} \frac{[\alpha\beta^\gamma - 1 + \gamma(\omega - 1)(\varepsilon_{comp}^{\gamma-1} - 1)]}{[(\alpha - 1) + \alpha\gamma(\beta - 1)]}$$

$$\eta_t = 1 - \frac{1}{\varepsilon_t^{\gamma-1}} \frac{[\alpha\beta^\gamma - 1 + \gamma(\omega - 1)(\varepsilon_{comp}^{\gamma-1} - 1)]}{[(\alpha - 1) + \alpha\gamma(\beta - 1)]} \tag{4.29}$$

The total compression ratio for the combined cycle is $\varepsilon_t = \varepsilon_{com}\varepsilon$. It should be noted that ε_t is different from the overall volume ratio $\varepsilon_o = \varepsilon_{com}\omega\varepsilon$.

The thermal efficiencies of the constant-pressure turbocharged and intercooled Otto and Diesel cycles are, respectively,

$$\eta_t = 1 - \frac{1}{\varepsilon_t^{\gamma-1}} \frac{[\alpha - 1 + \gamma(\omega - 1)(\varepsilon_{comp}^{\gamma-1} - 1)]}{(\alpha - 1)} \tag{4.30}$$

$$\eta_t = 1 - \frac{1}{\varepsilon_t^{\gamma-1}} \frac{[\beta^\gamma - 1 + \gamma(\omega - 1)(\varepsilon_{comp}^{\gamma-1} - 1)]}{\gamma(\beta - 1)} \tag{4.31}$$

4.3.1.3 Mean Effective Pressure

$$mep = \frac{W_{net}}{V_{1'} - V_{2'}} = \frac{W_{net}}{V_{1'}\left(1 - \frac{1}{\varepsilon}\right)} = \frac{\varepsilon W_{net}}{V_{1'}(\varepsilon - 1)}$$

Since $W_{net} = \eta_t q_{in}$ and the volume of air at point $1'$ per unit mass of air,

$$V_{1'} = c_v T_{1'}(\gamma - 1)/p_1$$

$$mep = \frac{\varepsilon q_{in} p_1}{c_v T_{1'}(\varepsilon - 1)(\gamma - 1)} \eta_t$$

$$q_{in} = c_v T_{1'} \varepsilon^{\gamma-1}[(\alpha - 1) + \alpha\gamma(\beta - 1)]$$

$$mep = \frac{\varepsilon p_1}{(\varepsilon - 1)c_v T_{1'}(\gamma - 1)} c_v T_{1'} \varepsilon^{\gamma-1} \eta_t[(\alpha - 1) + \alpha\gamma(\beta - 1)]$$

$$mep = \frac{\varepsilon^\gamma p_1}{(\varepsilon - 1)(\gamma - 1)} \eta_t[(\alpha - 1) + \alpha\gamma(\beta - 1)] \tag{4.32}$$

The *mep*s of the turbocharged and intercooled Otto and Diesel cycles are, respectively,

$$mep = \frac{\varepsilon^\gamma p_1}{(\varepsilon - 1)(\gamma - 1)} \eta_t(\alpha - 1) \tag{4.33}$$

$$mep = \frac{\gamma\varepsilon^\gamma p_1}{(\varepsilon - 1)(\gamma - 1)} \eta_t(\beta - 1) \tag{4.34}$$

4.3.2 Cycle with Variable-Pressure Turbocharging and Intercooling

The schematic diagram of an engine with variable-pressure turbocharging and inter-cooling is the same as in Figure 4.12 without the receiver. The $p - V$ and $T - s$ diagrams of the intercooled and non-intercooled cycles are shown in Figure 4.14. The shaded area $0 - 1 - 1' - 2' - 3' - 4 - 5' - 6'$ represents the intercooled cycle. Process $(0 - 1)$ is the compression process in the compressor, and process $(1 - 1')$ is the cooling process in the HE.

4.3.2.1 Cooling Process
The degree of cooling in the intercooler (HE) is, as before, given by

$$\omega = \frac{1}{1 - \epsilon(1 - 1/\varepsilon_{com}^{\gamma-1/\gamma})}$$

4.3.2.2 Thermal Efficiency

$$\eta_t = 1 - \frac{q_{out}}{q_{in}}$$

$$q_{in} = c_v T_{1'} \varepsilon^{\gamma-1}[(\alpha - 1) + \alpha\gamma(\beta - 1)]$$

$$q_{out} = q'_{out} + q''_{out}$$

$$q_{out} = c_p(T_{6'} - T_0) + c_p(T_1 - T_{1'})$$

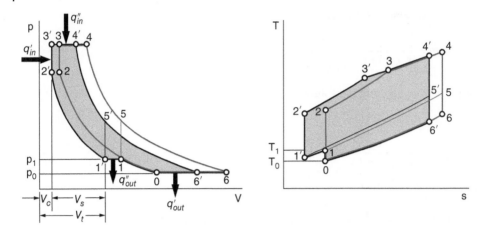

Figure 4.14 $p-V$ and $T-s$ diagrams for dual cycles with variable-pressure turbocharging without intercooling $(0-1-2-3-4-5-6)$ and with intercooling $(0-1-1'-2'-3'-4'-5'-6')$.

$$q_{out} = c_p T_0 \left(\frac{T_{6'}}{T_0} - 1 \right) + c_p T_1 \left(1 - \frac{T_{1'}}{T_1} \right)$$

$$q_{out} = c_p T_0 \left(\frac{V_{6'}}{V_0} - 1 \right) + c_p T_1 \left(1 - \frac{V_{1'}}{V_1} \right)$$

We previously defined $\rho = V_{6'}/V_0$; hence

$$q_{out} = c_p T_0 (\rho - 1) + c_p T_1 \left(1 - \frac{1}{\omega} \right)$$

$$q_{out} = c_p T_0 (\rho - 1) + c_p T_0 \varepsilon_{com}^{\gamma-1} \left(1 - \frac{1}{\omega} \right)$$

$$q_{out} = c_p T_0 \left[(\rho - 1) + \varepsilon_{com}^{\gamma-1} \left(1 - \frac{1}{\omega} \right) \right]$$

Using volume ratios, we can write the identity

$$V_{6'}/V_{2'} = V_{6'}/V_{3'}, \text{ or } (V_{6'}/V_0)(V_0/V_1)(V_1/V_{1'})(V_{1'}/V_{2'}) = (V_{6'}/V_{4'})(V_{4'}/V_{3'})$$

From the earlier definitions of the various volume ratios,

$$\rho \varepsilon_{com} \omega \varepsilon = \delta \beta \tag{4.35}$$

$$p_{6'} = p_{1'} \alpha \varepsilon^\gamma \left(\frac{1}{\delta} \right)^\gamma \text{ and } p_0 = \frac{p_1}{\varepsilon_{com}^\gamma} = \frac{p_{1'}}{\varepsilon_{com}^\gamma}, \text{ from which, since } p_0 = p_{6'}$$

$$\alpha \varepsilon^\gamma \varepsilon_{com}^\gamma = \delta^\gamma, \text{ or } \delta = \alpha^{1/\gamma} \varepsilon \varepsilon_{com}$$

Substituting for δ in Eq. (4.35) yields

$$\rho = \frac{\beta \alpha^{1/\gamma}}{\omega} \tag{4.36}$$

The heat rejected now becomes

$$q_{out} = c_p T_0 \left[\left(\frac{\beta \alpha^{1/\gamma}}{\omega} - 1 \right) + \varepsilon_{com}^{\gamma-1} \left(1 - \frac{1}{\omega} \right) \right]$$

$$q_{out} = \frac{c_p T_0}{\omega}[\beta\alpha^{1/\gamma} - \omega + \varepsilon_{com}^{\gamma-1}(\omega - 1)]$$

and the thermal efficiency is

$$\eta_t = 1 - \frac{\frac{c_p T_0}{\omega}[\beta\alpha^{1/\gamma} - \omega + \varepsilon_{com}^{\gamma-1}(\omega - 1)]}{c_v T_{1'}\varepsilon^{\gamma-1}[(\alpha - 1) + \alpha\gamma(\beta - 1)]}$$

Since $\omega = \dfrac{V_1}{V_{1'}} = \dfrac{T_1}{T_{1'}} = \dfrac{T_0 \varepsilon_{com}^{\gamma-1}}{T_{1'}}$ and $c_p/c_c = \gamma$,

$$\eta_t = 1 - \frac{\gamma}{\varepsilon_{com}^{\gamma-1}\varepsilon^{\gamma-1}}\frac{[\beta\alpha^{1/\gamma} - \omega + \varepsilon_{com}^{\gamma-1}(\omega - 1)]}{[(\alpha - 1) + \alpha\gamma(\beta - 1)]}$$

The final form for the thermal efficiency of the cycle with variable-pressure turbocharging and intercooling is

$$\eta_t = 1 - \frac{\gamma}{\varepsilon_t^{\gamma-1}}\frac{[\beta\alpha^{1/\gamma} - \omega + \varepsilon_{com}^{\gamma-1}(\omega - 1)]}{[(\alpha - 1) + \alpha\gamma(\beta - 1)]} \tag{4.37}$$

The thermal efficiencies of the Otto and Diesel cycles with variable-pressure turbocharging and intercooling are, respectively,

$$\eta_t = 1 - \frac{\gamma}{\varepsilon_t^{\gamma-1}}\frac{[\alpha^{1/\gamma} - \omega + \varepsilon_{com}^{\gamma-1}(\omega - 1)]}{(\alpha - 1)} \tag{4.38}$$

$$\eta_t = 1 - \frac{1}{\varepsilon_t^{\gamma-1}}\frac{[\beta - \omega + \varepsilon_{com}^{\gamma-1}(\omega - 1)]}{(\beta - 1)} \tag{4.39}$$

4.3.2.3 Mean Effective Pressure

$$mep = \frac{W_{net}}{V_{1'} - V_{2'}} = \frac{W_{net}}{V_{1'}\left(1 - \dfrac{1}{\varepsilon}\right)} = \frac{\varepsilon W_{net}}{V_{1'}(\varepsilon - 1)}$$

Since $W_{net} = \eta_t q_{in}$ and $V_{1'} = c_v T_{1'}(\gamma - 1)/p_1$

$$mep = \frac{\varepsilon q_{in} p_1}{(\varepsilon - 1)c_v T_{1'}(\gamma - 1)}\eta_t$$

$$q_{in} = c_v T_{1'}\varepsilon^{\gamma-1}[(\alpha - 1) + \alpha\gamma(\beta - 1)]$$

The *mep* of the cycle with variable-pressure turbocharging and intercooling is

$$mep = \frac{\varepsilon^\gamma p_1}{(\varepsilon - 1)(\gamma - 1)}\eta_t[(\alpha - 1) + \alpha\gamma(\beta - 1)] \tag{4.40}$$

The *meps* of the Otto and Diesel cycles with variable-pressure turbocharging and intercooling are, respectively,

$$mep = \frac{\varepsilon^\gamma p_1}{(\varepsilon - 1)(\gamma - 1)}\eta_t(\alpha - 1) \tag{4.41}$$

$$mep = \frac{\gamma\varepsilon^\gamma p_1}{(\varepsilon - 1)(\gamma - 1)}\eta_t(\beta - 1) \tag{4.42}$$

4.3.3 Cycle with Supercharging and Intercooling

A schematic diagram of a supercharged and intercooled engine is shown in Figure 4.15, featuring engine E, engine-driven compressor C, and heat exchanger HE.

A supercharged and intercooled dual cycle plotted in $p - V$ and $T - s$ coordinate systems is shown in Figure 4.16. The intercooled cycle comprises the dual cycle $(1 - 2 - 3 - 4 - 5)$, isentropic compression process $(0 - 1')$ in the compressor, and the constant-pressure cooling process $(1' - 1)$ in the HE.

4.3.3.1 Cooling Process

The HE effectiveness and degree of cooling in the intercooler are defined as before but with different numbering of the characteristic points of the cycle:

$$\epsilon = (T_{1'} - T_1)/(T_{1'} - T_0), \omega = V_{1'}/V_1 = T_{1'}/T_1.$$

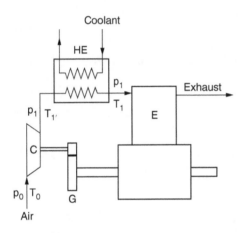

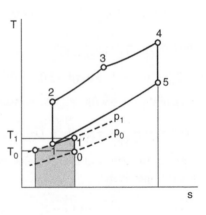

Figure 4.15 Schematic diagram of a supercharged engine with intercooling: E, engine; G, gearing; C, compressor; HE, heat exchanger/intercooler.

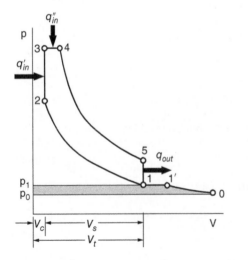

Figure 4.16 $p - V$ and $T - s$ diagrams of the supercharged and intercooled dual cycle; the shaded areas represents the work done by the compressor.

Knowing that $\varepsilon_{com}^{\gamma-1} = T_{1'}/T_0$, the same relationship for the degree of cooling (ω) is obtained:

$$\omega = 1/[1 - \epsilon(1 - 1/\varepsilon_{com}^{\gamma-1})]$$

4.3.3.2 Thermal Efficiency

The net work output W_{net} from the boosted cycle is equal to the work W_{dual} produced by the dual cycle $(1 - 2 - 3 - 4 - 5 -)$ minus the work W_{com} required by the compressor to raise the pressure of air from p_o to p_1(process $0 - 1'$).

The thermal efficiency of the boosted cycle is

$$\eta_t = \frac{q_{in} - q_{out}}{q_{in}} = \frac{W_{net}}{q_{in}} = \frac{W_{dual} - W_{com}}{q_{in}} = \frac{W_{dual}}{q_{in}} - \frac{W_{com}}{q_{in}}$$

$$\eta_t = \eta_{t(dual)} - \frac{\dfrac{\gamma R}{\gamma - 1}(T_{1'} - T_0)}{q_{in}},$$

or

$$\eta_t = \eta_{t(dual)} - \frac{\dfrac{\gamma R T_0}{\gamma - 1}(\varepsilon_{com}^{\gamma-1} - 1)}{c_v T_1 \varepsilon^{\gamma-1}[(\alpha - 1) + \alpha\gamma(\beta - 1)]}$$

The term $\eta_{t(dual)}$ is the thermal efficiency of the dual cycle with compression ratio ε.

Knowing $c_v = R/\gamma - 1$, $T_0 = T_1/\varepsilon_{com}^{\gamma}$, and accounting for the definitions of the degree of cooling $\omega = T_{1'}/T_1$ and the HE effectiveness $\epsilon = (T_{1'} - T_1)/(T_{1'} - T_o)$, the previous equation can be written as

$$\eta_t = \eta_{t(dual)} - \frac{\gamma\omega(\varepsilon_{com}^{\gamma-1} - 1)}{\varepsilon_{com}^{\gamma-1}\varepsilon^{\gamma-1}[(\alpha - 1) + \alpha\gamma(\beta - 1)]} \tag{4.43}$$

This equation shows that the thermal efficiency of the supercharged and intercooled cycle for a given heat input is always less than the efficiency of the natural-induction dual cycle with the same compression ratio. Using the equation for the thermal efficiency of the dual cycle and the equation for the degree of cooling, the cycle thermal efficiency can be written as

$$\eta_t = 1 - \frac{1}{\varepsilon^{\gamma-1}[(\alpha - 1) + \alpha\gamma(\beta - 1)]}\left\{(\alpha\beta^\gamma - 1) + \frac{\gamma\omega(\varepsilon_{com}^{\gamma-1} - 1)}{\varepsilon_{com}^{\gamma-1}}\right\},$$

or

$$\eta_t = 1 - \frac{1}{\varepsilon^{\gamma-1}[(\alpha - 1) + \alpha\gamma(\beta - 1)]}\left[(\alpha\beta^\gamma - 1) + \frac{\gamma(\varepsilon_{com}^{\gamma-1} - 1)}{\epsilon + (1 - \epsilon)\varepsilon_{com}^{\gamma-1}}\right] \tag{4.44}$$

Since $\varepsilon_{comp}^{\gamma-1} = (p_1/p_0)^{(\gamma-1)/\gamma}$, the thermal efficiency can also be written as

$$\eta_t = 1 - \frac{1}{\varepsilon^{\gamma-1}[(\alpha - 1) + \alpha\gamma(\beta - 1)]}\left\{(\alpha\beta^\gamma - 1) + \frac{\gamma[(p_1/p_o)^{(\gamma-1)/\gamma} - 1]}{\epsilon + (1 - \epsilon)(p_1/p_o)^{(\gamma-1)/\gamma}}\right\} \tag{4.45}$$

The thermal efficiencies of the supercharged and intercooled Otto and Diesel cycles are, respectively,

$$\eta_t = 1 - \frac{1}{(\alpha - 1)\varepsilon^{\gamma - 1}} \left[(\alpha - 1) + \frac{\gamma(\varepsilon_{com}^{\gamma - 1} - 1)}{\varepsilon + \varepsilon_{com}^{\gamma - 1}(1 - \epsilon)} \right] \tag{4.46}$$

$$\eta_t = 1 - \frac{1}{\gamma(\beta - 1)\varepsilon^{\gamma - 1}} \left[(\beta^\gamma - 1) + \frac{\gamma(\varepsilon_{com}^{\gamma - 1} - 1)}{\varepsilon + \varepsilon_{com}^{\gamma - 1}(1 - \epsilon)} \right] \tag{4.47}$$

4.3.3.3 Mean Effective Pressure

$$mep = \frac{W_{net}}{V_1 - V_2} = \frac{W_{net}}{V_1 \left(1 - \dfrac{1}{\varepsilon}\right)} = \frac{\varepsilon W_{net}}{V_1(\varepsilon - 1)}$$

Since $W_{net} = \eta_t q_{in} = \eta_t c_v T_1 \varepsilon^{\gamma - 1}[(\alpha - 1) + \alpha\gamma(\beta - 1)]$ and $V_1 = c_v T_1(\gamma - 1)/p_1$

$$mep = \frac{p_1 \varepsilon^\gamma}{(\varepsilon - 1)(\gamma - 1)} \eta_t[(\alpha - 1) + \alpha\gamma(\beta - 1)],$$

or

$$mep = \frac{p_0 \varepsilon_t^\gamma}{(\varepsilon - 1)(\gamma - 1)} \eta_t[(\alpha - 1) + \alpha\gamma(\beta - 1)] \tag{4.48}$$

4.4 Comparison of Boosted Cycles

The effect of cycle boosting with and without intercooling on the efficiency and *mep* of the dual cycle is shown in Figure 4.17. The compression ratio is kept constant for both the natural-induction and boosted dual cycle as well as the input heat and compressor pressure ratio. The intercooler effectiveness is kept constant at 0.5 in the intercooled cycles. As stated

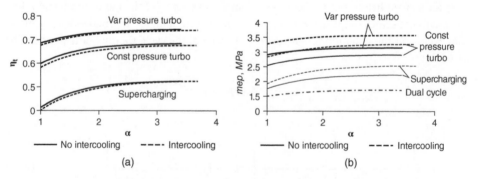

Figure 4.17 Comparison of the thermal efficiency and mean effective pressure of boosted dual cycles with and without intercooling ($q_{in} = 1625$ kJ/kg, $\varepsilon_{comp} = 2$, $\varepsilon = 9$, $\varepsilon_{tot} = 18$, $\epsilon = 0.5$, $p_0 = 0.1$ MPa, $T_0 = 298$ K).

earlier, the thermal efficiency for the dual cycle does not change when boosted using constant pressure turbocharging if the compression ratio is the same for both cases. Therefore, the reference line for comparison is the solid curve denoted by 'Const pressure turbo' in Figure 4.17a. Boosting improves the thermal efficiency only in the case of variable-pressure turbocharging (8–14% for the case shown). Supercharging causes the thermal efficiency to decrease significantly (23–31%). Intercooling leads to a slight decrease in thermal efficiency in all boosting schemes compared with straightforward boosting.

Cycle boosting in all its configurations has a significant effect on the *mep*, as can be seen in Figure 4.17b. Variable-pressure turbocharging gives the best results compared with

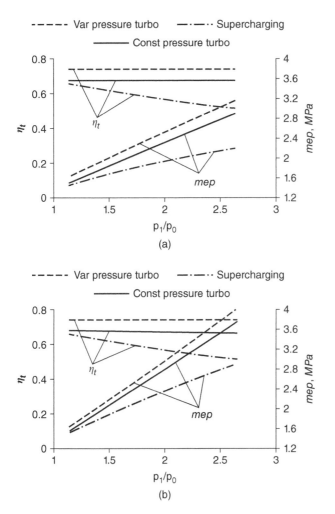

Figure 4.18 Effect of the pressure ratio in the compressor on the performance of boosted cycles without (a) and with (b) intercooling ($q_{in} = 1625$ kJ/kg, $\alpha = 2.5$, $\varepsilon_{comp} = var$, $\varepsilon = var$, $\varepsilon_{tot} = 18$, $\epsilon = 0.5$, $p_0 = 0.1$ MPa, $T_0 = 298$ K).

the natural-induction dual cycle (83–92%), followed by constant-pressure turbocharging (~69%), and then supercharging (16–29%). Unlike the case with thermal efficiency, inter-cooling significantly improves the *mep* in all cases. With intercooling, the *mep* increases further, with variable-pressure turbocharging by about 12–13%, with constant-pressure tur-bocharging by 10–12%, and with supercharging by 10–13%.

It should be noted that the values of the thermal efficiency and *mep* at $\alpha = 1$ are those of the Diesel cycle (heat input at constant pressure only). The performance of all boosted cycles improves as more heat is added to the dual cycle at constant volume (by increasing α), reaching its peak for the Otto cycle (heat input at constant volume only) when α reaches maximum at $\beta = 1$.

To reemphasise, the relationship between the pressure ratio $r_c = p_1/p_o$ and the compres-sion ratio (volume ratio V_o/V_1) in the compressor ε_{com} is as follows $r_c = p_1/p_o = \varepsilon_{com}^\gamma$, where p_1 is the pressure at the exit from the compressor and p_o is the ambient pressure.

The effect of the pressure ratio on the thermal efficiency and *mep* of the boosted dual cycle with and without intercooling is shown in Figure 4.18. The relative values of the configurations are essentially the same as observed earlier but exhibit different behaviour as the pressure ratio increases. The thermal efficiency remains almost unchanged for the variable-pressure and constant-pressure turbocharging with and without intercooling, whereas the *mep* increases sharply as p_1/p_o increases, with the rate of increase with intercooling being greater.

The intercooler effectiveness also has an effect on the performance of boosted and inter-cooled cycles. For example, increasing ε from 0.5 to 0.75 adds around 7% to the *mep* of the dual cycle with variable-pressure turbocharging and intercooling. The thermal efficiency, on the other hand, decreases by about 0.1–0.5%. For a perfect intercooler ($\varepsilon = 1.0$), the changes are about 15% and 1.0%, respectively.

Problems

4.1 Three boosting schemes are considered to increase the power output of a compres-sion ignition (CI) engine operating on the air-standard dual cycle: constant-pressure turbocharging, variable-pressure turbocharging, and supercharging. Determine the gain in thermal efficiency and *mep* for each scheme relative to the natural-induction dual-cycle engine. Take the following:

Ambient conditions	1 *bar*, 293 *K*
Ratio of specific heats of air	$\gamma = 1.4$
Pressure at start of compression in boosted engines	1.5 *bar*
Engine compression pressure for base engine	16.5
Overall pressure ratio for boosted engines	20
Pressure ratio at constant volume	$\alpha = 1.6$
Volume ratio at constant pressure	$\beta = 1.3$

4.2 Modify problem 4.1 by increasing the pressure at the start of compression to 1.8 bar while keeping the engine compression ratio constant. What effect will this change have on the thermal efficiency and mean effective pressure?

4.3 A turbocharged CI engine operates on a dual cycle with a forced-induction cycle at constant pressure with intercooling. The compression ratio in the compressor is $\varepsilon_{comp} = 1.8$, and the effectiveness of the intercooler is $\epsilon = 0.65$. Determine
(a) Degree of cooling in the cycle
(b) Thermal efficiency and mean effective pressure of the cycle if the compression ratio of the engine $\varepsilon = 14$, $\alpha = 1.5$, $\beta = 1.4$, $\gamma = 1.4$, $p_0 = 0.1$ MPa.

4.4 Compare the results in 4.3(b) with the results for the same engine without turbocharging and having an adjusted compression ratio to keep the maximum cycle pressure constant in both cases. All other data remain the same.

4.5 Performances of Otto and Diesel cycles are to be improved by using constant-pressure turbocharging. Determine the thermal efficiency and mean effective pressure for the data in the table:

Cycle	Parameter					
	ε	ε_{comp}	α	β	γ	p_0, kPa
Otto	10	1.5	1.7	–	1.4	100
Diesel	15	1.5	–	1.8	1.4	100

4.6 The following figure shows the p-V diagram of a dual cycle with intercooled turbocharging. The gases from the engine are supplied to the turbocharger partly at constant pressure and partly at variable pressure. Show that the thermal efficiency of the cycle is given by

$$\eta_t = 1 - \frac{\alpha\beta^\gamma - \alpha_T[1 + \gamma(\beta_T - 1)]}{\varepsilon^{\gamma-1}[\alpha - 1 + \gamma\alpha(\beta - 1)]}$$

where

$$\alpha_T = p_5/p_6, \beta_T = V_7/V_6 = V_7/V_1$$

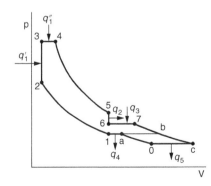

4.7 A CI engine operating on the air-standard dual cycle is to be boosted by variable-pressure turbocharging with intercooling. Fill the missing data in the table for these cycles: natural-induction (NI), turbocharged (TC), and turbocharged with intercooling (TC/IC).

Parameter	NI	TC	TC/IC
Ambient pressure, *bar*	1	1	1
Ambient temperature, K	288	288	288
Overall pressure ratio $\varepsilon_o = \varepsilon_{com}\omega\varepsilon$	17.5	20	23.2
Compressor compression ratio ε_{com}		1.4	2.16
Pressure ratio in compressor p_1/p_0			
Degree of cooling ω			1.16
Degree of cooling in degrees C			
Compression ratio in engine			
Pressure at end of compression			
Pressure ratio at constant volume, α	1.6	1.5	1.5
Volume ratio at constant pressure, β	1.43	1.5	1.5
Maximum pressure, *bar*			
Input heat, Q_{in}, kJ/kg			
Thermal efficiency			
Mean effective pressure			
Ratio of specific heats γ	1.4	1.4	1.4
Specific heat of air c_p, $kJ/kg. K$	1.005	1.005	1.005

5

Fuel-Air Cycles for Reciprocating Engines

The main difference between air-standard cycles (theoretical cycles) and fuel-air cycles is in the process of heat addition to the cycle. In the former, heat is added from an external source and is independent of the source. In the latter, the added heat is generated internally from the chemical reaction (combustion) of a mixture of a liquid or gaseous hydrocarbon fuel and air within the flammability limits. Another difference between the two types of cycles in this book is the assumption that the compression and expansion processes are polytropic not isentropic (adiabatic and reversible) as often assumed (Taylor, 1980; Haywood, 1988). We will first consider the case for the dual fuel-air cycle shown in Figure 5.1. On the surface, this cycle looks identical to the air-standard dual cycle discussed before; however, in reality it is one step closer to real compression ignition engine cycles. Once the performance equations are developed for this cycle, the constant-volume and constant-pressure cycles can be deduced as special cases.

5.1 Fuel-Air Cycle Assumptions

A number of assumptions are made for this cycle:

- Gas-exchange processes at the start and end of the cycle are ignored, i.e. there are no induction and exhaust processes and the cycle is a closed cycle.
- The working fluid trapped in the cylinder is air during the compression process 1–2 and products of complete combustion of the fuel during processes 2–3, 3–4, and 4–5.
- The compression process 1–2 is polytropic (heat to and from the surroundings is accounted for) with an averaged polytropic index n_1 $(n_1 < \gamma)$.
- The heat-addition profile in fuel-air cycles is exactly the same as in the theoretical (air-standard) cycle; however, the input heat is not taken from an external source but is generated internally as a result of the chemical reaction between the compressed air and the fuel in the cylinder. Additionally, the air-fuel ratio and thermal properties of the products of combustion are accounted for in the calculations.
- The expansion process 4–5 is polytropic, with an averaged polytropic index n_2 $(n_2 < \gamma)$.

Fundamentals of Heat Engines: Reciprocating and Gas Turbine Internal Combustion Engines, First Edition. Jamil Ghojel.
© 2020 John Wiley & Sons Ltd. This Work is a co-publication between John Wiley & Sons Ltd and ASME Press.
Companion website: www.wiley.com/go/JamilGhojel_Fundamentals of Heat Engines

5.2 Compression Process

This process starts at point 1 and ends at point 2 (Figure 5.1). During the compression process in actual engines, heat exchange between the cylinder gases and the surroundings is continuously changing with time, both in magnitude and direction. In computations of the compression stroke in fuel-air cycles, a single average polytropic index n_1 is assumed for the entire process. The value of n_1 is dependent mainly on engine speed (Sharaglazov et al., 2004) and can be estimated by the following empirical equation:

$$n_1 = 1.41 - \frac{100}{N}$$

where N is engine speed in revolutions per minute (rpm).

The pressure and temperature at the end of the compression process are given by

$$p_2 = p_1 \varepsilon^{n_1}$$

$$T_2 = T_1 \varepsilon^{n_1-1}$$

where ε is the compression ratio ($\varepsilon = V_1/V_2$).

Some indicative values of n_1, p_2, and T_2, given by Gavrilov et al. (2003), are shown in Table 5.1.

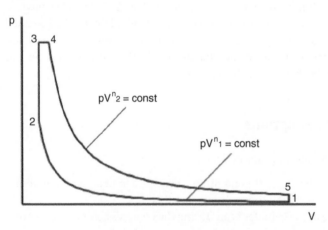

Figure 5.1 Dual fuel-air cycle.

Table 5.1 Typical values of the parameters of the compression process in real engines.

Engine type	n_1	p_2 (MPa)	T_2 (K)
SI engines	1.34–1.37	0.9–1.9	600–800
Gas engines[a]	1.36–1.39	1.0–2.0	650–800
Natural-induction CI engines	1.35–1.39	3.5–6.0	800–1000
Turbocharged CI engines	1.32–1.37	Up to 8.0	Up to 1100

a) Fuel: compressed natural gas (CNG) or liquefied petroleum gas (LPG).

5.3 Combustion Process

Consider the combustion process 2–3–4 for the mixed-combustion cycle shown in Figure 5.2a. In the ideal cycle configuration (air-standard cycle), heat is added from an external source partially at constant volume along process 2–3 and partially at constant pressure along process 3–4.

In the fuel-air cycle, we assume that combustion starts at 2 and the combustion products are formed instantaneously and maintained at the same composition during the entire heat-addition process along processes 2–3 and 3–4. In both the air-standard and fuel-air cycles, the portion of heat added to the cycle at constant volume is determined by the pressure-increase ratio $\alpha = p_3/p_2$, which is limited by the allowable maximum cycle pressure.

The first law of thermodynamics applied to processes 2–3 and 3–4 in Figure 5.2a for a closed system considering 1 kg of fuel is:

> *Released fuel energy*
>
> $\quad$ = *Change in internal energy of gases between 2 and 4*
>
> $\quad\quad$ + *work done by the gases between 3 and 4*

$$H_l = M_P U_{P4} - M_R U_{R2} + W_{3-4} \tag{5.1}$$

where

$\quad H_l$: lower heating value of the fuel, kJ/kg

$\quad U_{P4}$: specific internal energy of the combustion products at point 4, $kJ/kmole$

$\quad U_{R2}$: specific internal energy of the reactants (air) at point 2, $kJ/kmole$

$\quad M_P$: quantity of combustion products, $kmole$

$\quad M_R$: quantity of reactants, $kmole$

$\quad W_{3-4}$: work done per kilogram of fuel during expansion of the combustion products between points 3 and 4, kJ

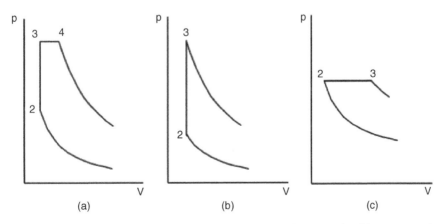

Figure 5.2 Combustion processes in fuel-air cycles: (a) dual-combustion (b) constant-volume (Otto); (c) constant-pressure (Diesel).

Work done by the combustion products during preliminary expansion at constant pressure is

$$W_{3-4} = p_4 V_4 - p_3 V_3 = p_4 V_4 - p_3 V_2$$

Since $p_3 = \alpha p_2$, where α is the pressure ratio during combustion at constant volume, $(p_3/p_2 = p_4/p_2)$,

$$W_{3-4} = p_4 V_4 - \alpha p_2 V_2$$

or, using the equation of state,

$$W_{3-4} = \overline{R}(M_P T_4 - \alpha M_R T_2) \tag{5.2}$$

where $\overline{R}$ is the universal gas constant.

Combining Eqs. (5.1) and (5.2), rearranging and dividing both sides by M_R yields the combustion equation

$$\frac{H_l}{M_R} + U_{R2} + \alpha \overline{R} T_2 = \mu_{th}(U_{P4} + \overline{R} T_4) \tag{5.3}$$

where $\mu_{th} = M_P/M_R$ is the theoretical coefficient of molar change.

The quantities of the combustion products (in *kmole*) can be found from the following reaction of a generic hydrocarbon fuel $C_x H_y$:

$$C_x H_y + \lambda \left(x + \frac{y}{4}\right)(O_2 + 3.7619 N_2) = x CO_2 + \frac{y}{2} H_2 O$$
$$+ \left(x + \frac{y}{4}\right)(\lambda - 1)O_2 + 3.7619\lambda \left(x + \frac{y}{4}\right) N_2 \tag{5.4}$$

This reaction is valid for complete combustion only with relative air-fuel ratio $\lambda \geq 1$. When $\lambda < 1$, the combustion is usually incomplete, and the air-fuel mixture is fuel-rich (insufficient air) and the heat of combustion is reduced as a result. This case will be discussed when considering practical cycles.

For 1 *kg* fuel, the amount of reactants and products (in *kmole/kg* fuel) are, respectively,

$$M_R = \frac{4.7619\lambda \left(x + \frac{y}{4}\right)}{(12x + y)} \tag{5.5}$$

$$M_P = \frac{1}{(12x + y)} \left[x + \frac{y}{2} + \left(x + \frac{y}{4}\right)(\lambda - 1) + 3.7619\lambda \left(x + \frac{y}{4}\right)\right] \tag{5.6}$$

The relative air-fuel ratio λ determines the amount of excess air in the reactants and the amounts of oxygen and nitrogen in the products. When $\lambda = 1$, the reaction becomes stoichiometric. Dissociation has been ignored in this analysis, but it could be accounted for if the reaction temperatures are high enough.

Example 5.1 Determine the amounts of reactants and products for the stoichiometric reaction ($\lambda = 1$) of the following two fuels:

- Compression-ignition engine surrogate fuel $C_{12}H_{21.6}$: $M_R = 0.5$ *kmole/kg fuel* and $M_P = 0.5329$ *kmole/kg fuel*
- Spark-ignition engine surrogate fuel $C_8 H_{18}$: $M_R = 0.5221$ *kmole/kg fuel* and $M_P = 0.5616$ *kmole/kg fuel*

The specific internal energy of the products is

$$U_P = \frac{1}{M_P}[M_{CO_2}U_{CO_2} + M_{H_2O}U_{H_2O} + M_{O_2}U_{O_2} + M_{N_2}U_{N_2}]$$

$M_{CO_2}, M_{H_2O}, M_{O_2}, M_{N_2}$ are the quantities (in $kmole$) of carbon dioxide, water vapour, oxygen, and nitrogen, respectively, in the products of complete combustion of 1 kg of fuel when $\lambda > 1$.

$U_{CO_2}, U_{H_2O}, U_{O_2}, U_{N_2}$ are the specific internal energies (in $MJ/kmole$) of carbon dioxide, water vapour, oxygen, and nitrogen, respectively. Specific internal energies of air and some common combustion products are given as temperature-dependent polynomials in Table 2.7 and in Table A.2.

The relationships between the combustion parameters can be found from the equations of state at 3 and 4:

$$\frac{p_4V_4}{p_2V_2} = \frac{M_P}{M_R}\frac{T_4}{T_2} = \mu_{th}\frac{T_4}{T_2}$$

By definition $p_4/p_2 = \alpha$ and $V_4/V_2 = \beta$; therefore, for the dual cycle,

$$\alpha\beta = \frac{M_P}{M_R}\frac{T_4}{T_2} = \mu_{th}\frac{T_4}{T_2} \tag{5.7}$$

5.3.1 Constant-Volume Combustion Cycle (Otto Cycle)

In the constant-volume cycle shown in Figure 5.2b, heat is added at constant volume only along process 2–3 ($\beta = 1$), and no work is done, i.e. $W_{3-4} = 0$. The combustion equation in this case is a modified version of Eq. (5.3) and is written as

$$\frac{H_l}{M_R} + U_{R2} = \mu_{th}U_{P3} \tag{5.8}$$

The relationships between the combustion parameters for the constant-volume cycle can be deduced from Eq. (5.7) by substituting 1 for β and replacing T_4 by T_3:

$$\alpha = \frac{M_P}{M_R}\frac{T_3}{T_2} = \mu_{th}\frac{T_3}{T_2} \tag{5.9}$$

5.3.2 Constant-Pressure Cycle (Diesel Cycle)

In the constant-pressure combustion cycle shown in Figure 5.2c, $\alpha = 1$, the internal energy of the gases increases at constant pressure along line 2–3, and the work component at constant pressure is retained. By accounting for the differences in the numbering of the characteristic points of the cycle, Eq. (5.3) is now reduced to

$$\frac{H_l}{M_R} + U_{R2} + \overline{R}T_2 = \mu_{th}(U_{P3} + \overline{R}T_3) \tag{5.10}$$

The relationship between the combustion parameters for the constant-pressure cycle for which $\alpha = 1$ is given by

$$\beta = \frac{M_P}{M_R}\frac{T_3}{T_2} = \mu_{th}\frac{T_3}{T_2} \tag{5.11}$$

Table 5.2 Typical values of the parameters of the expansion process in real engines.

Engine type	n_2	p_5 (MPa)	T_5 (K)
SI engines	1.23–1.30	0.35–0.6	1200–1700
Gas engines[a]	1.25–1.35	0.2–0.55	1100–1500
CI engines	1.18–1.28	0.2–0.5	1000–1200

a) Fuel: compressed natural gas (CNG) or liquefied petroleum gas (LPG).

5.4 Expansion Process

As stated earlier, the expansion process is polytropic with an index $1.23 < n_2 < 1.3$. The pressure and temperature at the end of the process for the dual cycle are given by

$$p_5 = \frac{p_4}{\delta^{n_2}} \tag{5.12}$$

$$T_5 = \frac{T_4}{\delta^{n_2-1}} \tag{5.13}$$

By definition, $\delta = V_5/V_4$ (expansion ratio), which is always less than ε for this cycle.

The pressure and temperature at the end of the expansion process for the constant-volume cycle, considering that $\delta = \varepsilon$, are given by

$$p_4 = \frac{p_3}{\varepsilon^{n_2}} \tag{5.14}$$

$$T_4 = \frac{T_3}{\varepsilon^{n_2-1}} \tag{5.15}$$

The pressure and temperature at the end of the expansion process for the constant-pressure cycle are given by

$$p_4 = \frac{p_3}{\delta^{n_2}} \tag{5.16}$$

$$T_4 = \frac{T_3}{\delta^{n_2-1}} \tag{5.17}$$

For this cycle, $\delta = V_4/V_3$ and is always less than ε.

Some indicative values of the expansion process are given in Table 5.2.

5.5 Mean Effective Pressure

The net work done in the dual-combustion cycle shown in Figure 5.1 is

$$W_{cyc} = W_{4-5} - W_{1-2} + W_{3-4}$$

where W_{4-5} is the polytropic expansion work, W_{1-2} is the polytropic compression work, and W_{3-4} is the work done by the gases expanding at constant pressure from 3 to 4:

$$W_{4-5} = \frac{p_4 V_4}{n_2 - 1}\left[1 - \left(\frac{V_4}{V_5}\right)^{n_2-1}\right] = \frac{(\alpha p_2)(\beta V_2)}{n_2 - 1}\left(1 - \frac{1}{\delta^{n_2-1}}\right)$$

$$W_{1-2} = \frac{p_2 V_2}{n_1 - 1}\left[1 - \left(\frac{V_2}{V_1}\right)^{n_1-1}\right] = \frac{p_2 V_2}{n_1 - 1}\left(1 - \frac{1}{\varepsilon^{n_1-1}}\right)$$

$$W_{3-4} = p_4 V_4 - p_3 V_2 = p_3 V_2 \left(\frac{p_4 V_4}{p_3 V_2} - 1\right) = \alpha p_2 V_2 (\beta - 1)$$

The final form of the equation of the net work for the dual cycle is

$$W_{cyc} = p_2 V_2 \left[\alpha(\beta - 1) + \frac{\alpha\beta}{n_2 - 1}\left(1 - \frac{1}{\delta^{n_2-1}}\right) - \frac{1}{n_1 - 1}\left(1 - \frac{1}{\varepsilon^{n_1-1}}\right)\right] \tag{5.18}$$

For the constant-volume cycle ($\beta = 1$):

$$W_{cyc} = p_2 V_2 \left[\frac{\alpha}{n_2 - 1}\left(1 - \frac{1}{\varepsilon^{n_2-1}}\right) - \frac{1}{n_1 - 1}\left(1 - \frac{1}{\varepsilon^{n_1-1}}\right)\right] \tag{5.19}$$

For the constant-pressure combustion cycle ($\alpha = 1$):

$$W_{cyc} = p_2 V_2 \left[(\beta - 1) + \frac{\beta}{n_2 - 1}\left(1 - \frac{1}{\delta^{n_2-1}}\right) - \frac{1}{n_1 - 1}\left(1 - \frac{1}{\varepsilon^{n_1-1}}\right)\right] \tag{5.20}$$

Another, more convenient, form of this equation can be obtained if we substitute for p_2 and V_2 from $p_2 = p_1 \varepsilon^{n_1}$, $V_2 = V_1/\varepsilon$ into Eq. (5.20):

$$W_{cyc} = (p_1 \varepsilon^{n_1})\left(\frac{V_1}{\varepsilon}\right)\left[\alpha(\beta - 1) + \frac{\alpha\beta}{n_2 - 1}\left(1 - \frac{1}{\delta^{n_2-1}}\right) - \frac{1}{n_1 - 1}\left(1 - \frac{1}{\varepsilon^{n_1-1}}\right)\right] \tag{5.21}$$

By definition, the cycle mean effective pressure (*mep*) is

$$mep = \frac{W_{cyc}}{(V_{max} - V_{min})} = \frac{W_{cyc}}{(V_1 - V_2)}$$

$$V_1 - V_2 = V_2\left(\frac{V_1}{V_2} - 1\right) = V_2(\varepsilon - 1) = \frac{V_1}{\varepsilon}(\varepsilon - 1)$$

$$mep = \frac{W_{cyc}}{(V_1 - V_2)} = \frac{W_{cyc}}{\frac{V_1}{\varepsilon}(\varepsilon - 1)}$$

Finally, the equation of the *mep* of the fuel-air dual cycle can be written as

$$mep = p_1 \frac{\varepsilon^{n_1}}{(\varepsilon - 1)}\left[\alpha(\beta - 1) + \frac{\alpha\beta}{n_2 - 1}\left(1 - \frac{1}{\delta^{n_2-1}}\right) - \frac{1}{n_1 - 1}\left(1 - \frac{1}{\varepsilon^{n_1-1}}\right)\right] \tag{5.22}$$

For the constant-volume cycle ($\beta = 1$):

$$mep = p_1 \frac{\varepsilon^{n_1}}{(\varepsilon - 1)}\left[\frac{\alpha}{n_2 - 1}\left(1 - \frac{1}{\varepsilon^{n_2-1}}\right) - \frac{1}{n_1 - 1}\left(1 - \frac{1}{\varepsilon^{n_1-1}}\right)\right] \tag{5.23}$$

For the constant-pressure cycle ($\alpha = 1$):

$$mep = p_1 \frac{\varepsilon^{n_1}}{(\varepsilon - 1)}\left[(\beta - 1) + \frac{\beta}{n_2 - 1}\left(1 - \frac{1}{\delta^{n_2-1}}\right) - \frac{1}{n_1 - 1}\left(1 - \frac{1}{\varepsilon^{n_1-1}}\right)\right] \tag{5.24}$$

The thermal efficiency for any of the above cycles is given by the general equation

$$\eta_{th} = \frac{W_{cyc}}{H_l} \tag{5.25}$$

5.6 Cycle Comparison

Figure 5.3 shows a comparison of the air-standard and fuel-air cycles on a $p - V$ coordinate system, and Table 5.3 summarises the operating and performance parameters. The introduction of fuel reaction with air and allowance for heat exchange during the compression and expansion processes results in significant differences between the two cycles. For the selected operating parameters, the maximum pressure, thermal efficiency, mean effective pressure, maximum temperature, and ratio of volume change at constant pressure decrease

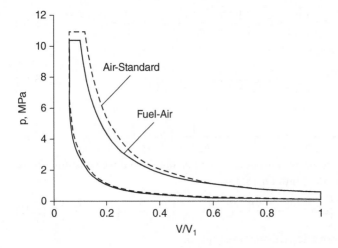

Figure 5.3 Comparison of the air-standard and fuel-air dual cycles.

Table 5.3 Operating and performance parameters of fuel-air and air-standard cycles.

	Fuel-air cycle	Air-standard cycle
Fuel	$C_{12}H_{21.6}$	—
Heat input	43 200 kJ/kg fuel	2110 kJ /kg air
Relative air-fuel ratio, λ	1.4	—
Ratio of pressure rise at constant volume, α	1.8	1.8
Ratio of volume increase at constant pressure, β	1.668	1.975
Compression ratio, ϵ	16	16
Compression index	$n_1 = 1.35$	$\gamma = 1.4$
Expansion index	$n_2 = 1.25$	$\gamma = 1.4$
Thermal efficiency, %	0.507	0.628
Mean effective pressure, MPa	1.577	2.068
Maximum temperature, K	2246	3211
Maximum pressure, MPa	10.32	10.91

by 5.4%, 19.3%, 23.7%, 30%, and 15.5%, respectively. The differences between the two cycles will vary if different values are assumed for the relative air-fuel ratio λ, ratio of pressure rise α, compression ratio ε, and polytropic exponents n_1 and n_2.

Problems

5.1 During the induction process of a four-stroke spark ignition (SI) engine, the fresh charge is heated by $15\,^\circ C$. Calculate the parameters of the compression stroke for the fuel-air cycle of this engine if the compression ratio is 8.6 and ambient conditions are $p_0 = 101.3\,kPa$ and $T_0 = 293\,K$. Take the coefficient and temperature of residual gases as 0.04 and $1050\,K$, respectively.

5.2 A four-cylinder, four-stroke CI engine with a total swept volume of $6.78\,l$ operates on the fuel-air dual cycle shown in the following figure. The state at each point of the cycle is given in brackets in terms of pressure (kPa), volume (m^3), and temperature (K), respectively. At an engine speed of $2000\,rpm$, the heat input into the cycle is $2.3\,kJ$ at constant volume and $1.2\,kJ$ at constant pressure. Take the mechanical efficiency as 0.8, and assume that the cycle work is equal to the indicated work.
  (a) Determine the unspecified volumes and temperatures
  (b) Determine the indicated work and indicated mean effective pressure of the cycle. What are the brake mean effective pressure and power developed by the engine? What is the rate of fuel consumption of the engine $(kg/s, kg/h)$ if the lower heating value (H_l) of the fuel is $43500\,kJ/kg$?

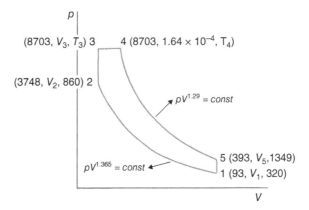

(Take $c_p = 1.1\,kJ/kg.K$, $c_v = 0.805\,kJ/kg.K$, $R = 0.287\,kJ/kg.K$)

5.3 Data are provided for three fuel-air cycles in the following table: dual cycle (compression ignition [CI] engine cycle), constant-volume cycle (Otto cycle), and constant-pressure cycle (Diesel cycle):

Parameter	Dual cycle	Otto cycle	Diesel cycle
x_C	0.869 565	0.869 565	0.869 565
x_H	0.130 435	0.130 435	0.130 435
x_O	0	0	0
Compression ratio, ϵ	16	8	16
Relative air-fuel ratio, λ	1.4	1	1.4
Ambient temp, K	298	298	298
Ambient pressure, kPa	100	100	100
n_1	1.38	1.38	1.35
n_2	1.25	1.25	1.25
H_1, MJ/kg	43.2	44	43.2
Ratio of pressure rise at constant volume, α	1.8		1

Determine the following for the dual cycle:
(a) Pressure and compression at the end of the compression process
(b) Quantity of reactants and products of combustion
(c) Combustion temperature
(d) Ratio of volume increase β
(e) Maximum cycle pressure

5.4 Repeat problem 5.3 for the Otto cycle. In (d), determine α.

5.5 Repeat problem 5.3 for the Diesel cycle.

6

Practical Cycles for Reciprocating Engines

The fuel-air cycles discussed in Chapter 5 differ significantly from the actual cycles of reciprocating piston engines. Actual cycles are usually obtained by measuring the pressure in the cylinder and plotting the results as an *indicator diagram*. They can also be reasonably accurately simulated by computer modelling using specialised software packages. Figure 6.1 shows the indicator diagrams of a natural-induction (NI), four-cylinder, four-stroke, compression-ignition (CI) engine operating at a speed of $2000\,rpm$ and developing $46\,kW$ of power. Figure 6.1a is the standard $p-\phi$ diagram, which is a plot of pressure versus crank-angle degrees. The $p-\phi$ diagram can be converted to a $p-V$ diagram (Figure 6.1b) using the kinematics of the reciprocating mechanism and engine dimensions. Figures 6.1a,b show the deviation of the heat-addition process (combustion process) in the actual cycle from the thermodynamic cycles investigated so far. Figure 6.1c is the blown-up lower part of the indicator diagram in Figure 6.1b. It shows the pressure profile at the end of the expansion process near bottom dead centre (BDC), which differs markedly from the isobaric (constant-pressure) heat-rejection process in the thermodynamic cycles. Figure 6.1c also shows a looping pressure profile between BDC and top dead centre (TDC), which is due to the pressure variation during the exhaust and induction processes. The area of the loop represents the work lost during the gas exchange, which reduces the net work produced by the cycle. This lost work is commonly known as the *pumping loss(es)*. It is apparent that the pumping losses for the engine whose indicator diagrams are shown in Figure 6.1 are quite small, which is due mainly to the optimal selection of valve timing.

Analysis of the processes taking place in actual four- and two-stroke, NI piston engines are presented in this chapter.

6.1 Four-Stroke Engine

Figures 6.2 and 6.3 show the four processes taking place in the four-stroke engine. The execution of the thermodynamic cycle in a four-stroke engine takes place over two revolutions of the crankshaft during which the piston traverses four strokes (a *stroke* is the distance between the TDC and BDC positions of the piston) and the crank rotates 720°. Each process

Fundamentals of Heat Engines: Reciprocating and Gas Turbine Internal Combustion Engines, First Edition. Jamil Ghojel.
© 2020 John Wiley & Sons Ltd. This Work is a co-publication between John Wiley & Sons Ltd and ASME Press.
Companion website: www.wiley.com/go/JamilGhojel_Fundamentals of Heat Engines

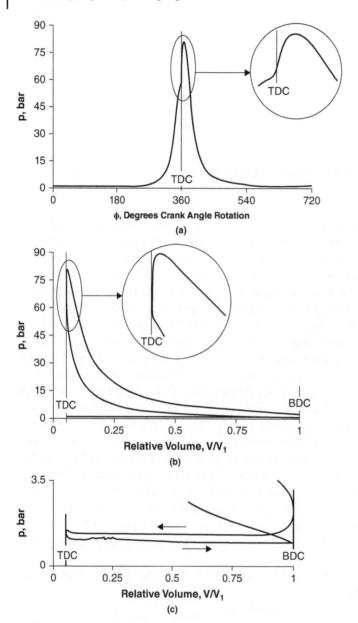

Figure 6.1 Indicator diagrams of a CI engine at partial load operating at 2000 *rpm*: (a) $p - \phi$ diagram; (b) $p - V$ diagram; (c) enlarged gas-exchange process showing end of expansion, exhaust, induction, and start of compression.

is represented by the shaded area in the engine schematic and by the solid line in the $p - V$ diagram immediately below it. The resulting diagram shown in Figure 6.3d is a schematic representation of the actual cycle with the gas-exchange process deliberately exaggerated to emphasise the effect of this process (see Figure 6.1c for comparison).

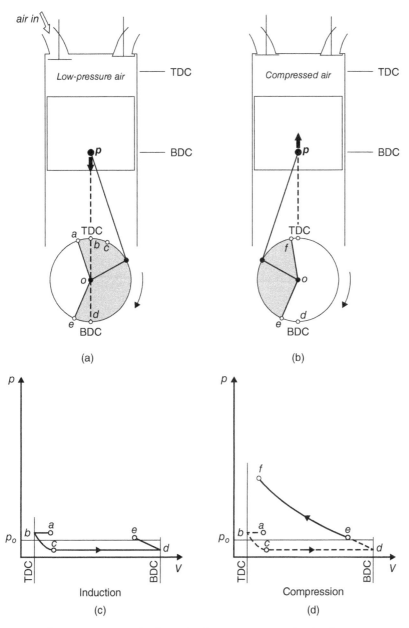

Figure 6.2 Four-stroke engine induction (a, c) and compression (b, d) processes: a – inlet valve opens, e – inlet valve closes/compression starts, f – compression ends.

6.1.1 The Induction Process $a - b - c - d - e$

This process is illustrated in Figures 6.2a,c. With the exhaust valve closed, the inlet valve opens at point a before the piston reaches TDC in its upward motion and closes at point e after the piston reaches BDC and reverses its direction upwards. The pressure of the exhaust

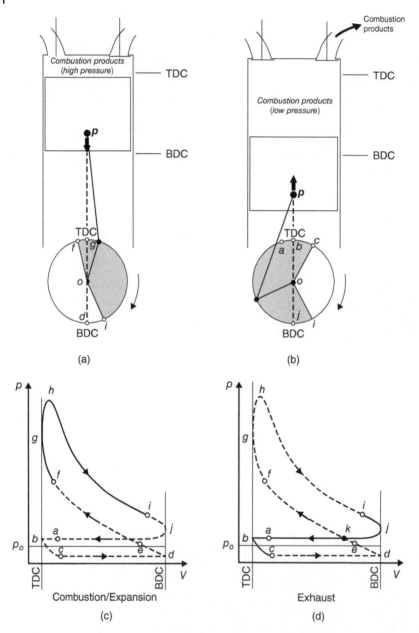

Figure 6.3 Four-stroke engine combustion/expansion (a, c) and exhaust (b, d) processes: *f* – combustion starts, *h*–pressure peaks, *i*– exhaust valve opens, *c* – exhaust valve closes.

gases at point **a**, which is slightly above the atmospheric pressure p_0, is maintained until the piston reaches TDC at point **b** and reverses its direction. During the downward motion of the piston the pressure drops below the atmospheric and air flows into the cylinder until the piston reaches BDC at point **d** then increases again with the upward motion of the piston from **d** to **e** to a value determined by the timing of the inlet valve closure. During this process the distance travelled by the piston exceeds one stroke which corresponds to a crank angle rotation greater than 180° (shaded area in Figure 6.2a).

6.1.2 The Compression Process $e - f$

The process is illustrated in Figures 6.2b,d. Both the inlet and exhaust valves are closed during this polytropic process, which starts at point e and ends when the combustion process starts at point f just before the piston reaches TDC. During this process, the piston travels less than one stroke, which corresponds to a crank angle rotation less than 180° (shaded area in Figure 6.2b). It should be noted that the pressure continues to increase beyond point f as a result of the combined effect of piston motion and combustion. The location of point f is determined by the ignition timing (in SI engines) and fuel injection timing (in CI engines).

6.1.3 The Combustion and Expansion Processes $f - g - h - i$

The combustion process is illustrated in Figures 6.3a,c. Combustion starts at point f, followed by a steep rise in the pressure to point h, and ends during the expansion of the gases with the downward motion of the piston (the exact end of the combustion process depends on the amount of fuel burned, i.e. engine load). The expansion process ends when the exhaust valve opens at point i before the piston reaches BDC. The piston travels less than a stroke during the expansion of the gases from TDC to BDC. The duration of this process is shown by the shaded area Figure 6.3a.

6.1.4 The Exhaust Process $i - j - a - b - c$

The exhaust process is shown in Figures 6.3b,d. As the exhaust valve opens at point i, the gas pressure starts decreasing to a value slightly above atmospheric at BDC. This pressure is maintained at that value during the upward motion of the piston and starts decreasing as the piston reverses its direction away from TDC, up to the moment the exhaust valve closes at point c. During this process, the distance travelled by the piston exceeds one stroke, which corresponds to a crank angle rotation greater than 180° (shaded area in Figure 6.3b). During the latter stages of this process, both the inlet and exhaust valves are open (process $a - b - c$). The period between the opening of the inlet valve at a and closing of the exhaust valve at c is known as *valve overlap,* during which the relatively high speed of the exhaust gases leaving the engine through the narrow exhaust valve throat is utilised to increase the pressure drop across the inlet valve throat, thereby increasing the mass of air inducted into the engine cylinder. The enclosed area $k - a - b - c - d - e - k$ represents the losses associated with the gas-exchange process during the induction and exhaust strokes and is known as the *pumping losses.* The pumping losses are usually lumped together with the mechanical losses when calculating engine performance parameters. The remaining area enclosed by the $p - V$ diagram is the net cycle work.

6.2 Two-Stroke Engine

The difference between the two-stroke engine and four-stroke engine is in the number of strokes it takes to execute the four processes of one thermodynamic cycle (induction, compression, combustion/expansion, and exhaust). In the two-stroke engine, the cycle is executed over one crankshaft revolution, during which the piston traverses two strokes only. In the four-stroke engine, two revolutions and four strokes are required to execute the thermodynamic cycle. Another difference is that poppet valves, operated by camshafts driven through gears by the engine crankshaft, are used in the four-stroke engine to control the flow of air into the cylinder and flow of exhaust gases out of the cylinder. On the other hand, small two-stroke engines utilise inlet and outlet ports with the opening and closing being controlled by the piston skirt. In large two-stroke engines, the induction is usually through ports and exhaust through poppet valves. The analysis of the operation of the two-stroke engine presented here and shown in Figures 6.4 and 6.5 is for an engine with inlet and outlet ports. Each process in Figures 6.4 and 6.5 is represented by the shaded area in the engine

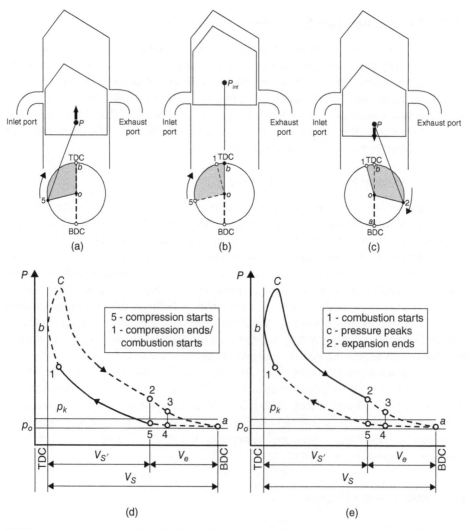

Figure 6.4 Two-stroke engine processes: compression 5 – 1 (a, b, d); combustion/expansion 1 – b – c – 2 (c, e).

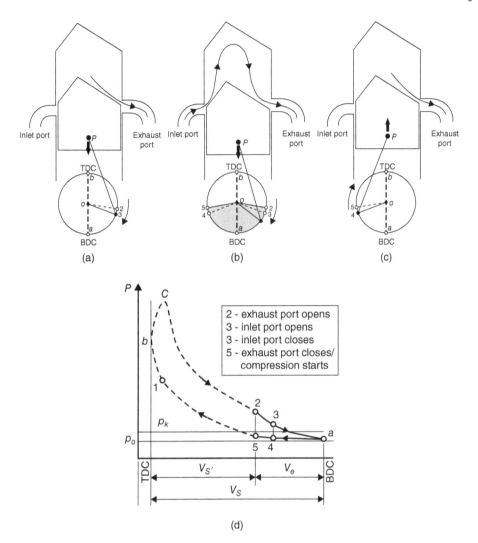

Figure 6.5 Two-stroke engine processes: exhaust only 2 – 3 (a); exhaust/induction with both ports open 2 – 3 – a – 4 (b, d); exhaust continues 4 – 5 exhaust port only open (c).

schematic and by the solid line in the $p - V$ diagram immediately below it. The processes of induction and exhaust (gas exchange) take place over volume change V_e, during which the piston does not produce any useful work, and the compression and combustion/expansion processes occur during volume change $V_{s'}$.

6.2.1 Compression Processes 5 – 1

The compression process (Figures 6.4a,b,d) starts at point **5** with both inlet and outlet ports closed during the upward motion of the piston and ends at point **1** before the piston reaches TDC. The duration of the compression process is well short of 180°, as shown by the shaded areas in Figures 6.4a,b, and shorter than in the case of the four-stroke engine.

6.2.2 Combustion and Expansion Processes 1 − b − c − 2

The combustion process, shown in Figures 6.4c,e, starts with the ignition of the fuel-air mixture at point **1**, accompanied by sharp increase in pressure to point **b** at TDC, and then continues to peak pressure at point **c** with the downward motion of the piston. The end of the combustion process (not shown) varies with engine operating conditions. The expansion process ends at point **2** just before the exhaust port opens. The shaded area in Figure 6.4c shows that the duration of the combustion/expansion process is much less than 180° and usually shorter than in the case of four-stroke engines.

6.2.3 Exhaust and Induction Processes 2 − 3 − a − 4 − 5

These two processes are illustrated in Figure 6.5. They take place simultaneously as the piston continues its downward motion past point **2** (end of expansion). The piston opens the exhaust port first at point **2**, and then the inlet port at point **3**.The piston reaches BDC at point **a** with both ports fully uncovered, and it reverses direction upward to close the inlet port first at point **4** and then the exhaust port at point **5**. The overlapping of the opening periods of the two ports facilitates the clearing of the trapped exhaust gases from the previous cycle and makes room for more air-fuel charge. This process is known as *scavenging* and is extremely important in two-stroke engines of the type under discussion. The combined exhaust and induction process is known as the *gas-exchange process*, and it occupies a significant portion of the swept volume represented by volume V_e, as shown in Figure 6.5d. This causes a reduction in specific cycle work compared with the four-stroke cycle.

6.3 Practical Cycles for Four-Stroke Engines

A relatively simple computational technique will be presented here that allows one to construct cycles that could be used as surrogates for actual internal combustion engine cycles and referred to in this book as a *practical cycles*. The cycles to be considered are for the compression ignition (CI) engine, spark ignition (SI) engine, and engine with constant-pressure combustion.

Cycle Assumptions

- The induction process starts at TDC and ends at BDC and takes place at constant pressure p_1. This pressure is estimated using Bernoulli's equation for the flow of air through the inlet manifold and inlet valve.
- The compression process is polytropic. It starts at BDC and ends at TDC.
- The heat-addition mode depends on the type of engine. In the CI engine, part of the heat is added at constant volume and the balance at constant pressure. In the SI engine, all the heat is added at constant volume. In the constant-pressure engine, all the heat is added at constant pressure.
- The added heat is not taken from an external source but is generated internally as a result of the chemical reaction between the compressed air and the fuel in the cylinder. Additionally, the air-fuel ratio and thermal properties of the products of combustion are accounted for in the calculations.

- The effect of residual gases is accounted for. Residual gases are combustion products that are trapped in the cylinder at the end of the exhaust process due, mainly, to the valve overlap and short time available for the completion of inlet and exhaust processes.
- Not all of the heating value of the fuel is utilised, as a result of the losses due to untimely ignition or incomplete combustion of the fuel, particularly in SI engines.
- The sharp corners of the cycles at the characteristic points are rounded off to mimic the processes near TDC and BDC in actual engine cycles.
- The expansion process is polytropic and starts at the end of the theoretical combustion process.
- Pumping losses are not treated separately but are lumped together, instead, with the mechanical losses and represented graphically by a horizontal line on the $p - V$ diagram with height p_1.
- The exhaust process starts at BDC and ends at TDC, and takes place at constant pressure p_1.

6.3.1 Compression Ignition Engine (CI Engine)

The practical cycle for a CI engine operating on the dual cycle is shown in Figure 6.6. This cycle is essentially a refinement of the fuel-air cycle discussed in Chapter 5. In its configuration as a dual cycle with heat addition partly at constant volume (process $2 - 3$) and partially at constant pressure (process $3 - 4$), it is computed as cycle $1 - 2 - 3 - 4 - 5$ (Figure 6.6a). To bring this cycle closer to the actual cycle, heat addition and rejection lines are rounded off as shown in Figures 6.6b to account for the losses during heat addition and rejection. The numbering system of the characteristic points of the standard and fuel-air cycles has been retained in Figure 6.6 for ease of comparison. The net computed cycle work is represented by area A, which is equal to the area $1 - 2 - 3 - 4 - 5$ minus the shaded areas. The pumping losses are not included in the computed cycle but will be accounted for as part of the total losses in practical cycles.

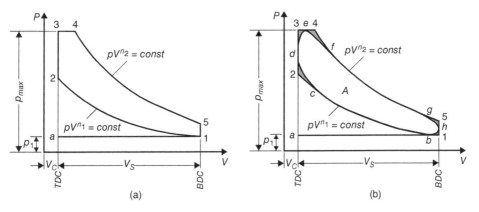

Figure 6.6 Practical cycle model of the compression ignition engine: (a) raw practical dual-combustion cycle; (b) cycle with rounding-off.

6.3.1.1 Induction Process

The induction process is represented by the isobaric line $a - 1$. The induction pressure and the pressure at the start of compression at point **1** are equal to p_1. Pressure losses in the induction manifold during the induction process reduce the amount of fresh charge admitted into the engine cylinder. These losses are estimated by assuming incompressible and steady flow of air into the engine and applying Bernoulli's equation in the induction manifold between the point just ahead of the air filter and the inlet valve throat in combination with the continuity equation between the valve throat and engine cylinder (Figure 6.7a). The equation for the pressure drop of the air as it enters the cylinder can be written as

$$\Delta p_{ind} = p_0 - p_1 = C \frac{N^2}{A_{min}^2} \tag{6.1}$$

where p_0 is the ambient pressure, C is a constant, N is the engine speed (*rpm*), and A_{min} is the cross-sectional area of the narrowest section in the induction manifold (inlet valve throat). Equation (6.1) shows that Δp_{ind} is directly proportional to the square of the engine speed and inversely proportional to the square of the cross-sectional area of the inlet valve seat. At high engine speeds, larger inlet valves or multiple smaller inlet valves with larger total area are required to maintain the same pressure drop. For a given engine speed, the pressure drop can be reduced by a similar strategy. The pressure drop during the induction process in four-stroke, NI, CI engines, according to experimental data by Dyechenko et al. (1974) lies within the range $\Delta p_{ind} = (0.10...0.25)p_0$, from which $p_1 = (0.75...0.9)p_0$.

In turbocharged (TC) engines, the air is supplied to the engine by a compressor with delivery pressure p_{comp}, and the pressure at the start of the compression process is higher than the atmospheric pressure (Figure 6.7b). Area B in this case is positive, i.e. there are no pumping losses during the gas-exchange process. The pressure drop in four-stroke, TC, CI engines during the induction process is $\Delta p_{ind} = p_{comp} - p_1 = (0.05 - 0.10)p_{comp}$ corresponding to the initial compression pressure $p_1 = (0.90 - 0.95)p_{comp}$.

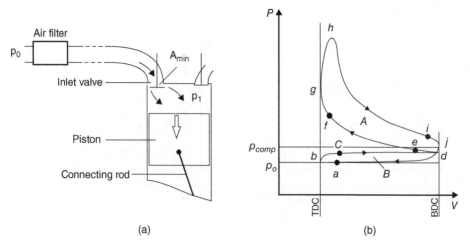

(a)   (b)

Figure 6.7 Pressure drop during induction process in four-stroke piston engines: (a) natural-induction scheme; (b) pressure drop in turbocharged SI engine.

The temperature at the end of the induction process can be found by applying the first law of thermodynamics to the induction process:

$$C_{pa}M_a(T_{in} + \Delta T) + C_{pg}M_{res}T_{res} = C_{pm}(M_a + M_{res})T_1 \tag{6.2}$$

where

C_{pa}, C_{pg}, C_{pm}: specific heats of air, combustion gases, and the air-gas mixture, respectively, $kJ/kmole. K$

T_{in}, T_{res}, T_1: temperature of the air at the inlet into the manifold, temperature of residual gases in the cylinder, and temperature of the air-gas mixture at the end of the induction process, respectively, K

ΔT: temperature increase of the inlet air during the induction process as it comes into contact with the hot surfaces of the induction system, K

M_a, M_{res}: quantity of inlet air and quantity of residual gases trapped in the cylinder during the induction process, respectively, $kmole$

The temperature at the end of the induction process from Eq. (6.2) is

$$T_1 = \frac{C_{pa}(T_{in} + \Delta T) + C_{pg}\gamma_{res}T_{res}}{C_{pm}(1 + \gamma_{res})} \tag{6.3}$$

The coefficient of residual gases γ_{res} is defined as

$$\gamma_{res} = \frac{M_{res}}{M_a}$$

At the temperatures encountered during the induction process, the specific heats of air, combustion products, and the mixture differ little from each other, and Eq. (6.3) can be simplified to

$$T_1 = \frac{T_{in} + \Delta T + \gamma_{res}T_{res}}{1 + \gamma_{res}} \tag{6.4}$$

T_{in} is equal to the ambient temperature for NI engines and to the compressor delivery temperature in TC engines. ΔT is the increase in temperature of the fresh charge during induction (for NI, SI, and CI engines $\Delta T = 10 - 25\ K$; for TC, CI engines $\Delta T = 5 - 10\ K$), γ_{res} is the coefficient of residual gases, and T_{res} is the temperature of residual gases (equal to the temperature of the exhaust gases). $T_1 = 310–350\ K$ for NI engines, and $T_1 = 320–400\ K$ for TC engines.

6.3.1.2 Compression Process

The temperatures at the end of the induction process and start of the polytropic compression process are the same and equal to T_1. The pressure and temperature at the end of the compression process $1 - 2$ in Figure 6.6 are, respectively,

$$p_2 = p_1\varepsilon^{n_1}$$

$$T_2 = T_1\varepsilon^{n_1-1}$$

The compression ratio ε is equal to V_1/V_2, and n_1 is the polytropic index of the compression process. Typical values of n_1, p_2, and T_2 can be found in Table 5.1.

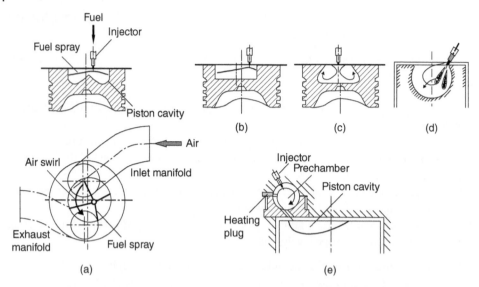

Figure 6.8 Fuel injection and combustion schemes in CI engines: (a, b, c, d) direct injection; (e) indirect injection.

6.3.1.3 Combustion Process

In actual CI engines, the thermal energy input into the cycle is generated inside the cylinder close to the TDC at the end of the compression process by direct injection of finely atomised fuel into an internally located combustion chamber. Combustion chambers come in different shapes and sizes, as shown in Figure 6.8a–d. Fuel is normally injected through a multi-hole nozzle at very high pressure and atomised, forming extremely small particles that penetrate the compressed air in the piston cavity; the fuel evaporates; mixes with air, forming a combustible mixture; and ignites at the appropriate moment in the cycle. Injector nozzles in high-speed engines typically have four or five holes, and air-fuel mixing is enhanced by giving the air in the cylinder a swirling motion. In large low-speed engines, it is preferable to enhance mixing by increasing the number of nozzle holes to 10 or more. The inducted air in CI engines is not throttled, and cycle input energy is controlled by varying the quantity of the injected fuel while maintaining a constant air mass. The combustion process takes place partially at constant volume and partially at constant pressure, with combustion duration increasing in proportion to the amount of fuel injected. Different combustion chambers can be used in direct injection engines, with the toroidal shape shown in Figure 6.8a being the most common. The indirect injection combustion system shown in Figure 6.8e was originally developed for high-speed CI engines but has fallen out of favour as the direct injection system improved significantly with the introduction of the common rail fuel injection system for passenger cars. Fuel economy in the indirect injection CI engine is inferior to a direct injection engine because of the increased heat losses through the relatively large surface area of the prechamber in the cylinder head and the cavity in the piston, in addition to gas-dynamic losses associated with the flow of high-velocity gases in narrow connecting passage(s).

The methodology for calculating the combustion process presented here is broadly based on the method described in Arkhangelsky et al. (1971, 1977) and can be used to model the combustion process in the practical cycles for CI, SI, and constant-pressure combustion engines. The practical cycle is essentially a refined fuel-air cycle with the combustion parameters determined from the energy equation. The energy (combustion) equation is written for the dual-combustion process $2 - 3 - 4$ in Figure 6.6 as follows:

Released fuel energy

$$= \textit{Change in internal energy of gases between 2 and 4}$$

$$+ \textit{work done by the gases between 3 and 4}$$

$$\varphi H_l = (M_P + M_{res})U_{P4} - M_R U_{R2} - M_{res}U_{P2} + W_{3-4} \tag{6.5}$$

where

φ: heat utilisation coefficient (of combustion) (≈ 0.8–0.9)

H_l: lower heating value of the fuel, kJ/kg

U_{P4}, U_{P2}: specific internal energy of the combustion products at points 4 and 2, respectively, $kJ/kmole$

U_{R2}: specific internal energy of the reactants (air) at point 2, $kJ/kmole$

M_P, M_R, M_{res}: quantities of the combustion products, reactants, and residual gases per kilogram of fuel, respectively, in $kmole$

W_{3-4}: work done per kilogram of fuel during expansion of the combustion products between points 3 and 4, kJ

The work done by the combustion products is

$$W_{3-4} = p_4 V_4 - p_3 V_3 = p_4 V_4 - p_3 V_2$$

Since $p_3 = \alpha \, p_2, p_3/p_2 = p_4/p_2$,

$$W_{3-4} = p_4 V_4 - \alpha p_2 V_2$$

or, using the equation of state

$$W_{3-4} = \overline{R} \left[(M_P + M_{res})T_4 - \alpha(M_R + M_{res})T_2 \right] \tag{6.6}$$

where $\overline{R}$ is the universal gas constant and α is the pressure rise during the constant-volume combustion process.

Substituting Eq. (6.6) into (6.5), rearranging, and dividing both sides by $(M_R + M_{res})$ yields

$$\frac{\varphi H_l}{M_R + M_{res}} + \frac{M_R U_{R2} + M_{res}U_{P2}}{M_R + M_{res}} + \alpha \overline{R} T_2 = \frac{M_P + M_{res}}{M_R + M_{res}}(U_{P4} + \overline{R} T_4)$$

Defining the coefficient of residual gases γ_{res} and the coefficient of molar change ψ as

$$\gamma_{res} = \frac{M_{res}}{M_R}, \psi = \frac{M_P + M_{res}}{M_R + M_{res}},$$

we finally obtain the combustion equation for the practical cycle for the CI engine:

$$\frac{\varphi H_l}{M_R(1 + \gamma_{res})} + \frac{U_{R2} + \gamma_{res}U_{P2}}{(1 + \gamma_{res})} + \alpha \overline{R} T_2 = \psi(U_{P4} + \overline{R} T_4) \tag{6.7}$$

In CI engines, $\gamma_{res} \approx 0.04$.

For a given fuel composition, relative air-fuel ratio (λ), and ratio of pressure rise at constant volume α, Eq. (6.7) can be solved for T_4 by iteration. When modelling actual engines, the value of the pressure rise ratio α is also needed as an input. The relationships between the combustion parameters can then be found from the equations of state at 2 and 4:

$$\frac{p_4 V_4}{p_2 V_2} = \frac{(M_P + M_{res})}{(M_R + M_{res})} \frac{T_4}{T_2} = \psi \frac{T_4}{T_2} \tag{6.8}$$

By definition, $p_4/p_2 = \alpha$, $V_4/V_2 = \beta$; therefore, for the dual cycle, Eq. (6.8) yields

$$\alpha\beta = \psi \frac{T_4}{T_2} \tag{6.9}$$

To find the quantities of the combustion products (in *kmole*), the method described in Chapter 5 for the fuel-air cycle can be used by assuming the following reaction of a generic hydrocarbon fuel $C_x H_y$ for $\lambda > 1$:

$$C_x H_y + \lambda \left(x + \frac{y}{4} \right) (O_2 + 3.7619 N_2)$$
$$= x CO_2 + \frac{y}{2} H_2 O + \left(x + \frac{y}{4} \right) (\lambda - 1) O_2 + 3.7619 \lambda \left(x + \frac{y}{4} \right) N_2$$

For 1 *kg* of fuel, the amount of reactants and products (in *kmole/kg fuel*) are, respectively,

$$M_R = \frac{4.7619 \lambda \left(x + \frac{y}{4} \right)}{(12 x + y)}$$

$$M_P = \frac{1}{(12 x + y)} \left[x + \frac{y}{2} + \left(x + \frac{y}{4} \right) (\lambda - 1) + 3.7619 \lambda \left(x + \frac{y}{4} \right) \right]$$

The relative air-fuel ratio λ determines the amount of excess air in the reactants and the amounts of oxygen and nitrogen in the products. The specific internal energy of the products is

$$U_P = \frac{1}{M_P} [M_{CO_2} U_{CO_2} + M_{H_2O} U_{H_2O} + M_{O_2} U_{O_2} + M_{N_2} U_{N_2}]$$

$M_{CO_2}, M_{H_2O}, M_{O_2}, M_{N_2}$ are the quantities (in *kmole*) of carbon dioxide, water vapour, oxygen, and nitrogen, respectively, in the products of complete combustion of 1 *kg* of fuel when $\lambda > 1$. $U_{CO_2}, U_{H_2O}, U_{O_2}, U_{N_2}$ are the specific internal energies (in *MJ/kmole*) of carbon dioxide, water vapour, oxygen, and nitrogen, respectively. Specific internal energies of air and common combustion products can be determined from the temperature-dependent polynomials in Table 2.7 or from Table A.2.

6.3.1.4 The Expansion Process

As stated earlier, the expansion process is polytropic, with an index $1.23 \leq n_2 \leq 1.3$. The pressure and temperature at the end of the process for the mixed-combustion cycle are given by

$$p_5 = \frac{p_4}{\delta^{n_2}} \tag{6.10}$$

$$T_5 = \frac{T_4}{\delta^{n_2-1}} \tag{6.11}$$

By definition, $\delta = V_5/V_4$ (expansion ratio), which is always less than ε for this cycle and n_2 is the polytropic index of the expansion process. Typical values of n_2, p_5, and T_5 are given in Table 5.2.

6.3.1.5 Cycle Work and Mean Effective Pressure

The work done by the practical dual cycle $1 - 2 - 3 - 4 - 5$ in Figure 6.6 is given by the same equation used for the fuel-air cycle, keeping in mind the differences in the values of the maximum cycle pressure and expansion ratio. To account for work losses at the characteristic points of the cycle represented by the shaded areas, a rounding-off coefficient σ is introduced. Thus, for the practical dual combustion cycle:

$$W_{cyc} = \sigma(p_1 \varepsilon^{n_1}) \left(\frac{V_1}{\varepsilon}\right) \left[\alpha(\beta - 1) + \frac{\alpha\beta}{n_2 - 1}\left(1 - \frac{1}{\delta^{n_2 - 1}}\right) - \frac{1}{n_1 - 1}\left(1 - \frac{1}{\varepsilon^{n_1 - 1}}\right)\right]$$

(6.12)

The mean effective pressure (*mep*) of the practical dual combustion cycle is

$$mep = p_1 \frac{\sigma \varepsilon^{n_1}}{(\varepsilon - 1)} \left[\alpha(\beta - 1) + \frac{\alpha\beta}{n_2 - 1}\left(1 - \frac{1}{\delta^{n_2 - 1}}\right) - \frac{1}{n_1 - 1}\left(1 - \frac{1}{\varepsilon^{n_1 - 1}}\right)\right]$$

(6.13)

Experimental results show that the values of the rounding-off coefficient σ fall within the range 0.90–0.97 for most engines. Lower to mid-range values yield good approximations of actual indicator diagrams of four-stroke internal combustion engines, and higher values yield good results in two-stroke engines (Dyechenko et al., 1974). If $\sigma = 1$, Eqs. (6.12) and (6.13) become identical in form to the equation for the fuel-air cycle.

Figure 6.9 demonstrates the applicability of the practical cycle to real engine operating conditions. The $p - V$ diagram, drawn to scale, gives an indication of the magnitude of cycle work losses at TDC and BDC. A rounding-off coefficient close to unity would be sufficient to obtain a good correlation between the two cycles.

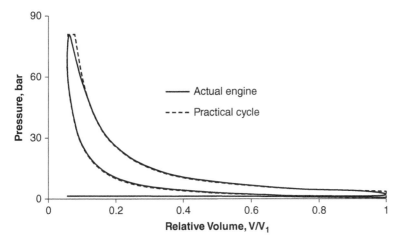

Figure 6.9 Calculated practical cycle and measured $p - V$ diagram for a CI engine (46 kW @ 2200 rpm).

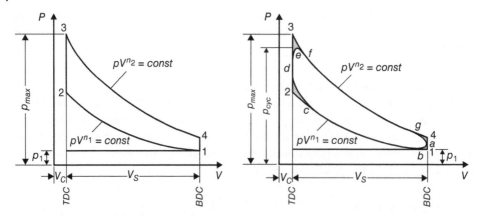

Figure 6.10 Practical cycle model of the SI engine.

6.3.2 Spark Ignition Engine (SI Engine)

A similar approach to the one previously discussed can be used to develop a model for the practical cycle of the SI engine. The model shown in Figure 6.10 assumes constant-volume heat addition along process 2–3; hence $\beta = 1$, and no work is done near TDC, i.e. the term W_{3-4} in Eq. (6.5) can be equated to zero.

The induction and compression processes are identical to the processes in the CI engine cycle in Section 6.3.1.

6.3.2.1 Combustion Process

In actual SI engines, the thermal energy input into the cycle is generated inside the cylinder close to the TDC at the end of the compression process. One of the biggest challenges engineers faced in the early stages of development of the piston engine was the design of a reliable fuel-metering system capable of delivering the precise amount of fuel at the precise time for optimal combustion. The carburettor reigned supreme as the fuel-metering system of choice in SI engines until it was superseded in the late 1980s by electronically controlled direct and indirect fuel injection systems. The preparation of the air-fuel mixture in most vehicles today occurs indirectly (outside the cylinder) by injecting finely atomised fuel into the inlet manifold as the inlet valve opens; the fuel is drawn into the combustion chamber by the incoming air (Figure 6.11). The subsequent ignition by a spark plug and combustion of the air-fuel mixture at almost constant volume deliver the required thermal energy, which can be controlled by varying the amount of air drawn into the cylinder by means of a throttle valve in the inlet manifold while maintaining the air-fuel ratio within the required range.

Accounting for the differences in the numbering of the characteristic points of this cycle as shown in Figure 6.10, the combustion equation for the SI engine can be written as

$$\frac{\varphi H_l}{M_R(1 + \gamma_{res})} + \frac{U_{R2} + \gamma_{res} U_{P2}}{(1 + \gamma_{res})} = \psi U_{P3} \tag{6.14}$$

For SI engines, $\gamma_{res} \approx 0.06$.

Equation (6.14) is valid when the air-fuel mixture is stoichiometric or lean ($\lambda \geq 1$) and the combustion is complete. SI engines operate part of the time with fuel-rich mixtures ($\lambda < 1$),

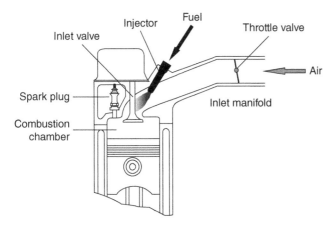

Figure 6.11 Fuel delivery and combustion scheme in an actual SI engine.

and there is not sufficient air for all the fuel to react, causing a reduction of the combustion heat from φH_l to $\varphi(H_l - \Delta H_l)$. The combustion equation for this case is

$$\frac{\varphi(H_l - \Delta H_l)}{M_R(1 + \gamma_{res})} + \frac{U_{R2} + \gamma_{res} U_{P2}}{(1 + \gamma_{res})} = \psi U_{P3} \tag{6.15}$$

The value of ΔH_l depends on the amount of products of incomplete combustion in the cylinder, such as hydrogen and carbon monoxide. The following equation can be used to estimate ΔH_l (Arkhangelsky et al., 1971):

$$\Delta H_l = 120(1 - \lambda)(A/F)_s^v \ MJ/kg \tag{6.16}$$

The relationship between the combustion parameters for the constant-volume cycle for which $\beta = 1$ is given by

$$\alpha = \psi \frac{T_3}{T_2} \tag{6.17}$$

The quantity of the combustion products (in *kmole*) depends on the value of the relative air-fuel ratio. If $\lambda > 1$, the following reaction of a hydrocarbon fuel $C_x H_y$ can be assumed, and the quantities of the combustion products and their thermodynamic properties (internal energy) determined as in the case of the CI engine:

$$C_x H_y + \lambda \left(x + \frac{y}{4}\right)(O_2 + 3.7619 N_2)$$
$$= xCO_2 + \frac{y}{2}H_2O + \left(x + \frac{y}{4}\right)(\lambda - 1)O_2 + 3.7619\lambda \left(x + \frac{y}{4}\right)N_2$$

Note that the products include $(x + y/4)(\lambda - 1)O_2$ *kmole* of unused oxygen, which is directly proportional to the value of λ decreasing to zero when $\lambda = 1$.

If $\lambda < 1$ and the combustion reaction is assumed to produce CO_2, CO, H_2O, H_2, and N_2, the combustion equation can be solved using Eq. (2.35a) to calculate the amount of products in conjunction with internal energy data from polynomials in Table 2.7 or Table A.2

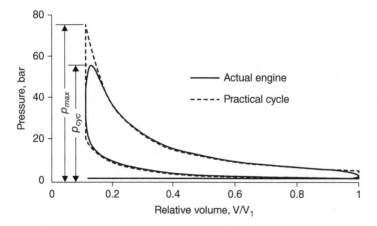

Figure 6.12 Calculated practical cycle and measured p-V diagram for a SI engine (15.8 *kW* @4000 *rpm*).

6.3.2.2 Expansion Process
The pressure and temperature at the end of the expansion process for the constant-volume cycle, considering that $\delta = \varepsilon$, are given by

$$p_4 = \frac{p_3}{\varepsilon^{n_2}} \tag{6.18}$$

$$T_4 = \frac{T_3}{\varepsilon^{n_2-1}} \tag{6.19}$$

6.3.2.3 Cycle Work and Mean Effective Pressure
The cycle work and *mep* can be obtained by substituting $\beta = 1$ in Eqs. (6.12) and (6.13):

$$W_{cyc} = \sigma(p_1\varepsilon^{n_1})\left(\frac{V_1}{\varepsilon}\right)\left[\frac{\alpha}{n_2-1}\left(1-\frac{1}{\varepsilon^{n_2-1}}\right)-\frac{1}{n_1-1}\left(1-\frac{1}{\varepsilon^{n_1-1}}\right)\right] \tag{6.20}$$

$$mep = p_1\frac{\sigma\varepsilon^{n_1}}{(\varepsilon-1)}\left[\frac{\alpha}{n_2-1}\left(1-\frac{1}{\delta^{n_2-1}}\right)-\frac{1}{n_1-1}\left(1-\frac{1}{\varepsilon^{n_1-1}}\right)\right] \tag{6.21}$$

Figure 6.12 compares the practical cycle and measured $p - V$ diagram of a low-output SI engine. Unlike the cycle for the CI engine, the computed maximum cycle pressure p_{max} is higher than the actual engine cycle p_{cyc} ($p_{cyc} \approx 0.85 p_{max}$).

6.3.3 Constant-Pressure Combustion Engine

There are no actual engines that operate on the constant-pressure cycle (in which the heat is wholly added at constant pressure). Low-speed CI engines, such as those used in marine propulsion and large diesel generators, come close. Nonetheless, if such an engine could exist, the model shown in Figure 6.13 would apply. The induction and compression processes are not shown here as they are identical to the processes for CI and SI engines.

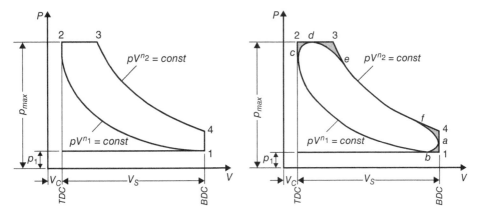

Figure 6.13 Practical cycle model for a low-speed CI engine.

6.3.3.1 Combustion Process

The combustion equation for this cycle can be derived from the equation for the dual-cycle model by equating α to unity and replacing U_{P4} and T_4 by U_{P3} and T_3 in Eq. (6.7):

$$\frac{\varphi H_l}{M_R(1+\gamma_{res})} + \frac{U_{R2}+\gamma_{res}U_{P2}}{(1+\gamma_{res})} + \overline{R}T_2 = \psi(U_{P3}+\overline{R}T_3) \tag{6.22}$$

The relationships between the parameters for the constant-volume cycle for which $\alpha = 1$ are

$$\beta = \frac{T_3}{T_2} \text{ and } \delta = \frac{\varepsilon}{\beta}, \text{where } \delta \text{ is the expansion ratio.}$$

These engines are most likely of the CI type, and the air-fuel mixture is always lean ($\lambda > 1$); hence the procedure to determine the mixture composition, find the internal energies, and solve Eq. (6.22) for T_3 is the same as for the CI engine.

6.3.3.2 Expansion Process

The pressure and temperature at the end of the expansion process are

$$p_4 = \frac{p_3}{\delta^{n_2}} \tag{6.23}$$

$$T_4 = \frac{T_3}{\delta^{n_2-1}} \tag{6.24}$$

The expansion parameters shown in Table 5.2 are applicable to this case if we take into account the numbering of the characteristic points of the cycle.

6.3.3.3 Cycle Work and Mean Effective Pressure

The net work done in the constant-pressure combustion cycle is

$$W_{cyc} = (p_1\varepsilon^{n_1})\left(\frac{V_1}{\varepsilon}\right)\left[\alpha(\beta-1)+\frac{\alpha\beta}{n_2-1}\left(1-\frac{1}{\delta^{n_2-1}}\right)-\frac{1}{n_1-1}\left(1-\frac{1}{\varepsilon^{n_1-1}}\right)\right] \tag{6.25}$$

And the *mep* is

$$mep = p_1\frac{\varepsilon^{n_1}}{(\varepsilon-1)}\left[(\beta-1)+\frac{\beta}{n_2-1}\left(1-\frac{1}{\delta^{n_2-1}}\right)-\frac{1}{n_1-1}\left(1-\frac{1}{\varepsilon^{n_1-1}}\right)\right] \tag{6.26}$$

6.4 Cycle Comparison

Figure 6.14 shows a comparison of the air-standard, fuel-air, and practical cycles of the dual-combustion configuration on $p - V$ coordinate systems, and Table 6.1 summarises the operating and performance parameters. The rounding-off losses in the practical cycle are not taken into account in this comparison.

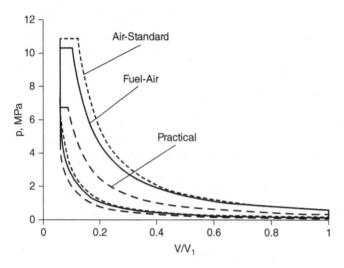

Figure 6.14 Comparison of the air-standard, fuel-air, and practical dual-combustion cycles (practical cycle: $\varphi = 0.9$, $\gamma_{res} = 0.04$).

Table 6.1 Operating and performance parameters of the dual-combustion air-standard, fuel-air, and practical cycles shown in Figure 6.14.

	Air-standard cycle	Fuel-air cycle	Practical cycle
Fuel	—	$C_{12}H_{21.6}$	$C_{12}H_{21.6}$
Heat input	2110 kJ/kg air	43200 kJ/kg fuel	43200 kJ/kg fuel
Relative air-fuel ratio, λ	—	1.4	1.4
Ratio of pressure rise, α	1.8	1.8	1.8
Ratio of volume increase, β	1.975	1.668	1.441
Compression ratio, ε	16	16	16
Compression index	$\gamma = 1.4$	$n_1 = 1.35$	$n_1 = 1.35$
Expansion index	$\gamma = 1.4$	$n_2 = 1.25$	$n_2 = 1.25$
Thermal efficiency, %	0.628	0.507	0.442
Mean effective pressure, MPa	2.068	1.577	0.929
Maximum temperature, K	3211	2246	2205
Maximum pressure, MPa	10.91	10.32	6.76

The thermal efficiency, *mep*, maximum temperature, and maximum pressure decrease significantly as we move from the air-standard cycle to the fuel-air cycle and finally to the practical cycle. The significant differences between the cycles are attributable to the assumptions made for each cycle, and these differences will change with the assumed values of the relative air-fuel ratio λ, ratio of pressure rise α, compression ratio ε, and polytropic indexes n_1 and n_2.

6.5 Cycles Based on Combustion Modelling (Wiebe Function)

Analytical functions approximating the burn rate (heat release) in IC engines are useful and cost-effective tools for more realistic engine cycle simulations beyond the cycles discussed so far. Most functions proposed to date are derivatives of the law of normal distribution of a continuous random variable, and the best known and most widely used in engine applications is the Wiebe function. A detailed review of this function's origins and applications is given by Ghojel (2010) based on the original work by Ivan Wiebe (1956, 1962). This function in its different forms can be used to describe the cumulative and rate of heat release at any instant during the combustion process in the four-stroke engine ($f - g - h$ ... in Figure 6.3c) and during the combustion process in the two-stroke engine ($1 - b - c$ ... in Figure 6.4e). The dots indicate that the exact end of the combustion is unknown due to its variability with engine operating conditions. The other processes – induction, compression, and exhaust – are the same as in the practical cycle.

6.5.1 The Wiebe Function

The cumulative heat released $x(t)$ at any instant in time is the fraction of the total heat released during the combustion process:

$$x(t) = \frac{q_C(t)}{q_T(t)} = \frac{m_{f, burnt}}{m_{f, total}} \tag{6.27}$$

where

$q_C(t)$, $m_{f, burnt}$: time-dependent cumulative heat released and fuel burnt at any instant from the start of combustion

$q_T(t)$, $m_{f, total}$: total heat released and fuel burnt during combustion

The Wiebe functions for the cumulative and rate of heat release as functions of time t are, respectively,

$$x(t) = 1 - e^{-6.908\left(\frac{t}{t_d}\right)^{m+1}} \tag{6.28}$$

$$w(t) = \frac{dx(t)}{dt} = \frac{6.908(m + 1)}{t_d}\left(\frac{t}{t_d}\right)^m e^{-6.908\left(\frac{t}{t_d}\right)^{m+1}} \quad 1/s \tag{6.29}$$

where t is the instantaneous time measured from start of combustion and t_d is the duration of the combustion process. Parameter m (often referred to as the *shape factor*) determines the time it takes for maximum rate of heat release to be reached and the shape of function (6.29). Indicator diagrams of piston engines are usually plotted as pressure versus crank-angle degrees; hence, the Wiebe functions are usually expressed in terms of crank-angle degrees (ϕ) as follows:

$$x(\phi) = 1 - e^{-6.908\left(\frac{\phi}{\phi_d}\right)^{m+1}} = 1 - exp\left[-6.908\left(\frac{\phi}{\phi_d}\right)^{m+1}\right] \tag{6.30}$$

$$w(\phi) = \frac{dx(\phi)}{d\phi} = \frac{6.908(m+1)}{\phi_d}\left(\frac{\phi}{\phi_d}\right)^{m} e^{-6.908\left(\frac{\phi}{\phi_d}\right)^{m+1}} \tag{6.31}$$

In these correlations, ϕ is the instantaneous crank angle measured from start of combustion, and ϕ_d is the duration of the combustion process. Figure 6.15 shows plots of Eqs. (6.30). and (6.31) for fixed combustion duration ϕ_d and variable shape factor m, and Figure 6.16 shows plots for constant combustion shape factor $m = 3$ and variable duration ϕ_d.

The cumulative heat release increases continuously and reaches a maximum at the end of the combustion process. The rate of heat release is bell shaped with a pronounced maximum point whose value and location are dependent on both the shape factor and the combustion duration.

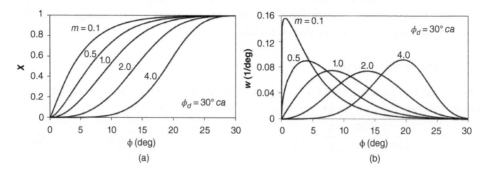

Figure 6.15 Wiebe cumulative heat release (a) and rate of heat release (b) for constant combustion duration $\phi_d = 30°$ and variable shape factor m.

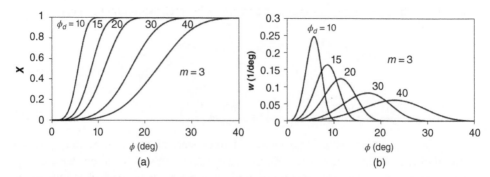

Figure 6.16 Wiebe cumulative heat release (a) and rate of heat release (b) for constant combustion shape factor $m = 3$ and variable duration ϕ_d.

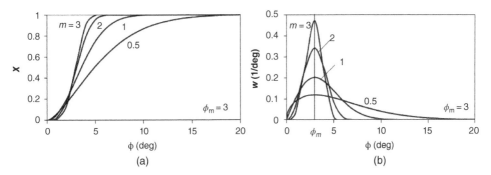

Figure 6.17 Wiebe cumulative heat release (a) and rate of heat release (b) for constant angle $\phi_m = 3$ and variable combustion shape factor m.

Equations (6.30) and (6.31) are the Wiebe functions most frequently used in combustion modelling studies. Other useful forms of the Wiebe functions can be written as

$$x(\phi) = 1 - exp\left[\left(-\frac{m}{m+1}\right)\left(\frac{\phi}{\phi_m}\right)^{m+1}\right] \tag{6.32}$$

and

$$w(\phi) = \frac{dx(\phi)}{d\phi} = \frac{m}{\phi_m}\left(\frac{\phi}{\phi_m}\right)^m e^{-\frac{m}{m+1}\left(\frac{\phi}{\phi_m}\right)^{m+1}}$$

The last equation is usually written as

$$w(\phi) = \frac{m}{\phi_m}\left(\frac{\phi}{\phi_m}\right)^m exp\left[\left(-\frac{m}{m+1}\right)\left(\frac{\phi}{\phi_m}\right)^{m+1}\right] \tag{6.33}$$

where ϕ_m is the angle at which the rate of heat release reaches its maximum value. For a given combustion duration, the time it takes (in seconds or crank-angle rotations) for the rate of heat release to reach a maximum is given by

$$t_m = t_d\left[\frac{m}{6.908(m+1)}\right]^{\frac{1}{m+1}} \tag{6.34}$$

$$\phi_m = \phi_d\left[\frac{m}{6.908(m+1)}\right]^{\frac{1}{m+1}} \tag{6.35}$$

Figure 6.17 shows plots of Eqs. (6.32) and (6.33) for constant heat-release rate peak angle ϕ_m and different values of the shape factor m.

6.5.2 Cycle Calculation Using Wiebe Function

Consider the $p-\theta$ and $p-V$ diagrams and Wiebe heat-release characteristics in Figures 6.18 and 6.19. In Figure 6.18a, the calculations are carried out incrementally from left to right with the crank angle θ increasing from 180° to 540° to include the compression, combustion, and expansion processes. The shaded small areas in Figure 6.18 represent the incremental pressure and volume change $1-2$ with the crank angle θ. Figure 6.19 shows the cumulative and rate of heat release as functions of angle ϕ whose origin is at the start

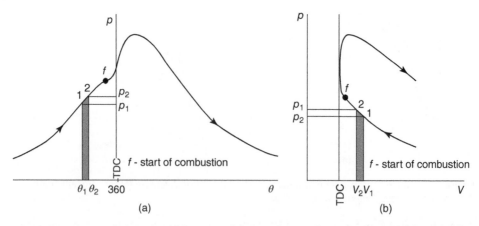

Figure 6.18 Schematic diagrams of pressure development during combustion: (a) $p - \theta$ diagram; (b) $p - v$ diagram.

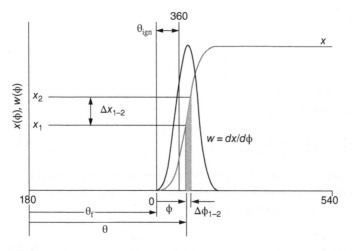

Figure 6.19 Schematic diagram of the application of the Wiebe function to combustion calculations.

of combustion at point f. The width of the elemental increment in Figure 6.18a is the same as in Figure 6.19 and usually taken equal to 1° ca, and the relationship between θ and ϕ can be written as $\phi = \theta - \theta_f = \theta - (360 - \theta_{ign})$, where θ_{ign} is the ignition angle (the angle at which the fuel ignites and the combustion process starts). The starting point of the Wiebe function ($\phi = 0$) is $\theta_f = (360 - \theta_{ign})$. In SI engines, the ignition angle corresponds closely to the spark timing. In CI engines, the ignition angle is smaller than the fuel injection timing, with the difference between them depending on the ignition lag or ignition delay (the time it takes for the diesel fuel to mix with air, evaporate, and ignite).

The first law of thermodynamics applied to a non-flow system (closed system) is usually written as

$$Q = \Delta U + W + Q_l \tag{6.36}$$

where

Q: heat transfer

ΔU: change of internal energy

W: work transfer

Q_l: heat losses

Applied to the increment $1 - 2$ shown by the shaded area in Figure 6.18b, the first law can be written as

$$q_{1-2} = c_{v(1-2)}(T_2 - T_1) + \int_{v_1}^{v_2} p dv + q_{l(1-2)} \tag{6.37}$$

where

q_{1-2}: heat from the combustion of fuel within the increment, kJ/kg

$c_{v(1-2)}$: average specific heat at a constant volume of the working fluid within increment $1-2$, $kJ/kg\ K$

$\int_{v_1}^{v_2} p dv$: specific work done by the working fluid over the incremental volume change Δv_{1-2}, kJ/kg

$q_{l(1-2)}$: incremental heat losses due to incomplete combustion, heat transfer to the cooling system, and dissociation, kJ/kg

v: specific volume of the working fluid, m^3/kg

Applying the equation of state to the first term on the right-hand side of Eq. (6.37), we can write

$$c_{v(1-2)}(T_2 - T_1) = \frac{c_{v(1-2)}}{R_{g(1-2)}}(p_2 v_2 - p_1 v_1)$$

Knowing that

$$c_p - c_v = R_g, \quad \gamma = \frac{c_p}{c_v}, \quad \frac{c_v}{R_g} = \frac{1}{\gamma - 1}, \quad \text{or} \quad \frac{c_{v(1-2)}}{R_{g(1-2)}} = \frac{1}{\gamma_{1-2} - 1}$$

we finally obtain

$$c_{v(1-2)}(T_2 - T_1) = \frac{1}{\gamma_{1-2} - 1}(p_2 v_2 - p_1 v_1) \tag{6.38}$$

The ratio of specific heats in real engines will vary with mixture composition and temperature. For an average pressure $(p_2 + p_1)/2$, the incremental work done between points 1 and 2 is

$$\int_{v_1}^{v_2} p dv = \frac{p_2 + p_1}{2}(v_2 - v_1) \tag{6.39}$$

The incremental work will be positive or negative depending whether the volume is increasing or decreasing.

The net incremental heat release, using the Wiebe cumulative function, is

$$q_{1-2} = q_z(x_2 - x_1) = q_z \Delta x_{1-2} \tag{6.40}$$

where

q_z: heat supplied to the cycle per kg of working fluid, kJ/kg

x_1: cumulative heat-release fraction to the beginning of increment $1-2$

x_2: cumulative heat-release fraction to the end of increment $1-2$

Δx_{1-2}: cumulative heat-release fraction over increment $1-2$

For the combustion of 1 kg of fuel,

$$q_z = \frac{H_l}{(1 + \gamma_{res})[\lambda(A/F)_s^g + 1]} \, kJ/kg \tag{6.41}$$

where

H_l: lower heating value of the fuel, kJ/kg or MJ/kg
γ_{res}: coefficient of residual gases
λ: relative air-fuel ratio (inverse of the equivalence ratio)
$(A/F)_s^g$: stoichiometric air-fuel ratio, $kg(air)/kg(fuel)$

The main component of the heat losses is the heat transfer by conduction and radiation from the gases to the cylinder walls and cylinder head and then to the cooling water. If the overall heat-transfer coefficient is known, the heat losses can be estimated from

$$q_{l(1-2)} = \frac{\Delta(hA)(T_g - T_w)}{m_{f(cyc)}(1 + \gamma_{res})[\lambda(A/F)_s^g + 1]} \left(\frac{1}{6N}\right) \Delta\theta \tag{6.42}$$

where

h: heat-transfer coefficient, W/m^2K
A: exposed heat-transfer surface, m^2
T_g: average gas temperature, K
T_w: average wall temperature (could be taken as average cooling water temperature), K
$m_{f(cyc)}$: fuel consumption per cycle, kg

Combining Eqs. (6.37), (6.38), (6.39), and (6.40), we obtain

$$q_z \Delta x_{1-2} - q_{l(1-2)} = \frac{1}{\gamma_{1-2} - 1}(p_2 v_2 - p_1 v_1) + \frac{p_2 + p_1}{2}(v_2 - v_1)$$

$$2q_z \Delta x_{1-2} - 2q_{l(1-2)} = \frac{2p_2 v_2}{\gamma_{1-2} - 1} - \frac{2p_1 v_1}{\gamma_{1-2} - 1} + p_2 v_2 - p_2 v_1 + p_1 v_2 - p_1 v_1$$

$$2q_z \Delta x_{1-2} - 2q_{l(1-2)} = p_2 \left(\frac{2v_2}{\gamma_{1-2} - 1} + v_2 - v_1\right) - p_1 \left(\frac{2v_1}{k\gamma_{1-2} - 1} + v_1 - v_2\right)$$

$$2q_z \Delta x_{1-2} - 2q_{l(1-2)} = p_2 \left[\left(\frac{2}{\gamma_{1-2} - 1} + 1\right)v_2 - v_1\right] - p_1 \left[\left(\frac{2}{\gamma_{1-2} - 1} + 1\right)v_1 - v_2\right]$$

Finally,

$$p_2 = \frac{[2q_z \Delta x_{1-2} - 2q_{l(1-2)}] + p_1 \left(\frac{\gamma_{1-2} + 1}{\gamma_{1-2} - 1}v_1 - v_2\right)}{\frac{\gamma_{1-2} + 1}{\gamma_{1-2} - 1}v_2 - v_1} \tag{6.43}$$

For the incremental change in cumulative heat-release fraction, either Eq. (6.30) or (6.32) can be used:

$$\Delta x_{1-2} = x_2 - x_1 = exp\left[-6.908\left(\frac{\phi_2}{\phi_d}\right)^{m+1}\right] - exp\left[-6.908\left(\frac{\phi_1}{\phi_d}\right)^{m+1}\right] \tag{6.44}$$

$$\Delta x_{1-2} = x_2 - x_1 = exp\left[\left(-\frac{m}{m+1}\right)\left(\frac{\phi_2}{\phi_m}\right)^{m+1}\right] - exp\left[\left(-\frac{m}{m+1}\right)\left(\frac{\phi_1}{\phi_m}\right)^{m+1}\right] \tag{6.45}$$

When building a database for the empirical parameters of the Wiebe function from heat-release analysis of experimental indicator diagrams, the angle at which the rate of heat release reaches its maximum (ϕ_m) is easier to measure than the angle at which combustion ends (ϕ_d). Therefore, it is preferable to use Eq. (6.45) when determining the incremental change in cumulative heat release.

The specific volume (volume per unit mass of reactants) at the start of compression (end of induction process) is

$$v_1 = \frac{V_1}{m_1} = \bar{R}\frac{M_1}{m_1}\frac{T_1}{p_1} \tag{6.46}$$

where

$$m_1 = m_R + m_{res} = (1 + \gamma_{res})m_R$$

The quantity of the air-fuel mixture in *kmole* is

$$M_1 = M_R + M_{res} = (1 + \gamma_{res})M_R$$

Hence,

$$\frac{M_1}{m_1} = \frac{M_R}{m_R}, \text{ and}$$

$$v_1 = \bar{R}\frac{M_R T_1}{m_R p_1}$$

$$M_R = \lambda(A/F)_s^v + \frac{1}{\mu_f} = \frac{\lambda(A/F)_s^g}{\mu_a} + \frac{1}{\mu_f} \text{ kmole}$$

$$m_R = \lambda(A/F)_s^g + 1 \text{ kg}$$

Finally, the specific volume can be written as

$$v_1 = \bar{R}\frac{T_1}{p_1}\frac{\left(\dfrac{\lambda(A/F)_s^g}{\mu_a} + \dfrac{1}{\mu_f}\right)}{[\lambda(A/F)_s^g + 1]} \text{ m}^3/\text{kg} \tag{6.47}$$

To calculate the instantaneous volume of the working fluid and the heat-transfer surface, reference is made to the piston-crank mechanism in Figure 6.20.

From the convention used for engine cycles, the volume at the start of compression is V_1 and at the end of compression is V_2 (also known as the *clearance volume V_c*), and the compression ratio is defined as

$$\varepsilon = \frac{V_1}{V_2} = \frac{V_1}{V_c}$$

from which

$$V_c = \frac{V_1}{\varepsilon}$$

Also

$$\varepsilon = \frac{V_1}{V_c} = \frac{V_1}{V_1 - V_s}$$

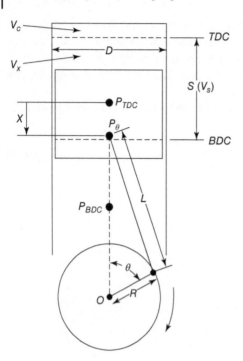

Figure 6.20 Piston-crank mechanism of the reciprocating engine.

From the last equation, we can determine V_s as

$$V_s = \left(\frac{\varepsilon - 1}{\varepsilon}\right) V_1$$

The piston travel as a function of the crank angle is

$$V_x = V_c + V_x = V_c + x A_p$$

where A_p is the cylinder cross-sectional area (or the piston surface area if that area is flat).

The exact relationship for piston displacement from the kinematics of the piston-crank mechanism is

$$x = R\left\{1 + \frac{1}{(R/L)} - \left(\cos\theta + \frac{1}{(R/L)}\sqrt{1 - (R/L)^2\sin^2\theta}\right)\right\} = R\sigma \qquad (6.48)$$

An approximate relationship for piston displacement of the following form is also often used:

$$x \approx R\left\{1 + \frac{(R/L)}{4} - \left(\cos\theta + \frac{(R/L)}{4}\cos2\theta\right)\right\} \approx R\sigma \qquad (6.49)$$

where
  R: crank radius
  L: length of connecting rod
  θ: crank-angle position
  σ: the value bounded by the external brackets in Eqs. (6.48) and (6.49)

$$V_x = V_c + x A_p = V_c + R A_p \sigma = V_c + \frac{A_p S}{2}\sigma = V_c + \frac{V_s}{2}\sigma$$

Substituting for V_c and V_s in the previous equation, we obtain

$$V_x = \frac{V_1}{\epsilon} \left(1 + \frac{\epsilon - 1}{2} \sigma \right)$$

This equation can be written in terms of specific volumes instead of absolute volumes as

$$v = \frac{v_1}{\epsilon} \left(1 + \frac{\epsilon - 1}{2} \sigma \right) \tag{6.50}$$

The instantaneous specific volume can now be found by substituting for the values of v_1 from Eq. (6.47) and σ into Eq. (6.50):

$$v = \frac{\overline{R} \, T_1}{\epsilon \, p_1} \frac{\left(\dfrac{\lambda (A/F)_s^g}{\mu_a} + \dfrac{1}{\mu_f} \right)}{[\lambda (A/F)_s^g + 1]} \left\{ 1 + \frac{\epsilon - 1}{2} \left[1 + \frac{(R/L)}{4} - \left(\cos\theta + \frac{(R/L)}{4} \cos 2\theta \right) \right] \right\} \tag{6.51}$$

The equation of state of the working fluid at the start of combustion at point f is

$$p_f V_f = \overline{R}(M_1 + M_r)T_f$$

The equation of state at any point during the combustion process, assuming negligible molar change for the working fluid, is

$$pV = \overline{R}(M_1 + M_r)T$$

The last two equations yield the temperature of the working fluid at any point during the combustion process if the pressure and volume are known:

$$T = p \frac{V T_f}{p_f V_f}$$

The previous equation can be rewritten in terms of the specific volumes as

$$T = p \frac{v T_f}{p_f v_f}$$

The temperature of the working fluid at the end of the first increment ($f - 2$) after the start of the combustion process can be determined from

$$T_2 = p_2 \frac{v_2 T_f}{p_f v_f} \tag{6.52}$$

All subsequent increments will be denoted $(1 - 2)$ and calculated sequentially using Eq. (6.43) for the pressure and Eq. (6.52) for the temperature after replacing f with 1. If precision is a prerequisite, calculation of p_2 and T_2 should be conducted in parallel, which will involve some iteration.

Equation (6.43) can be simplified by combining the first two terms in the numerator and accounting for all the losses by the factor φ:

$$p_2 = \frac{2 q_z \varphi \Delta x_{1-2} + p_1 \left(\dfrac{\gamma_{1-2} + 1}{\gamma_{1-2} - 1} v_1 - v_2 \right)}{\dfrac{\gamma_{1-2} + 1}{\gamma_{1-2} - 1} v_2 - v_1} \tag{6.53}$$

where φ is the coefficient of heat utilisation.

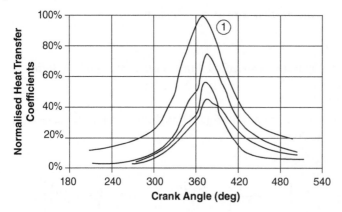

Figure 6.21 Comparison of estimated heat-transfer coefficients in reciprocating engines.

From Eqs. (6.43) and (6.53), for each increment we can write

$$x_{l(1-2)} = x_{1-2}(1 - \varphi) \tag{6.54}$$

At any instant of time (or crank-angle position), Eq. (6.54) can be written in terms of the cumulative heat-release fraction:

$$x_l = x_i(1 - \varphi) \tag{6.55}$$

and in terms of cumulative specific heat release,

$$q_l = q_i(1 - \varphi) \tag{6.56}$$

i.e. the losses at any point in the heat-cycle calculations constitute a fixed fraction of the total heat release.

The coefficient φ is essentially the ratio of the apparent (net) to the gross cumulative heat release. Analysis of experimentally obtained heat-release characteristics shows that, for a given engine, this ratio remains almost constant over the duration of the combustion process. In CI engines, the empirical coefficient φ changes between 0.7 and 0.9, depending on the type of engine and operating conditions (Sharaglazov et al., 2004). The justification for this approach to the calculation of heat losses can be seen by inspecting Figure 6.21, which shows a comparison between the predicted heat-transfer coefficients, normalised to the maximum value of plot 1 by four empirical models. The wide variation of the heat-transfer coefficient with the crank angle indicates the difficulty of accurately estimating heat-transfer losses and the likely wide variation of the predicted indicator diagrams.

6.6 Example of Wiebe Function Application

The methodology presented for the calculation of the combustion process using the Wiebe function together with the calculation method of the induction, compression, and expansion processes in the practical cycle will be applied to predict the indicator diagrams in real SI and CI engines.

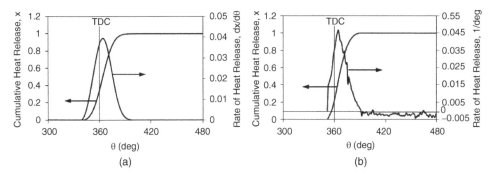

Figure 6.22 Heat-release characteristics for a SI engine (16 *kW* @ 4000 *rpm*): (a) Wiebe characteristics: $\varphi = 0.8$, $m = 2$, $\phi_m = 26$ *deg*; (b) experimental characteristics.

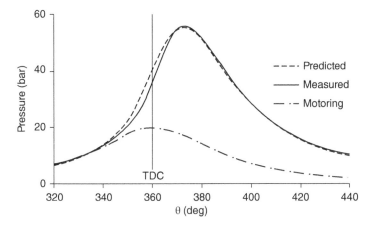

Figure 6.23 Calculated and predicted indicator diagrams for a SI engine (16 *kW* @ 4000 *rpm*).

6.6.1 SI Engine

The SI engine in this example is a small engine producing 16 *kW* at 4000 *rpm*, and the results are shown in Figures 6.22 and 6.23. The functions used in the example are Eqs. (6.32, 6.33), with the independent variable ϕ replaced by θ from $\phi = \theta - \theta_f$, where $\theta_f = 360 - \theta_{ign}$, from which $dx/d\theta = dx/d\phi$. The shape factor $m = 2$, and the angle at which maximum rate of heat release is reached $\phi_m = 26$. Figure 6.22a shows plots of the Wiebe functions, and Figure 6.22b shows the calculated heat-release characteristics for the engine from measured cylinder pressure data. Apart from the fluctuations of the experimental heat-release rate, caused by the small fluctuations of the measured pressure inside the cylinder, the experimental and predicted characteristics are quite similar. Comparison of the predicted and experimental indicator diagrams in Figure 6.23 shows relatively good agreement.

6.6.2 CI Engine

Wiebe functions (6.30, 6.31) are used in this case with shape factor $m = 0.2$ and combustion duration $\phi_d = 75$ *deg*. Comparison of the theoretical and calculated heat-release

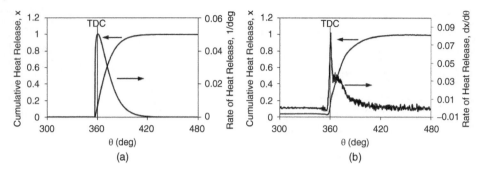

Figure 6.24 Heat-release characteristics for a CI engine (57.6 kW @ 2200 rpm): (a) Wiebe heat-release characteristics: $\varphi = 0.65$ $m = 0.2$, $\phi_d = 75$ deg; (b) experimental heat-release characteristics.

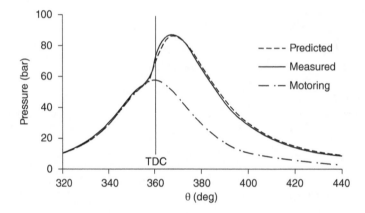

Figure 6.25 Calculated and predicted indicator diagrams for a CI engine (57.6 kW @ 2200 rpm).

characteristics are shown in Figure 6.24. There are significant differences between the theoretical (Wiebe) and experimental heat-release rate characteristic and the experimental results, indicating the presence of a second peak at a much lower value than the first, which is reached within very short period of time. However, despite this discrepancy, carefully selected Wiebe parameters can predict the cylinder pressure with reasonable accuracy, as shown in Figure 6.25.

6.7 Double Wiebe Models

Generally, heat-release characteristics in SI engines can be modelled using a single Wiebe function, as was shown earlier with reasonable results that represent a step up from the thermodynamic models discussed in Section 6.3.2. Analysis of the experimental data collected for the most widely used CI engines, known as direct injection CI engines (Figure 6.24b is a good example) indicates that a single Wiebe function does not represent the real state of affairs as far as the actual heat-release characteristics are concerned, particularly the heat-release rate. This can be seen by comparing Figures 6.24a,b, which clearly indicate the

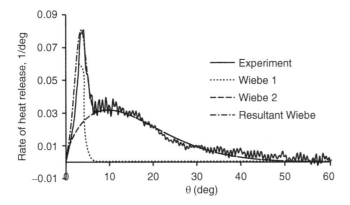

Figure 6.26 Double Wiebe representation of the experimental heat-release rate in a direct injection CI engine.

presence of two distinct combustion phases in direct injection CI engines: a rapid combustion phase and a diffusion combustion phase.

6.7.1 Rapid Combustion Phase

High-pressure fuel is injected into the cylinder cavity in the form of a finely atomised spray (see Figure 6.8). As the fuel penetrates the compressed air in the cylinder just before TDC, it starts evaporating; part of the injected fuel mixes with the hot air, forming a combustible mixture that eventually self-ignites (without any external source of heat, such as a spark plug) after a period of time known as *ignition lag* or *ignition delay*. The combustible mixture burns very rapidly over a short period of time, causing the sharp spike in heat-release rate shown in Figure 6.26 followed by equally rapid decrease in heat release rate as the initial mixture is burned out. The magnitude of the heat-release rate spike is affected by the ignition delay, amount of fuel injected during the ignition delay, and temperature of the compressed air at the start of fuel injection.

6.7.2 Diffusion Phase

This phase starts with continuing fuel injection and gradual formation of a combustible mixture, which causes the heat-release rate to increase again, albeit at a much-reduced rate compared with the first phase. The heat-release rate during this phase reaches a maximum then starts decreasing again, approaching zero at the end of the combustion process.

The combustion process described can be modelled using two Wiebe functions having the same origin of either of the following two forms of the cumulative heat-release fraction:

$$x(\phi) = A_1 \left[1 - e^{-6.908\left(\frac{\phi}{\phi_{d1}}\right)^{m_1+1}} \right] + A_2 \left[1 - e^{-6.908\left(\frac{\phi}{\phi_{d2}}\right)^{m_2+1}} \right] \tag{6.57}$$

$$x(\phi) = A_1 \left[1 - e^{-\frac{m_1}{m_1+1}\left(\frac{\phi}{\phi_{m1}}\right)^{m_1+1}} \right] + A_2 \left[1 - e^{-\frac{m_2}{m_2+1}\left(\frac{\phi}{\phi_{m2}}\right)^{m_2+1}} \right] \tag{6.58}$$

Table 6.2 Double Wiebe parameter values.

Parameter	Wiebe 1	Wiebe 2	Resultant Wiebe
m	2.2	0.6	—
ϕ_d	7	60	—
A_1	—	—	0.177
A_2	—	—	0.75

A_1 and A_2 are parameters that are dependent on engine and performance characteristics. Equation (6.57) and its first derivative are used in the example shown in Figure 6.26 to illustrate the application of a double-Wiebe function to a CI engine. The values of the function parameters are shown in Table 6.2.

The values in Table 6.2 selected for the Wiebe functions here are quite arbitrary, and the purpose of the discussion is to illustrate the methodology. However, a significant body of research was done in an attempt to establish a correlation between these parameters and engine design and operating conditions such as work by Watson et al. (1980), Ghojel (1974, 1982), Miyamoto et al. (1985), and Yasar et al. (2008).

6.8 Computer-Aided Engine Simulation

Heat-release characteristic models such as the Wiebe function, also known as *apparent* heat-release characteristics, represent the variation of the overall final thermal effect of the combustion process without delving into the details of the multiple processes involved in real cycle processes such as fuel delivery method and fuel evaporation, air-fuel mixture formation, partial oxidation, and so on, that vary spatially as well as temporally. Additionally, these models are not predictive, since the function parameters are usually unknown and can only be guessed or determined from measured pressure data. In cases where enough experimental data have been gathered to establish the relationship between the Wiebe parameters and engine operating conditions, it is possible to predict the heat-release characteristics for any altered conditions within the prescribed operating range (Woschni and Anisitis, 1974; Ghojel, 1982).

Fully predictive computer-aided engineering (CAE) tools are currently available for use by engine researchers, designers, manufacturers, and testing engineers. These tools are mostly 3-D computational fluid dynamic (CFD) packages that are capable of handling complex spatially and temporally resolved in-cylinder processes associated with combustion, such as turbulence, injection, sprays, mixing, auto-ignition, chemical kinetics, pollutant formation, and partial mixed combustion. They are particularly suited to modelling recent advanced combustion systems such as gasoline direct injection (GDI), homogeneous charge compression ignition (HCCI) engines, and common-rail direct injection diesel engines. They are now routinely used by engine researchers, designers, and testing engineers to predict pressure, heat release, and steady-state and transient engine

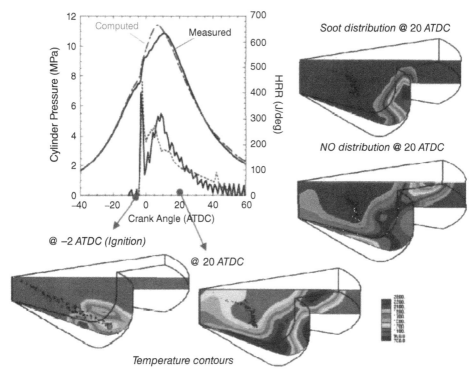

Figure 6.27 Some CFD modelling results for a direct injection compression ignition engine. Source: courtesy of ERC, University of Wisconsin.

performance and to compare various engine concepts and optimise component geometry and pollutant emissions.

The two categories of CAE tools and some of the better known software packages are:

- *Numerical engine modelling*: GT-POWER (Gamma Technologies, USA), WAVE (Ricardo, UK), AVL BOOST (GmbH, Austria), and PK Diesel-2/4T (Bauman State Technical University, Moscow, Russia)
- *Computational fluid dynamics (CFD) modelling*: FLUENT (Fluent Inc., USA), KIVA (Los Alamos National Laboratory, USA), AVL FIRE (GmbH, Austria), and Star-CD (CD-adapco, Siemens, Germany)

More information about these packages can be found in the technical literature and on the Web. Figure 6.27 shows a snapshot of CFD modelling capabilities combining pressure and heat-release prediction, in-cylinder temperature profile, and spatial soot and nitric oxide (NO) distributions. Judging by the correlation between the measured and computed pressure and heat-release profiles, model accuracy is not better than the results from models such as the Wiebe function. However, it should be noted, as mentioned earlier, that, unlike the Wiebe function, computer-based CFD modelling is fully predictive, and the results largely depend on the state of the art of the underlying theoretical sub-models (turbulence, chemical kinetics, and so on), which are continuously being refined and improved.

Problems

6.1 Solve the combustion equation for the CI engine cycle ($\lambda = 1.4$) to determine the maximum cycle temperature. The diesel fuel can be represented by $C_{12}H_{21}$, $T_2 = 973\ K$, $\varphi = 0.827$, $H_l = 42{,}200\ kJ/kg$, $\gamma_{res} = 0.033$, $\alpha = 1.8$, $\psi = 1.04$.

6.2 Find the characteristic points of the cycle and plot the $p - V$ diagram for the engine in problem 6.1. The following additional information is given:
$p_1 = 100\ kPa$, $T_1 = 334\ K$, $\varepsilon = 16.6$, $n_1 = 1.38$, $n_2 = 1.25$.

6.3 Solve the combustion equation for the SI engine cycle to determine the maximum cycle temperature. The source of energy is the stoichiometric reaction of isooctane (C_8H_{18}) in air. $T_2 = 693\ K$, $\varphi = 0.827$, $H_l = 44\ MJ/kg$, $\gamma_{res} = 0.06$, $\psi = 1.05$.

6.4 Find the characteristic points of the cycle and plot the p-V diagram for the engine in problem 6.3. The following additional information is given:
$p_1 = 100\ kPa$, $T_1 = 327\ K$, $\varepsilon = 8$, $n_1 = 1.36$, $n_2 = 1.28$.

6.5 The following information is provided for a natural-induction CI engine:

Cylinder diameter, D (m)	0.104
Piston stroke, S (m)	0.118
Compression ratio, ε	17.9
Swept volume, V_s (m^3)	1.000E-03
Clearance volume, V_c (m^3)	5.917E-05
Connecting-rod length, L (m)	1.059E-03
Connecting-rod to crank radius ratio, L/R	0.1818

If the $p - \theta$ diagram can be approximated by the function

$$p = \frac{\theta}{(a + b\theta + c\sqrt{\theta})}\ bar, \text{where } a = 5428.848, b = 14.757, c = -565.845,$$

(a) Plot the $p - V$ diagram to scale.
(b) Determine the *mep* of the cycle if the ambient pressure $p_1 = 0.1\ MPa$.

7

Work-Transfer System in Reciprocating Engines

The execution of the thermodynamic cycles in the engine cylinder discussed previously involves a series of energy-conversion processes from the chemical energy in the fuel to thermal energy by means of combustion, followed by mechanical work by means of the piston movement. This work is of little use unless converted to power and transmitted somehow to the end user. The most practical form of energy transmission is via rotary motion of the output shaft of the engine, which can then be used to produce variable power to run, for example, pumps, compressors, ship and aeroplane propellers, and motor vehicle wheels. This can be done efficiently by a crank slider mechanism that has been in use since the Industrial Revolution. In its current form, the mechanism comprises a piston, a connecting rod and a crank. Figure 7.1a illustrates the design for a six-cylinder inline reciprocating engine that also incorporates a flywheel and counterweights. The reciprocating (up and down) motion of the piston is converted to a rotary motion by the crank, and the flywheel prevents the piston from dwelling at the top and bottom dead centres and maintains uniform rotary motion. The counterweights help to reduce the imbalance caused by the forces generated by the rotary motion of the crank web and pin. Figure 7.1b shows a schematic diagram of a single crank without counterweights.

7.1 Kinematics of the Piston-Crank Mechanism

Consider the schematic diagram of the reciprocating (piston-crank) mechanism shown in Figure 7.2 and the definitions of the characteristic points. The instantaneous displacement x of the piston can be measured as the distance between the top dead centre (TDC) and the instantaneous position of the top of the piston, or the distance between the piston pin position P_{TDC} when the piston is at TDC and the instantaneous position of the pin thereafter. Either approach will yield the following relationship for the displacement:

$$x = L + R - (L \cos \beta + R \cos \theta) \tag{7.1}$$

The relationship between L and R can be written as

$$L \sin \beta = R \sin \theta$$

or

$$\sin \beta = \frac{R}{L} \sin \theta = \tau \sin \theta$$

Fundamentals of Heat Engines: Reciprocating and Gas Turbine Internal Combustion Engines, First Edition. Jamil Ghojel.
© 2020 John Wiley & Sons Ltd. This Work is a co-publication between John Wiley & Sons Ltd and ASME Press.
Companion website: www.wiley.com/go/JamilGhojel_Fundamentals of Heat Engines

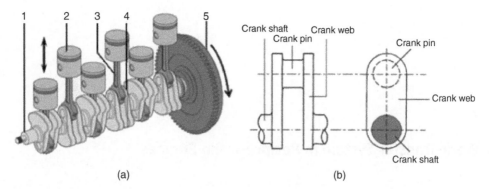

(a)   (b)

Figure 7.1 (a) Piston-connecting rod-crankshaft assembly of a six-cylinder inline engine: 1 – crankshaft, 2 – piston, 3 – connecting rod, 4 – crank web with counterweight, 5 – flywheel; (b) schematic diagram of a single crank without counterweights.

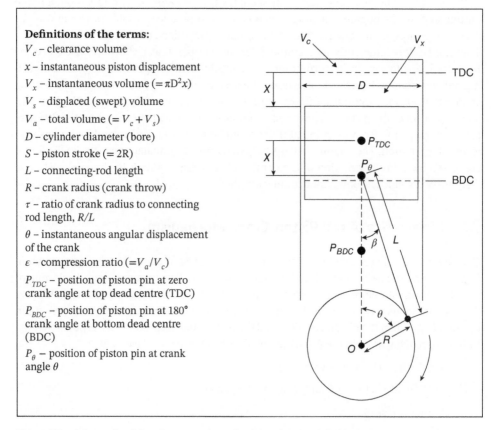

Definitions of the terms:

V_c – clearance volume

x – instantaneous piston displacement

V_x – instantaneous volume $(= \pi D^2 x)$

V_s – displaced (swept) volume

V_a – total volume $(= V_c + V_s)$

D – cylinder diameter (bore)

S – piston stroke $(= 2R)$

L – connecting-rod length

R – crank radius (crank throw)

τ – ratio of crank radius to connecting rod length, R/L

θ – instantaneous angular displacement of the crank

ε – compression ratio $(= V_a/V_c)$

P_{TDC} – position of piston pin at zero crank angle at top dead centre (TDC)

P_{BDC} – position of piston pin at 180° crank angle at bottom dead centre (BDC)

P_{θ} – position of piston pin at crank angle θ

Figure 7.2 Schematic of the piston-crank mechanism with the definitions used in this book.

Since

$$\cos\beta = \sqrt{1 - \sin^2\beta} = \sqrt{1 - \tau^2\sin^2\theta}$$

Eq. (7.1) can now be written as

$$x = L + R - (L\sqrt{1 - \tau^2\sin^2\theta} + R\cos\theta)$$

or

$$x = R\left[1 + \frac{1}{\tau} - \left(\cos\theta + \frac{1}{\tau}\sqrt{1 - \tau^2\sin^2\theta}\right)\right] \tag{7.2}$$

where $\tau = R/L$.

$$\sqrt{1 - \tau^2\sin^2\theta} = (1 - \tau^2\sin^2\theta)^{-1/2} \approx 1 - \frac{1}{2}\tau^2\sin^2\theta$$

$\cos 2\theta = \cos^2\theta - \sin^2\theta = 1 - 2\sin^2\theta$, from which

$$\sin^2\theta = \frac{1 - \cos 2\theta}{2}$$

Therefore, we can write the square root term in Eq. (7.2) as

$$\sqrt{1 - \tau^2\sin^2\theta} \approx 1 - \frac{1}{2}\tau^2\sin^2\theta = 1 - \frac{\tau^2}{4}(1 - \cos 2\theta)$$

Equation (7.2) can now be rewritten in a simpler, more manageable form as

$$x = R\left[1 + \frac{\tau}{4} - \left(\cos\theta + \frac{\tau}{4}\cos 2\theta\right)\right] \tag{7.3}$$

Equation (7.3) yields the instantaneous speed and acceleration of the piston as functions of crank angle rotation with a good degree of accuracy.

The instantaneous speed of the piston is

$$v = \dot{x} = \frac{dx}{dt} = \frac{dx}{d\theta}\frac{d\theta}{dt} = \omega\frac{dx}{d\theta}$$

where ω is the angular velocity of the crank ($d\theta/dt$).

$$v = \dot{x} = R\omega\left(\sin\theta + \frac{\tau}{2}\sin 2\theta\right) \tag{7.4}$$

The instantaneous acceleration of the piston is

$$a = \ddot{x} = \frac{dv}{dt} = \frac{dv}{d\theta}\frac{d\theta}{dt} = \omega\frac{dv}{d\theta}$$

$$a = \ddot{x} = R\omega^2(\cos\theta + \tau\cos 2\theta) \tag{7.5}$$

The piston displacement shown in Figure 7.3a changes from zero to a maximum measured from the position of the piston pin P_{TDC} when $\theta = 0$ (Figure 7.2). The maximum value of the displacement, reached at the position of the piston pin P_{BDC} when $\theta = 180$, is the stroke of the engine S, which is the distance travelled by the piston during half a revolution of the crank. In a two-stroke engine, where the completion of one thermodynamic cycle occurs over one revolution (360° crank angle rotation), the piston travels a distance equal to two strokes per cycle. In a four-stroke engine, where two revolutions (720°) is required to complete the thermodynamic cycle, the piston travels a distance equal to four strokes per cycle.

$$x = R\left[1 + \frac{\tau}{4} - \left(\cos\theta + \frac{\tau}{4}\cos 2\theta\right)\right]$$

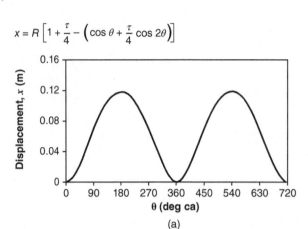

(a)

$$v = R\omega\left(\sin\theta + \frac{\tau}{2}\sin 2\theta\right)$$

(b)

$$a = R\omega^2(\cos\theta + \tau\cos 2\theta)$$

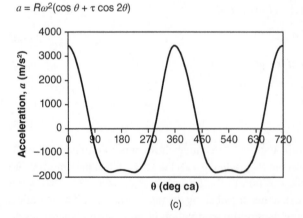

(c)

Figure 7.3 Kinematics of the reciprocating CI engine ($R = 59$ *mm*, $\tau = 0.325$, $\omega=209.4$ *rad/s*): displacement (a), speed (b), and acceleration (c) of the piston.

For a four-stroke engine, piston speed (Figure 7.3b) displays positive and negative values, reaching zero at $\theta = 0, 180, 360, 540$, and $720°$; maximum values at $\theta = 90$ and $450°$; and minimum values at $\theta = 270$ and $630°$. Piston acceleration also displays positive and negative values, as shown in Figure 7.3c.

7.2 Dynamics of the Reciprocating Mechanism

The piston–crank mechanism comprises physical components of specific shape and mass that affect the forces and moments generated in the engine. Hence, knowledge of the masses and their concentration is vital to analyse these forces and moments, most of which are undesirable, in order to minimise them as much as possible.

7.2.1 Mass-Distribution Scheme

The piston crank assembly can be replaced by a dynamically equivalent system of lumped masses connected by rigid, weightless links, as shown schematically in Figure 7.4. The subscripts used in this figure and in the equations are the following: r – connecting rod; c – crank; p – pin; pp – piston pin; cp – crank pin; cw – crank web; i – inertia.

7.2.1.1 Masses at the Piston Pin
The total reciprocating mass concentrated at the piston pin m_i (point A in Figure 7.4d) consists of the mass of the piston assembly m_p (includes the masses of the piston, piston pin, and piston rings; Figure 7.4a), and the mass of the connecting rod concentrated at its small end (at the piston pin at A) $m_{r.pp}$ (Figure 7.4b). The mass of the connecting rod concentrated at A is given by Eq. (7.6):

$$m_{r.pp} = m_r \left(\frac{l_{r.c}}{L} \right) \tag{7.6}$$

where m_r is the total mass of the connecting rod.

The total mass m_i at A executing a reciprocating motion is then

$$m_i = m_p + m_{r.pp} \tag{7.7}$$

7.2.1.2 Masses at the Crank Pin
The total rotating mass m_R at B consists of the mass of the crank pin m_{cp}, the mass equivalent to the crank web masses $2m_{cw}$ concentrated at the crank pin at B, and the mass of connecting rod $m_{r.cp}$ concentrated at its big end at B.

Masses m_{cp} and $2m_{cw}$ form a single dynamically equivalent mass m_c concentrated at B and given by

$$m_c = m_{cp} + 2m_{cw} \left(\frac{\rho}{R} \right) \tag{7.8}$$

where m_{cp} is the crank pin mass and m_{cw} is the mass of a single crank web.

The mass of the connecting rod concentrated at its big end at B is

$$m_{r.cp} = m_r \left(\frac{l_{r.pp}}{L} \right) \tag{7.9}$$

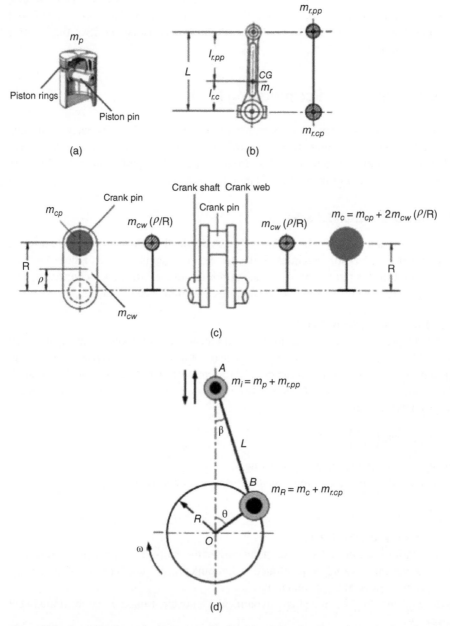

Figure 7.4 Dynamically equivalent mass distribution of the piston-crank mechanism: (a) piston assembly; (b) connecting rod; (c) crank; (d) equivalent dynamic system.

The total rotating mass concentrated at B is finally equal to

$$m_R = m_c + m_{r.cp} \tag{7.10}$$

7.2.2 Forces Acting on the Reciprocating Mechanism

The sign convention used for all forces acting on the piston-crank mechanism is shown in Figure 7.5. The force acting on the small end of the connecting rod is positive if acting toward the centre of rotation of the crankshaft (point O). The tangential force F_t producing a clockwise torque is positive, and force N acting along the crank web toward the centre O is also positive. The centrifugal force of the rotating crank mass always acts away from the centre O and is taken as negative in that direction.

7.2.2.1 Forces Acting on the Piston Pin

These forces include the gas pressure force F_g (usually plotted versus crank angle), and the inertia force F_i of the reciprocating mass acting on the piston pin at A (Figure 7.5a). The gas force acts on the piston surface and is always directed toward point O.

$$F_g = p_g A_p \tag{7.11}$$

where A_p is the piston top surface area.

The inertia force F_i resulting from reciprocating piston assembly mass (Eq. (7.7)) always acts opposite to the direction of motion of the piston and is given by Eq. (7.12).

$$F_i = -m_i a = -m_i R \omega^2 (\cos \theta + \tau \cos 2\theta) \tag{7.12}$$

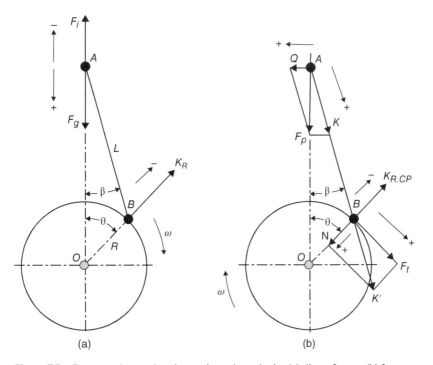

(a)   (b)

Figure 7.5 Forces acting at the piston pin and crank pin: (a) direct forces; (b) force components.

where a is the piston acceleration.

The sum of the gas and inertia forces F_p acts on the piston pin at A and could act toward O or away from O depending on the vector sum of the colinear forces F_g and F_i:

$$\overline{F}_p = \overline{F}_g + \overline{F}_i \tag{7.13}$$

Force F_p can be resolved into components K and Q (upper part of Figure 7.5b) as

$$\overline{F}_p = \overline{K} + \overline{Q} \tag{7.14}$$

Force K acts on the piston pin at the small end of the connecting rod (point B) and is given by

$$K = \frac{F_p}{\cos \beta} = \frac{F_p}{\sqrt{1 - \tau^2\sin^2\theta}} \tag{7.15}$$

Force Q acts on the cylinder wall perpendicular to the path of the reciprocating motion of the piston and is given by

$$Q = F_p \tan \beta = F_p \frac{\tau \sin \theta}{\sqrt{1 - \tau^2\sin^2\theta}} \tag{7.16}$$

Force Q is a variable force that presses against the cylinder walls of the engine and is known to cause so-called *piston slap*.

Figure 7.6 shows the forces F_g, K, and Q acting at point A. The shapes of the curves are determined to a large extent by the gas pressure force, which is directly proportional to the top surface area of the piston and combustion pressure. Combustion pressure and cylinder bore are usually greater in compression ignition (CI) engines compared to spark ignition (SI) engines of similar power output.

7.2.2.2 Forces Acting on the Crank Pin

For a rigid connecting rod, force K can be transferred unchanged along the connecting rod to act on the crank pin at B. The new force, denoted K' ($K' = K$), can now be resolved into a tangential force F_T and normal force N (lower part of Figure 7.5b):

$$\overline{K'} = \overline{F}_t + \overline{N} \tag{7.17}$$

Force F_T is given by

$$F_t = K' \sin(\theta + \beta) = F_p \frac{\sin(\theta + \beta)}{\cos \beta}$$

Knowing that $\cos\beta = \sqrt{1 - \sin^2\beta} = \sqrt{1 - \tau^2\sin^2\theta}$, and $\sin(\theta + \beta) = \sin \theta \cos \beta + \cos \theta \sin \beta = \sin \theta \sqrt{1 - \tau^2\sin^2\theta} + \cos \theta \, \tau \sin \theta,$ we can rewrite F_t as

$$F_t = F_p \left(\frac{\sin \theta \sqrt{1 - \tau^2\sin^2\theta} + \tau \sin \theta \cos \theta}{\sqrt{1 - \tau^2\sin^2\theta}} \right) \tag{7.18}$$

Force N acts along the crank radius toward O and is given by

$$N = K' \cos(\theta + \beta) = F_p \frac{\cos(\theta + \beta)}{\cos \beta}$$

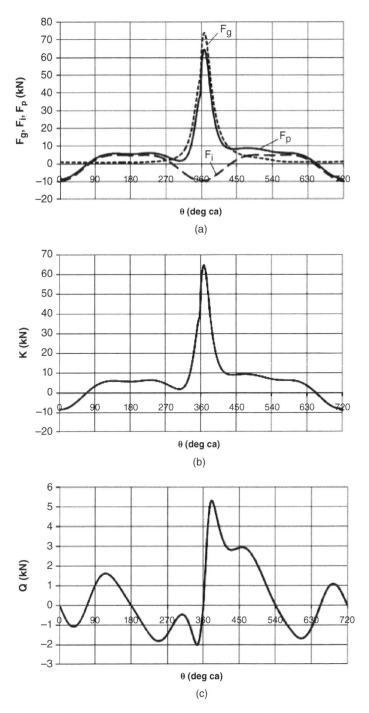

Figure 7.6 Forces acting on the piston pin at A in a CI engine ($D = 104$ mm, $p_{max} = 87$ bar, $R = 59$ mm, $\tau = 0.325$, $\omega = 209.4$ rad/s): (a) gas, inertia, and resultant forces; (b) and (c) components K and Q of force F_p.

Since $\cos(\theta + \beta) = \cos\theta\cos\beta - \sin\theta\sin\beta = \cos\theta\sqrt{1 - \tau^2\sin^2\theta} - \sin\theta\,\tau\sin\theta$, we obtain

$$N = F_p\left(\frac{\cos\theta\sqrt{1 - \tau^2\sin^2\theta} - \tau\sin^2\theta}{\sqrt{1 - \tau^2\sin^2\theta}}\right) \tag{7.19}$$

The inertia force (centrifugal force) of mass $m_{r.\,cp}$ at the big end of the connecting rod rotating at the set engine speed ω is $K_{R.\,cp}$. This constant force is always directed away from the centre of rotation O and taken as negative in that direction:

$$K_{R.cp} = -m_{r.cp}R\omega^2 \tag{7.20}$$

Figure 7.7 shows the forces K', N, and F_t acting on the crank pin at B plotted as functions of crank angle θ. Figures 7.6–7.7 are based on the data in Table 7.1 for one cylinder of a four-cylinder, four-stroke, CI engine. Similar plots based on data in the same table for a four-stroke SI engine can be found in Appendix B.

7.2.2.3 Forces and Moments Acting on the Crankshaft Supports at Point O

The forces acting at O and the moments resulting from these forces are shown in the bottom part of Figure 7.8. A constant inertia force (centrifugal force) K_R resulting from the rotating concentrated mass m_R $(=m_{r.\,cp} + m_c)$ at the crank pin B acts on the crank supports (bearings) in a direction away from the centre of rotation O:

$$K_R = K_{R.cp} + K_{R.c} = -m_{r.cp}R\omega^2 - m_c R\omega^2 = -R\omega^2(m_{r.cp} + m_c)$$
$$K_R = -m_R R\omega^2 \tag{7.21}$$

where $m_R = m_{r.\,cp} + m_c$.

To determine the forces and moments generated at O, we follow this procedure:

1. Force N is transferred along the rigid crank webs to O so that $N' = N$.
2. Two opposing forces F'_t and F''_t are applied at O ($F'_t = F''_t = F_t$).
3. The sum of vectors N' and F'_t yields vector K'' ($K'' = K' = K$).
4. Vector K'' is resolved into components Q' and $F_p{}'$ with origin at O ($Q' = Q$ and $F_p{}' = F_p$).
5. Opposing forces F_t and F''_t generate a moment about O, causing the crankshaft to rotate around its axis. For most engines, the crankshaft rotates in the clockwise direction and the torque generated in a clockwise direction is taken as positive.
6. Opposing forces Q and Q' cause the engine block to tilt relative to the axis of the crankshaft, clockwise or anticlockwise depending on the direction of force Q. This moment is taken as positive if it causes the engine block to tilt clockwise. In Figure 7.8 the moment is anticlockwise; hence, it is negative.

The moment generated by the opposing force F_t and F''_t is called the engine torque T. Using Eq. (7.18), the torque can be written as

$$T = F_t R = F_p R\left(\frac{\sin\theta\sqrt{1 - \tau^2\sin^2\theta} + \tau\sin\theta\cos\theta}{\sqrt{1 - \tau^2\sin^2\theta}}\right) \tag{7.22}$$

The tilting moment is equal to

$$M_{tilt} = -Qh = -QR\frac{\sin(\theta + \beta)}{\sin\beta}$$

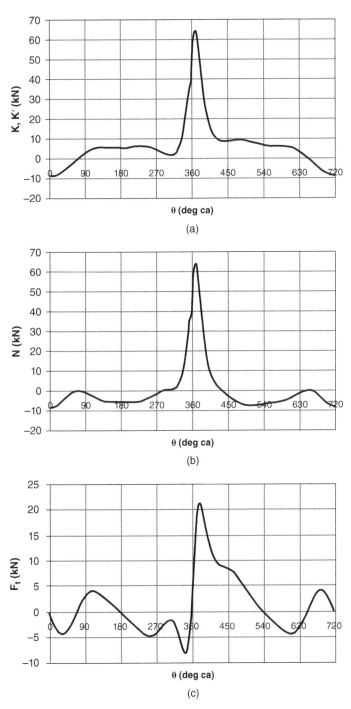

Figure 7.7 Forces acting on the crank pin at B in a CI engine ($D = 104$ mm, $p_{max} = 87$ bar, $R = 59$ mm, $\tau = 0.325$, $\omega = 209.4$ rad/s): (a) force K' transferred from A; (b) and (c) components N and F_t of force K'.

Table 7.1 Data for calculations of forces on piston and crank pin.

Engine type	CI	SI
Speed, N (rpm)	2000	5600
Compression ratio, ε	17	8.5
Bore, D (mm)	104	78
Stroke, S (mm) (=2R)	118	78
Conrod length, l_r (mm)	118	136.8
m_p (kg)	2.124	0.478
$m_{r.pp}$ (kg)	0.637	0.197
m_c (kg)	2.718	0.669
$m_{r.cp}$ (kg)	1.911	0.519
m_r (kg)	2.548	0.716

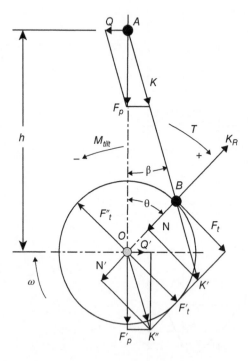

Figure 7.8 Forces and moments acting on the crankshaft supports at O.

Since $Q = F_p \tan \beta$,

$$M_{tilt} = -RF_p \frac{\sin(\theta + \beta)}{\cos \beta}$$ (7.23)

It was shown before that F_t can also be written as

$$F_t = F_p \frac{\sin(\theta + \beta)}{\cos \beta}$$

Hence,

$$M_{tilt} = -RF_t = -T \qquad (7.24)$$

In other words, the tilting moment is numerically equal to the torque but opposite in direction. The tilting moment acts on the fixed engine structure, causing it to alternately sway clockwise and anticlockwise and thus causing it to vibrate.

7.2.2.4 Resultant Forces Acting on the Crank Pin

The effect of the centrifugal forces resulting from the rotating masses of the connecting rod $(m_{r.\,cp})$ and the crank pin-web combination (m_c) on the crank pin and crankshaft are shown in Figure 7.9.

Force $\overline{F_{cp}}$, which is acting on the crank pin at B, is the vector sum of the tangential force $\overline{F_t}$ and the vector sum of the collinear vectors $\overline{N}$ and $\overline{K_{R.cp}}$ (Figure 7.9):

$$\overline{F_{cp}} = \overline{F_t} + \overline{N} + \overline{K_{R.cp}} \qquad (7.25)$$

$$F_{cp} = \sqrt{F_t^2 + (N + K_{R.cp})^2} \qquad (7.26)$$

It should be noted that N can be +ve or −ve, and $K_{R.\,cp}$ is always negative. The angle between F_{cp} and axis OB is ϕ, where

$$\tan \phi = \overline{F_t}/(\overline{N} + \overline{K_{R.cp}}) \qquad (7.27)$$

7.2.2.5 Polar Diagram

The magnitude and orientation of force F_{cp} for any crank-angle position θ can be determined graphically from the polar diagram of forces F_t and N in combination with force F_{cp}, as follows:

Figure 7.9 The resultant force F_{cp} acting on the crank pin at B.

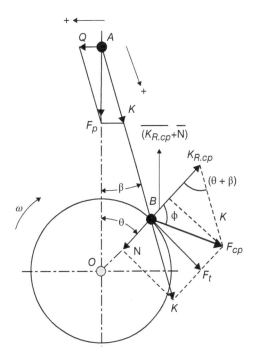

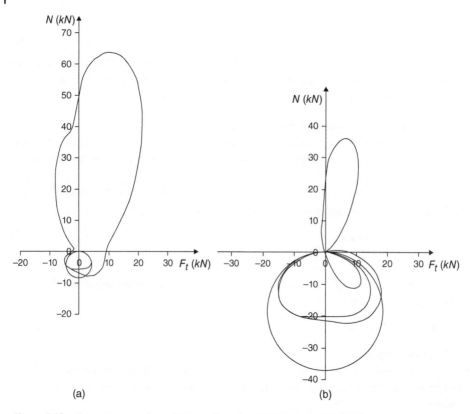

(a)   (b)

Figure 7.10   Force *N* versus force F_t for a CI engine: (a) 2000 *rpm*; (b) 4000 *rpm*.

1. Force N is plotted versus force F_t in Cartesian coordinates with origin O, as shown in Figure 7.10 for the CI engine data in Table 7.1 at two engine speeds with the assumption that $p - \theta$ diagram remains unchanged. The variation in the two plots is due to the variation of inertia forces with engine rotational speed.
2. For given values of θ, vectors N and F_t are added to find vector K acting at origin O_1, as shown in in Figure 7.11.
3. The crank pin centre O is located on the N axis at a distance equivalent to the magnitude of vector $K_{R.\,cp}$, which is drawn pointing downward in the negative direction.
4. Each $\overline{F_{cp}}$ force is the vector sum of vectors $\overline{K}$ and $\overline{K_{R.cp}}$.
5. The process is repeated for different crank-angle positions θ_1, θ_2, and so on to obtain a full polar plot as shown in Figure 7.12.

The distribution of forces F_{cp} on the crank pin can be estimated graphically and used to locate the optimum position for the lubricating oil hole, i.e. where the pressure forces acting on the crank pin are at their minimum (Arkhangelsky et al., 1977). After obtaining a polar diagram similar to the one shown in Figure 7.12, arcs with thicknesses proportional to the magnitude of forces F_{cp} for selected values of crank-angle position θ are drawn, as shown in Figure 7.13a:

Figure 7.11   Schematic diagram for the determination of F_{cp} from the polar diagram of K.

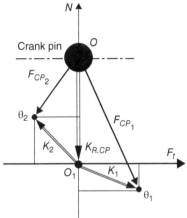

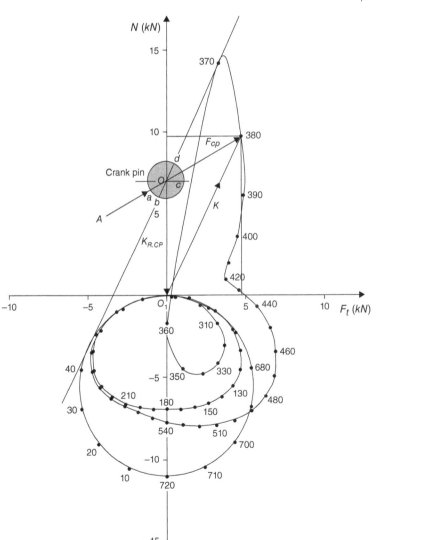

Figure 7.12   Polar diagram for the SI engine in Table 7.1.

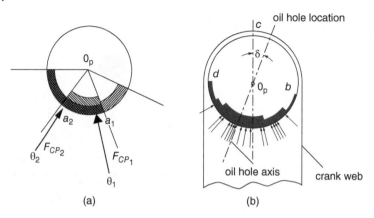

Figure 7.13 Construction of the crank pin wear diagram.

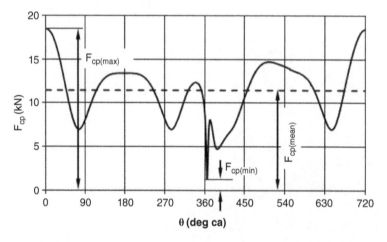

Figure 7.14 Force F_{cp} as a function of crank angle θ.

1. Each arc subtends an arbitrary angle (say, 60° on each side of points a_1 and a_2).
2. The cumulative sum of all the forces indicated by the arrows in Figure 7.13b yields the force concentration profile highlighted in black. The figure also shows the unloaded arc *bcd*.
3. The optimum position for drilling the lubrication hole is on the opposite side of the concentration of forces at angle δ relative to the vertical axis passing through the centre of the crank pin, O_p.

F_{cp} can also be plotted as a function of crank angle θ, as shown in Figure 7.14. Curve $F_{cp} = f(\theta)$ is the polar diagram in Figure 7.12 replotted in Cartesian coordinates. This plot is useful for determining the minimum force ($F_{cp(min)}$), maximum force ($F_{cp(max)}$), and mean force ($F_{cp(mean)}$) acting on the crank pin over one complete engine cycle.

7.2.2.6 Resultant Forces Acting on the Crankshaft Bearing Journals

Force F_c acts on the main crankshaft journal bearing adjacent to the crank at point O (Figure 7.15b, not drawn to scale), and it can be determined from the vector sum of force

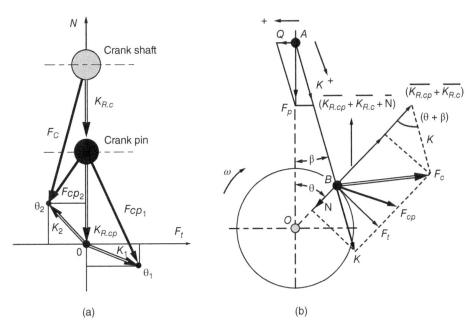

(a) (b)

Figure 7.15 Determination of the resultant force F_c acting on the crankshaft bearing at O.

F_{cp} (Eq. (7.26)) and the centrifugal force $K_{R.c}$ (Figure 7.15a). $K_{R.c}$ is the resulting force from the rotation of the combined mass m_c concentrated at B, which is dynamically equivalent to the masses of the two webs and crank pin of a single crank ($K_{R.c} = -m_c R\omega^2$):

$$\overline{F_c} = \overline{F_{cp}} + \overline{K_{R.c}} \tag{7.28}$$

Force F_c in combination with the polar diagram of force F_{cp} can be used to determine the wear diagram of the main crankshaft bearing using the same methodology described earlier for the wear diagram of the crank pin.

Force $\overline{F_c}$ can also be determined from Figure 7.15b as the vector sum of the tangential force $\overline{F_t}$ and the vector sum of the colinear vectors $\overline{N}$, $\overline{K_{R.cp}}$ and $\overline{K_{R.c}}$:

$$\overline{F_c} = \overline{F_t} + (\overline{N} + \overline{K_{R.cp}} + \overline{K_{R.c}}) \tag{7.29}$$

Hence,

$$F_c = \sqrt{F_t^2 + (N + K_{R.cp} + K_{R.c})^2} \tag{7.30}$$

The values in brackets under the square root sign are added algebraically before squaring the result.

Force F_c is a force that is applied on the crank pin at B, causing it to bend, and is transmitted via the two crank webs to the crankshaft supports (journal bearings) on both sides of the crank. For the case of multi-cylinder crankshafts, let us consider the shaft shown in Figure 7.16 with two cranks that have angle $\gamma = 120°$ between them. The reaction to forces $F_{c(i)}$ and $F_{c(i+1)}$ acting at the support between the adjacent cranks (i) and ($i+1$) can be determined from the following equilibrium conditions.

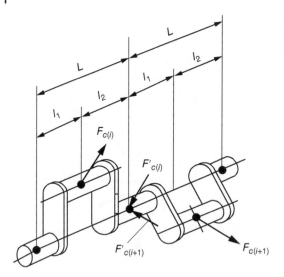

Figure 7.16 Forces acting on a crankshaft with two cranks at an angle of 120°.

For crank (i):

$$F'_{c(i)} = \frac{F_{c(i)}l_1}{L}$$

For crank ($i+1$):

$$F'_{c(i+1)} = \frac{F_{c(i+1)}l_2}{L}$$

The resultant force F_{cR} acting on the crankshaft journal bearing between crank (i) and crank ($i+1$) is the vector sum of forces $F'_{c(i)}$ and $F'_{c(i+1)}$:

$$\overline{F_{cR}} = \frac{(\overline{F_{c(i)}}l_1 + \overline{F_{c(i+1)}}l_2)}{L}$$

For a symmetrical crankshaft, $l_1 = l_2$, $F'_{c(i)} = -0.5F_{c(i)}$, and $F'_{c(i+1)} = -0.5F_{c(i+1)}$. Hence,

$$\overline{F_{cR}} = -0.5(\overline{F_{c(i)}} + \overline{F_{c(i+1)}})$$

The resultant vector of the forces acting on the main crank support $\overline{F_{cR}}$ can be found graphically by plotting the polar diagrams of cranks (i) and ($i+1$) separated by $\gamma = 120°$ and adding the vectors $\overline{F_{c(i)}}$ and $\overline{F_{c(i+1)}}$ at crank angles corresponding to the respective firing sequence of the two cylinders.

7.3 Multi-Cylinder Engines

The forces discussed so far relate to a single-cylinder, single-crank engine without counterweights. Most engines come in the form of inline or V-type multi-cylinder configurations. Figure 7.17a is an example of inline four-cylinder engine with possible firing sequence $1-3-2-4$ or $1-2-4-3$. Figure 7.17b is an example of an inline six-cylinder engine with four possible firing sequences: $1-5-3-6-2-4$, $1-2-4-6-5-3$, $1-4-2-6-3-5$, or $1-4-5-6-3-2$ (Bosch Automotive Handbook, 2004).

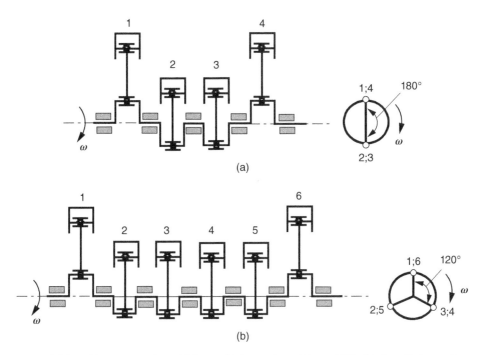

(a)

(b)

Figure 7.17 Inline-type engines: (a) four-cylinder engine with $\gamma = 180°$; (b) six-cylinder engine with $\gamma = 120°$.

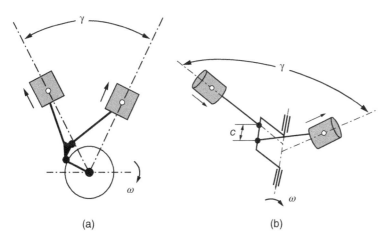

(a)   (b)

Figure 7.18 (a) V-engines with articulated connecting rod; (b) side-by-side connecting rods on one crank.

In V-type engines, two types of connecting rods can be used: an articulated rod with primary and secondary branches, as shown in Figure 7.18a; or two side-by-side connecting rods per cylinder pair, as shown in Figure 7.18b. In engine design practice, the articulated configuration is rarely used anymore (van Basshuysen and Schafer, 2007). The connecting rod configuration in Figure 7.18b in an eight-cylinder engine with two cylinder

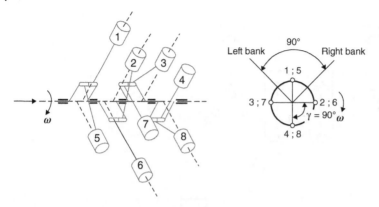

Figure 7.19 Eight-cylinder V 90° engine.

banks (four cylinders per bank) can be seen is Figure 7.19. The angle between the two banks is 90°, and there are three possible firing sequences: $1-6-3-5-4-7-2-8$, $1-5-4-8-6-3-7-2$, or $1-8-3-6-4-5-2-7$ (Bosch Automotive Handbook, 2004).

To determine the mean dynamic parameters in multi-cylinder engines, individual pressure profiles (indicator diagrams) for each cylinder are summed up, taking into account the angular phase shift of the processes, which is given by $\emptyset = 720/i$ for four-stroke engines and $\emptyset = 360/i$ for two-stroke engines (i is the number of cylinders), and the cylinder firing sequence. The firing sequence is usually selected to evenly distribute the loading stresses along the crankshaft. Figure 7.20 illustrates the methodology for summing up the cylinder pressure profiles in an eight-cylinder V-type engine with an angular phase shift of $\emptyset = 90°$ and firing sequence 1–8–3–6–4–5–2–7.

7.3.1 Torque in Multi-Cylinder Engines

The methodology for determining the mean torque of a multi-cylinder engine with equal firing intervals is shown for a four-stroke, four-cylinder inline SI engine in Figure 7.21. The torque curve for one cylinder is shown in Figure 7.21a. The other three cylinders will theoretically have similar curves but out of phase of each other by angle $\emptyset = 180$. Figures 7.21b,c,d show the torque curves for the second, third, and fourth cylinders, which are out of phase of the first cylinder by 180, 360, and 540°, respectively. In other words, the cycle in the second cylinder starts at 180°, the third cylinder at 360°, and the fourth cylinder at 540° relative to the first cylinder, with each cycle lasting 720°, as shown by the shaded areas. Summing up the four torque curves over one cycle duration (720°) results in the profile shown in Figure 7.21e, which displays four identical segments each having a duration of 180°. Figure 7.21f shows the resultant torque of the engine (shaded area) and its constituent components plotted over angle $\emptyset = 180°$.

The mean torque can be calculated from the shaded area in Figure 7.21f using the following equation:

$$T_m = \frac{1}{\emptyset} \int_{\theta_1}^{\theta_2} T d\theta \tag{7.31}$$

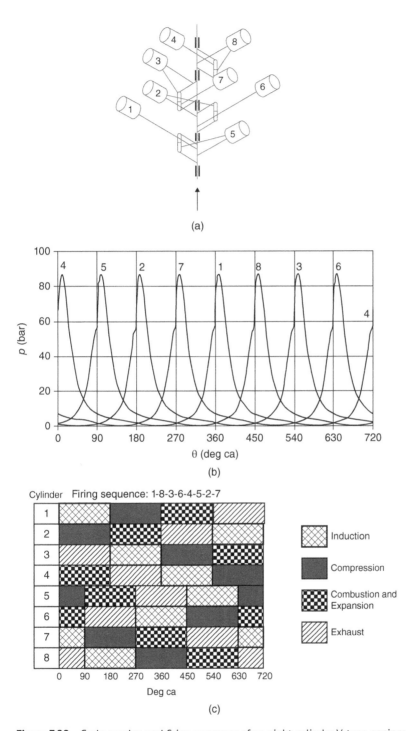

Figure 7.20 Cycle overlap and firing sequence of an eight-cylinder V-type engine: (a) cylinder-crank configuration; (b) pressure profiles; (c) process sequencing.

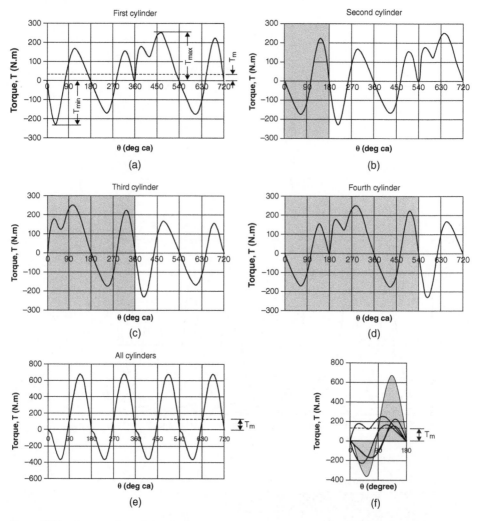

Figure 7.21 Construction of the resultant torque for a four-stroke, four-cylinder SI engine: Ø = 180°, engine output 60 kW at 5600 rpm, D = 78 mm, S = 78 mm.

For the single-cylinder configuration, $\theta_1 = 0$, $\theta_2 = 720$, $\emptyset = 720$, and the mean torque is found to be $T_{m(1)} \approx 32.7 \ N.m$. For the four-cylinder configuration, $\theta_1 = 0$, $\theta_2 = 180$, $\emptyset = 180$, and the mean torque is $T_{m(4)} \approx 130.8 \ N.m$. The "approximate" signs are indicative of the presence of errors caused by the numerical integration of areas under the torque curves.

The mean torque calculated using Eq. (7.31) should correspond to the indicated torque calculated from the actual thermodynamic cycle discussed in Chapter 8.

7.3.1.1 Torque Uniformity Factor (TUF)
The torque of a four-stroke single-cylinder engine exhibits a non-uniform periodic function when plotted versus the crank angle, as can be seen in Figure 7.21a. The torque uniformity

factor (TUF) is used to characterise the degree of fluctuations of the torque function $T = f(\theta)$ and is defined as

$$\mu = \frac{T_{max} - T_{min}}{T_m} \tag{7.32}$$

As expected, the mean torque increases fourfold when the number of cylinders is increased to four. Additionally, the torque curve becomes more uniform, and the magnitudes of the negative areas decrease. The TUF decreases from 12.56 to 9.46. Further examples of the effect of increasing the number of cylinders on the torque profile are shown in Figure 7.22. The figure shows full-cycle torque profiles spanning 720° and single-period profiles spanning a phase angle Ø for each engine configuration. The configurations are 2-cylinder, 3-cylinder, 5-cylinder, 6-cylinder, and 8-cylinder four-stroke inline SI engines operating at 5600 *rpm*. The four-cylinder variant is shown in Figure 7.21e,f. Increasing the number of cylinders reduces the TUF.

Mean engine torque increases directly with the increase in number of cylinders, and the torque profile becomes relatively smoother, with the negative areas under the torque curves becoming gradually smaller and disappearing completely in the case of the 8-cylinder engine. Table 7.2 shows the effect of increasing number of cylinders on the mean torque and the TUF.

Increasing the number of cylinders i of an engine, assuming all the cylinders perform identically, causes the mean torque T_m to increase linearly (Figure 7.23). The TUF changes with the increasing number of cylinders, with a general downward trend as shown in Figure 7.23. It decreases continuously to $i = 3$, increases somewhat at $i = 4$, and then starts decreasing again, reaching its lowest value at $i = 12$, indicating that the latter exhibits the smoothest torque curve.

7.3.2 Engine-Speed Fluctuations

The effect of the periodic change of engine torque is to cause the angular velocity of the crankshaft to fluctuate even when the engine is operating under steady-state conditions. Angular velocity fluctuations introduce additional inertial loadings of engine components. The coefficient of crankshaft imbalance (a distinction is made here between this imbalance and the imbalance caused by manufacturing irregularities) due to angular velocity fluctuations is defined as

$$\delta = \frac{\omega_{max} - \omega_{min}}{\omega_m} \tag{7.33}$$

Values of δ between 0.01 and 0.02 are common in automotive engine practice. Engine-speed fluctuations can be minimised by mounting a flywheel of appropriate dimensions on the crankshaft. The periodical engine torque is counterbalanced at each instance by a resistance moment and a moment produced by all the moving parts of the crank-piston mechanism reduced to an equivalent mass concentrated at the crank pin rotating about the axis of the crankshaft:

$$T = I_{tot} \frac{d\omega}{dt} + M_{res} \tag{7.34}$$

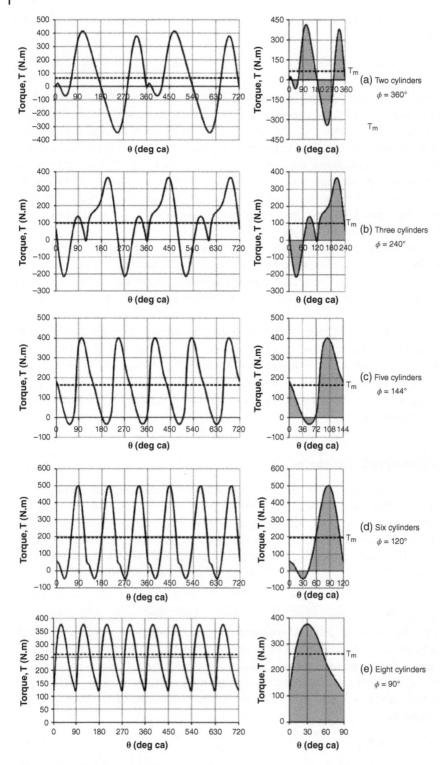

Figure 7.22 Effect of the number of cylinders on the torque profile and mean torque (four-stroke inline SI engine at 5600 *rpm*, *D* = 78 *mm*, *S* = 78 *mm*).

Table 7.2 Effect of the number of cylinders on the torque uniformity factor and mean torque.

No of cylinders	T_m, Nm	μ
1	32.7	12.56
2	65.3	9.46
3	98.0	5.02
4	130.7	5.81
5	163.4	2.66
6	196.0	1.84
8	261.4	0.999
12	392.1	0.532

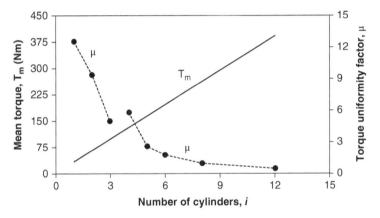

Figure 7.23 Effect of the number of cylinders on the mean torque and torque uniformity factor (four-stroke inline SI engine at 5600 *rpm*, *D* = 78 *mm*, *S* = 78 *mm*).

where

M_{res} is a resistance moment that includes the externally applied moment (engine load), the moment of internal frictional forces, and the torque required to drive the auxiliary equipment.

I_{tot} is the moment of inertia of all moving parts of the crank-piston mechanism, reduced to an equivalent rotating mass at the crank pin.

$d\omega/dt$ is the angular acceleration of the crankshaft.

Figure 7.24 shows the effect of torque fluctuations, relative to a mean value T_m, on the angular velocity of a multi-cylinder engine with firing interval $\phi = 720°/i$, where i is the number of cylinders. Under steady operating conditions, when $M_{res} = T_m$, the angular velocity will be driven by the difference between the instantaneous engine torque T and the mean torque T_m.

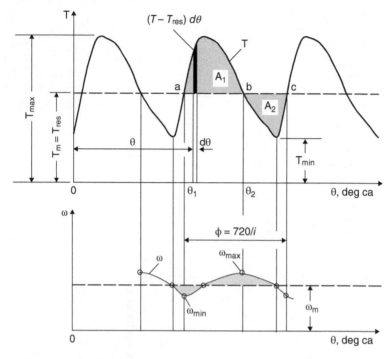

Figure 7.24 Fluctuations of the torque and angular velocity of a multi-cylinder engine operating under steady-state conditions.

Since $M_{res} = T_m$ and

$$\frac{d\omega}{dt} = \frac{d\omega}{d\theta}\frac{d\theta}{dt} = \omega\frac{d\omega}{d\theta} = \frac{1}{2}\frac{d\omega^2}{d\theta},$$

Eq. (7.34) can be rewritten as

$$(T - T_{res}) = \frac{1}{2}I_{tot}\frac{d\omega^2}{d\theta} \text{ or}$$

$$(T - T_{res})d\theta = \frac{1}{2}I_{tot}\,d\omega^2 \tag{7.35}$$

The torque surplus work represented by area A_1 above the mean torque – which is absorbed by the moving parts, causing the crankshaft to accelerate from ω_{min} to ω_{max} – can be determined by taking the integral of both sides of Eq. (7.35) between angles θ_1 and θ_2:

$$A_1 = \int_{\theta_1}^{\theta_2}(T - T_{res})d\theta = \frac{1}{2}I_{tot}\int_{\omega_{min}}^{\omega_{max}}d\omega^2 = I_{tot}\frac{(\omega_{max}^2 - \omega_{min}^2)}{2}$$

$$A_1 = I_{tot}\frac{(\omega_{max} + \omega_{min})}{2}(\omega_{max} - \omega_{min})$$

Assuming

$$\omega \approx \omega_m = \frac{(\omega_{max} + \omega_{min})}{2}$$

and making use of Eq. (7.33), we finally obtain

$$A_1 = I_{tot}\,\delta\omega^2 \text{ N.m} \tag{7.36}$$

The moment of inertia of the moving parts at the rated engine speed can be determined if the torque profile of the engine and the value of δ are known:

$$I_{tot} = \frac{A_1}{\delta\omega^2} \qquad (7.37)$$

The moment of inertia I_{tot}, in terms of engine rated speed N_r in revolutions per minute (*rpm*), is then

$$I_{tot} = \frac{900A_1}{\delta\pi^2 N_r^2} \qquad (7.38)$$

The moment of inertia of the flywheel I_f required to provide uniform angular velocity of the crankshaft for most engines is in the range of 75–90% of I_{tot}. For a simple flywheel in the form of a disk with heavy rim of diameter D_{rim},

$$I_f = \frac{m_f D_{rim}^2}{4} \; kg.m^2 \qquad (7.39)$$

The size of the flywheel decreases with an increasing number of cylinders as a result of decreasing torque surplus work (largest areas above the mean torque in Figure 7.22) and corresponding decrease in TUF, as shown in Figures 7.23.

7.4 Engine Balancing

Reciprocating piston engines are inherently unbalanced due to the presence of periodically varying forces and moments in the reciprocating mechanism, resulting in engine vibration, noise, and overloading of the supports. It is not practical to eliminate the imbalance completely, and the measures that need to be taken to reduce it depend to a great extent on the number of cylinders in the engine. The imbalance is due to:

- The unbalanced primary and secondary inertial forces of the reciprocating masses (F_{iI}, F_{iII}) and inertial forces of the rotating masses (K_R) of all cylinders:

$$F_i = F_{iI} + F_{iII} = (-m_i R\omega^2 \cos\theta) + (-m_i R\omega^2 \tau \cos 2\theta)$$
$$K_R = -m_R R\omega^2$$

- The moments resulting from the unbalanced inertial forces in individual cylinders along the crankshaft (M_{iI}, M_{iII}, and M_R).
- The periodically changing engine torque T and tilting moment M_{tilt} (see Figure 7.8):

$$T = F_p R \frac{\sin(\theta + \beta)}{\cos\beta}$$
$$M_{tilt} = -RF_p \frac{\sin(\theta + \beta)}{\cos\beta} = -T$$

7.4.1 Single-Cylinder Engine

In the single-cylinder engine shown in Figure 7.25, there are no longitudinal moments. The centrifugal inertial force resulting from the rotating masses K_R can be fully balanced

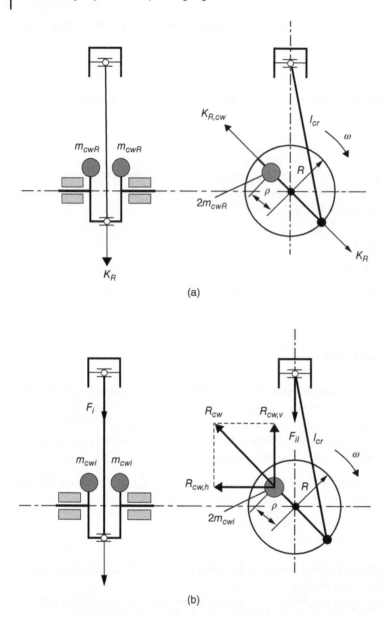

Figure 7.25 Balancing a single-cylinder engine: (a) centrifugal inertial force; (b) primary reciprocating inertial force.

by mounting a counterweight (m_{cwR}) on the extension of each crank web, as illustrated in Figure 7.25a. The required mass of the counterweight is then determined from

$$2m_{cwR}\rho\omega^2 = m_R R\omega^2$$
$$m_{cwR} = \frac{1}{2}\frac{R}{\rho}m_R \qquad\qquad (7.40)$$

Adding an additional m_{cwi} to the counterweight can generate a centrifugal force R_{cw}, the vertical component of which can eliminate the primary reciprocating inertial force F_{iI}. However, this introduces a large horizontal force $R_{cw,h}$, which is undesirable. One approach to reduce the magnitude of $R_{cw,h}$ is to aim at reducing F_{iI} by 50%. The additional mass to the counterweight is then given by

$$2m_{cwi} = 0.5m_i \frac{R}{\rho}$$

and the total mass of counterweight required is

$$m_{cw(t)} = (m_{cwR} + m_{cwi}) = \frac{R}{2\rho}(m_R + 0.5m_i) \tag{7.41}$$

7.4.2 Multi-Cylinder Engines

Multi-cylinder engines are typically classified as four-stroke or two-stroke inline- or V-engines. The cylinders in inline engines are grouped in a single bank (or block) with a common crank. In V-engines, the cylinders are grouped in two V-shaped banks that can have different angles and share a common crankshaft. Both these engine types can have crankshafts with different crank configurations. More complicated cylinder arrangements are used when compact design and cooling efficiency are of paramount importance. An example of this is the aircraft radial piston engine.

Multi-cylinder engines are generally considered to be balanced if

- Total primary reciprocating inertial forces $\sum F_{iI}$ and the moments they produce $\sum M_{iI}$ are equal to zero.
- Total secondary reciprocating inertial forces $\sum F_{iII}$ and the moments they produce $\sum M_{iII}$ are equal to zero.
- Total centrifugal inertial forces $\sum K_R$ and the moments they produce $\sum M_R$ are equal to zero.

7.4.2.1 Two-Cylinder Inline Engine

The previous example of a single-cylinder engine dealt with the reciprocating and rotating inertial forces only. In multi-cylinder engines, these forces generate transverse moments along the axis of the crankshaft by the forces between different cylinders, which could be balanced or unbalanced. Consider the two-cylinder engine with cranks at 180°, shown in Figure 7.26. With this crank arrangement, the reciprocating inertial forces of the first order are fully balanced, i.e. $\sum F_{iI} = 0$, but they generate a moment in the plane of the cranks (plane of paper):

$$\sum_{1}^{2} M_{iI} = F_{iI}a = m_i R\omega^2 \cos\theta\, a = Z_i \cos\theta\, a \tag{7.42}$$

where $Z_i = m_i \omega^2 R$.

This moment can be balanced by adding two counterweights located at the two outer webs of the two cranks at a distance ρ_1 from the crankshaft axis (not shown in Figure 7.26). The degree of balancing M_{iI} depends on the acceptable magnitude of the moment generated in the horizontal plane by the counterweights (see Section 7.4.1).

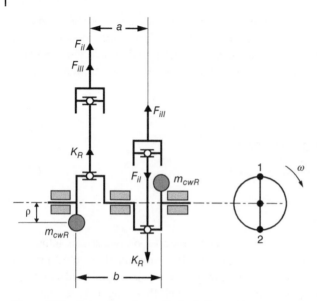

Figure 7.26 Balancing a two-cylinder inline engine.

The reciprocating inertial forces of the second order are unbalanced and equal to

$$\sum_1^2 F_{iII} = 2m_i R\omega^2 \tau \cos 2\theta = 2Z_i \tau \cos 2\theta \; a \tag{7.43}$$

These forces, which are usually left unbalanced, produce no moment, i.e. $M_{iII} = 0$. The centrifugal inertia forces K_R of the two cylinders are equal in magnitude and opposite in direction, producing a moment along the axis of the crankshaft; i.e. $\sum_1^2 K_R = 0$ and $\sum_1^2 M_R = m_R R\omega^2 a = Z_R a$, where $Z_R = m_R R\omega^2$. This moment can be fully balanced by adding two counterweight m_{cwR} located at the two outer webs of the two cranks at a distance ρ from the crankshaft axis. The value of this counterweight is

$$m_{cwR} = \frac{a}{b}\frac{R}{\rho} m_R \tag{7.44}$$

7.4.2.2 Two-Cylinder V-Engine

Two arrangements of a two-cylinder, V-engine with one crank were shown earlier, in Figure 7.18. The two connecting rods in Figure 7.18a are articulated, and they, along with the cylinder axes, are in one plane. The connecting rods in Figure 7.18b are separate, with the big ends side by side on a common crank pin. The connecting rods and cylinder axes are in parallel planes separated by a distance c. We will refer to Figure 7.27 for the following analysis; however, the general case will be considered first whereby the angle between the two cylinder banks $\gamma \neq 90°$. The reciprocating inertial forces of the first order for the left and right cylinders (F_{iIl}, F_{iIr}) when the left cylinder is in angular position θ are

$$F_{iIl} = -m_i R\omega^2 \cos\theta = -Z_i \cos\theta \tag{7.45}$$

$$F_{iIr} = -m_i R\omega^2 \cos[360° - (\gamma - \theta)] = -m_i R\omega^2 \cos(\gamma - \theta)$$

$$F_{iIr} = -Z_i \cos(\gamma - \theta) \tag{7.46}$$

Figure 7.27 Inertia forces in a two-cylinder V-engine.

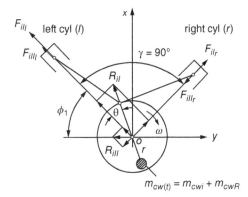

The resultant force R_{iI} is

$$\sum_1^2 F_{iI} = R_{iI} = \sqrt{F_{iIl}^2 + F_{iIr}^2 + 2F_{iIl}F_{iIr}\cos\gamma}\qquad(7.47)$$

The reciprocating inertial forces of the second order for the two cylinders are, respectively,

$$F_{iIIl} = -m_i R\omega^2\tau\cos 2\theta = -Z_i\tau\cos 2\theta$$
$$F_{iIIr} = -m_i R\omega^2\tau\cos 2[360^\circ - (\gamma - \theta)] = -m_i R\omega^2\tau\cos 2(\gamma - \theta)\qquad(7.48)$$

or,

$$F_{iIIr} = -Z_i\tau\cos 2(\gamma - \theta)$$

The resultant force is

$$\sum_1^2 F_{iII} = R_{iII} = \sqrt{F_{iIIl}^2 + F_{iIIr}^2 + 2F_{iIIl}F_{iIIr}\cos\gamma}\qquad(7.49)$$

Subscripts $i, I, II, l,$ and r denote, respectively, inertial, first order, second order, left cylinder, and right cylinder.

The inertial centrifugal force generated by the rotating masses at the crank pin (not shown in Figure 7.27) is always constant and directed along the crank axis away from the crankshaft axis and given by

$$\sum_1^2 K_R = K_R = -\acute{m}_R R\omega^2 = -Z_r$$

where

$\acute{m}_R = m_c + m_{r.cpl} + m_{r.cpr}.$

m_c is a mass that is dynamically equivalent to the masses of the crank pin and the crank webs.

$m_{r.cpl}$ is a mass which is dynamically equivalent to the mass of the connecting rod big end of the left cylinder.

$m_{r.cpr}$ is a mass which is dynamically equivalent to the mass of the connecting rod big end of the right cylinder.

The subscripts here are: c – crank; r. cp – connecting rod at crank pin; l – left cylinder; r – right cylinder. Force K_R can be fully balanced by two counterweights mounted at the ends of the crank webs at a distance r from O, the mass of each being given by

$$m_{cwR} = \frac{1}{2}\frac{R}{r}m_R \tag{7.50}$$

Equations (7.45–7.50) are obtained for the connecting rod-crank linkage shown in Figure 7.18a, in which forces R_{iI} and R_{iII} do not generate a turning moment in the longitudinal plane passing through the crankshaft axis. For the arrangement in Figure 7.18b, the reciprocating inertial forces as well as rotating inertial forces generate moments because of the distance c between the big ends of the two separate connecting rods. These moments are small and are usually ignored. Consequently, Eqs. (7.45–7.50) will be assumed valid for both arrangements shown in Figure 7.18.

For the common V arrangement in which $\gamma = 90^0$, the previous forces can be rewritten as

$$F_{iIl} = -m_i R\omega^2 \cos\theta = -Z_i \cos\theta \tag{7.51}$$

$$F_{iIr} = -m_i R\omega^2 \cos(90^0 - \theta) = -m_i R\omega^2 \sin\theta = -Z_i \sin\theta \tag{7.52}$$

The resultant force is

$$\sum_1^2 F_{iI} = R_{iI} = \sqrt{F_{iIl}^2 + F_{iIr}^2} = m_i R\omega^2 = Z_i \tag{7.53}$$

The angle between the resultant force and the axis of the left cylinder is

$$\tan^{-1}\frac{F_{iIr}}{F_{iIl}} = \theta$$

The resultant force is constant and directed along the crank radius of the left cylinder and can be fully balanced by two counterweights having a mass m_{cwi} each and given by

$$m_{cwi} = \frac{1}{2}\frac{R}{r}m_i \tag{7.54}$$

The resultant rotating inertial force is independent of the value of angle γ and is given, as before, by $\sum_1^2 K_R = K_R = -\dot{m}_R R\omega^2 = -Z_r$.

From Eqs. (7.44) and (7.50), the total mass $m_{cw(t)}$ mounted at the end of each crank web at a distance r from the centre of the crankshaft to balance the first-order inertial force and the rotating inertial force is

$$m_{cw(t)} = m_{cwR} + m_{cwi} = \frac{1}{2}\frac{R}{r}(m_i + m_R) \tag{7.55}$$

The reciprocating inertial forces of the second order for the left and right cylinders are, respectively,

$$F_{iIIl} = -m_i R\omega^2 \tau \cos 2\theta = -Z_i \tau \cos 2\theta \tag{7.56}$$

$$F_{iIIr} = -m_i R\omega^2 \tau \cos 2(90^0 - \theta) = m_i R\omega^2 \tau \cos 2\theta = Z_i \tau \cos 2\theta \tag{7.57}$$

The resultant force is

$$\sum_1^2 F_{iII} = R_{iII} = \sqrt{F_{iIIl}^2 + F_{iIIr}^2} = \sqrt{2}\, m_i R\omega^2 \tau \cos 2\theta = \sqrt{2}\, Z_i \tau \cos 2\theta \tag{7.58}$$

The angle between this force and the axis of the left cylinder is

$$\tan \phi_1 = \frac{F_{iIIr}}{F_{iII}} = -1$$

from which ϕ_1 is equal to $-45°$ or $135°$.

As before, the moments generated by the first- and second-order reciprocating inertial forces and the inertial rotating forces are all assumed equal to zero:

$$\sum_1^2 M_I = 0, \quad \sum_1^2 M_{II} = 0, \quad \sum_1^2 M_R = 0.$$

7.4.2.3 Balancing an Eight-Cylinder V-Engine

This example is for an engine with four throws and V angle $\gamma = 90°$. The cranks are arranged as shown in Figure 7.28, with cranks 1 and 4 in the same plane with opposite throws and cranks 2 and 3 at right angle to 1 and 4 with opposite throws. This engine can be treated as an assembly of four two-cylinder V-engines that are connected by a common crankshaft and numbered from 1 to 4 along the crankshaft in the z direction (Figure 7.28b). Figure 7.28a is a front view of the engine showing all four cranks and all eight connecting rods. For an angular displacement θ for crank 1, cranks 2, 3, and 4 will be displaced by $\theta + 90°, \theta + 270°$, and $\theta + 180°$, respectively.

7.4.2.3.1 Second-Order Inertial Forces
If we think of the two-cylinder V-engine discussed earlier as the first of the four pairs making up an eight-cylinder V-engine with the crankshaft shown in Figure 7.28b, the resultant of the second-order inertial forces for the first pair of cylinders is

$$R_{iII1} = \sqrt{2}\, m_i R\omega^2 \tau \cos 2\theta = \sqrt{2Z_i}\tau \cos 2\theta$$

R_{iII1}, R_{iII2}, R_{iII3}, and R_{iII4} are equal in magnitude but different in direction, as shown in Figure 7.28b. All four forces lie in the plane perpendicular to the plane of crank 1. The sum of the second-order inertial forces in eight cylinders is equal to the sum of the resultant forces for four cylinder pairs:

$$\sum_1^8 F_{iII} = \sum_{n=1}^{n=4} R_{iIIn} = 2(\sqrt{2}\, m_i R\omega^2 \tau \cos 2\theta) - 2(\sqrt{2}\, m_i R\omega^2 \tau \cos 2\theta) = 0$$

or

$$\sum_1^8 F_{iII} = \sum_{n=1}^{n=4} R_{iIIn} = 2(\sqrt{2Z_i}\tau \cos 2\theta) - 2(\sqrt{2Z_i}\tau \cos 2\theta) = 0 \tag{7.59}$$

and the sum of the moments generated by the forces in cranks 1 and 2 and cranks 3 and 4:

$$\sum_1^8 M_{iII} = (\sqrt{2Z_i}\tau \cos 2\theta a) - (\sqrt{2Z_i}\tau \cos 2\theta a) = 0 \tag{7.60}$$

7.4.2.3.2 First Order Inertial Forces
The resultant first-order inertial forces R_{iI1}, R_{iI2}, R_{iII3}, and R_{iII4} for the four cylinder pairs are equal in magnitude (Eq. (7.53)):

$$R_{iI1} = R_{iI12} = R_{iII13} = R_{iII14} = Z_i$$

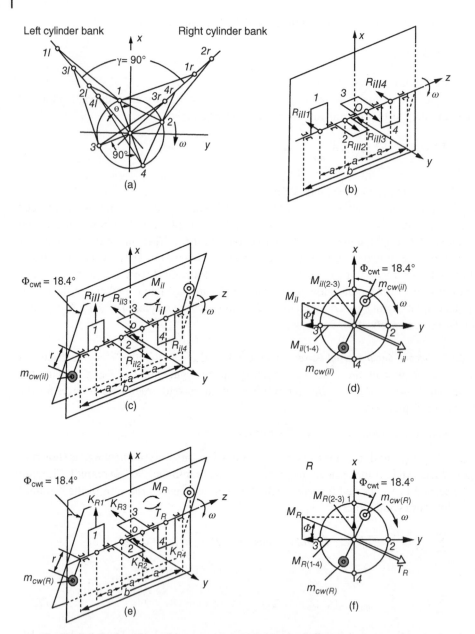

Figure 7.28 Balancing forces and moments in an eight-cylinder V-engine: (a) frontal view of the cranks, connecting rods, and piston displacements; (b) unbalanced second-order inertial forces; (c, d) balancing of first-order inertial forces; (e, f) balancing of rotating inertial forces.

However, forces R_{iI1} and R_{iI4} act in the plane of cranks 1 and 4 and cancel each other, while forces R_{iI2} and R_{iII3} act in the plane of cranks 2 and 3 and also cancel each other. The sum of the first-order inertial forces for all eight cylinders is then

$$\sum_1^8 F_{iI} = \sum_{n=1}^{n=4} R_{iIn} = 0 \tag{7.61}$$

Unlike the case with the second-order inertial forces, the resultant first-order inertial forces generate two couples acting in two perpendicular planes. The forces of the first and fourth cranks generate a couple in the plane of these cranks and equal in magnitude to $(3m_iR\omega^2 a)$. The forces of the second and third cranks generate a couple in the plane at a right angle to the first plane and equal in magnitude to $(m_iR\omega^2 a)$. The moments generated by the forces in cranks 1 and 4 and cranks 2 and 3 can be represented by vector $\overline{M_{iI(1-4)}}$ in the horizontal plane and vector $\overline{M_{iI(2-3)}}$ in the vertical plane, as shown in Figure 7.28d. These vectors are determined by the right-hand rule, with the thumb pointing in the direction of the vector and the remaining curled figures indicating direction of the moment. The resultant moment can now be determined by the vector addition of the two perpendicular vectors $\overline{M_{iI(1-4)}}$ and $\overline{M_{iI(2-3)}}$:

$$\sum_1^8 M_{iI} = \sqrt{(M_{iI(1-4)})^2 + (M_{iI(2-3)})^2} = \sqrt{10}\, m_i R\omega^2 a = \sqrt{10}\, Z_i a \tag{7.62}$$

The vector representing the resulting moment M_{iI} is at an angle ϕ determined from

$$\tan^{-1}\frac{M_{iI(2-3)}}{M_{iI(1-4)}} = \frac{1}{3} \approx 18.4° \tag{7.63}$$

Moment M_{iI} can be balanced by an opposing moment T_{iI} (Figure 7.28d) generated by two equal counterweights at both ends of the crankshaft mounted at a distance r from the axis of the crankshaft, as shown in Figure 7.28c. The mass of the counterweight $m_{cw(iI)}$ can be determined by equating M_{iI} and T_{iI}:

$$m_{cw(iI)}r\omega^2 b = \sqrt{10}\, m_i R\omega^2 a$$

from which

$$m_{cw(iI)} = \frac{a}{b}\frac{R}{r}\sqrt{10 m_i} \tag{7.64}$$

7.4.2.3.3 Rotating (Centrifugal) Inertial Forces These forces are shown in Figure 7.28e and are identical in orientation to the first-order inertial forces shown in Figure 7.28c. Based on the analysis of the forces acting in a two-cylinder V-engine, the centrifugal force in the first crank K_{R1} can be written as

$$K_{R1} = -(m_c + m_{r.cpl} + m_{r.cpr})R\omega^2$$

For identical arrangement and dimensions of the remaining three cylinder pairs,

$$K_{R2} = K_{R3} = K_{R4} = -(m_c + m_{r.cpl} + m_{r.cpr})R\omega^2 \tag{7.65}$$

The sum of these forces is

$$\sum_1^8 K_R = \sum_{n=1}^{n=4} K_{Rn} = 0 \tag{7.66}$$

However, these forces generate two moments: one in the plane of cranks 1 and 4 and equal to $3(m_c + m_{r.\,cpl} + m_{r.\,cpr})R\omega^2 a$ and the other in the plane of cranks 2 and 3 and equal to $(m_c + m_{r.\,cpl} + m_{r.\,cpr})R\omega^2 a$. The vector representation and summing of these two moments are shown in Figure 7.27f.

$$\sum_1^8 M_R = \sqrt{(M_{R(1-4)})^2 + (M_{R(2-3)})^2} = \sqrt{10}\,(m_c + m_{r.cpl} + m_{r.cpr})R\omega^2 a \tag{7.67}$$

The vector representing the resulting moment M_R is at an angle ϕ determined from

$$\tan^{-1}\frac{M_{R(2-3)}}{M_{R(1-4)}} = \frac{1}{3} \approx 18.4^\circ$$

This is the same angle for the sum of the first-order inertial forces given by Eq. (7.63). Moment M_R can be balanced by an opposing moment T_R (Figure 7.28f) generated by two equal counterweights at both ends of the crankshaft mounted at a distance r from the axis of the crankshaft, as shown in Figure 7.28e. The mass of the counterweight $m_{cw(R)}$ can be determined by equating T_R to M_R:

$$m_{cw(R)}r\omega^2 b = \sqrt{10}a(m_c + m_{r.cpl} + m_{r.cpr})R\omega^2$$

from which

$$m_{cw(R)} = \frac{a}{b}\frac{R}{r}\sqrt{10}(m_c + m_{r.cpl} + m_{r.cpr}) \tag{7.68}$$

Since the counterweights to balance M_{II} and M_R are in the same plane at an angle of 18.4° to the plane of cranks 1 and 4, a single counterweight of mass $m_{cw(t)}$ can be mounted at a distance r from the axis of the crankshaft to balance the combined moments of the first-order and rotating inertial forces:

$$m_{cw(t)} = m_{cw(II)} + m_{cw(R)} = \frac{a}{b}\frac{R}{r}\sqrt{10}(m_i + m_c + m_{r.cpl} + m_{r.cpr}) \tag{7.69}$$

7.4.2.4 Other Engine Configurations

The methodology outlined can be applied to any engine with any number of cylinders, crank configuration, and V angles. Tables B-1–B-4 show the unbalanced inertial forces and their moments for a selection of two- and four-stroke inline and V-engines with various crank configurations.

Problems

7.1 The instantaneous position x of the piston shown is approximated by
$$x = R\left[(1 - \cos\theta) + \frac{\tau}{4}(1 - \cos2\theta)\right]$$
If the angle subtended by the connecting rod to the vertical axis of the engine is β, derive the exact equation for x as a function of R, τ, and θ.

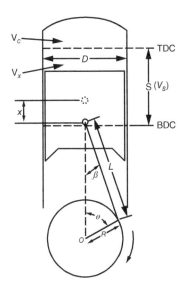

7.2 Plot x and $\dot{x}$ versus θ (0–720°) using the approximate and exact equations if $L = 150$ mm and $R = 50$ mm. Take the engine speed as 3000 *rpm*.

7.3 Piston pin offset is often used in reciprocating internal combustion engines to reduce the side forces on the cylinder walls and minimise the piston slap effect. The following figure shows two reciprocating mechanisms with piston pin offset (a) and without offset (b).

(a) What is the stroke S for the mechanism in (b)?

(b) Show that the stroke in mechanism (a) is given by

$$S = \overline{EA} - \overline{EA''} = R\left[\sqrt{\left(\frac{1}{\tau}+1\right)^2 - k^2} - \sqrt{\left(\frac{1}{\tau}-1\right)^2 - k^2}\right]$$

where $\tau = R/L$ and $k = e/R$

(c) Show that the instantaneous positions of the pistons for the mechanisms are given, respectively, by the equations

$$x = \overline{EA} - \overline{EA'} = R\left[\sqrt{\left(\frac{1}{\tau}+1\right)^2 - k^2} - \left(\cos\theta + \frac{1}{\tau}\cos\beta\right)\right]$$

$$x = \overline{OA} - \overline{OA'} = R\left[1 + \frac{1}{\tau} - \left(\cos\theta + \frac{1}{\tau}\sqrt{1 - \tau^2\sin^2\theta}\right)\right]$$

(d) Plot x versus θ (0–720°) for both mechanisms on the same coordinate system. Take $L = 160$ mm, $R = 60$ mm, and $e = 15$ mm.

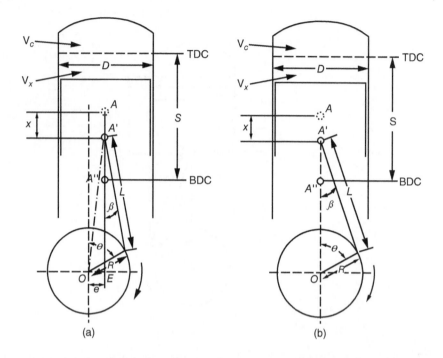

(a) (b)

7.4 A 3.6-*l* V6 SI-engine operates on a four-stroke cycle at 3800 *rpm*. The compression ratio is 10, the length of the connecting rod is 160 *mm*, and the ratio of cylinder diameter to piston stroke is 1. At this speed, combustion ends at 30° after TDC. Calculate

(a) Cylinder diameter, stroke length, and average piston speed
(b) Clearance volume of one cylinder
(c) Piston speed at the end of combustion
(d) Distance the piston travelled from TDC at the end of combustion
(e) Volume in the combustion chamber at the end of combustion

7.5 A six-cylinder CI engine for truck applications is operating at a speed of 2600 *rpm*. Diameter of engine cylinder: 120 *mm*; piston stroke: 120 *mm*; connecting rod length: 220 *mm*. Plot piston displacement, speed, and acceleration versus crank angle for the range 0–720°.

7.6 The engine in problem 7.5 has the equivalent mass distribution shown in the following table. Plot the forces acting on the piston and crank for one cylinder versus crank angle for the range 0–720°.

m_r	m_p	$m_{r.\,pp}$	$m_{r.\,cp}$	m_c	m_i	m_R
3.392 92	2.940 531	0.933 053	2.459 867	3.619 115	3.873 584	8.538 849

The indicator diagram for the engine cycle can be approximated by the function

$$p = 0.4095/(1 - 5.3852 \times 10^{-3}\theta + 7.2787 \times 10^{-6}\theta^2)$$

where p is the cylinder pressure in bar and θ is the crank angle.

7.7 A four-cylinder automotive SI engine is operating at a speed of 5600 *rpm*. Diameter of engine cylinder: 78 *mm*; piston stroke: 78 *mm*; connecting rod length: 136.8 *mm*. Plot piston displacement, speed, and acceleration versus crank angle for the range 0–720°.

7.8 For the engine in problem 7.7, plot the forces for one cylinder acting on the piston and crank pin versus crank angle for the range 0–720°. Equivalent mass distribution is given in the following table.

m_r	m_p	$m_{r.\,pp}$	$m_{r.\,cp}$	m_c	m_l	m_R
0.716	0.478	0.1969	0.5191	0.669	0.6749	1.1881

The indicator diagram for the cycle of this engine can be approximated by the function

$$p = \theta/(9754.79 + 25.8315\theta - 1003.56\sqrt{\theta})$$

where p is in bar, and θ is in degrees crank angle.

7.9 For the engine in problem 7.8, determine
(a) Torque versus crank angle for one cylinder for the range 0–720°
(b) Total torque for all four cylinders plotted versus crank angle for the range 0–720°
(c) Mean engine torque

8

Reciprocating Engine Performance Characteristics

8.1 Indicator Diagrams

Real engine cycles are routinely obtained experimentally in the form of plots of in-cylinder pressure p versus crank angle position θ and are widely used by engine manufacturers and researchers to assess the performance of engines. A $p - \theta$ plot or its derivative $p - V$ plot is known as the *indicator diagram*, an example of which is shown in Figure 6.1. To examine the engine performance parameters from first principles, we will make use of the actual cycles of naturally aspirated (NA) spark ignition (SI) and compression ignition (CI) engines shown in Figure 8.1. The work done by the SI engine and CI engine cycles shown in Figure 8.1 without rounding off the corners is, respectively,

$$W_{cyc(SI)} = (p_1 \varepsilon^{n_1}) \left(\frac{V_1}{\varepsilon} \right) \left[\frac{\alpha}{n_2 - 1} \left(1 - \frac{1}{\varepsilon^{n_2 - 1}} \right) - \frac{1}{n_1 - 1} \left(1 - \frac{1}{\varepsilon^{n_1 - 1}} \right) \right]$$

$$W_{cyc(CI)} = (p_1 \varepsilon^{n_1}) \left(\frac{V_1}{\varepsilon} \right) \left[\alpha(\beta - 1) + \frac{\alpha\beta}{n_2 - 1} \left(1 - \frac{1}{\delta^{n_2 - 1}} \right) - \frac{1}{n_1 - 1} \left(1 - \frac{1}{\varepsilon^{n_1 - 1}} \right) \right]$$

If we round off the corners, we obtain the indicated work for the two engines:

$$W_{i(SI)} = \sigma W_{cyc(SI)} \ Nm \ \text{or} \ J \tag{8.1}$$

$$W_{i(CI)} = \sigma W_{cyc(CI)} \ Nm \ \text{or} \ J \tag{8.2}$$

where σ is the rounding-off coefficient (empirical factor accounting for the losses in work near top dead centre [TDC] and bottom dead centre [BDC], as shown by the shaded areas). Experiments show that $\sigma = 0.92 - 0.97$ for four-stroke engines. For SI engines, σ is closer to the upper limit, whereas it approaches the lower limit in CI engines.

This work is represented for both engines by the area of rectangle $1 - a - c - f$, which is equal to the area enclosed by the rounded-off practical cycles in Figure 8.1. Area $f - c - d - e$ is equivalent to work lost because of the rounding off.

The mean effective pressure (*mep*) was previously defined as the average pressure that produces the same amount of work in a cycle with variable pressure for the same swept

Fundamentals of Heat Engines: Reciprocating and Gas Turbine Internal Combustion Engines, First Edition. Jamil Ghojel.
© 2020 John Wiley & Sons Ltd. This Work is a co-publication between John Wiley & Sons Ltd and ASME Press.
Companion website: www.wiley.com/go/JamilGhojel_Fundamentals of Heat Engines

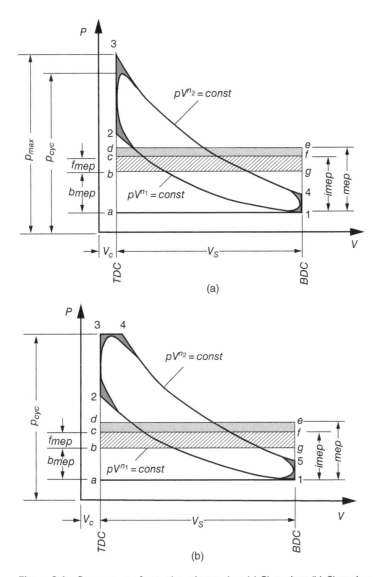

Figure 8.1 Parameters of actual engine cycles: (a) SI engine; (b) CI engine.

volume. The *mep* can be thought of as a hypothetical constant pressure that acts on the piston as it moves from TDC to BDC and the volume changes from minimum volume to maximum volume. For the cycles without rounding off:

$$mep_{(SI)} = p_1 \frac{\varepsilon^{n_1}}{(\varepsilon - 1)} \left[\frac{\alpha}{n_2 - 1} \left(1 - \frac{1}{\delta^{n_2 - 1}} \right) - \frac{1}{n_1 - 1} \left(1 - \frac{1}{\varepsilon^{n_1 - 1}} \right) \right]$$

$$mep_{(CI)} = p_1 \frac{\varepsilon^{n_1}}{(\varepsilon - 1)} \left[\alpha(\beta - 1) + \frac{\alpha\beta}{n_2 - 1} \left(1 - \frac{1}{\delta^{n_2 - 1}} \right) - \frac{1}{n_1 - 1} \left(1 - \frac{1}{\varepsilon^{n_1 - 1}} \right) \right]$$

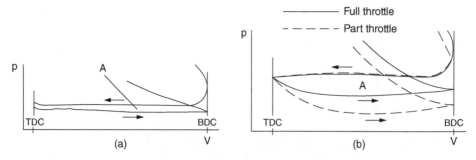

Figure 8.2 p-V indicator diagrams of pumping losses in naturally aspirated engines: (a) CI engine; (b) SI engine.

The indicated mean effective pressure (imep) of the cycles with rounding off can be defined in the same way as the indicated work:

$$imep_{(SI)} = \sigma mep_{(SI)} \tag{8.3}$$

$$imep_{(CI)} = \sigma mep_{(CI)} \tag{8.4}$$

Both $imep_{(CI)}$ and $imep_{(SI)}$ are represented by the height of the rectangle $1-a-c-f$ and equal to the height of rectangle $1-a-d-e$ (*mep*) less the height of rectangle $f-c-d-e$. Note that these two parameters are identical to the mean effective pressures defined by Eqs. (6.13) and (6.21) for the CI and SI engine cycles, respectively, in Chapter 6.

In the discussion of practical cycle models in Chapter 6, pumping losses were ignored. Examples of these losses represented by area A are shown in Figure 8.2; they are small in modern engines and dependent on the engine type and operating conditions. Pumping losses in CI engines are generally small and independent of the load (Figure 8.2a). In SI engines, pumping losses vary significantly with the throttle setting (Figure 8.2b).

In addition to pumping losses, there are other losses that include the work required to overcome friction in the piston-cylinder and connecting rod-crank assemblies and the work required to drive auxiliary equipment such as the oil pump, fuel pump, and alternator. These losses essentially characterise the efficiency of work (or power) transfer from the expanding gases in the cylinder to the crankshaft. The combined losses, collectively referred to as *frictional* or *mechanical losses*, reduce the indicator work W_i by the frictional work W_f, yielding the effective (brake) work W_b, i.e.

$$W_b = W_i - W_f \tag{8.5}$$

The brake work is represented by the area of rectangle $1-a-b-g$ and the frictional losses by the hatched rectangular area $g-b-c-f$ in Figure 8.1a,b, with the height of the rectangle defined as the frictional mean effective pressure (*fmep*).

The *fmep* reduces the *imep* to the brake mean effective pressure (*bmep*):

$$bmep = imep - fmep \tag{8.6}$$

The *bmep* is essentially the specific cycle work (work per unit volume, J/m^3) produced at the engine output shaft. If the cylinder swept volume (displacement) is V_s, then the *bmep* is

$$bmep = \frac{W_b}{V_s} \text{ in units of } \frac{Nm}{m^3} \rightarrow \frac{N}{m^2} \rightarrow Pa \tag{8.7}$$

The *bmep* is a useful parameter that can be used to compare the efficiency of converting heat into mechanical work in different engines irrespective of their physical size (displacement) or mode of combustion.

8.2 Indicated Parameters

Indicated parameters characterise the efficiency of the heat-to-work conversion process inside the cylinder. A very important parameter needs to be introduced at this stage: the volumetric efficiency η_v, defined as

$$\eta_v = \frac{actual\ mass\ of\ air\ in\ the\ cylinder\ at\ end\ of\ induction}{theoretical\ mass\ of\ air\ that\ could\ occupy\ the\ swept\ volume}$$

Mathematically,

$$\eta_v = \frac{m_a}{\rho_0 V_s} = \frac{\dot{m}_a}{\frac{2N}{j}\rho_a V_s} \tag{8.8}$$

where m_a is the actual mass of air trapped in the cylinder per cycle at the end of the induction stroke (kg), ρ_a is the density of the ambient air (kg/m^3) (strictly speaking, this should be the density of the air in the inlet manifold in NI engines and just after the compressor in turbocharged (TC) engines), V_s is the swept volume for one cylinder (m^3), $\dot{m}_a$ is the mass flow rate of air (kg/s), N is the engine speed (rev/sec), and j is the number of strokes per cycle.

The relative air-fuel ratio λ (the inverse of the equivalence ratio ϕ) is defined as

$$\lambda = \frac{actual\ air - fuel\ ratio\ (A/F)}{stoichiometric\ air - fuel\ ratio(A/F)_s}$$

The air-fuel ratios can be on a mass (gravimetric) basis or mole (volume) basis. λ can also be expressed as

$$\lambda = \frac{(A/F)^g}{(A/F)_s^g} = \frac{(m_a/m_f)}{(m_a/m_f)_s^g} = \frac{(\dot{m}_a/\dot{m}_f)}{(\dot{m}_a/\dot{m}_f)_s^g} \tag{8.9}$$

All of these ratios are mass-based. From Eq. (8.9), we obtain

$$\dot{m}_a = \dot{m}_f \lambda (\dot{m}_a/\dot{m}_f)_s^g\ kg/s \tag{8.10}$$

8.2.1 Indicated Work

$$W_i = (imep)V_s\ Nm \tag{8.11}$$

The swept volume $V_s = \pi D^2 S/4$, where D is the cylinder diameter (bore) and S is the piston stroke, both in metres.

8.2.2 Indicated Power

If N is the crankshaft rotational speed (rev/s) and j is the number of strokes per cycle (two for two-stroke engines and four for four-stroke engines), then $2N$ is the number of strokes per second and $2N/j$ the number of cycles per second. Indicated power becomes

$$\dot{W}_i = \frac{2N}{j}V_s(imep)\ W$$

For an engine with i cylinders,

$$\dot{W}_i = \frac{2N}{j} i V_s (imep) \ W \tag{8.12}$$

8.2.3 Indicated Specific Fuel Consumption

This parameter is defined as the amount of fuel consumed to produce a unit of power in a unit of time. If the rate of fuel consumption is denoted by $\dot{m}_f$,

$$isfc = \frac{\dot{m}_f}{\dot{W}_i} \tag{8.13}$$

The units of $isfc$ depend on the units of rate of fuel flow and indicated power. The units encountered in engine practice for $isfc$ include

$$kg/W.s, \ kg/kW.h, \ g/kW.h, \ g/hp.h, \ lbm/hp.h$$

Only the first three, based on SI units, will be used in this book.
From Eqs. (8.12) and (8.13),

$$isfc = \frac{\dot{m}_f}{\dfrac{2N}{j} i V_s (imep)} \tag{8.14}$$

From Eqs. (8.8) and (8.9),

$$\dot{m}_f = \frac{\eta_v \dfrac{2N}{j} \rho_a V_s}{\lambda (A/F)_s^g} \tag{8.15}$$

Substituting this value in Eq. (8.14) yields

$$isfc = \frac{\rho_a}{(A/F)_s^g} \frac{\eta_v}{\lambda (imep)} \ kg/J, \text{ or } kg/W.s \tag{8.16a}$$

where $imep$ is in units of Pa. For practical applications, $imep$ is usually given in units of MPa, and Eq. (8.16a) can be rewritten as

$$isfc = 3600 \frac{\rho_a}{(A/F)_s^g} \frac{\eta_v}{\lambda (imep)} \ kg/kW.h \tag{8.16b}$$

8.2.4 Indicated Efficiency

If the lower heating value of the fuel is H_l (in J/kg), the indicated efficiency can be written as

$$\eta_i = \frac{1}{(isfc)H_l} = \frac{1}{(isfc)H_l} \tag{8.17}$$

From Eqs. (8.16a) and (8.17),

$$\eta_i = \frac{(A/F)_s^g}{H_l} \frac{\lambda (imep)}{\rho_a \eta_v} \tag{8.18}$$

8.2.5 Indicated Mean Effective Pressure

For a known indicated efficiency, from Eq. (8.18),

$$imep = \frac{H_l}{(A/F)_s^g} \frac{\eta_i}{\lambda} \rho_a \ \eta_v \ Pa \tag{8.19}$$

8.2.6 Indicated Power

From Eqs. (8.12) and (8.19),

$$\dot{W}_i = \frac{H_l}{(A/F)_s^g} \frac{\eta_i}{\lambda} \frac{2N}{j} i V_s \rho_a \eta_v \ W \tag{8.20}$$

8.3 Brake Parameters

Brake parameters characterise the efficiency of work (power) transfer to the crankshaft. The power generated at the output from the crankshaft is known as the *effective* or *brake power* (W_b) and is represented by the area of rectangle $1 - a - b - g$ in Figure 8.1. The height of this rectangle is the *bmep*. The mechanical efficiency η_m is defined as the ratio of the brake work to the indicated work, brake power to indicated power, or *bmep* to *imep*:

$$\eta_m = \frac{W_b}{W_i} = \frac{\dot{W}_b}{\dot{W}_i} = \frac{bmep}{imep} \tag{8.21}$$

from which

$$W_b = \eta_m W_i, \ \dot{W}_b = \eta_m \dot{W}_i, bmep = \eta_m(imep)$$

Also,

$$\dot{W}_b = \frac{2N}{j} V_s(bmep) \ W \tag{8.22}$$

8.3.1 Brake-Specific Fuel Consumption

The brake-specific fuel consumption is defined as

$$bsfc = \frac{\dot{m}_f}{\dot{W}_b} \ kg/W.s \tag{8.23}$$

From Eqs. (8.15), (8.22), and (8.23),

$$bsfc = \frac{\rho_a}{(A/F)_s^g} \frac{\eta_v}{\lambda(bmep)} \ kg/W.s \tag{8.24}$$

From Eqs. (8.13) and (8.23),

$$\frac{isfc}{bsfc} = \frac{\dot{W}_b}{\dot{W}_i} = \eta_m, \text{from which}$$

$$bsfc = \frac{isfc}{\eta_m} \tag{8.25}$$

8.3.2 Brake Efficiency

If the lower heating value of the fuel is H_l, the brake efficiency can be written as

$$\eta_b = \frac{\dot{W}_b}{\dot{m}_f H_l} = \frac{1}{(bsfc)H_l} \tag{8.26}$$

From Eq. (8.24) and the right-hand side of Eq. (8.26),

$$\eta_b = \frac{(A/F)_s^g}{H_l} \frac{\lambda(bmep)}{\rho_a \eta_v}$$

(8.27)

Dividing Eqs. (8.18) by (8.27) yields

$$\frac{\eta_i}{\eta_b} = \frac{imep}{bmep} = \frac{1}{\eta_m}, \text{from which}$$

$$\eta_b = \eta_m \eta_i$$

(8.28)

8.3.3 Brake Mean Effective Pressure

From Eqs. (8.19) and (8.21),

$$bmep = \frac{H_l}{(A/F)_s^g} \frac{\eta_i}{\lambda} \eta_v \eta_m \rho_a \ N/m^2$$

(8.29)

8.3.4 Brake Power

For one cylinder:

$$\dot{W}_b = \frac{2N}{j} V_s(bmep) \ W$$

For a multi-cylinder engine with i cylinders:

$$\dot{W}_b = \frac{2N}{j} iV_s(bmep) \ W$$

(8.30)

From Eqs. (8.29) and (8.30),

$$\dot{W}_b = \frac{H_l}{(A/F)_s^g} \frac{\eta_i}{\lambda} \frac{2N}{j} iV_s \eta_v \eta_m \rho_a \ W$$

(8.31)

Engine speed is normally given in revolutions per min (*rpm*), and Eq. (8.31) can be rewritten in the following form:

$$\dot{W}_b = \frac{1}{30} \left(\frac{H_l}{(A/F)_s^g} \right) \left(\frac{\eta_i}{\lambda} \right) \left(\frac{1}{j} \right) (iV_s) \ (\eta_v) \ (\eta_m) \ (\rho_a) \ (N) \quad W$$

$$\begin{array}{c|c|c|c|c|c|c|c}
| & | & | & | & | & | & | & | \\
\hline
1 & 2 & 3 & 4 & 5 & 6 & 7 & 8
\end{array}$$

(8.32a)

Equation (8.32a) is valid for four- and two-stroke engines operating on liquid fuels. The fuels could be from petroleum or from renewable sources such as plants or waste oils and fats. The equation shows the dependence of power output on the following parameters:

(1) Fuel properties (heating value and stoichiometric air-fuel ratio)
(2) Combustion efficiency (characterised by the indicator efficiency and relative air-fuel ratio)
(3) Number of strokes per cycle (2 for two-stroke engine and 4 for four-stroke engine)

(4) Engine size (characterised by the product of the swept volume of one cylinder and the number of cylinders)
(5) Effectiveness of the gas-exchange process (indication of the amount of air induced per cycle) characterised by the volumetric efficiency η_v
(6) Mechanical effectiveness of power transmission to the crankshaft (accounts for pumping losses during gas exchange and frictional losses in the piston-crank mechanism by the mechanical efficiency η_m)
(7) Properties of the air in the inlet manifold, which are strongly dependent on the ambient conditions and the temperature of the induction manifold and can be significantly increased with forced induction (turbocharging)
(8) Engine rotational speed (usually limited by the allowable mechanical loading of engine components and frictional losses in the piston-crank mechanism)

The brake power output of a reciprocating internal combustion engine can be increased by varying each of the parameters in parentheses independently or in combinations. The dependence of power on these parameters varies widely. Changing parameter 1 by using alternative fuels affects power output only marginally. Improving 2, 5, and 6 is unlikely, as they seem to have reached their maximum potential. In certain cases, one parameter may have to be increased while reducing another; for example, for very large power outputs, engine displacement (4) can be increased significantly, which necessitates decreasing speed (8) to few hundred *rpm* to reduce mechanical losses and thermal loading. In compact engines, power can be increased by increasing the density (7) by turbocharging and/or supercharging and decreasing displacement (4). Using a two-stroke design instead of a four-stroke design should theoretically double the power output for the same displacement and speed; in practice, however, the increase is about 80%. For high-speed, high-performance engines, power is usually increased by increasing the number of cylinders i and reducing the swept volume V_s in addition to increasing engine speed (8). This allows the design of a very well-balanced, compact engine.

Piston engines predominantly operate on liquid fuels, but there are situations where gaseous fuels may be advantageous. Generally, piston engines can operate on gaseous fuels such as compressed natural gas (CNG) and liquefied petroleum gas (LPG). The brake power generated by such an engine can be shown to be given by the following equation:

$$\dot{W}_b = 89.7 \left(\frac{H_l}{M_r}\right) \left(\frac{p_2}{T_2}\right) \left(\frac{1}{j}\right) (iV_s)\eta_i(\eta_v)(\eta_m)(N) \ kW \tag{8.32b}$$

In this equation, the lower heating value H_l is in MJ/m^3 (1 kmole of gas is 22.4 m^3 at 273 K and 101.325 kPa), quantity of the reactants M_r in $kmole/kmole\ fuel$, the pressure p_2 and temperature T_2 at the end of compression process in MPa and degrees K, the swept volume V_s in litres, and engine speed N in *rpm*.

8.4 Engine Design Point and Performance

Internal combustion (IC) engines have many applications as stationary or mobile power sources. In stationary applications, they can be found in electricity generation and pumping stations (liquid and gas) in a variety of sizes and outputs. Mobile applications include air, sea (surface and subsurface), and land transportation. Operating regimes in terms of engine power and speed vary significantly depending on each specific application. An indication of the versatility of the IC engine is its capability to cater to all these varying operating regimes economically and reliably.

8.4.1 Design Point Calculations

In Chapter 6, different engine cycle models were introduced that allowed the determination of cycle work with varying degrees of accuracy in comparison with actual cycles. Any of those models in conjunction with the engine brake parameters discussed earlier enable the engine designer to determine the performance of the engine at a single operating point known as the *rated point* or *nominal point*. The parameters at this point are referred to as the *rated power, rated speed, rated torque, rated fuel consumption,* and so on. With the exception of the computer-aided simulation software packages briefly discussed in Section 6.10, there are no theoretical models that are capable of fully predicting engine performance at speeds and loads other than the rated values. Only testing of a prototype engine in a well-equipped laboratory can determine the full range of engine performance characteristics.

An example of a design point calculation methodology for a CI engine is given in Appendix C. The cycle work in the example is calculated from the practical cycle discussed in Section 6.3.1. The other engine performance parameters are calculated using the performance parameter equations in Section 8.3.

8.4.2 Engine Performance Characteristics

Engine performance is dependent on the amount of fuel burned and the corresponding thermal energy release. This energy can be varied in the SI engine by means of the throttle in the induction manifold, which controls the amount of combustible air-fuel mixture entering the cylinder during the induction process. The process is known as *throttling*. Maximum power output is obtained at full throttle; lower power settings are referred to as *partial throttle* settings. In the CI engine, there is no throttling, and power output is regulated by increasing or decreasing the amount of fuel injected into the compressed air at the end of the compression process by means of a control lever in the fuel pump. The terms *throttling, full throttle,* and *part throttle* will be used in this book for both engine types since they are widely used in this context in engine practice.

Engine test results obtained by manufacturers and researchers are usually presented in the form of speed and load characteristics:

1. *Speed characteristics*: These are variable-speed test results of the engine at constant throttle positions. Parameters such as brake power, torque, *bsfc*, rate of fuel consumption, and volumetric efficiency are plotted versus engine speed.
2. *Power characteristics*: These are variable-power test results of the engine at constant speed. Parameters such as *bsfc*, rate of fuel consumption, and relative air-fuel ratio λ are plotted versus brake power or *bmep*.

Figure 8.3 illustrates examples of speed characteristics of SI and CI engines showing variation of brake power and *bsfc* with engine speed at full- and partial-throttle positions.

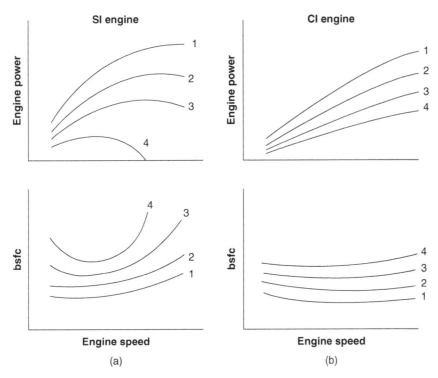

Figure 8.3 Variable-speed characteristics: (a) SI engine; (b) CI engine. (1 – full throttle; 2,3,4 – part throttle).

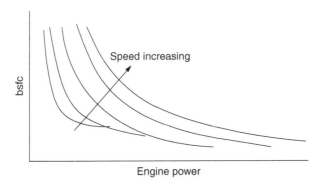

Figure 8.4 Variable-power characteristics of a typical automotive SI engine.

Figure 8.4 is an example of load characteristics showing *bsfc* versus brake power at several constant engine speeds. As Figure 8.3a shows, the throttling action in the SI engine causes a sharp increase in specific fuel consumption and a corresponding sharp decrease in power at the lower scale of partial throttling. The reason behind this is the increase in mechanical losses (decrease in mechanical efficiency) and decrease in volumetric efficiency with throttling.

Table 8.1 Examples of engine test standards.

Standard	Ambient pressure and temperature	Test conditions
SAE (Society of Automotive Engineers – International)	760 $mmHg$, 15.6 °C	Engine tested without muffler, fan, flywheel, alternator, and water pump
CUNA (Italian automotive standards)	760 $mmHg$, 15 °C	Engine tested without muffler
DIN (German industrial standards)	760 $mmHg$, 20 °C	Engine tested with all standard auxiliary equipment

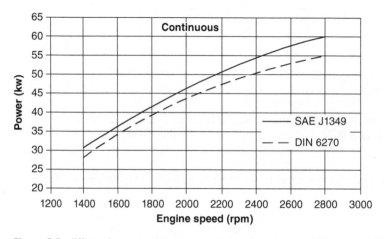

Figure 8.5 Effect of test conditions on engine brake power at different speeds.

When prototype engines are tested to determine performance characteristics, tests are carried out in accordance with set test standards, examples of which are shown in Table 8.1. Depending on the test standards, different results are obtained for the same engine. For example, the Society of Automotive Engineers (SAE) Gross rating is 10–25% higher than the DIN (net) rating, and the CUNA rating is 5–10% higher than the DIN rating. Figure 8.5 shows the continuous power output of a CI engine tested without the cooling fan (SAE) and with the cooling fan (DIN). The difference is the power consumption of the cooling fan. Engine brake power at any atmospheric conditions other than the test conditions can be determined in terms of the measured power according to the test standard from

$$\dot{W}_b = \dot{W}_{b(standard)} \left(\frac{P_{actual}}{P_{standard}} \right) \sqrt{\frac{T_{actual}}{T_{standard}}} \tag{8.33}$$

Both the pressure and temperature of atmospheric air tend to decrease with altitude. However, the rate of decrease of air pressure (and hence density) is much greater. As a result, engine brake power decreases by about 1% for every 100 m increase in altitude.

8.5 Off-Design Performance

Design point calculations based on the simplified cycle models discussed earlier allowed the determination of engine rated performance parameters such as brake power, torque, bsfc, and so on. Full engine characteristics are difficult to estimate because of the variability of the various design and performance parameters involved with engine speed. Taking brake power Eq. (8.32a) as an example, it can be seen that for a given engine, total displacement, and fuel type, there are five variables – η_i, λ, η_v, η_m, and ρ_o – that can change significantly with engine speed (Figure 8.6). The prediction of the behaviour of these parameters as engine speed deviates from the nominal is made difficult by the fact that they are dependent on ambient conditions (temperature, pressure, humidity), design characteristics (turbocharged, multi-valve, flexible valve timing), and the intended application of the engine (automotive, marine, aviation, industrial).

Additional complications arise due to the interdependencies between some parameters. A good example of this is η_i and λ. Eq. (8.32a) indicates that engine power output can be increased by increasing the ratio η_i/λ. However, as Figure 8.7 shows, this could be at the expense of reducing the indicated efficiency η_i. In SI engines, the fuel-delivery system is usually set to operate rich ($\lambda < 1$) for maximum power and lean ($\lambda > 1$) for maximum efficiency (Figure 8.7a). The shaded are is the range within which a typical SI engine would normally operate: $0.85 \le \lambda \le 1.15$. The CI engine displays different characteristics, as can be seen in Figure 8.7b. Because of the way air and fuel are mixed in diesel engines, there is a

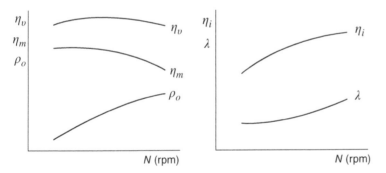

Figure 8.6 Variation of engine parameters with engine speed at full throttle.

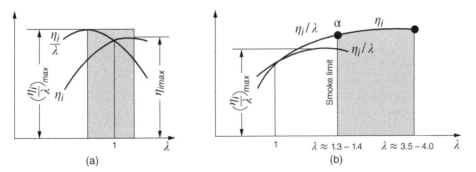

Figure 8.7 Indicated efficiency η_i and the ratio η_i/α versus the relative air-fuel ratio (λ) for (a) SI and (b) CI engines.

minimum value of $\lambda = \lambda_a$ below which the engine should not be operated (point a). This point is known as the *smoke limit* because any further enrichment of the air-fuel mixture below this point causes a sharp increase in soot formation in the combustion chamber, and dense black smoke appears in the exhaust. Since the maximum value of the ratio η_i/λ occurs at a relative air-fuel ratio value below λ_a, CI engines normally operate below their potential peak power. The value of the relative air-fuel ratio at the smoke limit is $\lambda_a = 1.3-1.4$, and the lean limit ($\lambda = 3.5-4$) is much higher than for SI engines; the total operating range is shown by the shaded area.

8.5.1 Speed Characteristics

Attempts were made in the past to predict off-design behaviour of engines using empirical correlations based on engine test data. An example for automotive engines is the power versus speed correlation given here (Kolchin and Demidov, 1984):

$$\left(\frac{\dot{W}_b}{\dot{W}_{b(nom)}}\right) = a\left(\frac{N}{N_{nom}}\right) + b\left(\frac{N}{N_{nom}}\right)^2 + c\left(\frac{N}{N_{nom}}\right)^3 \tag{8.34}$$

The subscript *nom* defines the rated or nominal power and engine speed obtained by the design-point calculations. It should be noted that the rated power of the CI engine occurs at the rated speed and is equal to the maximum power that can be generated (Figure 8.3b). In most SI engines, the maximum power occurs at a speed below the rated speed, as shown in Figure 8.3a; hence, when applying Eq. (8.34), we assume that $W_{b(nom)} = W_{b(max)}$. The recommended constants a, b, and c are given in Table 8.2.

For the brake-specific fuel consumption, the following correlation is suggested:

$$\left(\frac{bsfc}{(bsfc)_{nom}}\right) = a + b\left(\frac{N}{N_{nom}}\right) + c\left(\frac{N}{N_{nom}}\right)^2 \tag{8.35}$$

The recommended constants a, b, and c are given in Table 8.3.

Other performance characteristics, such as torque, rate of fuel consumption, *bmep*, and mechanical efficiency can be estimated based on the brake power and *bsfc*. Another important parameter that needs to be estimated beforehand is the frictional mean effective pressure (*fmep*), which characterises the frictional losses in the moving parts of the recip-rocating mechanism (mainly piston-cylinder friction) and the pumping losses. Frictional losses change almost linearly with mean piston speed C_p (Obert, 1973; Arkhangelsky et al., 1971); therefore, we can write

$$fmep = a + bC_p \; MPa \; (C_p \; in \; m/s) \tag{8.36}$$

Table 8.2 Constants for the empirical Eq. (8.34).

Engine Type		a	b	c
SI engines		1	1	−1
CI engines	Direct injection	0.87	1.13	−1
	Prechamber	0.6	1.4	−1
	Swirl chamber	0.7	1.3	−1

Table 8.3 Constants for the empirical Eq. (8.35).

Engine Type	a	b	c
SI engines	1.2	−1.2	1
CI engines	1.55	−1.55	1

Table 8.4 Constants for the empirical Eq. (8.36).

Engine type		a	b
SI engines	$\dfrac{S}{D} > 1$	0.05	0.0155
	$\dfrac{S}{D} < 1$	0.04	0.0135
CI engines	Direct injection	0.105	0.012
	Indirect injection	0.105	0.0138

According to Arkhangelsky et al. (1971), the values of a and b can be taken as shown in Table 8.4.

The mean piston speed can be written in terms of the engine speed N (rpm) and piston stroke S (m) as follows:

$$C_p = SN/30 \ m/s \tag{8.37}$$

If the *imep* is known,

$$bmep = imep - fmep$$

If the brake power is known,

$$bmep = \frac{120\dot{W}_b}{iV_dN} \ N/m^2 \text{ for four} - \text{stroke engine} \tag{8.38a}$$

$$bmep = \frac{60\dot{W}_b}{iV_dN} \ N/m^2 \text{ for two} - \text{stroke engine} \tag{8.38b}$$

In these equations, $\dot{W}_b$ is in *watts*, iV_d is in m^3, and N is in *rpm*.

$$\eta_m = \frac{bmep}{bmep + fmep} \tag{8.39}$$

Engine torque, based on the equation $\dot{W}_b = \omega T$, can be written as

$$T = \frac{30\dot{W}_b}{\pi N} \ N.m \ (W_b \text{ in } watts, N \text{ in } rpm) \tag{8.40}$$

Figure 8.8 shows plots of the measured and calculated values of $\dot{W}_b$, η_m, and T versus rotational speed for a SI engine with a rated power of 90 kW at 4200 *rpm*.

8.5.2 Load Characteristics

Load characteristics are more difficult to evaluate empirically, and there are no known attempts to devise empirical equations for that purpose. On the basis of analysis of

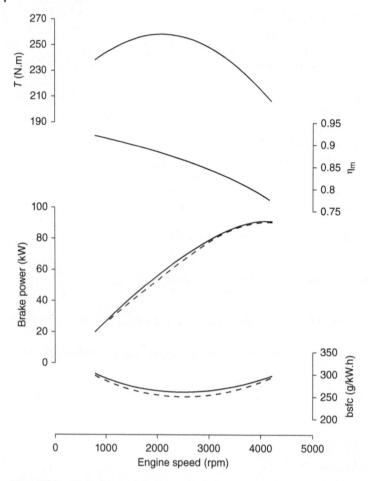

Figure 8.8 SI engine performance characteristics: $S = 95.25$ *mm*, $D = 86.36$ *mm*, $iV_d = 3.69$ *l* (solid line: measured; dashed line: calculated).

published data, two correlations are given that can be used as a qualitative representation of typical load characteristics of SI and CI engines.

8.5.2.1 SI Engines

The following derived correlation for the SI engine is rather cumbersome, but it yields characteristics close to actual engine experimental data for the speed range 800–3200 *rpm* and brake power range 10–120 *kW*:

$$bsfc = a + \frac{b}{N} + c \ln(\dot{W}_b) + \frac{d}{N^2} + e \ln(\dot{W}_b)^2 + \frac{f \ln(\dot{W}_b)}{N} + \frac{g}{N^3} + h \ln(\dot{W}_b)^3$$

$$+ \frac{i \ln(\dot{W}_b)^2}{N} + \frac{j \ln(\dot{W}_b)}{N^2} \tag{8.41}$$

The constant coefficients for this correlation are given in Table 8.5.

Table 8.5 Constant coefficients for Eq. (8.41) (SI engine).

Constant	Value	Constant	Value
a	$-4.76\mathrm{E}+02$	f	$1.41\mathrm{E}+06$
b	$-4.56\mathrm{E}+05$	g	$-7.93\mathrm{E}+11$
c	$1.97\mathrm{E}+03$	h	$6.03\mathrm{E}+01$
d	$2.08\mathrm{E}+09$	i	$2.60\mathrm{E}+05$
e	$-6.58\mathrm{E}+02$	j	$8.55\mathrm{E}+07$

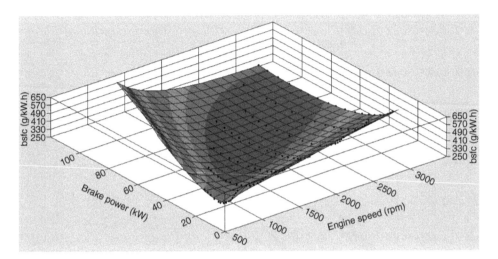

Figure 8.9 3-D representation of the load characteristics of an SI engine: $N = 800-3200\,rpm$, $\dot{W}_b = 10-120\,kW$.

Figures 8.9 and 8.10 show the characteristics, plotted using Eq. (8.41), in 2-D and 3-D formats. Engine fuel economy (*bsfc*) exhibits minima that decrease in magnitude with increasing speed and power. As the engine speed increases, the fuel-consumption curves (*bsfc* = $f(\dot{W}_b)$) become flatter; *bsfc* is at its lowest point at the higher ends of speed and load.

A fuel map is a carpet plot representing engine performance in terms of the brake power (or torque or *bmep*) versus engine speed at constant *bsfc* contours. The fuel map derived from data in Figure 8.9 is shown in Figure 8.11. As the figure shows, the gap between the contours gets narrower with decreasing *bsfc* toward the centre of the concentric profiles. The curve labelled Maximum Power is achieved by regulating the throttle position to obtain the maximum power at each engine speed. The power at maximum speed is the engine rated (nominal) power. Fuel maps are very useful for determining the best power/speed combinations for maximum engine fuel efficiency (lowest *bsfc*).

8.5.2.2 CI Engines
On the basis of analysis of a wide range of published experimental data, a correlation for the *bsfc* as a function of engine power and speed is arrived at for direct-injection CI engines

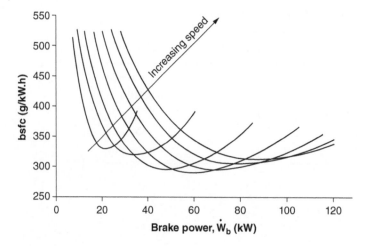

Figure 8.10 2-D load characteristics of the SI engine in Figure 8.9.

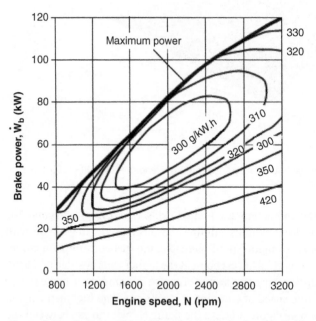

Figure 8.11 Fuel map for an SI engine derived using Eq. (8.41).

within the brake power range 20–320 *kW* and speed range 600–2800 *rpm*:

$$bsfc = a + b(N) + c\left(\frac{1}{\dot{W}_b}\right) + d(N^2) + e\left(\frac{1}{\dot{W}_b{}^2}\right) + f\left(\frac{N}{\dot{W}_b}\right) + g(N^3)$$

$$+ h\left(\frac{1}{\dot{W}_b{}^3}\right) + i\left(\frac{N}{\dot{W}_b{}^2}\right) + j\left(\frac{N^2}{\dot{W}_b}\right)$$

(8.42)

Table 8.6 Constant coefficients for Eq. (8.42) (CI engine).

Constant	Value	Constant	Value
a	3.36E + 02	f	− 6.15E − 01
b	− 1.67E − 01	g	− 3.63E − 09
c	− 3.75EE + 03	h	− 1.74EE + 06
d	5.33E − 05	i	− 2.42EE + 00
e	1.56EE + 05	j	2.40E − 03

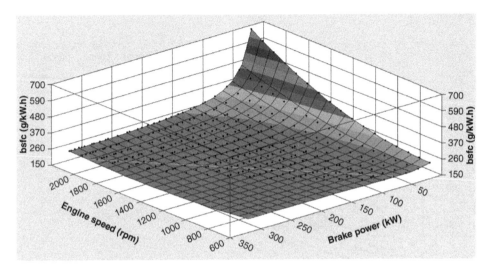

Figure 8.12 3-D representation of the power characteristics for a CI engine: $N = 600-2800\ rpm$, $\dot{W}_b = 20-320\ kW$.

As in the case of SI engines, this equation is not universal and can be used only as a qualitative representation of power characteristics of diesel engines. The characteristics of a newly designed engine can only be obtained by testing a prototype engine under controlled laboratory conditions.

The constant coefficients for Eq. (8.42) for a generic engine generating $320\ kW$ at $2200\ rpm$ are given in Table 8.6.

Figure 8.12 shows a 3-D plot of the function $bsfc = f(\dot{W}_b, N)$ for CI engines in the power range $20-320\ kW$ and speed range $600-2800\ rpm$. The plot shows that $bsfc$ changes little over most of the operating range, but increases rapidly with increasing speed at low power, and with decreasing power at high speed.

Power characteristics in the form $bsfc = f(N)$ at constant power and $bsfc = f(\dot{W}_b)$ at constant speed are shown in Figure 8.13. Figure 8.13b is the most widely used form by researchers and engine manufacturers to present engine performance. In addition to $bsfc$, the rate of fuel consumption and relative air-fuel ratio are sometimes included as dependent variables on the graph.

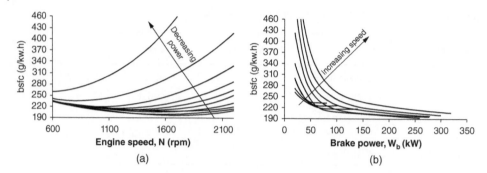

Figure 8.13 2-D power characteristics of a CI engine for the ranges $N = 800$–$2200\ rpm$, $\dot{W}_b = 20$–$320\ kW$.

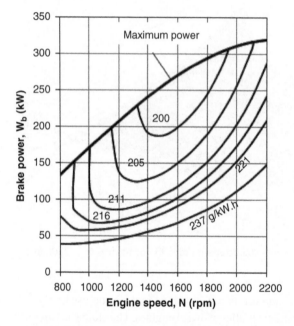

Figure 8.14 Fuel map for the generic CI engine derived from Eq. (8.42)

The carpet plot (fuel map) for the generic CI engine is shown in Figure 8.14. As in the case of the SI engine, the power-speed range gets narrower as *bsfc* decreases. The Maximum Power curve is achieved in the case of CI engines by regulating the amount of fuel delivered by the fuel pump to obtain the maximum power at each engine speed. The power at maximum speed is the engine rated or nominal power. The same conclusions about fuel economy, made at a glance from the 3-D image, can be made by inspecting Figures 8.13 and 8.14.

Fuel maps are sometimes presented in tabular form, as shown in Table 8.7. This abridged tabular version is for the power characteristics $\dot{m}_f = f(\dot{W}_b, N)$ in which *bsfc* is replaced by the fuel rate $\dot{m}_f$ (*kg/h*). This form of the fuel map is particularly useful as input in computer programs designed to simulate performance of on- and off-road vehicles operating under specific travel routes (Ghojel et al., 1990; Ghojel, 1991, 1992, 1993; Ghojel and Watson, 1995).

Table 8.7 CI engine fuel rate map.

$\dot{W}_b(kW)\rightarrow$	40	80	120	160	200	240	280	320
N (rpm)	Fuel rate (kg/h)							
600	9.7	18.7						
700	9.5	18.1						
800	9.4	17.7	26.5					
900	9.4	17.3	25.8					
1000	9.5	17.1	25.3	33.8				
1100	9.7	17.0	25.0	33.1				
1200	9.9	17.0	24.7	32.7	40.7			
1300	10.2	17.1	24.7	32.4	40.2			
1400	10.6	17.4	24.7	32.3	39.9	47.7		
1500	11.1	17.7	24.9	32.4	39.9	47.4		
1600	11.7	18.2	25.3	32.6	40.0	47.4		
1700	12.3	18.7	25.7	33.0	40.3	47.7	55.1	
1800	13.1	19.4	26.4	33.5	40.8	48.1	55.4	
1900	13.9	20.2	27.1	34.2	41.5	48.7	56.1	
2000	14.7	21.0	28.0	35.1	42.3	49.6	56.9	
2100	15.7	22.0	28.9	36.1	43.3	50.7	58.0	
2100	16.7	23.1	30.1	37.3	44.5	51.9	59.3	66.7

Problems

8.1 Measurements show that the mechanical efficiency of a 3.2-litre, 4-stroke, 6-cylinder SI engine operating at full throttle at 6200 *rpm* is 70%. The indicated mean effective pressure (in bar) of this engine is given by the following equation:

$$imep = 0.95 \left(\frac{\varepsilon^{n_1}}{\varepsilon - 1}\right)\left[\frac{\alpha}{n_2 - 1}\left(1 - \frac{1}{\delta^{n_2-1}}\right) - \frac{1}{n_1 - 1}\left(1 - \frac{1}{\varepsilon^{n_1-1}}\right)\right]$$

where $\varepsilon = 10.5$, $\alpha = 4.5$, $n_1 = 1.365$, $n_2 = 1.26$.

Calculate:

(a) Brake mean effective pressure
(b) Brake power and torque of the engine
(c) Rate of fuel consumption, given that the brake-specific fuel consumption is 280 g/kW-h
(d) Thermal efficiency of the engine if the lower heating value of the fuel is 44 MJ/kg

8.2 A four-cylinder racing engine of 2.5 *l* displacement has a compression ratio of 12/1. When tested with a dynamometer, a torque of 290 N. m was obtained at 5000 *rpm*, and at the peak speed of 6750 *rpm* the torque was 250 N. m. The minimum fuel

consumption was 17.2 ml/s at a speed of 5000 rpm. Calculate the maximum $bmep$ in bar, the minimum $bsfc$ in $kg/kW \cdot h$, and the brake thermal efficiency at maximum torque. Compare the latter with the air standard efficiency.
(Take the specific gravity of the fuel as 0.735, $H_l = 44{,}200\ kJ/kg$, brake power $\dot{W}_b = (2N/j)iV_s(bmep)$, and air standard thermal efficiency $\eta_t = 1 - 1/\varepsilon^{\gamma-1}$.)

8.3 A four-stroke four-cylinder SI automotive engine is to be designed to produce 55 kW of power at 5500 rpm. The initial design and operating conditions for the engine are presented in the following table. The design process will be broken into several steps for ease of solution.

Ambient temperature, K	288
Ambient pressure, MPa	0.1
Temperature increase during induction ΔT, K	10
Temperature of the residual gases, K	950
Coefficient of residual gases	0.06
Polytropic compression index n_1	1.36
Polytropic expansion index n_2	1.28
Compression ratio ε	8
Relative air-fuel ratio λ	0.9
Gravimetric fuel analysis, kg	$x_C = 0.855, x_H = 0.145$
Molar mass of fuel	114
Lower heating value of the fuel H_l, kJ/kg	44 000
Heat utilization coefficient, φ	0.82
Volumetric efficiency, η_v	0.925

As a first step in the design, determine
(a) Air required for the combustion process per kg of fuel
(b) Products of combustion

8.4 As a next step of the design of the engine in problem 8.3, determine
(a) Pressure and temperature at the end of the induction process
(b) Pressure and temperature at the end of the compression process

8.5 As a next step of the design of the engine in problem 8.3, determine
(a) Combustion temperature
(b) Maximum cycle pressure (both theoretical and actual)
(c) Pressure and temperature at the end of the expansion process

8.6 As a next step of the design of the engine in problem 8.3, determine
(a) Theoretical and actual indicated mean effective pressure imep
(b) Mechanical efficiency (η_m)
(c) Brake mean effective pressure ($bmep$)

(d) Indicated specific fuel consumption (*isfc*)
(e) Indicated efficiency (η_i)
(f) Brake-specific fuel consumption (*bsfc*)
(g) Brake efficiency (η_b)

8.7 As final step of the design of the engine in problem 8.3, determine
(a) Engine displaced volume and volume per cylinder (in litres)
(b) Piston diameter and stroke
(c) Mean piston speed

Part III

Gas Turbine Internal Combustion Engines

Introduction III: History and Classification of Gas Turbines

Gas turbines are air-breathing, internal combustion heat engines. There are two types of gas turbine cycles: closed and open. Only open cycle gas turbine engine will be considered in this book, and the configurations that will be discussed are highlighted in Chart III.1.

The development of the steam turbine late in the nineteenth century, beginning with the introduction of the first practical engine by Charles Parsons in 1884, was a logical evolution from piston steam engines to rotating heat engines. Similarly, the development of the gas turbine internal combustion engine, starting with the experimental engines of Rene Armengaud and Charles Lemale as early as 1903, was another evolutionary step to replace the steam turbine by eliminating the bulky boiler, condenser, and circulating water. It was also envisaged that with its superior rotary power generation, the gas turbine would replace the piston engine and its cumbersome reciprocating piston-crank mechanism. The gas turbine has been successful in this endeavour only partially. The steam turbine still reigns supreme in electric power generation (~80% of all electric power in the United States), and the piston engine is still dominant in land transport, industrial, and agricultural machinery and remains a serious competitor in low-range power generation. However, the gas turbine has been very successful in other areas such as base, midrange, and peak load electric power generation, aircraft propulsion (subsonic and supersonic), ship propulsion, and pump and compressor drives for liquid and gas pipelines. It is also an ideal heat engine for use in combined cycle and cogeneration applications, operating with impressive thermal efficiencies. Efforts to use gas turbines in cars, trucks, and buses have not been successful to date due to high initial cost and high fuel consumption at idle, in addition to the noise associated with high flow rates of air in the engine.

Historically, the development of aircraft propulsion is of relatively recent origin, starting with Frank Whittle in the UK successfully testing a kerosene-fuelled turbojet engine in 1937 and Hans von Ohain designing a hydrogen-fuelled gas turbine engine that was used in the first flight of a turbojet aircraft in 1939 in Germany just before the outbreak of World War II.

The gas turbine is a continuous-flow engine that can run at high speeds and deliver large amounts of air; therefore, for a given air-fuel ratio, high specific power outputs can be obtained compared with the piston engine. Yet this feature also makes it difficult to operate the gas turbine at a temperature comparable to the temperature in the piston engine. In the piston engine, the combustion process is intermittent, and engine components are exposed to high combustion temperatures only briefly followed by a cooling period between consecutive firing processes. In the gas turbine, engine components are continuously exposed

Fundamentals of Heat Engines: Reciprocating and Gas Turbine Internal Combustion Engines, First Edition. Jamil Ghojel.
© 2020 John Wiley & Sons Ltd. This Work is a co-publication between John Wiley & Sons Ltd and ASME Press.
Companion website: www.wiley.com/go/JamilGhojel_Fundamentals of Heat Engines

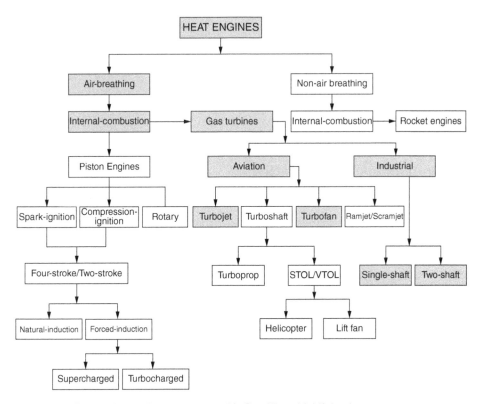

Chart III.1   Gas turbine engine types covered in Part III are highlighted.

to the high-temperature fluid flow, and metallurgical considerations limit the maximum permissible value to about 1800 K. If high-temperature operation is not essential (industrial turbines), the combustion temperature can be controlled by increasing the air-fuel ratio and thereby diluting the combustion products and reducing the turbine entry temperature, with a consequent reduction in thermal efficiency. If high-temperature operation is required in order to increase thermal efficiency and reduce engine size (aircraft engines), critical components such as the blades can be cooled by compressed air at the expense of increased engine cost. The gas turbine engines considered in this book and highlighted in the chart are of the simple type without an air bleed for blade cooling.

As in Part II, the subject matter is presented on the basis of thermodynamic cycles, increasing in complexity gradually from ideal air-standard cycles to more realistic cycles that can be used to model actual engines to a reasonable degree of accuracy. Chapter 15 presents a comprehensive overview of off-design calculations of gas turbines using two methods. Introduced first is the classical method based on the performance characteristics of the turbine(s) and compressor. The second method, called the thermo-gas-dynamic method, employs thermodynamic and gas dynamic relations in lieu of component performance characteristics, making it more universal.

9

Air-Standard Gas Turbine Cycles

Ideal or theoretical cycles are thermodynamically the simplest means to investigate the maximum potential of gas turbines. More complex cycles can then be built around the ideal cycle, up to practical cycles that can be used to predict performance of different types of real gas-turbine engines. Since the working fluid in these cycles is air, they are also known as *air-standard cycles*.

9.1 Joule-Brayton Ideal Cycle

The configuration of a simple single-shaft gas turbine is shown in Figure 9.1a, and the ideal air-standard Joule-Brayton cycle in T-s coordinates is shown in Figure 9.1b. The air-standard cycle is based on the following assumptions:

- The working fluid is air.
- Air behaves likes a perfect gas.
- The compression and expansion processes are isentropic and adiabatic.
- The combustion process is replaced with the addition of heat from an external source at constant pressure equal to the compressor delivery pressure.
- All pressure losses in the combustion chamber, inlet, and exit ducts are ignored.
- No heat is lost or gained in the flow ducts and combustion chamber.

The thermal efficiency of the ideal gas-turbine cycle from the energy balance is

$$\eta_{th} = 1 - \frac{q_{out}}{q_{in}} = 1 - \frac{c_p(T_4 - T_1)}{c_p(T_3 - T_2)} = 1 - \frac{T_1\left(\frac{T_4}{T_1} - 1\right)}{T_2\left(\frac{T_3}{T_2} - 1\right)} \tag{9.1}$$

$$\frac{p_2}{p_1} = \left(\frac{T_2}{T_1}\right)^{\gamma/(\gamma-1)}$$

$$\frac{p_3}{p_4} = \left(\frac{T_3}{T_4}\right)^{\gamma/(\gamma-1)}$$

Since $\dfrac{p_2}{p_1} = \dfrac{p_3}{p_4}$

$$\frac{T_2}{T_1} = \frac{T_3}{T_4}, \text{ or } \frac{T_4}{T_1} = \frac{T_3}{T_2}, \text{ or } \frac{T_4}{T_1} - 1 = \frac{T_3}{T_2} - 1$$

Fundamentals of Heat Engines: Reciprocating and Gas Turbine Internal Combustion Engines, First Edition. Jamil Ghojel.
© 2020 John Wiley & Sons Ltd. This Work is a co-publication between John Wiley & Sons Ltd and ASME Press.
Companion website: www.wiley.com/go/JamilGhojel_Fundamentals of Heat Engines

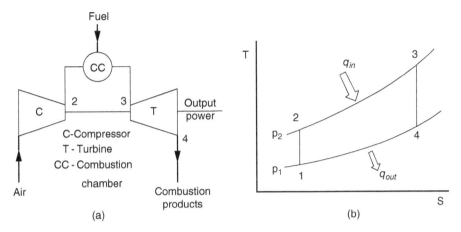

Figure 9.1 Simple gas turbine cycle: (a) engine schematic; (b) T-s diagram of Brayton air-standard cycle.

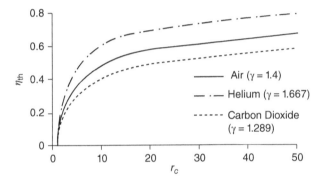

Figure 9.2 Thermal efficiency of the Brayton cycle for three working fluids.

Hence, Eq. (9.1) can be rewritten as

$$\eta_{th} = 1 - \frac{T_1}{T_2} = 1 - \frac{1}{T_2/T_1} = 1 - \frac{1}{(p_2/p_1)^{(\gamma-1)/\gamma}}$$

The compressor pressure ratio $r_c = p_2/p_1$; the thermal efficiency can now be written as

$$\eta_{th} = 1 - \frac{1}{r_c^{(\gamma-1)/\gamma}} \tag{9.2}$$

Figure 9.2 shows the thermal efficiency as a function of the pressure ratio for air, carbon dioxide, and helium. The efficiency increases rapidly with the pressure ratio, with diminishing effect at higher values. The efficiency also increases with an increasing ratio of specific heats γ.

The net specific output work (kJ/kg) is

$$w = w_t - w_c = c_p(T_3 - T_4) - c_p(T_2 - T_1)$$

$$w = c_p T_1 \left(\frac{T_3}{T_1} - \frac{T_4}{T_1} \right) - c_p T_1 \left(\frac{T_2}{T_1} - 1 \right)$$

The nondimensional work term $w/c_p T_1$ (called *specific output work* henceforward) can be written as

$$\frac{w}{c_p T_1} = \left(\frac{T_3}{T_1} - \frac{T_4}{T_1}\right) - \left(\frac{T_2}{T_1} - 1\right) = \frac{T_3}{T_1}\left(1 - \frac{T_4}{T_3}\right) - \left(\frac{T_2}{T_1} - 1\right)$$

$$\frac{w}{c_p T_1} = \frac{T_3}{T_1}\left(1 - \frac{1}{(p_3/p_4)^{(\gamma-1)/\gamma}}\right) - \left(\left(\frac{p_2}{p_1}\right)^{(\gamma-1)/\gamma} - 1\right)$$

$$\frac{w}{c_p T_1} = \frac{T_3}{T_1}\left(1 - \frac{1}{r_c^{(\gamma-1)/\gamma}}\right) - (r_c^{(\gamma-1)/\gamma} - 1)$$

If we define $a = T_{max}/T_{min} = T_3/T_1$ and $b = r_c^{(\gamma-1)/\gamma}$, specific output work can be written as

$$\frac{w}{c_p T_1} = a\left(1 - \frac{1}{b}\right) - b + 1 \tag{9.3}$$

Figure 9.3 shows the specific output work as a function of the pressure ratio in the compressor r_c and temperature ratio T_3/T_1.

The specific output work increases with the pressure ratio at a given temperature ratio, reaching a maximum at a specific r_c. With increasing temperature ratio, the output work increases and the pressure ratio for maximum output work increases (see the dashed maxima line). The maximum output work can be found by differentiating Eq. (9.3) with respect to $r_c^{(\gamma-1)/\gamma}$ and equating the result to zero, which yields

$$(r_{c,w=max})^{(\gamma-1)/\gamma} = \sqrt{\frac{T_3}{T_1}} \tag{9.4}$$

The effect of the pressure ratio on the cycle output work can further be illustrated by the T-s diagram for the Brayton cycle with different pressure ratios while the turbine inlet temperature is maintained constant, as shown in Figure 9.4.

When $r_c \approx 1$, $w = 0$ (the isentropic compression and expansion processes take place with hardly any temperature change, and heat added is equal to heat rejected). On the other hand, for the extreme case when the isentropic compression and expansion processes take place between T_1 and T_3, the pressure ratio is at its maximum and

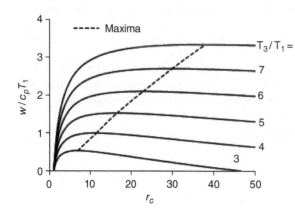

Figure 9.3 Specific output work versus pressure ratio and temperature ratio T_3/T_1.

Figure 9.4 T-s diagrams of the Brayton cycle with different pressure ratios.

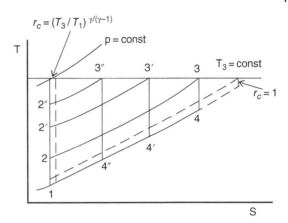

given by $r_c = (T_3/T_1)^{\gamma/(\gamma-1)}$, and $w = 0$ (compressor work is equal to turbine work). The isentropic compression and expansion processes for cycle 1-2-3-4 between the two extremes discussed is

$$r_c^{(\gamma-1)/\gamma} = \frac{T_2}{T_1} = \frac{T_3}{T_4}$$

From Eq. (9.4) for maximum output work,

$$\frac{T_2}{T_1} = \frac{T_3}{T_4} = \sqrt{\frac{T_3}{T_1}} = \sqrt{a}$$

or

$$\frac{T_2}{T_1}\frac{T_3}{T_4} = \frac{T_3}{T_1}$$

or

$$T_2 = T_4 \tag{9.5}$$

This means the specific output work is a maximum when the compressor delivery temperature is equal to the turbine exit temperature.

The temperature ratio a has a significant effect on the specific output work, with the latter increasing linearly with a at a constant pressure ratio r_c (Figure 9.5). The rate of increase of

Figure 9.5 Effect of temperature ratio on the specific output work at two constant compressor pressure ratios.

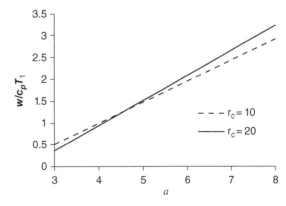

w/c_pT with a rises with increasing r_c. For a constant compressor inlet temperature T_1, the thermal efficiency changes significantly with increasing turbine inlet temperature T_3.

9.2 Cycle with Heat Exchange (Regeneration)

The thermal efficiency of the simple gas-turbine cycle can be increased by using a heat exchanger to heat the air entering the combustion chamber. The condition for achieving this is that the turbine exit temperature is greater than the compressor delivery temperature ($T_4 > T_2$). Heat can be transferred from the exhaust gases leaving the turbine to the compressed air leaving the compressor in a heat exchanger as shown in Figure 9.6. Ideally, the temperature of the compressed air leaving the heat exchanger at point x may be equal to the temperature of the gas leaving the turbine at point 4, yielding a heat-exchanger effectiveness of 1.0. This means the external heat source, such as fuel burning in the combustion chamber, will need to contribute toward increasing the temperature from T_x to T_3 only and not from T_2 to T_3.

The thermal efficiency of the cycle for this case is

$$\eta_{th} = \frac{w_{net}}{q_{in}} = \frac{w_t - w_c}{q_{in}} \tag{9.6}$$

$$w_t = c_p(T_3 - T_4)$$

$$w_c = c_p(T_2 - T_1)$$

$$q_{in} = c_p(T_3 - T_x)$$

For ideal regeneration (effectiveness = 1.0), $T_4 = T_x$ and, hence, $q_{in} = w_t$ and

$$\eta_{th} = 1 - \frac{w_c}{w_t} = 1 - \frac{c_p(T_2 - T_1)}{c_p(T_3 - T_4)} = 1 - \frac{T_1\left(\dfrac{T_2}{T_1} - 1\right)}{T_3\left(1 - \dfrac{T_4}{T_3}\right)}$$

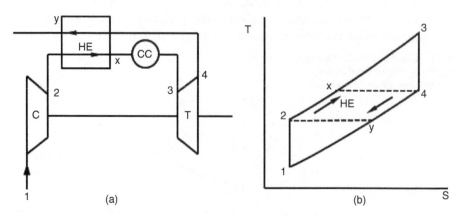

Figure 9.6 Cycle with heat exchange: (a) engine layout; (b) cycle T-s diagram (C – compressor, HE – heat exchanger, CC – combustion chamber, T – turbine).

$$\eta_{th} = 1 - \frac{T_1 \, [(p_2/p_1)^{(\gamma-1)/\gamma} - 1]}{T_3 \, [1 - (p_1/p_2)^{(\gamma-1)/\gamma}]}$$

or

$$\eta_{th} = 1 - \frac{T_1 \, [r_c^{(\gamma-1)/\gamma} - 1]}{T_3 \, [1 - r_c^{-(\gamma-1)/\gamma}]}$$

Multiplying the denominator and numerator of the second term on the right-hand side by $r_c^{(\gamma-1)/\gamma}$ yields

$$\eta_{th} = 1 - \frac{T_1 \, [r_c^{(\gamma-1)/\gamma} - 1] \, r_c^{(\gamma-1)/\gamma}}{T_3 \, [1 - r_c^{-(\gamma-1)/\gamma}] \, r_c^{(\gamma-1)/\gamma}} = 1 - \frac{T_1}{T_3} \frac{[r_c^{(\gamma-1)/\gamma} - 1] r_c^{(\gamma-1)/\gamma}}{[r_c^{(\gamma-1)/\gamma} - 1]}$$

Finally,

$$\eta_{th} = 1 - \frac{T_1}{T_3} r_c^{(\gamma-1)/\gamma} \tag{9.7}$$

By definition, $a = T_{max}/T_{min} = T_3/T_1$ and $b = r_c^{(\gamma-1)/\gamma}$; hence, Eq. (9.7) becomes

$$\eta_{th} = 1 - \frac{b}{a} \tag{9.8}$$

Figure 9.7a shows that, for a given temperature ratio T_3/T_1 (with $T_1 = \text{const}$), the thermal efficiency of a heat-exchange cycle decreases with increasing pressure ratio in the compressor. For a given compressor pressure ratio, the thermal efficiency of the cycle increases with increasing temperature ratio T_3/T_1. At $a = 6$ and $r_c = 20$, the thermal efficiency increases by 5.7% compared with the Brayton cycle. Heat exchange has no effect on the specific output work, and the curves in Figure 9.7b are plotted using Eq. (9.3).

As we have seen earlier, maximum output work is obtained when the temperature equality $T_4 = T_2$ is achieved and

$$(r_{c,w=max})^{(\gamma-1)/\gamma} = \sqrt{\frac{T_3}{T_1}}$$

from which

$$\frac{T_1}{T_3} = \frac{1}{(r_{c,w=max})^{2(\gamma-1)/\gamma}}$$

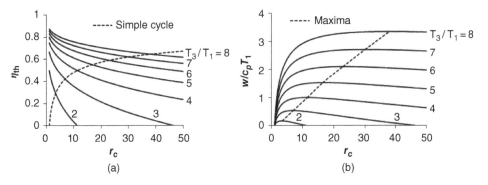

Figure 9.7 Cycle with heat exchange: (a) thermal efficiency; (b) specific output work.

Substituting for T_1/T_3 in Eq. (9.7), the efficiency corresponding to maximum output work becomes

$$\eta_{th(w=max)} = 1 - \frac{(r_{c(w=max)})^{(\gamma-1)/\gamma}}{(r_{c(w=max)})^{2(\gamma-1)/\gamma}} = 1 - \left(\frac{1}{r_{c(w=max)}}\right)^{(\gamma-1)/\gamma} \tag{9.9}$$

Equation (9.9) is equivalent to the thermal efficiency of the Brayton cycle, and it indicates that the thermal efficiencies of the cycle with heat exchange are equal to the thermal efficiencies of the Brayton cycle at the pressure ratios corresponding to the maxima of the output work at constant temperature ratios (points of intersection of the constant T_3/T_1 lines with the efficiency curve of the Brayton cycle in Figure 9.7a). The gain in thermal efficiency due to the heat exchange lies above the Brayton thermal efficiency curve.

Heat exchangers can be used as long as $T_4 > T_2$. When T_4 becomes equal to T_2, the heat-exchange cycle efficiency becomes equal to the Brayton cycle efficiency. When T_4 becomes less than T_2, the heat exchanger cools the air leaving the compressor, thus reducing cycle efficiency.

9.3 Cycle with Reheat

Inspection of Eq. (9.6) shows that for a give compressor work and input heat, the thermal efficiency of the Brayton cycle can be increased by increasing the turbine work. This can be achieved by dividing the expansion process into high-pressure and low-pressure stages and reheating the gas in between, along the process 4–5 (Figure 9.8). The reheating can be effected by providing additional heat in combustion chamber CC2 at a constant pressure $p_4 = p_5$. The vertical distance between any pair of constant pressure lines on the T-s diagram increases with increasing entropy; therefore

$$(T_3 - T_4) + (T_5 - T_6) > (T_3 - T_{4'})$$

This means the turbine expansion work increases to

$$w_t = c_p[(T_3 - T_4) + (T_5 - T_6)] \tag{9.10}$$

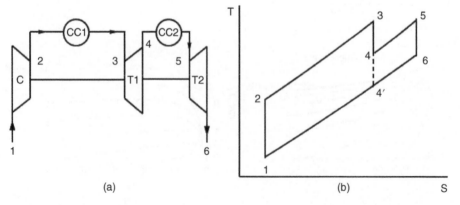

Figure 9.8 Cycle with reheat: (a) engine layout; (b) cycle T-s diagram (C – compressor, CC1 and CC2 – combustion chambers, T1 – HP turbine, T2 – LP turbine).

If the gas is reheated to T_3 ($T_5 = T_3$), we can write

$$w_t = c_p[2T_3 - (T_4 + T_6)] \tag{9.11}$$

And since $p_5 = p_4, p_6 = p_{4'}$,

$$\frac{T_5}{T_6} = \frac{T_3}{T_6} = \left(\frac{p_5}{p_6}\right)^{(\gamma-1)/\gamma} = \left(\frac{p_4}{p_6}\right)^{(\gamma-1)/\gamma}$$

$$\frac{T_4}{T_{4'}} = \left(\frac{p_4}{p_6}\right)^{(\gamma-1)/\gamma}$$

from which

$$\frac{T_5}{T_6} = \frac{T_3}{T_6} = \frac{T_4}{T_{4'}} \text{ and } T_4 = \frac{T_3 T_{4'}}{T_6}$$

and the turbine expansion work can now be written as

$$w_t = c_p\left[2T_3 - \left(\frac{T_3 T_{4'}}{T_6} + T_6\right)\right] \tag{9.12}$$

The maximum turbine work can be found by differentiating Eq. (9.12) with respect to T_6, assuming $T_3 = $ const. and equating the result to zero:

$$\frac{dw_t}{dT_6} = c_p\left[0 - \left(\frac{-T_3 T_{4'}}{T_6{}^2} + 1\right)\right] = 0$$

from which

$$T_6{}^2 = T_3 T_{4'} \tag{9.13}$$

Since

$$T_4 = \frac{T_3 T_{4'}}{T_6}, \text{ or } T_3 T_{4'} = T_4 T_6,$$

it follows from Eq. (9.13) that

$$T_6 = T_4$$

$$\frac{T_3/T_4}{T_5/T_6} = \frac{T_3/T_4}{T_3/T_6} = 1$$

$$\frac{T_3/T_4}{T_5/T_6} = \frac{(p_3/p_4)^{(\gamma-1)/\gamma}}{(p_5/p_6)^{(\gamma-1)/\gamma}} = \frac{(p_3/p_4)^{(\gamma-1)/\gamma}}{(p_4/p_6)^{(\gamma-1)/\gamma}} = 1$$

The last two terms yield

$$\frac{p_3}{p_4} = \frac{p_4}{p_6}, \text{ or } p_4{}^2 = p_3 p_6, \text{ or } p_4 = \sqrt{p_3 p_6}$$

This indicates that maximum expansion work in the turbine is obtained by reheat when

- $T_3 = T_5$
- $T_4 = T_6$
- The high-pressure and low-pressure expansion ratios are equal.

Hence the net output work is

$$w = w_t - w_c$$

$$w = c_p(2T_3 - 2T_4) - c_p(T_2 - T_1)$$

$$\frac{w}{c_p T_1} = \left[\left(2\frac{T_3}{T_1} - 2\frac{T_4}{T_1} \right) - \left(\frac{T_2}{T_1} - 1 \right) \right]$$

$$\frac{w}{c_p T_1} = \left[\left(2\frac{T_3}{T_1} - 2\frac{T_3/T_1}{T_3/T_4} \right) - \left(\frac{T_2}{T_1} - 1 \right) \right]$$

In terms of pressure ratios, the specific output work becomes

$$\frac{w}{c_p T_1} = \left[2\frac{T_3}{T_1} - \frac{2(T_3/T_1)}{(p_3/p_4)^{(\gamma-1)/\gamma}} - \left(\frac{p_2}{p_1} \right)^{(\gamma-1)/\gamma} + 1 \right]$$

Since $p_4 = \sqrt{p_3 p_6}$ and $p_6 = p_1, p_3 = p_2$,

$$\frac{w}{c_p T_1} = \left[2\frac{T_3}{T_1} - \frac{2(T_3/T_1)}{r_c^{(\gamma-1)/2\gamma}} - r_c^{(\gamma-1)/\gamma} + 1 \right] \tag{9.14}$$

Using the following definitions again,

$$a = T_{max}/T_{min} = \frac{T_3}{T_1} \text{ and } b = r_c^{(\gamma-1)/\gamma}$$

$$\frac{w}{c_p T_1} = \left[2a - \frac{2a}{\sqrt{b}} - b + 1 \right] \tag{9.15}$$

The thermal efficiency of the reheat cycle is

$$\eta_{th} = \frac{w}{q_{in}}$$

where q_{in} is the input heat into combustion chambers CC1 and CC2.

$$q_{in} = c_p[(T_3 - T_2) + (T_5 - T_4)] = c_p(2T_3 - T_4 - T_2)$$

$$q_{in} = c_p T_1 \left(\frac{2T_3}{T_1} - \frac{T_4}{T_1} - \frac{T_2}{T_1} \right) = c_p T_1 \left(\frac{2T_3}{T_1} - \frac{T_3/T_1}{T_3/T_4} - \frac{T_2}{T_1} \right)$$

Since $p_4 = \sqrt{p_3 p_6}$, $p_6 = p_1$, and $p_3 = p_2$, q_{in} can be written as

$$q_{in} = c_p T_1 \left[\frac{2T_3}{T_1} - \frac{T_3/T_1}{(p_3/p_4)^{(\gamma-1)/\gamma}} - (p_2/p_1)^{(\gamma-1)/\gamma} \right]$$

or

$$q_{in} = c_p T_1 \left[\frac{2T_3}{T_1} - \frac{T_3/T_1}{r_c^{(\gamma-1)/2\gamma}} - r_c^{(\gamma-1)/\gamma} \right]$$

or

$$q_{in} = c_p T_1 \left(2a - \frac{a}{\sqrt{b}} - b \right) \tag{9.16}$$

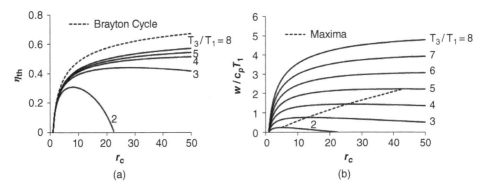

Figure 9.9 Cycle with reheat: (a) thermal efficiency; (b) specific output work.

Finally, the efficiency can be written as

$$\eta_{th} = \frac{2a - \dfrac{2a}{\sqrt{b}} - b + 1}{2a - \dfrac{a}{\sqrt{b}} - b} \tag{9.17}$$

Figure 9.9 shows the plots of the efficiency η_{th} defined by Eq. (9.17) and output work $w/c_p T_1$ defined by Eq. (9.15) for the reheat cycle. Generally, both increase with increasing compressor pressure ratio r_c and temperature ratio T_3/T_1. Reheat has a significant positive effect on the specific work at the expense of efficiency, which does increase with increasing temperature ratio T_3/T_1 but remains below the Brayton cycle efficiency curve. The reason only five efficiency curves are shown in Figure 9.9a is that the curves are so close to each other for ratios $T_3/T_1 = 5 - 8$ that they are almost graphically indistinguishable. The efficiency and specific output work curves exhibit maxima at specific compressor pressure ratios. With increasing temperature ratio, these maxima tend to occur at higher compressor pressure ratios.

9.4 Cycle with Intercooling

Another method to increase the net output work is to reduce compression work while keeping the turbine work constant ($w = w_t - w_c$). This can be achieved by dividing the compression process into low-pressure (C1) and high-pressure (C2) stages and cooling the air between the stages (Figure 9.10).

Intercooling reduces compression work because

$$(T_2 - T_1) + (T_4 - T_3) < (T_{2'} - T_1)$$

Analytical expressions for specific output work and thermal efficiency can be deduced in a similar way to the reheat cycle. As for reheat, the conditions for optimum compression work with intercooling are: $p_4/p_3 = p_2/p_1$, $p_2{}^2 = p_1 p_4$, $T_3 = T_1$, $T_2 = T_4$. The net output work is

$$w = w_t - w_c$$

$$w = c_p(T_5 - T_6) - c_p[(T_2 - T_1) + (T_4 - T_3)]$$

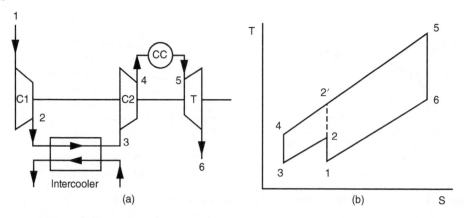

Figure 9.10 Cycle with intercooling: (a) engine layout, (b) cycle T-s diagram (C1 – LP compressor, C – HP compressor, CC – combustion chamber).

$$w = c_p(T_5 - T_6) - c_p(2T_2 - 2T_1)$$

$$w = c_p T_1 \left[\left(\frac{T_5}{T_1} - \frac{T_6}{T_1} \right) - 2 \left(\frac{T_2}{T_1} - 1 \right) \right]$$

$$w = c_p T_1 \left[\left(\frac{T_5}{T_1} - \frac{T_5/T_1}{T_5/T_6} \right) - 2 \left(\frac{T_2}{T_1} - 1 \right) \right]$$

In terms of pressure ratios, the specific output work becomes

$$\frac{w}{c_p T_1} = \left[\frac{T_5}{T_1} - \frac{T_5/T_1}{(p_5/p_6)^{(\gamma-1)/\gamma}} - 2 \left(\frac{p_2}{p_1} \right)^{(\gamma-1)/\gamma} + 2 \right]$$

Since $p_2 = \sqrt{p_1 p_4}$ and $p_6 = p_1$, $p_4 = p_5$ and $r_c = p_4/p_1 = p_5/p_6$ (total pressure ratio),

$$\frac{w}{c_p T_1} = \left[\frac{T_5}{T_1} - \frac{T_5/T_1}{r_c^{(\gamma-1)/\gamma}} - 2 r_c^{(\gamma-1)/2\gamma} + 2 \right] \tag{9.18}$$

Using

$$a = T_{max}/T_{min} = \frac{T_5}{T_1} \text{ and } b = r_c^{(\gamma-1)/\gamma},$$

$$\frac{w}{c_p T_1} = a \left(1 - \frac{1}{b} \right) - 2\sqrt{b} + 2 \tag{9.19}$$

The thermal efficiency of the cycle is

$$\eta_{th} = \frac{w}{q_{in}}$$

where q_{in} is the input heat in CC:

$$q_{in} = c_p(T_5 - T_4)$$

$$q_{in} = c_p T_1 \left(\frac{T_5}{T_1} - \frac{T_4}{T_1} \right)$$

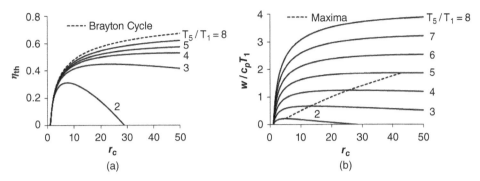

Figure 9.11   Cycle with intercooling: (a) thermal efficiency; (b) specific output work.

Rewriting this equation in terms of pressure ratios, and knowing that $p_6 = p_1$ and

$$q_{in} = c_p T_1 \left[\frac{T_5}{T_1} - (p_4/p_1)^{(\gamma-1)/\gamma} \right]$$

$$q_{in} = c_p T_1 \left[\frac{T_5}{T_1} - r_c^{(\gamma-1)/2\gamma} \right]$$

$$q_{in} = c_p T_1 (a - \sqrt{b})$$

we can finally write the efficiency as

$$\eta_{th} = \frac{a \left(1 - \frac{1}{b} \right) - 2\sqrt{b} + 2}{a - \sqrt{b}} \tag{9.20}$$

Figure 9.11 shows the plots for Eqs. (9.19) and (9.20). The specific output work increases over the entire range of r_c and T_5/T_1, while the thermal efficiency decreases compared with the simple cycle. This pattern of behaviour is similar to the cycle with reheat.

9.5   Cycle with Heat Exchange and Reheat

Adding reheat or heat exchange leads to a significant increase in the specific output work of the cycle. However, while adding a heat exchange also causes an increase in the thermal efficiency, particularly at lower compressor pressure rations, adding reheat to the simple cycle causes a decrease in thermal efficiency of the base cycle. It seems logical then to expect an improvement in efficiency and output work if both reheat and heat exchange are added to the Brayton cycle, as shown in Figure 9.12.

The net output work for this cycle is the same as the reheat cycle:

$$w = w_t - w_c = c_p(T_3 - T_4) + c_p(T_5 - T_6) - c_p(T_2 - T_1)$$

For $T_3 = T_5$ and $T_4 = T_6$,

$$w = c_p(2T_3 - 2T_4) - c_p(T_2 - T_1)$$

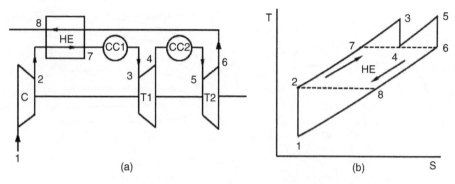

Figure 9.12 Cycle with heat exchange and reheat: (a) engine layout; (b) cycle T-s diagram (C – compressor, HE – heat exchanger, CC1 and CC2 – combustion chambers, T1 – HP turbine, T2 – LP turbine).

As before, $a = T_{max}/T_{min} = T_3/T_1 = T_5/T_1$ and $b = r_c^{(\gamma-1)/\gamma}$. Hence,

$$\frac{w}{c_p T_1} = \left(2a - \frac{2a}{\sqrt{b}} - b + 1 \right) \tag{9.21}$$

The heat input is

$$q_{in} = c_p(T_3 - T_7) + c_p(T_5 - T_4)$$

Since $T_7 = T_4 = T_6$,

$$q_{in} = 2c_p(T_3 - T_4)$$

$$q_{in} = 2c_p T_1 \left(\frac{T_3}{T_1} - \frac{T_4}{T_1} \right) = 2c_p T_1 \left(\frac{T_3}{T_1} - \frac{T_3/T_1}{T_3/T_4} \right) = 2c_p T_1 \left(\frac{T_3}{T_1} - \frac{T_3/T_1}{(p_3/p_4)^{(\gamma-1)/\gamma}} \right)$$

Also, since $p_4 = \sqrt{p_3 p_6}$,

$$q_{in} = 2c_p T_1 \left(\frac{T_3}{T_1} - \frac{T_3/T_1}{r_c^{(\gamma-1)/2\gamma}} \right)$$

Finally, the input heat can be written as

$$q_{in} = c_p T_1 \left(2a - \frac{2a}{\sqrt{b}} \right)$$

The thermal efficiency is

$$\eta_{th} = \frac{w}{q_{in}}$$

or

$$\eta_{th} = \frac{2a\left(1 - \dfrac{1}{\sqrt{b}} \right) - b + 1}{2a\left(1 - \dfrac{1}{\sqrt{b}} \right)} \tag{9.22}$$

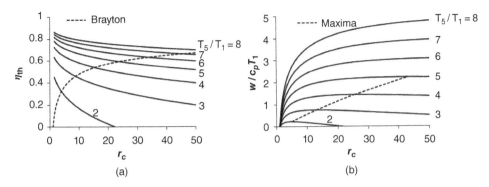

Figure 9.13 Cycle with heat exchange and reheat: (a) thermal efficiency; (b) specific output work.

The efficiency curves for this cycle, shown in Figure 9.13a, are similar to the curves for the heat-exchange cycle, while the specific output work curves (Figure 9.13b) remain unchanged with the addition of a heat exchanger to the reheat cycle. It is evident that the thermal efficiency increases compared with both the Brayton cycle and the Brayton cycle with reheat alone. At $a = 6$ and $r_c = 20$, the thermal efficiency increases by 17.6% compared with the Brayton cycle and 37.3% compared with the Brayton cycle with reheat.

9.6 Cycle with Heat Exchange and Intercooling

The addition of a heat exchanger to a cycle that has intercooling can increase the thermal efficiency of the latter, as was the case with the cycle with reheat and heat exchange. The specific output work of the resulting cycle (Figure 9.14) is the same as for the cycle with intercooling alone and is given by

$$w = w_t - w_c = c_p(T_5 - T_6) - c_p(T_2 - T_1) - c_p(T_4 - T_3)$$

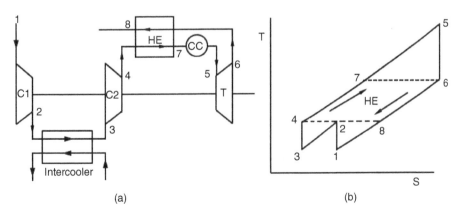

Figure 9.14 Cycle with heat exchange and intercooling: (a) engine layout, (b) cycle T-s diagram (C1 – LP compressor, C2 – HP compressor, HE – heat exchanger, CC – combustion chambers, T – turbine).

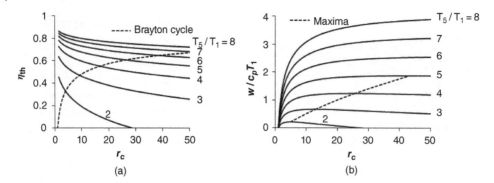

Figure 9.15 Cycle with heat exchange and intercooling: (a) thermal efficiency; (b) specific output work.

For $a = T_{max}/T_{min} = T_5/T_1$ and $b = r_c^{(\gamma-1)/\gamma}$

$$\frac{w}{c_p T_1} = a\left(1 - \frac{1}{b}\right) - 2\sqrt{b} + 2 \qquad (9.23)$$

$$q_{in} = c_p(T_5 - T_7) = c_p(T_5 - T_6)$$

The heat input, considering that $T_7 = T_6$ and $T_4 = T_8$ (heat exchanger effectiveness of 1.0), is

$$q_{in} = c_p T_1\left(\frac{T_5}{T_1} - \frac{T_6}{T_1}\right) = c_p T_1\left(\frac{T_5}{T_1} - \frac{T_5/T_1}{T_5/T_6}\right) = c_p T_1\left(\frac{T_5}{T_1} - \frac{T_5/T_1}{(p_5/p_6)^{(\gamma-1)/\gamma}}\right)$$

$$q_{in} = c_p T_1\left(\frac{T_5}{T_1} - \frac{T_5/T_1}{r_c^{(\gamma-1)/\gamma}}\right)$$

For $a = T_{max}/T_{min} = T_5/T_1$, $b = r_c^{(\gamma-1)/\gamma}$, and $r_c = p_4/p_1 = p_5/p_6$, the heat input is

$$q_{in} = c_p T_1\left(a - \frac{a}{b}\right)$$

The thermal efficiency can now be written as

$$\eta_{th} = \frac{a\left(1 - \frac{1}{b}\right) - 2\sqrt{b} + 2}{a\left(1 - \frac{1}{b}\right)} \qquad (9.24)$$

Figure 9.15 shows the plots for thermal efficiency and specific output work of the Brayton cycle with combined intercooling and heat exchange. It is evident that the thermal efficiency increases compared with both the simple Brayton cycle and the Brayton cycle with intercooling alone. As an example of the changes, for $a = 6$ and $r_c = 20$, the thermal efficiency increases by 20% compared with the simple Brayton cycle and 29.4% compared with the Brayton cycle with intercooling.

9.7 Cycle with Heat Exchange, Reheat, and Intercooling

To take advantage of heat exchange, reheat, and intercooling simultaneously, all three can be combined with the Brayton cycle to obtain the cycle shown in Figure 9.16. The net output

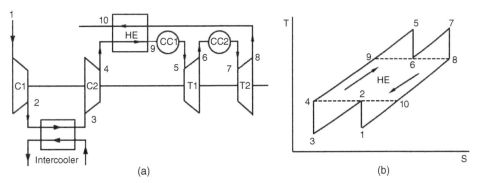

Figure 9.16 Cycle with heat exchange, reheat, and intercooling: (a) engine layout; (b) cycle T-s diagram (C1 – LP compressor, C2 – HP compressor, HE – heat exchanger, CC1 and CC2 – combustion chambers, T1 – HP turbine, T2 – LP turbine).

work of the cycle is

$$w = w_t - w_c = c_p[(T_5 - T_6) + (T_7 - T_8)] - c_p[(T_2 - T_1) + (T_4 - T_3)]$$

Now, $T_1 = T_3$, $T_2 = T_4 = T_{10}$, $T_9 = T_6 = T_8$, $T_5 = T_7$, $p_7 = p_6 = \sqrt{p_5 p_8}$, $p_2 = p_3 = \sqrt{p_4 p_1}$; hence,

$$w = c_p(2T_5 - 2T_6) - c_p(2T_2 - 2T_1)$$

$$w = c_p T_1 \left(2\frac{T_5}{T_1} - 2\frac{T_6}{T_1} \right) - c_p T_1 \left(2\frac{T_2}{T_1} - 2 \right)$$

$$w = c_p T_1 \left(2\frac{T_5}{T_1} - 2\frac{T_5/T_1}{T_5/T_6} \right) - c_p T_1 \left(2\frac{T_2}{T_1} - 2 \right)$$

$$w = c_p T_1 \left(2\frac{T_5}{T_1} - 2\frac{T_5/T_1}{(p_5/p_6)^{(\gamma-1)/\gamma}} \right) - c_p T_1 \left[2\left(\frac{p_2}{p_1}\right)^{(\gamma-1)/\gamma} - 2 \right]$$

$$w = c_p T_1 \left(2\frac{T_5}{T_1} - 2\frac{T_5/T_1}{(p_5/p_8)^{(\gamma-1)/2\gamma}} \right) - c_p T_1 \left[2\left(\frac{p_4}{p_1}\right)^{(\gamma-1)/2\gamma} - 2 \right]$$

For $a = T_{max}/T_{min} = T_5/T_1$, $b = r_c^{(\gamma-1)/\gamma}$, and $r_c = p_5/p_1$

$$w = c_p T_1 \left(2a - 2\frac{a}{\sqrt{b}} \right) - c_p T_1 (2\sqrt{b} - 2)$$

$$\frac{w}{c_p T_1} = \left(2a - \frac{2a}{\sqrt{b}} - 2\sqrt{b} + 2 \right) \tag{9.25}$$

The heat input is the same as in the Brayton cycle with reheat and heat exchange and can be written as

$$\frac{q_{in}}{c_p T_1} = \left(2a - \frac{2a}{\sqrt{b}} \right)$$

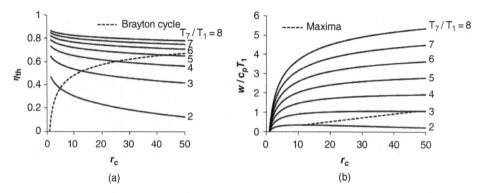

Figure 9.17 Cycle with heat exchange, reheat, and intercooling: (a) thermal efficiency; (b) specific output work.

The thermal efficiency is then

$$\eta_{th} = \frac{\left(a - \dfrac{a}{\sqrt{b}}\right) - \sqrt{b} + 1}{\left(a - \dfrac{a}{\sqrt{b}}\right)} \tag{9.26}$$

Figure 9.17 shows the plots for the thermal efficiency and specific output work as functions of compressor pressure ratio and at different temperature rations.

9.8 Cycle Comparison

Qualitatively, compared to the Brayton cycle, the following are observed:

- Adding a heat exchanger increases thermal efficiency without altering specific output work.
- Adding reheat increases specific output work, but thermal efficiency decreases.
- Adding intercooling has a similar effect to adding reheat.
- Combining heat exchange and reheat with the Brayton cycle increases both efficiency and specific output work.
- Combining heat exchange and intercooling with the Brayton cycle increases both efficiency and specific output work.
- Combining heat exchange, reheat, and intercooling with the Brayton cycle increases both efficiency and specific output work to a greater extent.

Table 9.1 shows the percentage changes of efficiency and specific output work of different cycle configurations compared to the Brayton cycle. It is evident that best performance is obtained when reheat, intercooling, and heat exchange are combined with the simple Brayton cycle.

Figure 9.18 summarises the performance of the different theoretical cycles discussed at temperature ratio $a = 6$ and compressor pressure ratio $r_c = 20$.

Figure 9.19 compares the thermal efficiency and specific output work of the theoretical cycles at a constant temperature ratio $a = 5$ as they vary with the compressor pressure

Table 9.1 Comparison of all theoretical cycles ($a = 6$, $r_c = 20$).

Cycle	Configuration	Efficiency	% Change	Sp. work	% Change
Brayton	I	0.575	0.00	2.097	0.00
Heat exchange	II	0.607	5.68	2.097	0.00
Reheat	III	0.492	−14.37	2.824	34.68
Intercooling	IV	0.533	−8.45	2.382	13.60
Reheat + HE	II + III	0.676	17.55	2.824	34.68
Intercooling + HE	II + IV	0.690	20.06	2.382	13.60
Reheat + intercooling + HE	II + III + IV	0.744	29.43	3.109	48.28

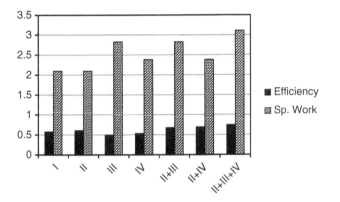

Figure 9.18 Comparison of all theoretical cycles.

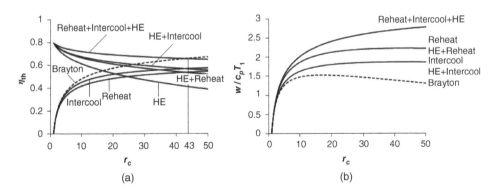

Figure 9.19 Comparison of all theoretical cycles ($a = 5$, $r_c = var$).

ratio r_c. It is clearly seen that all modified Brayton cycles show improvement of specific work over the entire range of r_c (Figure 9.19b). As for the thermal efficiency, not all modifications to the simple Brayton cycle show a positive effect. The cycles with reheat and intercooling are inferior to the Brayton cycle over the entire range of r_c, and the curves are similar to the latter. The remaining cycles exhibit a different curve shape with a downward trend as r_c increases, and the improvement in efficiency occurs at the lower end of r_c. The cycle with reheat, intercooling, and heat exchange is the best performer up to $r_c \approx 43$.

It should be borne in mind that most of the cycle modifications discussed are rarely implemented in real engines, as they are complicated, expensive, and not cost-effective in the long run.

Problems

9.1   A simple gas turbine engine operates on the air-standard ideal cycle with a compressor pressure ratio of 4, between the temperature limits of 20 and 870 °C. Assuming constant specific heat for air, sketch the T-s diagram and determine from first principles
(a) Compressor work
(b) Turbine work
(c) Thermal efficiency

9.2   In a cycle with heat exchange shown next, the temperatures at points 1 and 4 are, respectively, 300 and 700 K.

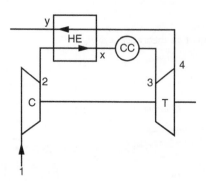

If the pressure ratio in the compressor is 6.5 and the fluid is air, determine
(a) The thermal efficiency of the regenerative cycle
(b) The percentage gain in efficiency compared with the simple cycle

9.3   Repeat the solution for the cycle in Problem 9.1 if the compressor and turbine isentropic efficiencies are, respectively, 0.82 and 0.92.

9.4   Repeat the solution for the cycle in Problem 9.3 if a heat exchanger of effectiveness 0.7 is added.

9.5 Repeat the solution for the cycle in Problem 9.3 if the compressor is replaced by a two-stage compressor with an ideal intercooler in between so that the compressor work is minimised.

9.6 Repeat the solution for the cycle in Problem 9.3 if the turbine is replaced by a two-stage turbine with an ideal reheater in between.

9.7 Repeat the solution for the cycle in Problem 9.6 if the compressor is replaced by a two-stage compressor with an ideal intercooler in between so that the compressor work is minimised.

10

Irreversible Air-Standard Gas Turbine Cycles

In this chapter, ideal air-standard cycles will be made more realistic by accounting for cycle irreversibility due to an increase in entropy in compressors, turbines, diffusers, and nozzles, and pressure losses in inlet and outlet ducts, combustion chambers, and reheaters. Additionally, other losses related to fluid flow and temperature gradients in heat exchangers cause a decrease in heat-transfer effectiveness between hot and cold fluid streams. The resulting cycle is referred to in this book as an *irreversible air-standard cycle*. The thermodynamic irreversible cycle in T-s coordinates and the engine it represents are shown in Figure 10.1. The engine comprises a compressor, a combustion chamber, and a turbine, as shown in Figure 10.1a. This configuration can be used as a gas generator in two-shaft gas turbine, turbojet, and turbofan engines. It is also used as a stand-alone single-shaft gas turbine engine for power generation. In analysing this cycle, we assume that

- The working fluid is air that behaves like a perfect gas (c_p = const.).
- The compression and expansion processes are irreversible (accompanied by an increase in entropy).
- The combustion process is replaced by heat addition from an external source (fuel mass and effect of properties of the combustion products are ignored).
- Pressure losses in the combustion chamber and exit and inlet ducts are accounted for (losses can be represented by fixed pressure drops).
- Heat is not lost or gained in any component of the gas turbine engine.

If a heat exchanger is added to the cycle, the deviation from the ideal case for this components will be accounted for by its effectiveness. The compression and expansion processes are shown as irreversible processes 1–2 and 3–4, respectively, in the T-s diagram in Figure 10.1b. The irreversibilities, accompanied by increased entropy in the compressor and turbine, are normally accounted for by their respective isentropic efficiencies. The pressure losses in the combustion chamber, inlet duct, and exit duct are represented by the small pressure drops Δp_{23}, Δp_{1a}, and Δp_{4a}, respectively. The losses are usually given as fractions or percentages of the compressor delivery pressure or inlet pressure. The inlet duct acts as a nozzle, causing a decrease in pressure from p_a to p_1 and a slight increase in the speed of air flow into the compressor.

Fundamentals of Heat Engines: Reciprocating and Gas Turbine Internal Combustion Engines, First Edition. Jamil Ghojel.
© 2020 John Wiley & Sons Ltd. This Work is a co-publication between John Wiley & Sons Ltd and ASME Press.
Companion website: www.wiley.com/go/JamilGhojel_Fundamentals of Heat Engines

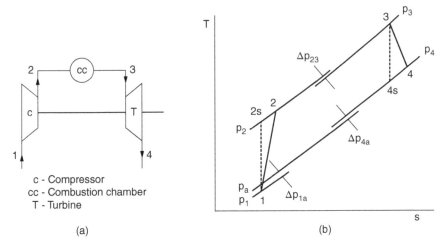

Figure 10.1 Irreversible cycle of a single-shaft gas turbine: (a) engine schematic; (b) T-s diagram of the cycle.

10.1 Component Efficiencies

10.1.1 Compressor Isentropic Efficiency

The compression process in the compressor is shown in Figure 10.2. The process from 1 to 2 is the actual (irreversible) process and from 1 to 2s is the isentropic process. The isentropic efficiency of the compressor is defined either as the ratio of the isentropic specific work w_s to the actual (irreversible) specific work w or as the ratio of the isentropic enthalpy change $(h_{2s} - h_1)$ to the actual enthalpy change $(h_2 - h_1)$:

$$\eta_c = \frac{w_s}{w} = \frac{h_{2s} - h_1}{h_2 - h_1} = \frac{c_p(T_{2s} - T_1)}{c_p(T_2 - T_1)}$$

For a perfect gas, $c_p = $ const.; therefore, the isentropic efficiency of the compressor can be written in terms of the temperatures:

$$\eta_c = \frac{(T_{2s} - T_1)}{(T_2 - T_1)} \tag{10.1}$$

The actual temperature change in the compressor as a result of the compression process is

$$\Delta T_{12} = T_2 - T_1 = \frac{T_1}{\eta_c}\left(\frac{T_{2s}}{T_1} - 1\right)$$

$$T_2 - T_1 = \frac{T_1}{\eta_c}(r_c^{(\gamma-1)/\gamma} - 1) \tag{10.2}$$

where r_c is the pressure ratio p_2/p_1 in the compressor.

10.1.2 Turbine Isentropic Efficiency

The isentropic efficiency of the turbine can be defined in a similar way in terms of the isentropic (3–4s) and irreversible (3–4) expansion processes, as shown in Figure 10.3.

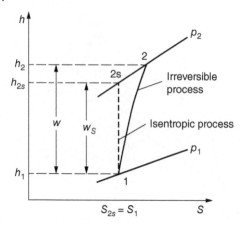

Figure 10.2 Isentropic and irreversible compression processes in a compressor.

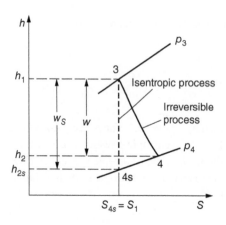

Figure 10.3 Isentropic and irreversible expansion processes in a turbine.

The isentropic efficiency of the turbine is either the ratio of the actual (irreversible) work w to the isentropic work w_s, or the ratio of actual enthalpy change $(h_3 - h_4)$ to the isentropic enthalpy change $(h_3 - h_{4s})$:

$$\eta_t = \frac{w}{w_s} = \frac{h_3 - h_4}{h_3 - h_{4s}} = \frac{c_p(T_3 - T_4)}{c_p(T_3 - T_{4s})}$$

or, for a perfect gas with $c_p = \text{const.}$,

$$\eta_t = \frac{(T_3 - T_4)}{(T_3 - T_{4s})} \tag{10.3}$$

The temperature change in the turbine as a result of the expansion process is

$$\Delta T_{34} = T_3 - T_4 = \eta_t T_3 \left(1 - \frac{T_{4s}}{T_3}\right)$$

$$T_3 - T_4 = \eta_t T_3 \left(1 - \frac{1}{r_t^{(\gamma-1)/\gamma}}\right) \tag{10.4}$$

where r_t is the pressure ratio p_3/p_4 in the turbine $(r_t \neq r_c)$.

10.1.3 Polytropic (Small-Stage) Compressor Efficiency

The isentropic compressor and turbine efficiencies defined previously are not constant and vary with the number of stages in the compressor (variation in pressure ratio). This is due to the fact that the vertical distance between any two constant-pressure lines increases with an increase in entropy. Therefore it is incorrect to assume constant isentropic efficiencies for the compressor and turbine over the wide range of pressure ratios encountered in some applications. The concept of polytropic or small-stage efficiency of compression and expansion is useful, since it can be assumed to be constant and independent of the pressure ratio.

Consider an axial compressor comprising several infinitesimally small stages. The polytropic compression efficiency η_{pc} is defined as the constant isentropic efficiency for each small stage (Figure 10.4).

For an infinitesimal isentropic compression of a perfect gas,

$$\eta_{pc} = \frac{dT_s}{dT} = const. \tag{10.5}$$

For any isentropic compression

$$\frac{T^\gamma}{p^{(\gamma-1)}} = const. \text{ or } \frac{T}{p^{(\gamma-1)/\gamma}} = const.$$

In differential form, the process can be written as

$$\frac{dT}{T} = \frac{\gamma - 1}{\gamma} \frac{dp}{p}$$

For the incremental isentropic process shown in Figure 10.4

$$\frac{dT_s}{T} = \frac{\gamma - 1}{\gamma} \frac{dp}{p}$$

Combining this equation with Eq. (10.5) yields

$$\eta_{pc} \frac{dT}{T} = \frac{\gamma - 1}{\gamma} \frac{dp}{p}$$

Figure 10.4 Infinitely small multistage compression.

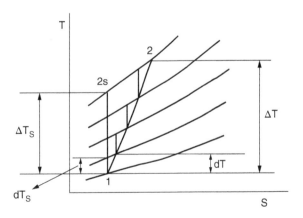

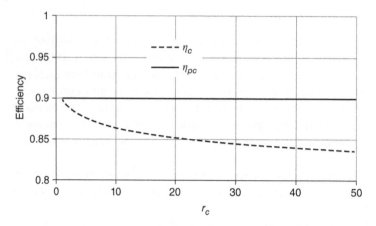

Figure 10.5 Variation of compressor isentropic efficiency with the compressor pressure ratio at a constant polytropic efficiency of 0.9.

Integrating between points 1 and 2, keeping in mind that η_{pc} is by definition constant, we obtain

$$\eta_{pc} = \frac{\ln(p_2/p_1)^{(\gamma-1)/\gamma}}{\ln(T_2/T_1)} \qquad (10.6)$$

This equation enables us to calculate η_{pc} from the known values of p and T at the inlet and outlet of the compressor. Rearranging Eq. (10.6) yields

$$\frac{T_2}{T_1} = \left(\frac{p_2}{p_1}\right)^{(\gamma-1)/\gamma\eta_{pc}}$$

The overall isentropic efficiency from Figure 10.4 is

$$\eta_c = \frac{\Delta T_s}{\Delta T} = \frac{T_{2s} - T_1}{T_2 - T_1} = \frac{\dfrac{T_{2s}}{T_1} - 1}{\dfrac{T_2}{T_1} - 1} = \frac{(p_2/p_1)^{(\gamma-1)/\gamma} - 1}{(p_2/p_1)^{(\gamma-1)/\gamma\eta_{pc}} - 1}$$

$$\eta_c = \frac{r_c^{(\gamma-1)/\gamma} - 1}{r_c^{(\gamma-1)/\gamma\eta_{pc}} - 1} \qquad (10.7)$$

This equation enables us to determine the variation of η_c with the compressor pressure ratio for a given polytropic efficiency η_{pc}. As Figure 10.5 shows, the compressor isentropic efficiency decreases continuously with increasing pressure ratio at the constant polytropic efficiency of 0.9. The drop in the isentropic efficiency is about 6.7% when the pressure ratio increases to 50.

The temperature change in the compressor work can now be written as

$$\Delta T_{12} = T_2 - T_1 = T_1(r_c^{(\gamma-1)/\gamma\eta_{pc}} - 1) \qquad (10.8)$$

10.1.4 Polytropic (Small-Stage) Turbine Efficiency

Consider an axial turbine comprising several infinitely small stages. The polytropic expansion efficiency η_{pt} is defined as the constant isentropic efficiency for each small stage (Figure 10.6).

Figure 10.6 Infinitely small multistage expansion.

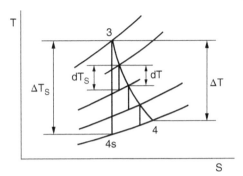

For an infinitesimal isentropic expansion of a perfect gas, the polytropic efficiency is

$$\eta_{pt} = \frac{dT}{dT_s} = const.$$

For an isentropic expansion,

$$\frac{T_s^\gamma}{p^{(\gamma-1)}} = const., \quad or \quad \frac{T_s}{p^{(\gamma-1)/\gamma}} = const.$$

In differential form,

$$\frac{dT_s}{T} = \frac{\gamma-1}{\gamma}\frac{dp}{p}, \quad or \quad \frac{1}{\eta_{pt}}\frac{dT}{T} = \frac{\gamma-1}{\gamma}\frac{dp}{p}$$

Integrating between points 3 and 4, we obtain

$$\eta_{pt} = \frac{ln(T_4/T_3)}{ln(p_4/p_3)^{(\gamma-1)/\gamma}} \tag{10.9}$$

This equation enables us to calculate η_{pt} from the known values of p and T at the inlet and outlet of the turbine. Rearranging Eq. (10.9) yields

$$\frac{T_4}{T_3} = \left(\frac{p_4}{p_3}\right)^{\eta_{pt}(\gamma-1)/\gamma}$$

The overall isentropic efficiency from Figure 10.6 is

$$\eta_t = \frac{\Delta T}{\Delta T_s} = \frac{T_3 - T_4}{T_3 - T_{4s}} = \frac{1 - \dfrac{T_4}{T_3}}{1 - \dfrac{T_{4s}}{T_3}} = \frac{1 - (p_4/p_3)^{\eta_{pt}(\gamma-1)/\gamma}}{1 - (p_4/p_3)^{(\gamma-1)/\gamma}}$$

$$\eta_t = \frac{r_t^{\eta_{pt}(\gamma-1)/\gamma} - 1}{r_t^{(\gamma-1)/\gamma} - 1} \tag{10.10}$$

For a given polytropic turbine efficiency ($\eta_{pt} = 0.85$), the turbine isentropic efficiency increases continuously with increasing pressure ratio (Figure 10.7). The increase is about 7% at the maximum expansion ratio shown.

The temperature change in the turbine can now be written as

$$\Delta T_{34} = T_3 - T_4 = T_3(1 - 1/r_t^{\eta_{pt}(\gamma-1)/\gamma}) \tag{10.11}$$

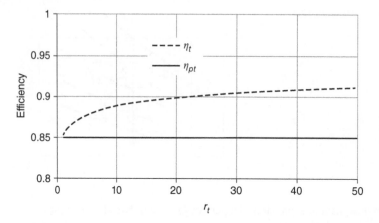

Figure 10.7 Variation of turbine isentropic efficiency with the turbine pressure ratio at constant polytropic efficiency of 0.85.

10.2 Simple Irreversible Cycle

The simple irreversible cycle is the same cycle defined earlier for the single-shaft gas turbine, with all losses accounted for except the inlet pressure loss, as shown in Figure 10.8. As a result, the compressor inlet pressure p_1 is taken as equal to the ambient pressure p_a.

The actual temperature change in the compressor and turbine, accounting for process irreversibility, can be written as per Eqs. (10.8) and (10.11), respectively:

$$\Delta T_{12} = T_2 - T_1 = T_1(r_c^{(\gamma-1)/\gamma\eta_{pc}} - 1) \ \text{ or } \ T_2/T_1 = r_c^{(\gamma-1)/\gamma\eta_{pc}}$$

$$\Delta T_{34} = T_3 - T_4 = T_3\left(1 - \frac{1}{r_t^{\eta_{pt}(\gamma-1)/\gamma}}\right) \ \text{ or } \ T_3/T_4 = r_t^{\eta_{pt}(\gamma-1)/\gamma}$$

The compressor specific work can be written as

$$w_c = c_p(T_2 - T_1)$$

$$w_c = c_p T_1(r_c^{(\gamma-1)/\gamma\eta_{pc}} - 1), \text{ or }$$

$$\frac{w_c}{c_p T_1} = (r_c^{(\gamma-1)/\gamma\eta_{pc}} - 1) \tag{10.12}$$

The turbine specific work can be written as

$$w_t = c_p(T_3 - T_4)$$

$$w_t = c_p T_3\left(1 - \frac{1}{r_t^{\eta_{pt}(\gamma-1)/\gamma}}\right), \text{ or }$$

$$\frac{w_t}{c_p T_1} = \frac{T_3}{T_1}\left(1 - \frac{1}{r_t^{\eta_{pt}(\gamma-1)/\gamma}}\right) \tag{10.13}$$

Figure 10.8 The simple air-standard irreversible cycle.

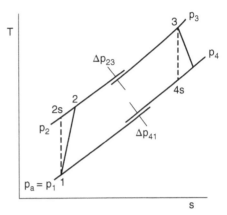

The pressure loss in the combustion chamber Δp_{23} is usually given as a fraction x of the compressor delivery pressure, and the pressure loss in the exhaust duct Δp_{41} is given as a fraction y of the compressor inlet pressure:

$$\Delta p_{23} = xp_2, \Delta p_{41} = yp_1$$

The pressure ratio in the turbine can be expressed in terms of the compressor pressure ratio thus:

$$r_t = \frac{(p_2 - \Delta p_{23})}{(p_1 + \Delta p_{41})} = \frac{p_2 - xp_2}{p_1 + yp_1} = \frac{p_2(1-x)}{p_1(1+y)}$$

$$r_t = r_c\left(\frac{1-x}{1+y}\right)$$

Substituting for r_t in Eq. (10.13) we obtain

$$\frac{w_t}{c_pT_1} = \frac{T_3}{T_1}\left\{1 - \frac{1}{\left[r_c\left(\dfrac{1-x}{1+y}\right)\right]^{\eta_{pt}(\gamma-1)/\gamma}}\right\} \tag{10.14}$$

Let

$$z = \left(\frac{1-x}{1+y}\right)^{(\gamma-1)/\gamma}$$

Substituting for z in Eq. (10.14) yields

$$\frac{w_t}{c_pT_1} = \frac{T_3}{T_1}\left[1 - \frac{1}{z^{\eta_{pt}}r_c^{\eta_{pt}(\gamma-1)/\gamma}}\right] \tag{10.15}$$

(if pressure losses are ignored, $z = 1$).

The net non-dimensional specific output work w can be written as

$$\frac{w}{c_pT_1} = \frac{T_3}{T_1}\left(1 - \frac{1}{z^{\eta_{pt}}r_c^{\eta_{pt}(\gamma-1)/\gamma}}\right) - (r_c^{(\gamma-1)/\gamma\eta_{pc}} - 1)$$

If $a = T_{max}/T_{min} = T_3/T_1$ and $b = r_c^{(\gamma-1)/\gamma}$, the net work in its final form is

$$\frac{w}{c_pT_1} = a\left(1 - \frac{1}{(zb)^{\eta_{pt}}}\right) - b^{1/\eta_{pc}} + 1 \tag{10.16}$$

The heat input is

$$q_{in} = c_p(T_3 - T_2) = c_p T_1 \left(\frac{T_3}{T_1} - \frac{T_2}{T_1} \right) = c_p T_1 \left(\frac{T_3}{T_1} - r_c^{(\gamma-1)/\gamma\eta_{pc}} \right)$$

$$q_{in} = c_p T_1 (a - b^{1/\eta_{pc}})$$

The thermal efficiency is

$$\eta_{th} = \frac{w}{q_{in}}$$

$$\eta_{th} = \frac{a \left(1 - \dfrac{1}{(zb)^{\eta_{pt}}} \right) - b^{1/\eta_{pc}} + 1}{a - b^{1/\eta_{pc}}} \tag{10.17}$$

When the isentropic efficiencies are used, the specific output work and thermal efficiency for the simple irreversible cycle without pressure losses can be written as

$$\frac{w}{c_p T_1} = \left[a\eta_t \left(1 - \frac{1}{b} \right) - \frac{1}{\eta_c}(b - 1) \right] \tag{10.18}$$

$$\eta_{th} = \frac{a\eta_t \left(1 - \dfrac{1}{b} \right) - \dfrac{1}{\eta_c}(b - 1)}{a - \dfrac{b - 1}{\eta_c} - 1} \tag{10.19}$$

To assess the importance of the different parameters affecting the thermal efficiency and specific output work, the following cycles are compared at
$a = T_3/T_1 = 5$:

- Cycle 1: The Brayton cycle
- Cycle 2: Irreversible cycle with isentropic efficiencies ($\eta_c = \eta_t = 0.9$) and no pressure losses ($\Delta p_{23} = 0$, $\Delta p_{41} = 0$, $z = 1$)
- Cycle 3: Irreversible cycle with polytropic efficiencies ($\eta_{pc} = \eta_{pt} = 0.9$) and no pressure losses ($\Delta p_{23} = 0$, $\Delta p_{41} = 0$, $z = 1$)
- Cycle 4: Irreversible cycle with polytropic efficiencies ($\eta_{pc} = \eta_{pt} = 0.9$) and pressure losses ($\Delta p_{23} = 0.06 p_2$, $\Delta p_{41} = 0.04 p_1$, $z = 0.972$)

The results can be seen in Figure 10.9. The introduction of losses reduces the thermal efficiency and output work significantly in all cycles in comparison with the Brayton cycle. For example, for the cycle with isentropic efficiencies without pressure losses (2), the decrease in thermal efficiency and specific output work is between 16.8% and 43.2% over the entire compressor pressure range used.

Comparing cycles 2 and 3 shows that replacing the isentropic efficiencies with the polytropic efficiencies (with other losses ignored) has little effect on the output work, with the change ranging between −4.2% and +1.9%. However, the efficiency varies significantly, with the change ranging between 1.0% and +12.5%. When cycles 2 and 4 are compared, the output work changes between −10.9% and −32.8%. The efficiency changes to a greater extent, with the change being between −32.6% and +12.5%. As pressure losses play a significant role in actual gas turbine engines, all the cycles to be analysed under the name 'irreversible cycles' will include the effect of pressure losses in the combustion chamber and exit duct in addition to the use of the polytropic efficiencies in the compression and expansion processes.

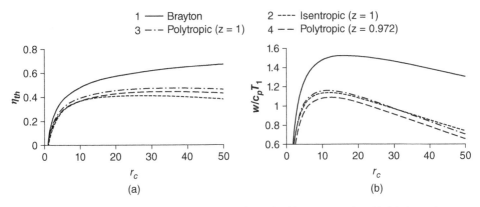

Figure 10.9 Comparison of air-standard cycles with and without losses ($a = 5$): (a) thermal efficiencies; (b) specific output work.

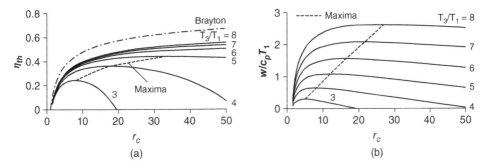

Figure 10.10 Irreversible cycle with pressure losses and irreversibilities accounted for ($\eta_{pc} = \eta_{pt} = 0.9$, $\Delta p_{23} = 0.06p_2$, $\Delta p_{41} = 0.04p_1$, $z = 0.972$).

The full characteristics of the irreversible cycle determined by Eqs. (10.16) and (10.17) are shown in Figure 10.10 where $\eta_{pc} = \eta_{pt} = 0.9$, $\Delta p_{23} = 0.06p_2$, $\Delta p_{41} = 0.04p_1$, and $z = 0.972$. The thermal efficiency of the Brayton cycle is also shown for comparison.

For a given constant compressor inlet temperature T_1 (ambient temperature, in this case), both maximum cycle temperature T_3 and compressor pressure ratio r_c affect the performance of the cycle. For each temperature ratio a (proportional to T_3 at a fixed T_1), the efficiency and specific output work reach maximum values at different pressure ratios (Figure 10.10). As a increases, the pressure ratio at which the peak occurs increases. Thermal efficiencies peak at higher pressure ratios compared to specific output work (see maxima lines). As a is increased above 5, the gain in efficiency becomes marginal, but the specific output work increases markedly (Figure 10.10b). This is particularly important for gas turbine applications where high specific output work, and hence compactness, is an essential design requirement and outweighs efficiency considerations (e.g. aircraft engines).

The shape of the performance characteristics of the irreversible gas turbine cycle can be explained by inspecting Figure 10.11 which depicts the turbine work, compressor work, output work, heat input, and thermal efficiency plotted versus the compressor pressure ratio at $a = T_3/T_1 = 4$.

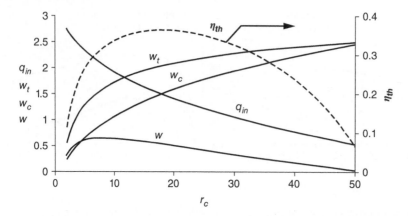

Figure 10.11 Work components, input heat, and efficiency of the irreversible simple cycle at $a = T_3/T_1 = 4$.

After reaching its peak, the net specific output work starts falling because the rate of increase of compressor work is greater than the rate of increase of turbine work with increasing r_c. At constant a (constant T_3 if T_1 is fixed), increasing r_c is accompanied by a decrease of input heat as a result of the higher compressor delivery temperature; however, this does not compensate for the decrease in net output work, and the efficiency eventually starts decreasing as r_c increases further. If T_3/T_1 is increased, all indicators except compressor work increase, and the peak of the efficiency curve shifts towards higher r_c.

10.3 Irreversible Cycle with Heat Exchange (Regenerative Irreversible Cycle)

The addition of a heat exchanger to the Brayton cycle increases the thermal efficiency but has no effect on the specific output work. The irreversible cycle with heat exchange in which irreversibilities in the compressor and turbine are accounted for in addition to the effectiveness of the heat exchanger is shown in Figure 10.12. The effect of irreversibilities in the compressor and turbine are expressed in terms of the polytropic efficiencies η_{pc} and η_{pt}. The losses in the heat exchanger are expressed in terms of the effectiveness of a counter-flow heat exchanger ϵ. The pressure losses in the combustion chamber and exit duct are as for the simple cycle.

The turbine, compressor, and net output work are the same as in the air-standard ideal cycle:

$$\frac{w_c}{c_p T_1} = (b^{1/\eta_{pc}} - 1)$$

$$\frac{w_t}{c_p T_1} = \left(a - \frac{a}{(zb)^{\eta_{pt}}} \right)$$

$$\frac{w}{c_p T_1} = a \left(1 - \frac{1}{(zb)^{\eta_{pt}}} \right) - b^{1/\eta_{pc}} + 1$$

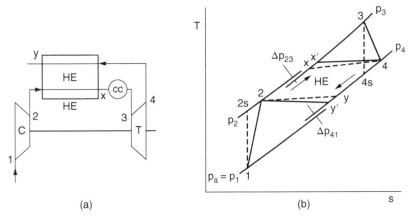

Figure 10.12 Irreversible cycle with heat exchange: (a) engine schematic; (b) T-s diagram of the cycle.

The input heat is

$$q_{in} = c_p(T_3 - T_x)$$

where T_x is the air temperature at the exit from the heat exchanger. The heat-exchanger effectiveness in terms of the cold fluid is

$$\epsilon = \frac{\text{Cold side HE exit temperature} - \text{Compressor exit temperature}}{\text{Turbine exit temperature} - \text{Compressor exit temperature}}$$

$$\epsilon = \frac{T_x - T_2}{T_4 - T_2}$$

Hence,

$$q_{in} = c_p[T_3 - T_2 - \epsilon(T_4 - T_2)]$$

$$q_{in} = c_p T_1 \left[\frac{T_3}{T_1} - \frac{T_2}{T_1} - \epsilon \left(\frac{T_4}{T_1} - \frac{T_2}{T_1} \right) \right] = c_p T_1 \left[\frac{T_3}{T_1} - \frac{T_2}{T_1}(1 - \epsilon) - \epsilon \left(\frac{T_4}{T_1} \right) \right]$$

$$q_{in} = c_p T_1 \left[\frac{T_3}{T_1} - \frac{T_2}{T_1}(1 - \epsilon) - \epsilon \left(\frac{T_3/T_1}{T_3/T_4} \right) \right]$$

$$q_{in} = c_p T_1 \left\{ \frac{T_3}{T_1} - [(1 - \epsilon)(r_c^{(\gamma-1)/\gamma\eta_{pc}})] - \epsilon \left[\frac{(T_3/T_1)}{r_t^{\eta_{pt}(\gamma-1)/\gamma}} \right] \right\}$$

Using the previous definitions of a, b, and z, we obtain

$$q_{in} = c_p T_1 \left[a - (1 - \epsilon)b^{1/\eta_{pc}} - \frac{\epsilon a}{(zb)^{\eta_{pt}}} \right]$$

$$\frac{q_{in}}{c_p T_1} = \left[a - (1 - \epsilon)b^{1/\eta_{pc}} - \frac{\epsilon a}{(zb)^{\eta_{pt}}} \right] \tag{10.20}$$

The thermal efficiency is therefore

$$\eta_{th} = \frac{a\left(1 - \dfrac{1}{(zb)^{\eta_{pt}}}\right) - b^{1/\eta_{pc}} + 1}{a - (1 - \epsilon)b^{1/\eta_{pc}} - \dfrac{\epsilon a}{(zb)^{\eta_{pt}}}} \tag{10.21}$$

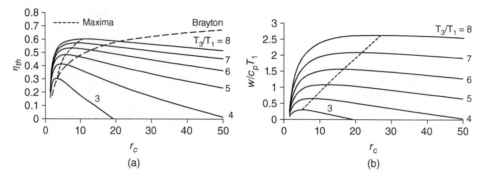

Figure 10.13 Irreversible cycle with heat exchange: (a) thermal efficiency; (b) output work for the actual heat-exchange cycle ($\epsilon = 0.8$, $\eta_{pc} = \eta_{pt} = 0.9$, $\Delta p_{23} = 0.06 p_2$, $\Delta p_{41} = 0.04 p_1$, $z = 0.972$).

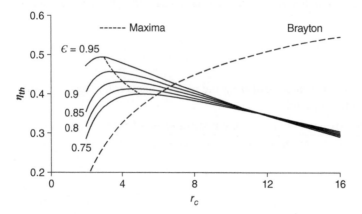

Figure 10.14 Effect of heat-exchanger effectiveness ϵ on the thermal efficiency of the irreversible cycle with heat exchange at $a = 4$ ($\eta_{pc} = \eta_{pt} = 0.9$, $\Delta p_{23} = 0.06 p_2$, $\Delta p_{41} = 0.04 p_1$, $z = 0.972$).

Figure 10.13 shows the specific output work and thermal efficiency of this cycle as functions of the temperature ratio a and compressor pressure ratio r_c. Efficiency gains occur at low pressure ratios, and output work increases significantly at much higher pressure ratios. The output work curves retain the shape we have seen for the ideal Brayton cycle with heat exchange, with peak values shifting towards the higher values of r_c (Figure 9.7). The efficiency curves, on the other hand, are very different from the curves for the Brayton cycle and exhibit a peak for each value of a with a small increase in the corresponding value of r_c as a increases (Figure 10.13a).

An increase in the effectiveness of the heat exchanger at constant temperature ratio $a = 4$ causes an increase in the efficiency at pressure ratios below 12, with peaks moving towards lower pressure ratios (Figure 10.14). At pressure ratios above 12, increasing the effectiveness causes the heat exchanger to cool the air leaving the compressor instead of heating it, and causes the thermal efficiency to decrease. This can happen when the compressor delivery temperature becomes higher than the turbine exit temperature. Heat exchanger effectiveness has no effect on the specific output work.

10.4 Irreversible Cycle with Reheat

The T-s diagram for the irreversible reheat cycle is shown in Figure 10.15. The losses accounted for are the irreversibilities in the compressor and turbine (expressed in terms of the polytropic efficiencies η_{pc} and η_{pt}) and the pressure losses in the combustion chamber (Δp_{23}) and exit duct (Δp_{61}).

The net output work is

$$w = c_p[(T_3 - T_4) + (T_5 - T_6)] - c_p(T_2 - T_1)$$

The same assumptions made in the ideal cycle with reheat are made here: $T_3 = T_5$, $T_4 = T_6$, and $p_4 = \sqrt{p_3 p_6}$. Hence,

$$w = c_p T_1 \left[2\left(\frac{T_3}{T_1} - \frac{T_4}{T_1}\right)\right] - c_p T_1 \left(\frac{T_2}{T_1} - 1\right)$$

$$\frac{w}{c_p T_1} = \left[2\left(\frac{T_3}{T_1} - \frac{T_3/T_1}{T_3/T_4}\right)\right] - \left(\frac{T_2}{T_1} - 1\right)$$

$$\frac{w}{c_p T_1} = \left[2\left(\frac{T_3}{T_1} - \frac{T_3/T_1}{(p_3/p_4)^{\eta_{pt}(\gamma-1)/\gamma}}\right)\right] - [(p_2/p_1)^{(\gamma-1)/\gamma \eta_{pc}} - 1]$$

$$\frac{w}{c_p T_1} = \left[2\left(\frac{T_3}{T_1} - \frac{T_3/T_1}{r_t^{\eta_{pt}(\gamma-1)/2\gamma}}\right)\right] - [r_c^{(\gamma-1)/\gamma \eta_{pc}} - 1]$$

Using the previous definitions of a, b, and z, we obtain

$$\frac{w}{c_p T_1} = 2a\left(1 - \frac{1}{(zb)^{\eta_{pt}/2}}\right) - b^{1/\eta_{pc}} + 1 \tag{10.22}$$

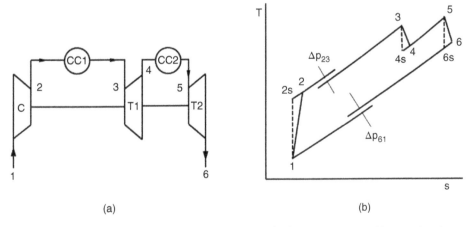

(a)   (b)

Figure 10.15   Irreversible cycle with reheat and losses in the compressor, turbine, combustion chamber, and exit duct.

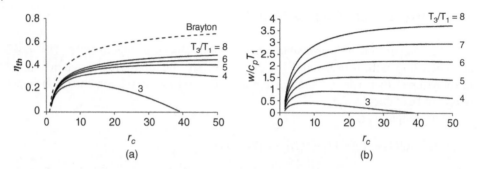

Figure 10.16 Irreversible cycle with reheat: (a) thermal efficiency; (b) output work ($\eta_{pc} = \eta_{pt} = 0.9$, $\Delta p_{23} = 0.06 p_2$, $\Delta p_{61} = 0.04 p_1$, $z = 0.972$).

The input heat is

$$q_{in} = c_p(T_3 - T_2) + c_p(T_5 - T_4)$$

$$q_{in} = c_p T_1 \left[\left(\frac{T_3}{T_1} - \frac{T_2}{T_1} \right) + \left(\frac{T_5}{T_1} - \frac{T_4}{T_1} \right) \right]$$

$$q_{in} = c_p T_1 \left[2 \left(\frac{T_3}{T_1} \right) - \frac{T_2}{T_1} - \frac{T_4}{T_1} \right] = c_p T_1 \left[2 \left(\frac{T_3}{T_1} \right) - \frac{T_2}{T_1} - \frac{T_3/T_1}{T_3/T_4} \right]$$

$$q_{in} = c_p T_1 \left[2 \left(\frac{T_3}{T_1} \right) - r_c^{(\gamma-1)/\gamma\eta_{pc}} - \frac{T_3/T_1}{r_t^{\eta_{pt}(\gamma-1)/2\gamma}} \right]$$

$$\frac{q_{in}}{c_p T_1} = 2a - b^{1/\eta_{pc}} - \frac{a}{(zb)^{\eta_{pt}/2}} \tag{10.23}$$

The thermal efficiency is therefore

$$\eta_{th} = \frac{2a \left(1 - \dfrac{1}{(zb)^{\eta_{pt}/2}} \right) - b^{1/\eta_{pc}} + 1}{2a - b^{1/\eta_{pc}} - \dfrac{a}{(zb)^{\eta_{pt}/2}}} \tag{10.24}$$

Figure 10.16 shows the thermal efficiency and the specific output work of this cycle as functions of temperature ratio a and compressor pressure ratio r_c. Unlike the output work, the efficiency changes little with r_c at temperature ratios above 5. The maxima of efficiency and output work shift significantly towards high values of r_c with increasing temperature ratio a.

10.5 Irreversible Cycle with Intercooling

The T-s diagram for the irreversible reheat cycle is shown in Figure 10.17. The losses accounted for are the same as for the reheat cycle.

The net output work is

$$w = c_p(T_5 - T_6) - c_p[(T_2 - T_1) + (T_4 - T_3)]$$

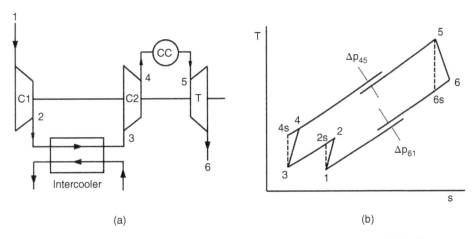

Figure 10.17   Irreversible cycle with intercooling: (a) engine schematic, (b) cycle T-s diagram.

Assuming that $T_3 = T_1$, $T_4 = T_2$, and $p_3 = \sqrt{p_4 p_1}$,

$$w = c_p T_1 \left(\frac{T_5}{T_1} - \frac{T_6}{T_1} \right) - c_p T_1 \left[\left(\frac{T_2}{T_1} - 1 \right) + \left(\frac{T_4}{T_1} - \frac{T_3}{T_1} \right) \right]$$

$$\frac{w}{c_p T_1} = \left(\frac{T_5}{T_1} - \frac{T_5/T_1}{T_5/T_6} \right) - \left(2\frac{T_2}{T_1} - 2 \right)$$

$$\frac{w}{c_p T_1} = \left(\frac{T_5}{T_1} - \frac{T_5/T_1}{(p_5/p_6)^{\eta_{pt}(\gamma-1)/\gamma}} \right) - 2(p_2/p_1)^{(\gamma-1)/\gamma \eta_{pc}} + 2$$

$$\frac{w}{c_p T_1} = \left(\frac{T_5}{T_1} - \frac{T_5/T_1}{r_t^{\eta_{pt}(\gamma-1)/\gamma}} \right) - 2r_c^{(\gamma-1)/2\gamma \eta_{pc}} + 2$$

Using the previous definitions of a, b, and z, we obtain

$$\frac{w}{c_p T_1} = a \left(1 - \frac{1}{(zb)^{\eta_{pt}}} \right) - 2b^{1/2\eta_{pc}} + 2 \tag{10.25}$$

The heat input is

$$q_{in} = c_p (T_5 - T_4)$$

$$q_{in} = c_p T_1 \left(\frac{T_5}{T_1} - \frac{T_4}{T_1} \right) = c_p T_1 \left(\frac{T_5}{T_1} - \frac{T_2}{T_1} \right)$$

$$q_{in} = c_p T_1 \left[\frac{T_5}{T_1} - r_c^{(\gamma-1)/2\gamma \eta_{pc}} \right]$$

$$\frac{q_{in}}{c_p T_1} = a - b^{1/2\eta_{pc}} \tag{10.26}$$

The thermal efficiency is therefore

$$\eta_{th} = \frac{a \left(1 - \frac{1}{(zb)^{\eta_{pt}}} \right) - 2b^{1/2\eta_{pc}} + 2}{a - b^{1/2\eta_{pc}}} \tag{10.27}$$

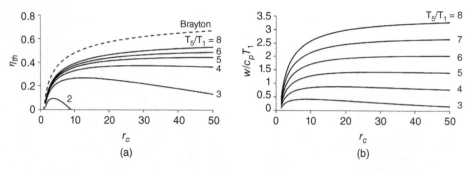

Figure 10.18 Irreversible cycle with intercooling: (a) thermal efficiency; (b) output work ($\eta_{pc} = \eta_{pt} = 0.9$, $\Delta p_{45} = 0.06p_4$, $\Delta p_{61} = 0.04p_1$, $z = 0.972$).

The plots for the thermal efficiency and specific output work are shown in Figure 10.18. The shapes of the plots are similar to the plots for the reheat cycle, but the values for output work are lower.

10.6 Irreversible Cycle with Heat Exchange and Reheat

Schematic diagrams of the engine and cycle T-s diagram are shown in Figure 10.19. The losses taken into account are losses due to irreversible processes in the compressor and turbine, losses in the heat exchanger, pressure losses in the combustion chamber, and pressure losses in the exhaust duct. We have already seen that there is nothing to gain from using reheat or intercooling without heat exchange in an ideal cycle. The same applies to irreversible cycles: adding reheat or intercooling to a heat-exchange cycle will cause the specific output work to increase markedly without loss of efficiency.

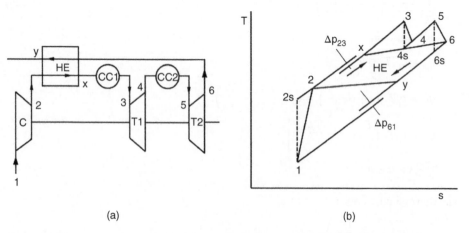

Figure 10.19 Irreversible cycle with heat exchange and reheat: (a) engine schematic, (b) cycle T-s diagram.

Net specific output work is the same as for the irreversible reheat cycle:

$$\frac{w}{c_p T_1} = 2a\left(1 - \frac{1}{(zb)^{\eta_{pt}/2}}\right) - b^{1/\eta_{pc}} + 1 \tag{10.28}$$

The input heat is

$$q_{in} = c_p(T_3 - T_x) + c_p(T_5 - T_4) = c_p(2T_3 - T_x - T_4)$$

Heat-exchanger effectiveness in terms of the cold fluid is

$$\epsilon = \frac{T_x - T_2}{T_6 - T_2}$$

$$T_x = T_2 + \epsilon(T_6 - T_2)$$

$$q_{in} = c_p[2T_3 - T_2 - \epsilon(T_6 - T_2) - T_4] = c_p[2T_3 - T_2(1 - \epsilon) - T_4(1 + \epsilon)]$$

$$q_{in} = c_p T_1 \left[2\frac{T_3}{T_1} - \frac{T_2}{T_1}(1 - \epsilon) - \frac{T_4}{T_1}(1 + \epsilon)\right]$$

$$q_{in} = c_p T_1 \left[2\frac{T_3}{T_1} - \frac{T_2}{T_1}(1 - \epsilon) - \frac{T_3/T_1}{T_3/T_4}(1 + \epsilon)\right]$$

$$q_{in} = c_p T_1 \left[2\frac{T_3}{T_1} - (1 - \epsilon)\left(\frac{p_2}{p_1}\right)^{(\gamma-1)/\eta_{pc}\gamma} - \frac{T_3/T_1}{(p_3/p_4)^{\eta_{pt}(\gamma-1)/\gamma}}(1 + \epsilon)\right]$$

$$q_{in} = c_p T_1 \left[2\frac{T_3}{T_1} - (1 - \epsilon)r_c^{(\gamma-1)/\eta_{pc}\gamma} - \frac{T_3/T_1}{r_t^{\eta_{pt}(\gamma-1)/2\gamma}}(1 + \epsilon)\right]$$

$$q_{in} = c_p T_1 \left[2a - (1 - \epsilon)b^{1/\eta_{pc}} - \frac{a}{(zb)^{\eta_{pt}/2}}(1 + \epsilon)\right]$$

$$\frac{q_{in}}{c_p T_1} = 2a - (1 - \epsilon)b^{1/\eta_{pc}} - \frac{a}{(zb)^{\eta_{pt}/2}}(1 + \epsilon) \tag{10.29}$$

The thermal efficiency is therefore

$$\eta_{th} = \frac{2a\left(1 - \frac{1}{(zb)^{\eta_{pt}/2}}\right) - b^{1/\eta_{pc}} + 1}{2a - (1 - \epsilon)b^{1/\eta_{pc}} - \frac{a}{(zb)^{\eta_{pt}/2}}(1 + \epsilon)} \tag{10.30}$$

The resulting efficiency and output work are plotted as functions of a and r_c as shown in Figure 10.20.

For the temperature ratios investigated, thermal efficiency curves display peaks in the lower range of compressor pressure ratio. The efficiencies up to $r_c \approx 25$ are higher for all temperature ratios compared with the Brayton cycle. Most curves for the specific output work also exhibit maxima within the investigated range of compressor pressure ratios with the points of the maxima shifting towards the higher range of r_c.

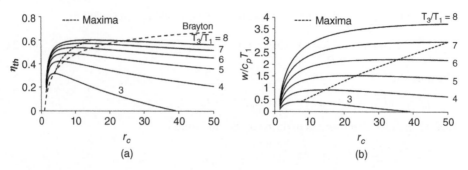

Figure 10.20 Irreversible cycle with heat exchange and reheat: (a) thermal efficiency; (b) output work ($\epsilon = 0.8$, $\eta_{pc} = \eta_{pt} = 0.9$, $\Delta p_{23} = 0.06 p_2$, $\Delta p_{61} = 0.04 p_1$, $z = 0.972$).

10.7 Irreversible Cycle with Heat Exchange and Intercooling

This cycle is shown in Figure 10.21. The addition of heat exchange to the cycle with intercooling should have the effect we saw for the ideal cycle with intercooling and heat exchange.

The net specific output work is the same as the intercooled cycle:

$$\frac{w}{c_p T_1} = a \left(1 - \frac{1}{(zb)^{\eta_{pt}}} \right) - 2b^{1/2\eta_{pc}} + 2$$

The input heat is

$$q_{in} = c_p (T_5 - T_x)$$

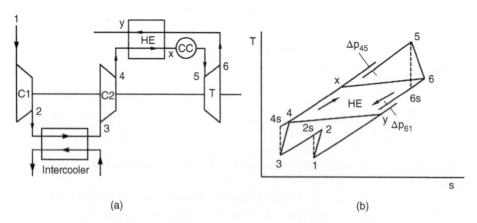

(a)   (b)

Figure 10.21 Irreversible cycle with heat exchange and intercooling: (a) engine schematic; (b) cycle T-s diagram.

Heat-exchanger effectiveness in terms of the cold fluid is

$$\epsilon = \frac{T_x - T_4}{T_6 - T_4}$$

$$T_x = T_4 + \epsilon(T_6 - T_4)$$

$$q_{in} = c_p[T_5 - T_4 - \epsilon(T_6 - T_4)] = c_p[T_5 - T_4(1 - \epsilon) - \epsilon T_6]$$

$$q_{in} = c_p T_1 \left[\frac{T_5}{T_1} - \frac{T_4}{T_1}(1 - \epsilon) - \epsilon \frac{T_6}{T_1} \right]$$

$$q_{in} = c_p T_1 \left[\frac{T_5}{T_1} - \frac{T_2}{T_1}(1 - \epsilon) - \epsilon \frac{T_5/T_1}{T_5/T_6} \right]$$

$$q_{in} = c_p T_1 \left[\frac{T_5}{T_1} - (1 - \epsilon)\left(\frac{p_2}{p_1}\right)^{(\gamma-1)/\eta_{pc}\gamma} - \epsilon \frac{T_5/T_1}{(p_5/p_6)^{\eta_{pt}(\gamma-1)/\gamma}} \right]$$

$$q_{in} = c_p T_1 \left[\frac{T_5}{T_1} - (1 - \epsilon)r_c^{(\gamma-1)/\eta_{pc}\gamma} - \epsilon \frac{T_5/T_1}{r_t^{\eta_{pt}(\gamma-1)/\gamma}} \right]$$

$$\frac{q_{in}}{c_p T_1} = \left[a - (1 - \epsilon)b^{1/2\eta_{pc}} - \frac{\epsilon a}{(zb)^{\eta_{pt}}} \right]$$

The thermal efficiency is therefore

$$\eta_{th} = \frac{a\left(1 - \dfrac{1}{(zb)^{\eta_{pt}}}\right) - 2b^{1/2\eta_{pc}} + 2}{a - (1 - \epsilon)b^{1/2\eta_{pc}} - \dfrac{\epsilon a}{(zb)^{\eta_{pt}}}} \tag{10.31}$$

The plots for the efficiency and specific output work as functions of a and r_c are shown in Figure 10.22. The comments made on the plots for the cycle with heat exchange and reheat apply to these plots, noting that the maxima of specific output work curves move further towards higher r_c values.

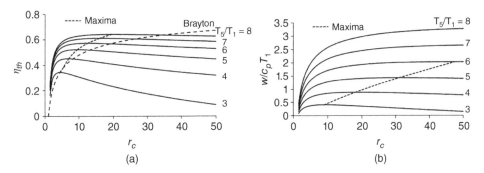

Figure 10.22 Irreversible cycle with heat exchange and intercooling: (a) thermal efficiency; (b) output work ($\epsilon = 0.8$, $\eta_{pc} = \eta_{pt} = 0.9$, $\Delta p_{45} = 0.06p_4$, $\Delta p_{61} = 0.04p_1$, $z = 0.972$).

10.8 Irreversible Cycle with Heat Exchange, Reheat, and Intercooling

The cycle with combined reheat, intercooling, and heat exchange is shown in Figure 10.23. The losses taken into account are losses due to irreversible processes in the compressor and turbine, losses in the heat exchanger, pressure losses in the combustion chamber, and pressure losses in the exhaust duct.

The specific turbine work is

$$w_t = c_p(T_5 - T_6) + c_p(T_7 - T_8) = 2(T_5 - T_6)$$

The specific compressor work

$$w_c = c_p(T_2 - T_1) + c_p(T_4 - T_3) = 2(T_2 - T_1)$$

The net specific output work is

$$w = 2c_p[(T_5 - T_6) - (T_2 - T_1)] = 2c_p T_1 \left[\left(\frac{T_5}{T_1} - \frac{T_6}{T_1} \right) - \left(\frac{T_2}{T_1} - 1 \right) \right]$$

$$w = 2c_p T_1 \left[\left(\frac{T_5}{T_1} - \frac{T_5/T_1}{T_5/T_6} \right) - \left(\frac{T_2}{T_1} - 1 \right) \right]$$

$$w = 2c_p T_1 \left[\left(\frac{T_5}{T_1} - \frac{T_5/T_1}{(p_5/p_6)^{\eta_{pt}(\gamma-1)/\gamma}} \right) - \left(\frac{p_2}{p_1} \right)^{(\gamma-1)/\gamma\eta_{pc}} + 1 \right]$$

$$w = 2c_p T_1 \left[\left(a - \frac{a}{r_t^{\eta_{pt}(\gamma-1)/2\gamma}} \right) - r_c^{(\gamma-1)/2\eta_{pc}\gamma} + 1 \right]$$

$$\frac{w}{c_p T_1} = 2a \left(1 - \frac{1}{(zb)^{\eta_{pt}/2}} \right) - 2b^{1/2\eta_{pc}} + 2 \tag{10.32}$$

The heat input is the same as in the actual cycle with reheat and heat exchange:

$$\frac{q_{in}}{c_p T_1} = 2a - (1 - \epsilon)b^{1/2\eta_{pc}} - \frac{a}{(zb)^{\eta_{pt}/2}}(1 + \epsilon) \tag{10.33}$$

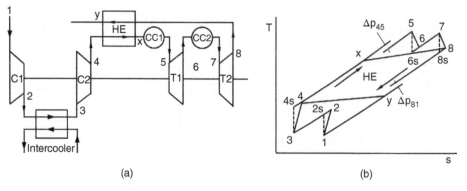

(a)   (b)

Figure 10.23 Irreversible cycle with heat exchange, reheat, and intercooling: (a) engine schematic; (b) cycle T-s diagram.

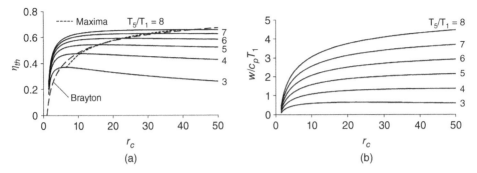

Figure 10.24 Irreversible cycle with heat exchange, reheat, and intercooling: (a) thermal efficiency; (b) output work ($\epsilon = 0.8$, $\eta_{pc} = \eta_{pt} = 0.9$, $\Delta p_{45} = 0.06 p_4$, $\Delta p_{81} = 0.04 p_1$, $z = 0.972$).

The thermal efficiency is

$$\eta_{th} = \frac{2a\left(1 - \dfrac{1}{(zb)^{\eta_{pt}/2}}\right) - 2b^{1/2\eta_{pc}} + 2}{2a - (1-\epsilon)b^{1/2\eta_{pc}} - \dfrac{a}{(zb)^{\eta_{pt}/2}}(1+\epsilon)} \tag{10.34}$$

Figure 10.24 shows the plots for efficiency and specific output work for this cycle. Comparison of these plots with the plots for the previous two cycles indicates that

- The curves are similar
- Efficiency and specific output work values are higher over the entire range of compressor pressure ratio and for all temperature ratios
- The maxima for both efficiency and specific output work shift towards higher r_c values
- Only two specific work curves (temperature ratios 3 and 4) reach their peaks within the r_c range

10.9 Comparison of Irreversible Cycles

The performance of the various cycle configurations discussed in this chapter are summarised in Table 10.1 and Figure 10.25. Calculations are conducted for the temperature ratio $a = 6$ and compressor pressure ratio $r_c = 20$. The performances of the different configurations in comparison with the simple cycle indicate behaviour similar to the ideal cycles in Chapter 9, albeit with much lower values of the efficiency and specific output work.

The thermal efficiencies and output work as functions of the pressure ratio for the configurations under discussion are shown graphically in Figures 10.26 and 10.27 for $a = 6$. With respect to thermal efficiency, irreversible cycles are superior to the Brayton cycle only at the lower range of pressure ratios ($r_c < 24$ for $a = 6$), as shown in Figure 10.26. Two cycles (II + IV and II + III + IV) show improved thermal efficiency compared with the irreversible simple cycle over the entire investigated range of pressure ratios. A further increase in a will increase the thermal efficiency further. The implementation of such cycles with very high turbine inlet temperatures is impractical due to cost, design complexity, and metallurgical limitations of the materials used in the combustion chamber and turbine blades.

Table 10.1 Comparison of all irreversible cycles ($a = 6$, $r_c = 20$).

Cycle	Configuration	Efficiency	% Change	Sp. work	% Change
Simple irreversible cycle[a]	I	0.457	0.00	1.561	0.00
Heat exchange	II	0.487	6.55	1.561	0.00
Reheat	III	0.405	−11.34	2.140	37.11
Intercooling	IV	0.439	−4.35	1.931	23.74
Reheat + HE	II + III	0.519	13.43	2.14	37.11
Intercooling + HE	II + IV	0.568	24.24	1.932	23.74
Reheat + intercooling + HE	II + III + IV	0.593	29.60	2.511	60.85

a) Configuration I: $\eta_{pc} = \eta_{pt} = 0.9$, $\Delta p_{23} = 0.06 p_2$, $\Delta p_{41} = 0.04 p_1$, $z = 0.972$.

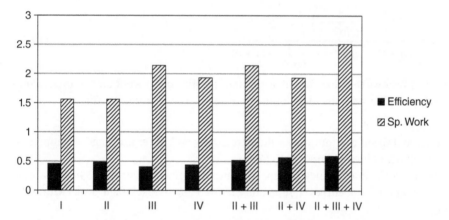

Figure 10.25 Comparison of irreversible cycles ($a = 6$, $r_c = 20$).

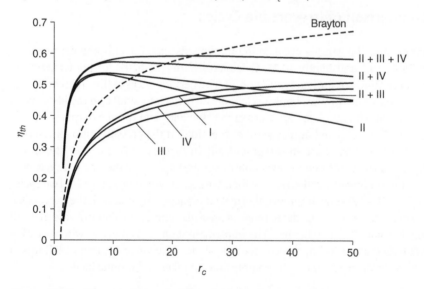

Figure 10.26 Thermal efficiency of irreversible cycles compared with the Brayton cycle ($a = 6$).

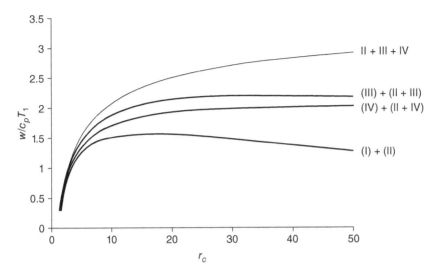

Figure 10.27 Specific output work of irreversible cycles compared with the simple cycle ($a = 6$).

The meanings of the cycle configurations are given in Table 10.1. Direct comparison of the air-standard ideal and air-standard irreversible cycles of different configurations are not presented here; anyone interested in such a comparison can refer to the thermal efficiency and specific work equations summarised in Appendix D.

Problems

10.1 Show that the specific work and thermal efficiency for the simple irreversible cycle without pressure losses can be written in terms of the isentropic efficiencies as follows:

$$\frac{w}{c_p T_1} = a\eta_t \left(1 - \frac{1}{b}\right) - \frac{1}{\eta_c}(b-1)$$

$$\eta_{th} = \frac{a\eta_t \left(1 - \frac{1}{b}\right) - \frac{1}{\eta_c}(b-1)}{a - \frac{1}{\eta_c}(b-1) - 1}$$

where

$$a = T_3/T_1, b = r_c^{(\gamma-1)/\gamma}$$

10.2 Deduce the equations for the simple irreversible cycle with pressure losses in terms of the isentropic efficiencies of the compressor and turbine.

10.3 A gas turbine engine operates on the irreversible cycle with heat exchange. Air at 0.1 MPa and 293 K is compressed with a pressure ratio of 12:1 in the compressor, and the temperature ratio in the engine $a = 4$. The fluid is air with $\gamma = 1.4$. Calculate the specific work and thermal efficiency of the cycle for the following conditions:

Pressure losses in the combustion chamber	6% of compressor delivery pressure
Pressure losses in the inlet duct	4% of compressor inlet pressure
Compressor polytropic efficiency	0.9
Turbine polytropic efficiency	0.9
Heat exchanger effectiveness	0.65

What is the effect on the specific work and efficiency of the cycle of increasing HE effectiveness to 0.8?

10.4 Deduce the equations for the specific work and thermal efficiency for the irreversible cycle with intercooling in terms of the isentropic efficiencies of the compressor and turbine.

10.5 Modify Problem 10.3 by adding a reheating process to the cycle.

10.6 Modify Problem 10.3 by adding an intercooling process to the cycle.

10.7 Modify Problem 10.3 by adding both reheating and intercooling processes to the cycle.

11

Practical Gas Turbine Cycles

The air-standard irreversible cycle discussed in Chapter 10 is a model of the gas turbine cycle that enables us to determine cycle thermal efficiency and output work by means of mathematical equations that are functions of the maximum cycle temperature, the compressor pressure ratio, component efficiencies, pressure losses in the combustion chamber and exit duct, and heat-exchanger effectiveness (if an exchanger is present). A more realistic model will need to incorporate additional features such as

- Mechanical efficiency of the rotating components
- Temperature-dependent thermodynamic properties of air and combustion products (the working fluids can no longer be treated as perfect gases with constant specific heat at constant pressure)
- Fuel properties
- Variation of combustion temperature with air-fuel ratio and combustion efficiency

Since it is impractical to model these features with simple mathematical equations as was the case earlier, it will be more appropriate to use step-by-step calculations of the processes in a gas turbine engine. The ensemble of these calculations will be referred to in this book as the *practical gas turbine cycle*, or simply the *practical cycle*, and will serve as the basis for the *design-point* and *off-design* performance calculations to be discussed in Chapters 12, 13, and 15, respectively.

11.1 Simple Single-Shaft Gas Turbine

The practical cycle of the simple single-shaft gas turbine shown in Figure 11.1 will be considered in this chapter, followed by other configurations when considering design-point calculations in Chapters 12 and 13. This cycle is similar in appearance to the irreversible air-standard cycle in Chapter 10, but the processes are significantly different as more realistic thermodynamic properties of the fluids in the engine are incorporated into the calculations.

Fundamentals of Heat Engines: Reciprocating and Gas Turbine Internal Combustion Engines, First Edition. Jamil Ghojel.
© 2020 John Wiley & Sons Ltd. This Work is a co-publication between John Wiley & Sons Ltd and ASME Press.
Companion website: www.wiley.com/go/JamilGhojel_Fundamentals of Heat Engines

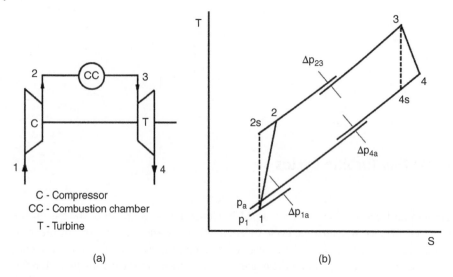

Figure 11.1   Basis for design-point calculations of the single-shaft gas turbine.

11.2   Thermodynamic Properties of Air

If air is assumed to be an ideal gas, the specific heat at constant pressure is a function of temperature only and can be approximated by correlation (11.1), which is a modified version of a correlation by Rivkin (1987) for the temperature range from −50 to 1500 °C. Alternatively, correlation (2.38) and Table 2.6 in Chapter 2 can be used for higher temperatures based on the JANAF data.

$$C_{pa} = \sum_{n=1}^{8} B_n \left(\frac{T}{1000} \right)^{n-1} kJ/kmole.K \tag{11.1}$$

where C_{pa} is the molar specific heat at constant pressure, B_n is the polynomial coefficient (Table 11.1), and T is the absolute temperature in degrees K.

$$c_{pa} = \frac{C_{pa}}{M_a} \; kJ/kg.K \tag{11.2}$$

$$\gamma_a = \frac{1}{\left(1 - \dfrac{R_a}{c_{Pa}} \right)} \tag{11.3}$$

where
  M_a: molecular mass of air (28.97 kg/kmole)
  R_a: gas constant for air (0.287 kJ/kg. K)

The specific heats from correlations (2.38) and (11.1) together with the calculated ratio of specific heats of air γ_a from Eq. (11.3) are plotted versus temperature in Figure 11.2. The plot shows that γ_a decreases quickly with increasing temperature up to about 2000 K, deviating significantly from the value of 1.4, which is commonly assumed in theoretical cycles.

Table 11.1 The values of coefficient B_n in Eq. (11.1).

B_n	Value
B_1	29.438 265
B_2	−1.610 822
B_3	−11.991 744
B_4	68.828 384
B_5	−98.239 929
B_6	64.883 505
B_7	−20.909 380
B_8	2.665 240

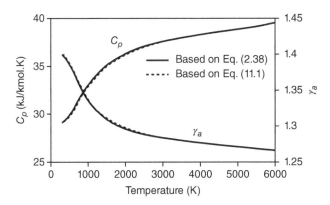

Figure 11.2 Molar specific heat of air at constant pressure and ratio of specific heats as functions of temperature.

11.3 Compression Process in the Compressor

The temperature equivalent of the compressor work can be calculated from either of the following equations, depending on the compressor efficiency used:

$$\Delta T_{12} = T_2 - T_1 = \frac{T_1}{\eta_c}[r_c^{(\gamma_a-1)/\gamma_a} - 1] \tag{11.4}$$

$$\Delta T_{12} = T_2 - T_1 = T_1[r_c^{(\gamma_a-1)/\gamma_a\eta_{pc}} - 1] \tag{11.5}$$

Knowing the compressor pressure ratio r_c, inlet temperature T_1, and compression process efficiency, Eq. (11.4) or (11.5) can be solved iteratively for T_2 by finding the value of γ_a from Eqs. (11.1), (11.2), and (11.3) using a mean temperature for the compression process: $T_m = (T_1 + T_2)/2$.

11.3.1 Power to Drive the Compressor

The turbine power required to drive the compressor, expressed as the power $\dot{W}_c$ absorbed by the compressor, taking into account mechanical losses, is

$$\dot{W}_c = \frac{\dot{m}_a}{\eta_m} c_{pa} \Delta T_{12} \tag{11.6}$$

where

$\dot{m}_a$: mass flow rate of air
η_m: mechanical efficiency of the compressor-turbine unit
c_{pa}: specific heat at constant pressure of the air in the compressor

11.4 Combustion Process

The energy balance in the combustion chamber in terms of heat addition and corresponding temperature rise can be written as

$$\eta_{com} H_l \dot{m}_f = \dot{m}_a c_{pm}(T_3 - T_2) \tag{11.7}$$

where

η_{com}: combustion efficiency
H_l: lower heating value of the fuel
$\dot{m}_f$: mass flow rate of the fuel
$\dot{m}_a$: mass flow rate of the air
c_{pm}: mean specific heat of the air

For the proper design of a gas turbine engine, temperature T_3, referred to as the *combustion temperature* or *turbine entry temperature* (TET), must be estimated as accurately as possible. Knowing $(T_3 - T_2)$ for a designated fuel, the mass flow of the fuel $\dot{m}_f$, specific fuel consumption, and thermal efficiency of the cycle can be estimated.

11.4.1 Combustion Chamber Design

The combustion process and the design of the combustion chamber play a critical role in the performance of gas turbine engines. It is reported that the reason for the delay in the development of the jet engine by Britain during WWII was due to the difficulties encountered by Sir Frank Whittle in developing the liquid fuel combustor for his gas turbine engine W1, which powered the Gloster Pioneer in its first flight in May 1941. The first successful test flight of the first airplane powered by a jet engine, designed by Hans von Ohain in Germany, took place in August 1939. The combustor in this engine had a less complicated design and burned gaseous hydrogen.

Combustion chambers for gas turbines are designed in different configurations depending on the engine application. In stationary industrial applications or power generation, a single large combustor may suffice (Figure 11.3); whereas in applications where compactness of the engine is important, such as in aviation gas turbines, the combustor is usually designed as a single chamber of annular shape or as a cluster of a number of small chambers arranged circumferentially around the turbine axis (Figure 11.4).

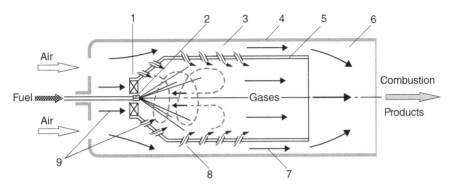

Figure 11.3 Combustion chamber for a stationary gas turbine: 1 – swirl vanes; 2 – fuel injector; 3 – annular space; 4 – casing; 5 – flame tube; 6 – mixing chamber; 7 – tertiary air; 8 – secondary air; 9 – primary air.

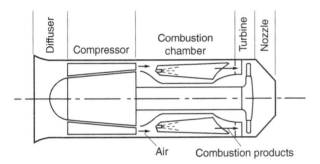

Figure 11.4 Annular combustion chamber in a turbojet engine.

The schematic diagram of a typical combustor for an industrial gas turbine is shown in Figure 11.3. The compressed air is ducted to the combustion chamber directly from the compressor or from the outlet of the heat exchanger if the engine is operating on a regenerative cycle (heat-exchange cycle). The overall air-fuel ratio is significantly higher than in reciprocating engines and could be as high as $100:1$ ($\lambda = 6.71$) on a mass basis if kerosene is used as a fuel. A homogeneous air-fuel mixture with such a ratio is well outside the flammability limits of hydrocarbon fuels (see Table 11.2) and will not ignite; therefore, different amounts of air are introduced into different zones of the combustor in order to form a stratified air-fuel mixture (non-homogeneous mixture) as follows:

- 15–20% of the total air flow (primary air) is introduced around the fuel jet: partly through the swirl vanes (1) surrounding the fuel injector (2) and partly through small holes upstream of the flame tube (5), to form a stoichiometric mixture for high-temperature, rapid ignition and flame formation.
- 30% of the air (secondary air) flows through holes (8) to stabilise the flame and provide recirculation of the hot combustion products with incoming relatively cool air and fuel.
- The remaining air (tertiary air) continues to the mixing chamber (dilution zone) (6) to be mixed with the gases from the flame tube and cool the final combustion products to the required design TET.

Table 11.2 Flammability limits for some hydrocarbon fuels at standard atmospheric conditions.

Fuel	Stoichiometric A/F ratio	Lean flammability limit		Rich flammability limit	
		F/A ratio, % volume	Relative A/F ratio, λ	F/A ratio, % volume	Relative A/F ratio, λ
Methane	17.39	4.4	2.366	16.4	0.634
Hydrogen	34.78	4	10.411	75	0.555
Propane	15.8	2.1	1.984	10.1	0.412
Gasoline	15.04	1.4	1.206	8	0.211
Diesel fuel	15.14	0.6	1.876	7.5	0.150
Kerosene Jet A-1	14.9	0.7	1.653	5	0.231

Because of the high air-fuel ratio, gas turbine combustors produce very little carbon monoxide and unburned hydrocarbons. The emission of sulfur compounds will depend on the presence of sulfur in the fuel. The main pollutants emitted are oxides of nitrogen (NO_x), which form as a result of the reaction of nitrogen with oxygen at high temperatures.

11.4.2 Thermodynamic Properties of the Combustion Products

When calculating the combustion process in the combustor, the thermodynamic properties of the species in the combustion products are of vital importance. The molar enthalpies of several combustion products can be found in the form of polynomials in Chapter 2 or in the tables in Appendix A. Data for four species of complete combustion – CO_2, H_2O, O_2, and N_2 – are reproduced here for ease of reference in the form of tenth-order polynomials:

$$\Delta H_g = \sum_{n=0}^{10} B_n \left(\frac{T}{1000} \right)^n kJ/kmol.K \tag{11.8}$$

The polynomial coefficients are shown in Table 11.3.

Two other properties are important in gas turbine cycle calculations: the specific heat at constant pressure c_{pg} and the ratio of specific heats $\gamma_g = c_{pg}/c_{vg}$. The values of c_{pg} and γ_g for the combustion gases can be determined analytically by considering the composition of the combustion products of any hydrocarbon fuel, with or without dissociation, using JANAF data. For the temperatures and pressures encountered in gas turbine engine practice, dissociation can generally be ignored, and the combustion products will be limited to CO_2, H_2O, O_2, and N_2 only. Detailed analysis of combustion reactions with varying numbers of species within the temperature range 800–2500 K and pressure range 1–100 atm (Ghojel, 1994) show that:

- γ_g is mainly a function of the air-fuel ratio and temperature, and the pressure and number of species in the products have little effect.
- The analysis of fixed combustion products (no dissociation) yields a good approximation of γ_g.
- The mean molecular weight of the combustion products varies with the air-fuel ratio within a narrow range and can be taken as equal to the value for air in the case of combustion in gas turbines in which the air-fuel mixture is very lean.

Table 11.3 Coefficients of polynomial (11.8) for the enthalpies of the gaseous products of complete combustion ($T = 298 - 6000\ K$).

Coefficient	CO_2	CO	H_2O	N_2	O_2
			ΔH_g, kJ/kmole		
B_0	$-8.75035E + 00$	$-8.79549E + 00$	$-9.97897E + 00$	$-8.96855E + 00$	$-8.09901E + 00$
B_1	$1.98233E + 01$	$3.09993E + 01$	$3.40028E + 01$	$3.21683E + 01$	$2.54662E + 01$
B_2	$3.81012E + 01$	$-9.24622E + 00$	$-4.60987E + 00$	$-1.16998E + 01$	$5.20982E + 00$
B_3	$-2.34710E + 01$	$1.79305E + 01$	$1.09847E + 01$	$1.96777E + 01$	$3.26310E + 00$
B_4	$1.02904E + 01$	$-1.41821E + 01$	$-5.75463E + 00$	$-1.47654E + 01$	$-5.66051E + 00$
B_5	$-3.19713E + 00$	$6.61689E + 00$	$1.56920E + 00$	$6.65939E + 00$	$3.51907E + 00$
B_6	$6.94534E - 01$	$-1.96865E + 00$	$-2.24252E - 01$	$-1.93166E + 00$	$-1.22789E + 00$
B_7	$-1.02394E - 01$	$3.77856E - 01$	$9.34966E - 03$	$3.63304E - 01$	$2.60641E - 01$
B_8	$9.69029E - 03$	$-4.53299E - 02$	$1.82025E - 03$	$-4.28545E - 02$	$-3.34816E - 02$
B_9	$-5.26987E - 04$	$3.09266E - 03$	$-2.71534E - 04$	$2.88203E - 03$	$2.39941E - 03$
B_{10}	$1.24152E - 05$	$-9.16141E - 05$	$1.12781E - 05$	$-8.43160E - 05$	$-7.37573E - 05$

- Equations for γ_g can be derived for any fuel as a function of the relative air-fuel ratio λ (or the equivalence ratio ϕ) and the average gas temperature, in the following relatively simple forms:

$$\gamma_g = a + \frac{b}{T} + \frac{c}{T^2} + \frac{d}{\lambda} \tag{11.9a}$$

$$\gamma_g = a + \frac{b}{T} + \frac{c}{T^2} + d\,\phi \tag{11.9b}$$

where

a, b, c, d: fuel-dependent constant coefficients (Table 11.4)

λ: relative air-fuel ratio defined as the actual air/fuel ratio divided by the stoichiometric air/fuel ratio

ϕ: the equivalence ratio, defined as the stoichiometric air/fuel ratio divided by the actual air/fuel ratio or actual fuel/air ratio, divided by the stoichiometric fuel/air ratio ($\phi = 1/\lambda$)

T: average gas temperature, K

Equation (11.9a) yields results that correlate well with the direct calculation method based on the JANAF data of c_p and derived data of c_v.

Table 11.4 Coefficients of Eq. (11.9a) for different types of hydrocarbon fuels ($\lambda > 1$).

Fuel type	a	b	c	d
Kerosene $C_{11}H_{21.34}$	1.250349	96.83762	$-15\,252.8$	$-3.97E-02$
Light diesel $C_{12}H_{26}$	1.2500391	97.48906	$-15\,422.9$	$-3.97E-02$
Methane CH_4	1.2482057	101.3199	$-16\,423.5$	$-4.01E-02$
Propane C_3H_8	1.2494339	98.75799	$-15\,754.3$	$-3.99E-02$
Iso-octane C_8H_{18}	1.249931	97.71575	$-15\,482.1$	$-3.98E-02$

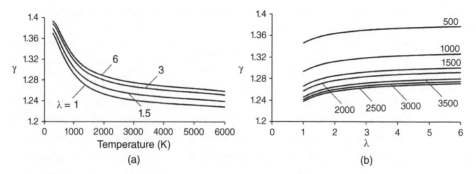

Figure 11.5 Ratio of specific heats of the combustion products of kerosene ($C_{11}H_{21.34}$): (a) versus temperature at constant λ; (b) versus λ at constant temperature.

Knowing γ_g, c_{pg} can be determined from

$$c_{pg} = \frac{\overline{R}}{M_g} \frac{\gamma_g}{(\gamma_g - 1)} \; kJ/kg\,K \tag{11.10}$$

where $\overline{R}$ is the universal gas constant (8.3143 $kJ/kmol.\,K$) and M_g the mean molecular mass of the products.

Figure 11.5 shows the variation of the ratio of specific heats γ_g with the relative air-fuel ratio λ and average temperature of the products of complete combustion calculated using specific heat data of the species in the products. Equation (11.9a) can also be used to plot similar graphs for any fuel with known correlation coefficients (Table 11.4).

11.4.3 Combustion Temperature

The gas turbine combustion process can be represented by the single steady-state flow system shown in Figure 11.6. For the constant-pressure combustion process ($p_2 \approx p_3$) without heat or work transfer, the energy equation can be written

$$H_p - H_R = 0 \tag{11.11}$$

H_p and H_R are the enthalpies of products and reactants, respectively, per kmole of fuel burned.

The enthalpy of the reactants per kmole of fuel is the sum of the enthalpy of the air (H_a) and the enthalpy of the fuel (H_f):

$$H_R = H_a + H_f$$

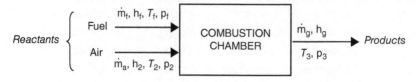

Figure 11.6 Combustion system in gas turbines.

$$H_R = \sum_R n_R(\Delta h_R + \Delta h_{fR}) + c_{pf}(T_f - 298) + \Delta h_{ff} \tag{11.12}$$

For each component of air (usually O_2 and N_2), n_R is the number of moles, Δh_R is the enthalpy at T_2 relative to the enthalpy at the reference temperature of 298 K, and Δh_{fR} is the enthalpy of formation at 298 K and 0.1 MPa. Δh_{fR} is usually taken as equal to zero for naturally occurring elements such as O_2 and N_2. T_f is the initial temperature of the fuel, c_{pf} is the average specific heat of the fuel for the temperature range between 298 K and T_f, and Δh_{ff} is the enthalpy of formation of the fuel at 298 K and 0.1 MPa.

The enthalpy of the products per kmole of fuel is

$$H_P = \sum_P n_P(\Delta h_P + \Delta h_{fP}) \tag{11.13}$$

where, for each component of the combustion products, n_P is the number of moles, Δh_P is the enthalpy at T_3 relative to the enthalpy at the reference temperature of 298 K, and Δh_{fp} is the enthalpy of formation at 298 K and 0.1 MPa.

For any hydrocarbon fuel of composition C_xH_y, the general reaction with excess air and complete combustion can be written as

$$C_xH_y + \lambda\left(x + \frac{y}{4}\right)(O_2 + 3.7619N_2) = xCO_2 + \frac{y}{2}H_2O + \left(x + \frac{y}{4}\right)(\lambda - 1)O_2$$
$$+ 3.7619\lambda\left(x + \frac{y}{4}\right)N_2 \tag{11.14}$$

The relative air-fuel ratio λ determines the amount of excess air in the reactants and the amounts of oxygen and nitrogen in the products. When $\lambda = 1$, the reaction becomes stoichiometric. Dissociation has been ignored in this analysis, but it could be accounted for readily, if warranted, by deciding on the reaction scheme appropriate for gas turbines. For the temperatures and pressures encountered in gas turbine engine practice, dissociation can generally be ignored, and the combustion can be treated as complete combustion with the products consisting of CO_2, H_2O, O_2, and N_2 only.

Based on (11.14) the enthalpies of the reactants and products are

$$H_R = \lambda\left(x + \frac{y}{4}\right)(\Delta h_{O_2} + 3.7619\Delta h_{N_2}) + c_{pf}(T_f - 298) + \Delta h_{ff} \tag{11.15}$$

$$H_P = \left[x(\Delta h_{CO_2} + \Delta h_{fCO_2}) + \frac{y}{2}(\Delta h_{H_2O} + \Delta h_{fH_2O}) + \left(x + \frac{y}{4}\right)(\lambda - 1)\Delta h_{O_2}\right.$$
$$\left. + 3.7619\lambda\left(x + \frac{y}{4}\right)\Delta h_{N_2}\right] \tag{11.16}$$

To determine the combustion temperature T_3, Eq. (11.11) must be solved, which requires prior knowledge of the following

- Compressor exit temperature T_2
- Temperature, specific heat, and the enthalpy of formation of the fuel
- Relative air-fuel ratio λ

To simplify the analysis, we assume that the fuel is delivered in a liquid phase to the combustion chamber at the ambient temperature 298 K and pressure 0.1 MPa, as a result of which the enthalpy term for the fuel in Eq. (11.15) becomes equal to zero.

The temperature at the exit from the combustion chamber for any fuel of known composition can be determined using any of the four methods presented next.

11.4.3.1 Method 1

Using the enthalpies of the reactants and products calculated from Eq. (11.8) or the tables in Appendix A, H_R is determined from Eq. (11.15) for the required relative air-fuel ratio λ and reactants temperature T_2, and then Eq. (11.16) is solved iteratively for T_3 until H_P becomes equal to H_R. The calculated final temperature T_3 is the adiabatic flame temperature, which is the maximum theoretical temperature attainable in the combustion chamber when the pressure is assumed constant and heat losses from the combustion chamber are ignored. It will be taken as the TET in cycle calculations.

11.4.3.2 Method 2

A more direct approach to finding T_3 for a specific fuel involves compiling the values of H_R and H_P versus temperature in the range 298–4000 K for several values of λ in the range 1–8 and fitting the data to nonlinear regression models with two independent variables. The best fit obtained for both the reactants and products from the analysis of several hydrocarbon fuels is of the form shown in the following equations:

$$H_R = a + bT_2 + c\lambda + dT_2^2 + e\lambda^2 + fT_2\lambda + gT_2^3 + h\lambda^3 + iT_2\lambda^2 + jT_2^2\lambda \tag{11.17a}$$

$$H_P = a + bT_3 + c\lambda + dT_3^2 + e\lambda^2 + fT_3\lambda + gT_3^3 + h\lambda^3 + iT_3\lambda^2 + jT_3^2\lambda \tag{11.17b}$$

The values of the coefficients in these equations for dodecene $C_{12}H_{24}$ (a surrogate for Jet A-1 fuel) are given in Table 11.5 and the coefficients for a number of hydrocarbon fuels are given in Appendix A.

Equations (11.17a) and (11.17b) can be solved in a spreadsheet by arranging them in a table as shown in Table 11.6, assigning values in the reactants block (say, $T_2 = 298\ K$ and $\lambda = 1$) and the same value of λ in the products block, and then iterating for T_3 until $X = Y$.

A 3-D graphical representation of the enthalpy of the products for dodecene is shown in Figure 11.7. The plot for the reactants has a similar shape. Both the average temperature

Table 11.5 The coefficients for Eqs. (11.17a) and (11.17b) for dodecene $(C_{12}H_{24})$ $(T = 298 – 4000\ K, p_2 = 0.1\ MPa, \lambda = 1 – 8)$.

	Products H_P	Reactants H_R
a	$-4.22509E + 04$	$2.42286E + 03$
b	$-6.39574E + 00$	$-8.15967E + 00$
c	$-5.31346E + 03$	$-5.31346E + 03$
d	$5.26732E – 03$	$4.47710E – 03$
e	$-1.98725E – 06$	$-4.15375E – 05$
f	$1.60557E + 01$	$1.60557E + 01$
g	$-7.81303E – 07$	$-6.95719E – 07$
h	$1.26494E – 07$	$2.65266E – 06$
i	$9.71552E – 11$	$2.28862E – 09$
j	$5.45679E – 04$	$5.45679E – 04$

Table 11.6 Spreadsheet solution for Method 2.

Reactants		Products	
T_2	298	T_3	?
λ	1	λ	1
H_R	X	H_P	Y

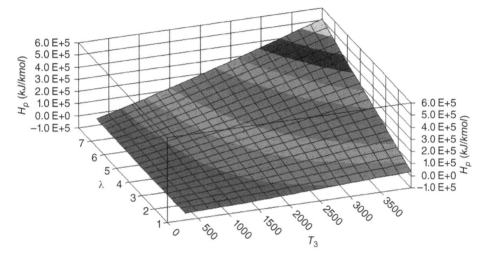

Figure 11.7 3-D graphical representation of the enthalpy of the products of combustion of dodecene (Eq. (11.17b)).

T_3 and the overall relative air-fuel ratio λ have a significant effect on the enthalpy of the combustion products. At low values of λ, the temperature has little effect on H_P. At high values of λ, H_P increases sharply as T_3 increases. The reason is that the concentrations of both O_2 and N_2 increase significantly with increasing λ, and their enthalpies increase with increasing T_3. The accuracy of this method depends on the accuracy of the fitted models with two independent variables.

11.4.3.3 Method 3

The calculations using the previous method can be repeated for several combinations of T_2 and λ to compile enough data for a regression model of the type $T_3 = f(T_2, \lambda)$. The model obtained by analysing the reactions of several fuels is

$$T_3 = a + \frac{b}{\lambda} + cT_2 + \frac{d}{\lambda^2} + eT_2{}^2 + f\frac{T_2}{\lambda} + \frac{g}{\lambda^3} + hT_2{}^3 + i\frac{T_2{}^2}{\lambda} + j\frac{T_2}{\lambda^2} \tag{11.18}$$

The coefficients in Eq. (11.18) for dodecene are given in Table 11.7 and its 3-D plot is shown in Figure 11.8. Data for other gaseous and liquid fuels are given in Appendix A.

For a given initial temperature of the reactants T_2, the combustion temperature increases steadily as the air-fuel mixture becomes richer (λ decreases). The higher the value of T_2, the

Table 11.7 Coefficients for Eq. (11.18) for dodecene combustion without dissociation ($p_2 = 0.1$ MPa, $\lambda = 1 - 8$).

Coefficient	$C_{12}H_{24}$
a	$-1.26729E + 01$
b	$3.25940E + 03$
c	$9.37338E - 01$
d	$-1.06247E + 03$
e	$1.47381E - 04$
f	$-9.14949E - 01$
g	$1.83875E + 02$
h	$-7.40568E - 08$
i	$1.67612E - 04$
j	$2.27784E - 01$

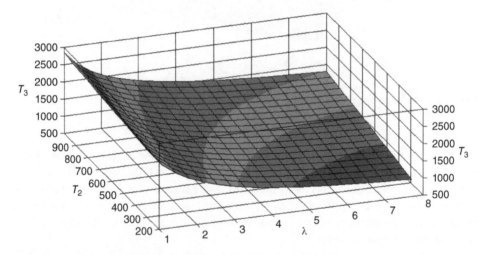

Figure 11.8 3-D graphical representation of Eq. (11.18) for dodecene.

greater the rate of increase of T_3 with decreasing λ. Typically, hydrocarbon fuels ignite at λ values close to the stoichiometric, resulting in high adiabatic flame temperatures exceeding $2000\,K$ (~$2400\,K$ for dodecene). Since such temperatures are not desirable because of metallurgical considerations for the turbine components, particularly the blades, air-fuel ratios in gas turbine combustors are much higher than in piston engines, reaching value as high as 5. A homogeneous air-fuel mixture with such a ratio is well outside the flammability limits of hydrocarbon fuels (see Table 11.2 The solution to this dilemma, as explained in Section 11.4.1, is to introduce different amounts of air into different zones of the combustor in order to stabilise the flame following ignition and control the resulting high temperatures

Table 11.8 Effect of calculation method on the value of the predicted combustion temperature T_3 of dodecene ($T_2 = 298\ K$, $p_2 = 0.1\ MPa$).

	T_3 (K)		Difference
Relative air/fuel ratio λ	Method 1	Methods 2 and 3	%
1	2425	2466	−1.69%
2	1524	1553	−1.90%
3	1165	1175	−0.86%
4	970	969	−0.10%
5	848	840	−0.94%
6	765	751	−1.83%

by means of a tertiary air stream. In aircraft gas turbine engines, where high temperatures are essential to increase thermal efficiency, the turbine blades are cooled internally to withstand the higher gas temperatures instead of cooling the combustion products by dilution.

Comparison of the combustion temperatures predicted using Methods 1–3 is given in Table 11.8. The temperatures obtained from Methods 2 and 3 are almost identical, since they share the same H_R and H_P data obtained from Eqs. 11.17a and 11.17b. The difference between the values predicted by Method 1 (direct method) and Methods 2 and 3 is less than 2%.

11.4.3.4 Method 4

Combustion temperature data for hydrocarbon fuels are usually presented in published literature as charts of $T_3 = \psi(T_2, \phi)$ or $T_3 = \psi(T_2, f)$, where ϕ is the equivalence ratio and f is the fuel/air ratio defined as follows

$$\phi = \frac{1}{\lambda} \text{ and } f = \left(\frac{F}{A}\right)_{actual} = \frac{1}{\lambda\left(\dfrac{A}{F}\right)_{stoich}}$$

A chart in the form $T_3 = \psi(T_2, \lambda)$ based on current analysis using Eq. (11.18) is shown in Figure 11.9 (T_2 ranges from $300 - 1000\ K$ in increments of $100\ K$). For mixtures close to the stoichiometric ($\lambda \approx 1$), the temperature curves for a comprehensive chart with a wide range of reactant temperatures T_2 at small increments will be closely clustered and may be difficult to read unless the chart is significantly magnified.

The combustion temperature charts for dodecene as $T_3 = \psi(T_2, \phi)$ and $T_3 = \psi(T_2, f)$ are shown in Figures 11.10 and 11.11. The following relationships are used for the plots:

$$\phi = \frac{1}{\lambda} \text{ and } f = \frac{1}{\lambda\left(\dfrac{A}{F}\right)_{stoich}} = \frac{1}{\lambda(14.714)}$$

In determining the combustion temperatures so far, we have ignored losses such as incomplete combustion, reactants pressure, pressure losses in the combustor, and heat losses to the environment. Paradoxically, these losses have a negligible effect on the efficiency of the

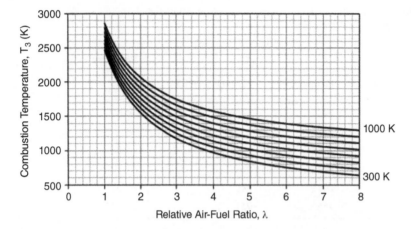

Figure 11.9 Combustion temperature chart: combustion temperature versus relative air/fuel ratio at constant reactant temperatures for dodecene ($p_2 = 0.1$ *MPa*).

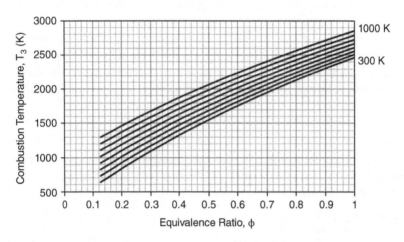

Figure 11.10 Combustion temperature chart: combustion temperature versus equivalence ratio at constant reactant temperatures for dodecene ($p_2 = 0.1$ *MPa*).

actual combustion process, defined as

$$\eta_{com} = \frac{actual\ temperature\ change\ in\ the\ combustor}{theoretical\ temperature\ change\ in\ the\ combustor}$$

$$\eta_{com} = \frac{(T_3 - T_2)_{actual}}{(T_3 - T_2)_{theoretical}} \tag{11.19}$$

Measured combustion efficiencies are equal to 98–99% (Saravanamuttoo et.al., 2001) and for aircraft engines close to 99% at sea level (Hill and Peterson, 1992) or even close to 100% under high thrust conditions at sea level (Cumpsty, 2003). In light of these numbers, the theoretical temperature rise can be taken as the actual temperature rise for engineering calculations.

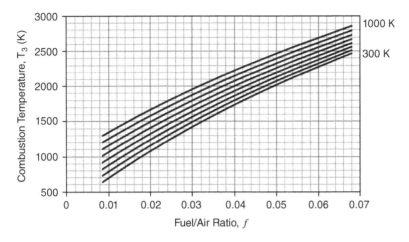

Figure 11.11 Combustion temperature chart: combustion temperature versus fuel/air ratio at constant reactant temperatures for dodecene ($p_2 = 0.1$ MPa).

11.4.4 Effect of Dissociation on the Combustion Temperature

It was stated earlier that dissociation is usually ignored in gas turbine calculations due to the highly lean air-fuel mixtures used and the fact that average combustion temperatures are below the dissociation threshold. However, combustion temperatures and, consequently, TETs have been rising steadily in pursuit of higher efficiencies and more compact designs, particularly in aviation applications, which may merit accounting for dissociation in these cases. An example is presented here of the effect of dissociation when applied to dodecene, assuming a reaction involving the formation of 11 species at a pressure of 1.0 MPa (see Chapter 2 for more details of this reaction). The form of the correlation for the function $T_3 = \psi(T_2, \lambda)$ is the same as correlation (11.18) for the reaction without dissociation, albeit with different coefficients, as shown in Table 11.9.

Table 11.9 Coefficients for Eq. (11.18) for dodecene combustion with dissociation ($p_2 = 1.0$ MPa, $\lambda = 1 - 8$).

Coefficient	$C_{12}H_{24}$
a	$6.47592E + 01$
b	$2.66484E + 03$
c	$9.23342E - 01$
d	$-2.73958E + 02$
e	$-1.43039E - 05$
f	$-1.95037E - 01$
g	$-2.56925E + 02$
h	$2.67779E - 08$
i	$-4.90025E - 05$
j	$-2.27349E - 01$

Table 11.10 Effect of dissociation on the combustion temperature of dodecene ($T_2 = 500$ K).

	T_3 (K)		Difference
λ	W/T dissociation $p_2 = 0.1$ MPa	With dissociation $p_2 = 1.0$ MPa	%
1	2563	2437	4.92
1.1	2423	2336	3.59
1.2	2301	2238	2.74
1.5	2012	1981	1.54
2	1691	1675	0.95
3	1333	1325	0.60
4	1138	1137	0.09
5	1016	1020	−0.39
6	932	940	−0.86

Table 11.10 shows the effect of dissociation on the combustion temperature at initial reactants temperature $T_2 = 500$ K. Generally, the differences are small, with the maximum difference occurring at the stoichiometric mixture ($\lambda = 1$). The reason is that the maximum temperature occurs at $\lambda = 1$ for the reaction without dissociation and at around $\lambda = 0.95$ for the reaction with dissociation. The data in Table 11.10 are a confirmation that temperatures predicted from reactions comprising four species without dissociation are adequate for gas turbine calculations.

11.5 Expansion Process in the Turbine

The temperature equivalent of the turbine work can be calculated from

$$\Delta T_{34} = T_3 - T_4 = \eta_t T_3 \left[1 - \frac{1}{r_t^{(\gamma_g - 1)/\gamma_g}} \right] \tag{11.20}$$

based on the definition of the turbine isentropic efficiency, or from

$$\Delta T_{34} = T_3 - T_4 = T_3 \left[1 - \frac{1}{r_t^{\eta_{pt}(\gamma_g - 1)/\gamma_g}} \right] \tag{11.21}$$

based on the definition of the turbine polytropic efficiency.

If the pressure losses in the inlet duct are ignored, i.e. $p_1 = p_a$, the turbine pressure ratio (expansion ratio) can be calculated as was done for the irreversible cycle in Chapter 10 as

$$r_t = r_c \left(\frac{1-x}{1+y} \right) \tag{11.22}$$

where r_c is the compressor pressure ratio, $x = \Delta p_{23}/p_2$, and $y = \Delta p_{4a}/p_a = \Delta p_{41}/p_1$.

The temperature ratio in terms of the turbine expansion ratio r_t is

$$\frac{T_3}{T_4} = \left[r_c \left(\frac{1-x}{1+y} \right) \right]^{\eta_{pt}(\gamma_g - 1)/\gamma_g}$$

The unknown temperature T_4 in Eq. (11.20) or (11.21) is calculated iteratively as follows:

1. Assume a value for T_4, and calculate $T_m = (T_3 + T_4)/2$.
2. Calculate γ_g using Eq. (11.9a) for the mean temperature T_m.
3. Use the calculated γ_g in Eq. (11.20) or (11.21) to calculate ΔT_{34}.
4. If this value is not equal to the difference between T_3 and the assumed temperature T_4 in step 1, assume another value for T_4 and repeat the procedure until equality is achieved.

11.5.1 Total Turbine Power

$$\dot{W}_t = \dot{m}_g c_{pg} \Delta T_{34} \tag{11.23}$$

where

$\dot{m}_g$: mass flow rate of the combustion gases $(= m_a + m_f)$
c_{pg}: specific heat of the combustion gases at constant pressure

Net power output is

$$\dot{W}_{net} = \dot{W}_t - \dot{W}_c \tag{11.24}$$

11.5.2 Specific Fuel Consumption

This is a measure of the mass of fuel used to produce a unit of power over a period of time:

$$sfc = \frac{\dot{m}_f}{\dot{W}_{net}} \tag{11.25}$$

The units of the specific fuel consumption (*sfc*) depend on the units of the mass flow rate and power. The most widely used unit is $kg/kW.h$, where $\dot{m}_f$ is in kg/h and $\dot{W}_{net}$ is in kW.

11.5.3 Cycle Thermodynamic Efficiency

This is defined as the ratio of net power output to the rate of the heat supply to the engine:

$$\eta_{th} = \frac{\dot{W}_{net}}{\dot{m}_f H_l} \tag{11.26}$$

where H_l is the lower heating value of the fuel in kJ/kg.
   Combining Eqs. (11.25) and (11.26) gives

$$\eta_{th} = \frac{1}{(sfc)H_l} \tag{11.27}$$

which indicates that for a given fuel, the thermal efficiency is inversely proportional to the specific fuel consumption.

Problems

11.1 Calculate the compressor delivery temperature for the compressor pressure ratios 2, 10, 20, 30, 40, and 50 using both constant and temperature-dependent specific heats of air, and plot the results (you can use Eqs. (11.1) and (11.5)). Take the compressor inlet conditions as 0.1 *MPa* and 298 *K* and $\eta_{pc} = 0.86$.

11.2 Using the results from Problem 10.1, determine the effect on the compressor power of using temperature-dependent specific heats, if the mass flow of air is 20 *kg/s* and the mechanical efficiency is 0.98.

11.3 The compressor pressure ratio in a single-shaft gas turbine is 20:1, and compressor inlet conditions are the atmospheric pressure 0.1 *MPa* and temperature 293 *K*. Iso-octane (C_8H_{18}) is delivered to the combustion chamber at atmospheric temperature and burned at a relative air-fuel ratio $\lambda = 3.387$. Calculate the combustion temperature of the fuel (adiabatic flame temperature) using Method 1 from Section 11.4.3.1. Assume a constant heat capacity for air during compression ($\eta_{pc} = 0.85$), and use the following enthalpies of formation: $\Delta h_{fCO_2} = -393500 \, kJ/kg$, $\Delta h_{fH_2O} = -241820 \, kJ/kg$, $\Delta h_{ffuel} - 250105 \, kJ/kg \, fuel$.

11.4 Check the result in Problem 11.3 using correlation (11.18) for octane and explain the discrepancy, if any. Use the same correlation to estimate the adiabatic flame temperatures for methane (CH_4), propane (C_3H_8), and dodecane ($C_{12}C_{25}$).

11.5 Air at 0.1 *MPa* and 293 *K* is compressed in a single-shaft gas turbine engine with a pressure ratio of 12 : 1 in the compressor. The compressed air reacts with light diesel fuel ($C_{12}H_{26}$) in the combustion chamber and burns completely without dissociation at a maximum temperature of 1490 *K*. For the following data, calculate the turbine exit temperature using the temperature-dependent specific heat of the combustion products.

Pressure losses in the combustion chamber	5% of compressor delivery pressure
Pressure losses in the combustion chamber	4% of compressor delivery pressure
Compressor polytropic efficiency	0.9
Turbine polytropic efficiency	0.9
Relative air-fuel ratio, λ	2.944
Ratio of specific heat of air, γ	1.4

Equation (11.9a) can be used to calculate the specific heat of the products of combustion.

12

Design-Point Calculations of Aviation Gas Turbines

The purpose of the design point calculations of aviation gas turbine engines (or aero engines) is to determine the airflow rate, specific fuel consumption, thermal efficiency, and propulsive efficiency for a given thrust demand at a specified cruise conditions. The input data include the ambient conditions at cruise altitude, compressor pressure ratio, turbine inlet temperature (or air-fuel ratio in the combustion chamber [CC]), and component efficiencies. Based on these calculations, engine components such as the compressor, combustion chamber and turbine can be designed. During the process of design optimisation, special attention is given to specific thrust (thrust per mass flow rate), which affects engine weight, volume, and frontal area. Designers will also endeavour to decrease the specific fuel consumption to reduce weight of the engine and fuel consumption. Three types of aircraft gas turbine engines will be discussed in this chapter, namely, the simple turbojet engine, the unmixed-flow turbofan engine and the mixed-flow turbofan engine.

12.1 Properties of Air

Aircraft designers and operators need to have reliable data on the environment within which the gas turbine engine will operate, in order to predict its performance. The relevant data include pressure, temperature, density, and speed of sound of the atmospheric air, which are all altitude dependent. Additionally, as aircraft are intended to fly at high speeds, the properties of air entering the engine tend to undergo significant change within the engine components. In this section, the effect of altitude above sea level on the properties of air will be discussed, together with the effect of air speed on the properties within the engine components.

12.1.1 International Standard Atmosphere (ISA)

The actual properties of atmospheric air are difficult to predict due to the effects of altitude, weather pattern, heat transfer, and air mass motion. To mimic the actual atmosphere, the International Civil Aviation Organization (ICAO) devised an idealised dry quiescent atmosphere known as the International Standard Atmosphere that can be used to standardise aircraft instruments and estimate the expected flight conditions at different altitudes.

Fundamentals of Heat Engines: Reciprocating and Gas Turbine Internal Combustion Engines, First Edition. Jamil Ghojel.
© 2020 John Wiley & Sons Ltd. This Work is a co-publication between John Wiley & Sons Ltd and ASME Press.
Companion website: www.wiley.com/go/JamilGhojel_Fundamentals of Heat Engines

ISA does not change by geographical location or by season, and it uses the following data for its standard reference point at sea level:

$$p_0 = 101.325 \ kPa$$

$$\rho_0 = 1.225 \ kg/m^3$$

$$T_0 = 288.15 \ K$$

$$a_0 = 340 \ m/s$$

$$R = 278 \ J/kg.K$$

$$g = 9.80665 \ m/s^2$$

Temperature vs. altitude:

$$T = T_0 - 6.5\frac{H}{1000} \ K, \ \ H \leq 11{,}000 \ m$$

H is the altitude in m. The temperature remains constant for $H > 11{,}000 \ m$

Pressure vs. altitude:

$$p_T = p_0\left(1 - 0.0065\frac{H}{T_0}\right)^{5.2561} \ kPa, \ \ H \leq 11{,}000 \ m$$

$$p = p_T \exp\left[-\frac{g}{RT_T}(H - H_T)\right] \ kPa, \ \ H > 11{,}000 \ m$$

Subscript T denotes properties at the tropopause ($H_T = 11{,}000 \ m$):

$$p_T = 22.632 \ kPa, T_T = 216.65 \ K$$

Density:

Density is calculated assuming ideal gas behaviour from

$$\rho = \frac{p}{RT} \ kg/m^3$$

Speed of sound:

The speed of sound in the atmosphere can be calculated as a function of the ratio of specific heats γ_a, gas constant R (=287 J/kg. K), and air temperature:

$$a = \sqrt{\gamma_a RT} \ m/s$$

The ratio of specific heats γ_a from Eq. (11.3) (see Chapter 11) can be used in the calculations. Since the temperature of the air remains constant at altitudes above the tropopause, the speed of sound also remains unchanged at 295.16 m/s above $H = 11\ 000 \ m$. Alternatively, the following correlation fitted to speed of sound data versus altitude H (in metres) can be used:

$$a = aH^2 + bH + c \ m/s, \ \ H \leq 11{,}000 \ m$$

where $a = -2.79E - 08, b = -3.80E - 03, c = 340.3207$.

Real atmospheric conditions are usually expressed in terms of ISA values, and they can be higher or lower than the standard value. Actual temperatures are then $ISA \mp \Delta T$, where $\Delta T = T_{ISA} - T_{act}$. As an example, If the actual temperature at altitude 7000 m

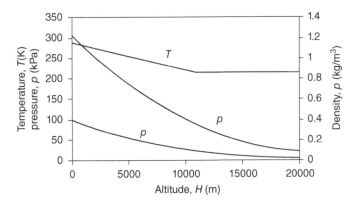

Figure 12.1 Plots of ISA pressure, temperature, and density of air versus altitude.

is $T_{act} = -40.5\ ^\circ C$ (232.55 K) and the ISA temperature $T_{ISA} = -30.5\ ^\circ C$ (242.65 K), the atmospheric temperature is then expressed as ISA + 10. If the actual temperature is $-25.5\ ^\circ C$, the atmospheric temperature is expressed as ISA – 5. Flying in excessive ISA-plus temperatures will negatively impact aircraft performance.

Figure 12.1 shows plots of standard atmospheric pressure, temperature, and density up to an altitude of 20 000 m.

12.1.2 Stagnation Properties

Aviation gas turbines power both civil and military aircraft that travel at high speeds, and the air enters the engine at a speed equal to the forward speed of the aircraft. As a result of the ram effect, the thermodynamic properties of the air at the inlet into the compressor will differ significantly from the static properties of the ambient air. The air properties at the inlet into the compressor are known as the *total* or *stagnation* properties.

If it is imagined that the high-speed air entering the diffuser in a turbojet engine shown in Figure 12.2 is brought to rest adiabatically and isentropically and without work at the entry into the compressor, the thermodynamic properties of the air will change in proportion to the flow velocity.

The energy equation for the diffuser in which the pressure increases at the expense of velocity as the fluid with initial velocity of C_1 is brought to rest at station 2 is

$$h_1 + \frac{C_1^2}{2} = h_{2t} + 0$$

By definition, the stagnation, or total, enthalpy h_{2t} is the enthalpy that a gas stream of enthalpy h_1 and velocity C_1 would possess when brought to rest adiabatically and without work transfer:

$$h_{2t} = h_1 + \frac{C_1^2}{2} \tag{12.1}$$

For a perfect gas, $h = c_p T$; therefore Eq. (12.1) can be rewritten as

$$T_{2t} = T_1 + \frac{C_1^2}{2c_p} \tag{12.2}$$

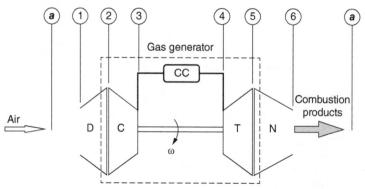

D-Diffuser   C-Compressor   CC-Combustion chamber
T - Compressor turbine   N - Nozzle

Figure 12.2   Schematic diagram of a turbojet engine.

The stagnation temperature can now be defined as the temperature that a gas stream of temperature T_1 and velocity C_1 would possess when brought to rest adiabatically and without work transfer.

The stagnation pressure p_{2t} can be defined as the pressure that a gas stream of pressure p_1, temperature T_1, and velocity C_1 would possess when brought to rest adiabatically and reversibly (isentropically) without work transfer:

$$p_{2t} = p_1 \left(\frac{T_{2t}}{T_1} \right)^{\gamma/(\gamma-1)} \tag{12.3}$$

$$p_{2t} = p_1 \left(1 + \frac{C_1^2}{2C_p T_1} \right)^{\gamma/(\gamma-1)} \tag{12.4}$$

The energy equation for process $2 - 3$ in the compressor in Figure 12.1 with work transfer can be written as

$$h_2 + \frac{C_2^2}{2} + W_c = h_3 + \frac{C_3^2}{2}$$

Applying the concept of stagnation properties to the process, we obtain

$$W_c = c_p(T_3 - T_2) + \frac{1}{2}(C_3^2 - C_2^2) = \left(c_p T_3 + \frac{C_3^2}{2} \right) - \left(c_p T_2 + \frac{C_2^2}{2} \right)$$

$$W_c = c_p \left(T_3 + \frac{C_3^2}{2c_p} \right) - c_p \left(T_2 + \frac{C_2^2}{2c_p} \right)$$

$$W_c = c_p(T_{3t} - T_{2t}) \tag{12.5}$$

Similarly, for the adiabatic expansion process $4 - 5$ in the turbine,

$$h_4 + \frac{C_4^2}{2} = h_5 + \frac{C_5^2}{2} + W_t$$

or

$$W_t = c_p(T_{4t} - T_{5t}) \tag{12.6}$$

Process 3 – 4 is a steady-flow process in the combustion chamber with heat transfer but without work transfer:

$$h_3 + \frac{C_3{}^2}{2} + Q = h_4 + \frac{C_4{}^2}{2}, \text{ or}$$

$$Q = c_p(T_{4t} - T_{3t}) \tag{12.7}$$

Thus, if stagnation temperatures are employed, there is no need to refer explicitly to the kinetic energy term in the energy equations for the compressor, combustion chamber, and turbine. This conclusion is very useful since the measured temperature of a flowing stream using a thermocouple, for example, is essentially the stagnation temperature and not the static temperature.

For air at $T = 25°$ C (298 K) and $p = 101.325$ kPa flowing at 20 m/s, the stagnation temperature and pressure are, respectively, 298.2 K and 101.56 kPa, which are only slightly higher than the static values. Therefore, for stationary industrial gas turbines with small air inlet flow velocities, the stagnation properties can be assumed to be equal to the static properties. For turbojet engines, the differences between static and stagnation properties could be significant. For example, for an aircraft flying at a speed of 850 km/h (236 m/s) at an altitude of 10 000 m, the static ISA air temperature is −50°C (223.15 K), and the pressure is 26.436 kPa. The stagnation temperature in this case is 250.81 K (−22.19° C), which indicates an increase in the air temperature at the inlet into the compressor of 27.66 °C. The corresponding stagnation pressure is 39.82 kPa. The effect of air speed on the stagnation temperature and pressure for the same initial conditions at an altitude of 10 000 m is shown in Figure 12.3. The temperature increases gradually with flow speed, whereas the stagnation pressure initially increases gradually up to 600 m/s and then starts increasing rapidly thereafter. The inset in Figure 12.3 shows the extreme stagnation values that can be attained at very high flow speeds. These are conditions similar to the ones encountered by the leading edge of a space craft re-entering the earth's atmosphere.

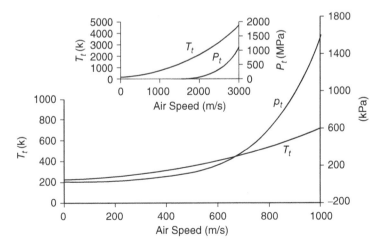

Figure 12.3 Stagnation temperature and pressure at an altitude of 10 000 m as functions of air speed (initial conditions: $T = 223.1$ K, $p = 26.436$ kPa, $\gamma = 1.4$).

12.2 Simple Turbojet Engine

Figure 12.4 shows the configuration of a simple turbojet engine with the main stations numbered, and the corresponding $T - s$ diagram. The engine comprises, in addition to the gas generator, two more major components: a diffuser D and a nozzle N (Figure 12.4a). The diffuser is mounted in front of the compressor, and the nozzle is mounted behind the turbine. The air enters the diffuser at speed C_1, which is equal to the forward speed of the aircraft. The gases leaving the turbine expand further in the nozzle before they are exhausted into the atmosphere at high speed C_6, generating a propulsive thrust that provides the forward motion of the aircraft. It should be noted that the temperatures at all stations but two are labelled as stagnation temperatures in the $T - s$ diagram: at diffuser entry 1 and nozzle exit 6 at the throat. The subscripts of the temperature and pressure are arranged as a combination of a digit followed by the letter t. The digit denotes the number of the station on the engine configuration, and the letter t denotes the total or stagnation property.

The losses in the diffuser and nozzle (assumed to be adiabatic devices) with no work or heat transfer can be accounted for by their respective isentropic efficiencies η_i and η_n. Similarly, the losses in the compressor and turbine are accounted for by their respective isentropic efficiencies η_c and η_t or polytropic efficiencies η_{pc} and η_{pt}. The losses in the combustion chamber are represented by the pressure drop Δp_{34}. There is no afterburner in the configuration, no power take-off to operate auxiliary equipment, and no air bleeding for blade-cooling purposes. Also, the cycle shown in T-s coordinates in Figure 12.4b is drawn as an open cycle, which is a distinguishing feature of real gas turbine cycles.

12.2.1 Intake (Diffuser)

The isentropic efficiency η_i of diffuser D is defined as

$$\eta_i = \frac{T_{2s} - T_1}{T_{2t} - T_1} \tag{12.8}$$

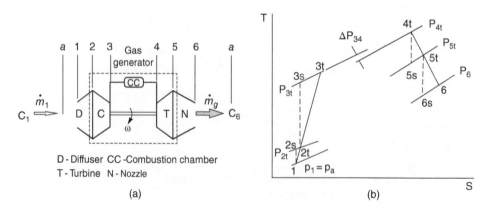

(a)   (b)

Figure 12.4 Simple turbojet engine cycle with $T - s$ diagram.

The intake pressure ratio in the diffuser (isentropic process $1-2$) is

$$\frac{P_{2t}}{P_1} = \left(\frac{T_{2s}}{T_1}\right)^{\gamma_a/(\gamma_a-1)} \tag{12.9}$$

where γ_a is temperature-dependent ratio of specific heats for air (see Chapter 11). From Eqs. (12.2) and (12.8),

$$T_{2s} - T_1 = \eta_i \frac{C_1^2}{2c_{pa}}, \text{ or}$$

$$\frac{T_{2s}}{T_1} = 1 + \eta_i \frac{C_1^2}{2c_{pa}T_1} \tag{12.10}$$

Therefore, the intake pressure ratio is

$$\frac{P_{2t}}{P_1} = \left[1 + \eta_i \frac{C_1^2}{2c_{pa}T_1}\right]^{\gamma_a/(\gamma_a-1)} \tag{12.11}$$

The Mach number at intake inlet conditions is

$$M_1 = \frac{C_1}{\sqrt{\gamma_a R_a T_1}} \tag{12.12}$$

where $\gamma_a R_a = c_{pa}(\gamma_a - 1)$

From Eqs. (12.11) and (12.12)

$$\frac{P_{2t}}{P_1} = \left[1 + \eta_i \frac{\gamma_a - 1}{2} M_1^2\right]^{\gamma_a/(\gamma_a-1)} \tag{12.13}$$

From Eqs. (12.2) and (12.12) and $\gamma_a R_a = c_{pa}(\gamma_a - 1)$,

$$\frac{T_{2t}}{T_1} = 1 + \frac{\gamma_a - 1}{2} M_1^2 \tag{12.14}$$

These equations are valid only when the flow in the intake is subsonic ($M_1 < 1$) or sonic ($M_1 = 1$).

12.2.2 Compressor

The temperature equivalent of the compressor work (process $2-3$) is

$$\Delta T_{2-3} = T_{3t} - T_{2t} = \frac{T_{2t}}{\eta_c}[r_c^{(\gamma_a-1)/\gamma_a} - 1], \text{ or}$$

$$\Delta T_{2-3} = T_{3t} - T_{2t} = T_{2t}[r_c^{(\gamma_a-1)/\gamma_a \eta_{pc}} - 1]$$

$$P_{2t} = r_c P_{1t}$$

The power absorbed by the compressor is

$$\dot{W}_c = \dot{m}_1 c_{pa} \Delta T_{2-3} = \dot{m}_1 c_{pa}(T_{3t} - T_{2t}) \tag{12.15}$$

12.2.3 Combustion Chamber

The rate of heat liberated during the combustion process is the heat added to the cycle, and it can be found by writing the energy equation without work transfer and no change in kinetic energy:

$$\dot{Q}_{in} = \dot{m}_g c_{pg} T_{4t} - \dot{m}_1 c_{pa} T_{3t} \tag{12.16}$$

where, $\dot{m}_g = \dot{m}_1 + \dot{m}_f$, c_{pa} is the specific heat of the air at station 3, c_{pg} is the specific heat of the combustion products at station 4, and $\dot{m}_f$ is the mass flow rate of the fuel in the combustion chamber.

If the mass flow of the fuel is ignored, the energy equation can be written as

$$\dot{Q}_{in} = \dot{m}_1 c_{pm}(T_{4t} - T_{3t})$$

where the mean specific heat c_{pm} is taken as the average between the values for the air at compressor delivery temperature and for the gases at turbine entry temperature:

$$c_{pm} = \frac{c_{pa} + c_{pg}}{2}$$

12.2.4 Turbine

In the turbojet engine, the turbine generates just enough power to drive the compressor and overcome the frictional losses in the compressor-turbine unit. The temperature equivalent of the turbine work is

$$\Delta T_{45} = T_{4t} - T_{5t} = \eta_t T_{4t} \left[1 - \frac{1}{r_t^{(\gamma_g-1)/\gamma_g}} \right] \tag{12.17}$$

or,

$$\Delta T_{45} = T_{4t} - T_{5t} = T_{4t} \left[1 - \frac{1}{r_t^{\eta_{pt}(\gamma_g-1)/\gamma_g}} \right] \tag{12.18}$$

The turbine power required to drive the compressor is

$$\dot{W}_t = \dot{m}_g c_{pg} \Delta T_{45} = \dot{m}_g c_{pg}(T_{4t} - T_{5t}) \tag{12.19}$$

Since $\dot{W}_t = \dot{W}_c/\eta_m$, where η_m is the mechanical efficiency of the gas generator,

$$\dot{W}_t = \dot{m}_g c_{pg}(T_{4t} - T_{5t}) = \frac{\dot{m}_1 c_{pa}(T_{3t} - T_{2t})}{\eta_m}$$

and

$$\Delta T_{45} = T_{4t} - T_{5t} = \frac{1}{\eta_m} \frac{\dot{m}_1}{\dot{m}_g} \frac{c_{pa}}{c_{pg}}(T_{3t} - T_{2t}) \tag{12.20}$$

Depending on the turbine efficiency used, Eq. (12.20) can be rewritten

$$\Delta T_{45} = \frac{1}{\eta_m} \frac{\dot{m}_1}{\dot{m}_g} \frac{c_{pa}}{c_{pg}}(T_{3t} - T_{2t}) = \eta_t T_{4t} \left[1 - \frac{1}{r_t^{(\gamma_g-1)/\gamma_g}} \right] \tag{12.21}$$

or

$$\Delta T_{45} = \frac{1}{\eta_m} \frac{\dot{m}_1}{\dot{m}_g} \frac{c_{pa}}{c_{pg}}(T_{3t} - T_{2t}) = T_{4t}\left[1 - \frac{1}{r_t^{\eta_{pt}(\gamma_g-1)/\gamma_g}}\right] \tag{12.22}$$

where r_t is the pressure ratio in the turbine.

Equation (12.20) can be used to determine the temperature drop ΔT_{45} in the turbine, and either Eq. (12.21) or (12.22) can be used to determine the pressure ratio in the turbine:

$$r_t = \left[\frac{1}{1 - \dfrac{\Delta T_{45}}{\eta_t T_{4t}}}\right]^{\gamma_g/(\gamma_g-1)}$$

or

$$r_t = \left[\frac{1}{1 - \dfrac{\Delta T_{45}}{T_{4t}}}\right]^{\gamma_g/\eta_{pt}(\gamma_g-1)}$$

12.2.5 Nozzle

The isentropic efficiency of the adiabatic converging nozzle in Figure 12.2a is defined as

$$\eta_n = \frac{T_{5t} - T_6}{T_{5t} - T_{6s}} \tag{12.23}$$

For the expansion process in the nozzle (isentropic process $5-6$),

$$\frac{T_{5t}}{T_{6s}} = \left(\frac{p_{5t}}{p_6}\right)^{(\gamma_g-1)/\gamma_g}$$

Combining this equation with Eq. (12.23) yields

$$\Delta T_{56} = T_{5t} - T_6 = \eta_n T_{5t}\left\{1 - \left[\frac{1}{(p_{5t}/p_6)^{(\gamma_g-1)/\gamma_g}}\right]\right\} \tag{12.24}$$

The temperature and pressure ratios in the nozzle can be written in terms of the Mach number, as was the case with the diffuser. The Mach number in the nozzle throat is

$$M_6 = \frac{C_6}{\sqrt{\gamma_g R_g T_6}}$$

where $\gamma_g R_g = c_{pg}(\gamma_g - 1)$

For a nozzle exit speed C_6, the energy equation for the flow between stations 5 and 6 is

$$h_5 = h_6 + \frac{C_6^2}{2}$$

For a perfect gas, the equation can be rewritten as

$$T_{5t} - T_6 = \frac{C_6^2}{2c_{pg}}, \text{ or}$$

$$\frac{T_{5t}}{T_6} = 1 + \frac{C_6^2}{2c_{pg}T_6} = 1 + \frac{\gamma_g - 1}{2}M_6^2 \tag{12.25}$$

From Eq. (12.23),

$$\frac{T_{6s}}{T_{5t}} = 1 - \frac{1}{\eta_n}\left(1 - \frac{T_6}{T_{5t}}\right)$$

Also

$$\frac{p_6}{p_{5t}} = \left(\frac{T_{6s}}{T_{5t}}\right)^{\gamma_g/(\gamma_g-1)} = \left[1 - \frac{1}{\eta_n}\left(1 - \frac{T_6}{T_{5t}}\right)\right]^{\gamma_g/(\gamma_g-1)}$$

Combining this equation with Eq. (12.25) yields

$$\frac{p_6}{p_{5t}} = \left[1 - \frac{1}{\eta_n}\left(1 - \frac{1}{1 + \dfrac{\gamma_g - 1}{2}M_6^{\;2}}\right)\right]^{\gamma_g/(\gamma_g-1)} \tag{12.26}$$

Two distinct cases can be identified, depending on the state of the gases in the nozzle throat at station 6 in Figure 12.2: sonic flow and subsonic flow.

12.2.5.1 Sonic Flow

If the flow in the nozzle throat is sonic, $M_6 = 1$ and the gases at the nozzle throat at station 6 are at the critical conditions of p_c and T_c. These gases expand from the pressure at station 5 to the critical pressure p_c ($p_c > p_a$) at station 6 first, and then to the ambient pressure p_a:

$$C_6 = C_c = \sqrt{\gamma_g R_g T_c} \tag{12.27}$$

The isentropic nozzle efficiency is

$$\eta_n = \frac{T_{5t} - T_c}{T_{5t} - T_{cs}} \tag{12.28}$$

where T_c is the critical temperature and T_{cs} is the temperature of the gases at the end of the isentropic expansion when the flow is sonic. The temperature ratio, from Eq. (12.25) becomes

$$\frac{T_{5t}}{T_c} = \frac{\gamma_g + 1}{2} \tag{12.29}$$

Knowing that $p_6 = p_c$ and $M_6 = 1$, the pressure ratio from Eq. (12.26) becomes

$$\frac{p_c}{p_{5t}} = \left[1 - \frac{1}{\eta_n}\left(\frac{\gamma_g - 1}{\gamma_g + 1}\right)\right]^{\gamma_g/(\gamma_g-1)} \tag{12.30}$$

or

$$\frac{p_{5t}}{p_c} = \frac{1}{\left[1 - \dfrac{1}{\eta_n}\left(\dfrac{\gamma_g - 1}{\gamma_g + 1}\right)\right]^{\gamma_g/(\gamma_g-1)}} \tag{12.31}$$

Equation (12.31) is used to determine the conditions in the nozzle throat. If the calculated pressure ratio p_{5t}/p_a is greater than the critical pressure ratio ($p_{5t}/p_a > p_{5t}/p_c$), then the nozzle is said to be *choked*, and the critical speed, temperature, and pressure are found from Eqs. (12.27), (12.29), and (12.30) respectively.

The critical density is

$$\rho_6 = \rho_c = \frac{p_c}{R_g T_c} \tag{12.32}$$

The nozzle throat area is

$$A_6 = \frac{\dot{m}_g}{\rho_c C_c} \tag{12.33}$$

Since the gas in the nozzle expands first to the critical pressure p_c and then to the ambient pressure p_a, the thrust is made up of two components: the momentum thrust and the pressure thrust:

$$F = (\dot{m}_g C_c - \dot{m}_1 C_1) + (p_c - p_1)A_6 \quad N \tag{12.34}$$

where the mass flow rate is in *kg/s* and the speeds are in *m/s*.
The specific fuel consumption (*sfc*) is

$$sfc = \frac{\dot{m}_f}{F} \tag{12.35}$$

The units of the *sfc* depend on the units of $\dot{m}_f$ and *F*: $kg/N.\,s$, $kg/kN.\,s$, or $kg/kN.\,h$.

12.2.5.2 Subsonic Flow
If the gases in the nozzle are expanded to the atmospheric pressure ($p_6 = p_a$) and the pressure ratio p_{5t}/p_a is less than the critical pressure ratio p_{5t}/p_c, the nozzle is said to be *unchoked*, with $M_6 < 1$ and the temperature and pressure ratios in the nozzle is given by Eqs. (12.25) and (12.26). The pressure at the throat of the nozzle in Eq. (12.26) is replaced by the ambient pressure $p_a = p_1$. The flow velocity in the nozzle throat can be determined from either of the following equations:

$$C_6 = \sqrt{2c_{pg}(T_{5t} - T_6)}$$

or

$$C_6 = M_6 \sqrt{c_{pg} T_6 (\gamma_g - 1)}$$

The gas density in the nozzle throat is

$$\rho_6 = \frac{p_a}{R_g T_6} \tag{12.36}$$

The nozzle throat area is

$$A_6 = \frac{\dot{m}_g}{\rho_5 C_6} \tag{12.37}$$

Since the gases in the nozzle expand to p_a, the pressure component of the thrust is equal to zero, reducing the thrust equation to

$$F = \dot{m}_g C_6 - \dot{m}_1 C_1 \quad N \tag{12.38}$$

where the mass flow rate is in *kg/s* and the velocities are in *m/s*.
The *sfc* is

$$sfc = \frac{\dot{m}_f}{F} \tag{12.39}$$

The units of the *sfc* depend on the units of $\dot{m}_f$ and *F*, as was shown previously.

12.2.6 Engine Performance

The overall efficiency is defined as the ratio of the useful power provided to the aircraft to the rate of energy input:

$$\eta_o = \frac{C_1 F}{\dot{m}_f H_l} = \frac{1}{sfc} \frac{C_1}{H_l} \tag{12.40}$$

C_1 is the aircraft forward speed, H_l is the lower heating value of the fuel (the water in the products remains in vapour form), and $\dot{m}_f$ is the fuel flow rate. The right-hand side of Eq. (12.40) shows that for a given fuel and constant flight speed, the overall efficiency is inversely proportional to the *sfc*.

The propulsion (or propulsive) efficiency is defined as the ratio of the power provided to the aircraft by the momentum thrust to the power generated by the change in kinetic energy of the gases flowing through the engine:

$$\eta_p = \frac{C_1(\dot{m}_g C_6 - \dot{m}_1 C_1)}{\frac{1}{2}(\dot{m}_g C_6^2 - \dot{m}_1 C_1^2)} \tag{12.41}$$

C_6 is the velocity of the jet in the nozzle throat, C_1 is the aircraft forward speed, and $\dot{m}_1$ and $\dot{m}_g$ ($\dot{m}_g = \dot{m}_1 + \dot{m}_f$) are the mass flow rates of the air and combustion gases, respectively.

If the mass flow rate of the fuel is ignored ($\dot{m}_g \approx \dot{m}_1$), the last equation is reduced to

$$\eta_p = \frac{C_1(C_6 - C_1)}{\frac{1}{2}(C_6^2 - C_1^2)} = \frac{C_1(C_6 - C_1)}{\frac{1}{2}(C_6 - C_1)(C_6 + C_1)} = \frac{2C_1}{(C_6 + C_1)}, \text{ or, finally}$$

$$\eta_p = \frac{2}{(1 + C_6/C_1)} \tag{12.42}$$

The gas velocity in the nozzle throat C_6 is replaced by the critical velocity C_c when the flow in the nozzle throat becomes sonic (the nozzle is choked).

The thermal efficiency is the ratio of the overall efficiency to the propulsion efficiency:

$$\eta_{th} = \frac{\eta_o}{\eta_p} \tag{12.43}$$

12.3 Performance of Turbojet Engine – Case Study

The performance characteristics presented are for a specific example based on the data in Table 12.1. The engine is operated at subsonic cruise conditions at an altitude of 10 000 *m* for five values of the maximum-to-minimum temperature ratio T_{4t}/T_{2t} (4, 5, 6, 7, and 8). Considering that the stagnation compressor entry temperature T_{2t} under the assumed flight conditions is 251.0 *K*, the corresponding turbine entry temperatures T_{4t} are 1004, 1255, 1506, 1757, and 2008 *K*. The data presented are for critical properties at the nozzle throat (choked nozzle).

Figures 12.5 and 12.6 show some representative data with respect to the main stations and components of the simple turbojet engine when operated at compressor pressure ratios r_c=10 and 36, turbine entry temperature $T_{4t} = 1757$ *K*, and relative air-fuel ratio $\lambda = 1.93$. These show the wide variations of the temperature, pressure, and specific heat of the gases

Table 12.1 Data for the calculation of the turbojet engine performance characteristics

	Fuel: $C_{12}H_{24}$					
Turbine entry temperature, K	$T_{max} = T_{4t}$	1004	1255	1506	1757	2008
Diffuser isentropic efficiency	η_i	0.9	0.9	0.9	0.9	0.9
Compressor polytropic efficiency	η_{pc}	0.85	0.85	0.85	0.85	0.85
Turbine polytropic efficiency	η_{pt}	0.88	0.88	0.88	0.88	0.88
Mechanical efficiency	η_m	0.98	0.98	0.98	0.98	0.98
Nozzle isentropic efficiency	η_n	0.9	0.9	0.9	0.9	0.9
Aircraft forward speed, km/h	C_1	862.5	862.5	862.5	862.5	862.5
Flight altitude, m	H	10 000	10 000	10 000	10 000	10 000
Ambient pressure, kPa	p_1	26.428	26.428	26.428	26.428	26.428
Ambient temperature, K	T_1	223.2	223.2	223.2	223.2	223.2
Flight Mach Number	M_1	0.8	0.8	0.8	0.8	0.8
Compressor inlet temperature, K	T_{2t}	251	251	251	251	251
Temperature Ratio	T_{4t}/T_{2t}	4	5	6	7	8
Combustion chamber pressure losses, kPa	Δp_{34}	4	4	4	4	4
Air flow rate, kg/s	$\dot{m}_1$	1.0	1.0	1.0	1.0	1.0
Relative air-fuel ratio	λ	variable	variable	variable	variable	variable
Fuel flow rate, kg/s	$\dot{m}_f$	variable	variable	variable	variable	variable
Gas flow rate, kg/s	$\dot{m}_g$	variable	variable	variable	variable	variable
Lower heating value, kJ/kg	H_l	44 340	44 340	44 340	44 340	44 340

(air and combustion products) within the flow passages in the turbojet. In calculating the latter, a gas constant of 0.287 $kJ/kg\,K$ is assumed. The compressor pressure ratio has much greater effect on the pressure and temperature profiles than on the specific heat and ratio of specific heats. However, the effect of temperature on the specific heat is significant along the flow path of gases in the turbojet.

Figures 12.7–12.9 show the general performance characteristics of the simple turbojet engine. The parameters investigated include the specific thrust f, sfc, and propulsion efficiency. The following comments summarise the performance characteristics of the uninstalled turbojet engine:

- At a given compressor pressure ratio, the specific thrust increases significantly with turbine entry temperature. Thus, increasing T_{4t}/T_{2t} allows for the design of a more compact engine, which is essential to reduce weight and drag.
- The thrust curves exhibit maxima at different compressor pressure ratios, with the location of the peak values shifting towards higher r_c with increasing T_{4t}.
- At a constant r_c, the sfc increases with increasing T_{4t}.

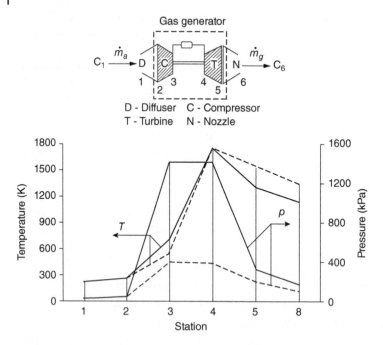

Figure 12.5 Variation of temperature T and pressure p in the simple turbojet engine: $T_{4t} = 1757\ K$, $\lambda = 1.93$, $r_c = 10$ (dashed lines), $r_c = 36$ (solid lines).

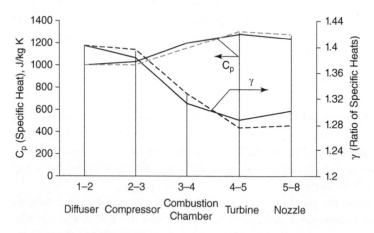

Figure 12.6 Variation of specific heat c_p and ratio of specific heats γ in the simple turbojet engine: $T_{4t} = 1757\ K$, $\lambda = 1.93$, $r_c = 10$ (dashed lines), $r_c = 36$ (solid lines).

- Only the curve for the lowest temperature ratio $T_{4t}/T_{2t} = 4$ ($T_{4t} = 1004\ K$) exhibits a minimum *sfc* at $r_c = 16$.
- For a given turbine inlet temperature, the propulsive efficiency increases with increasing compressor pressure ratio, with the rate of increase becoming greater at lower turbine inlet temperatures. At constant r_c, the propulsive efficiency decreases with increasing turbine inlet temperature (increasing T_{4t}/T_{2t}).

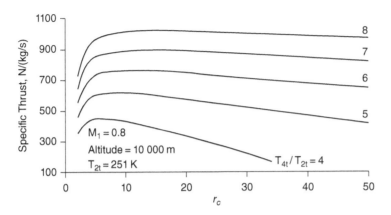

Figure 12.7 Specific thrust of the turbojet engine at $H = 10\,000\,m$, compressor inlet temperature $T_{2t} = 251\,K$, and inflow air Mach number $M_1 = 0.8$.

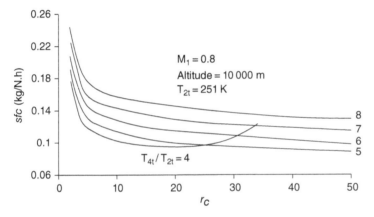

Figure 12.8 Specific fuel consumption of the turbojet engine at $H = 10\,000\,m$, compressor inlet temperature $T_{2t} = 251\,K$, and inflow air Mach number $M_1 = 0.8$.

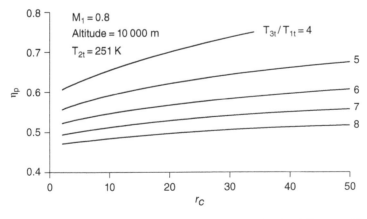

Figure 12.9 Propulsive efficiency of the turbojet engine for different turbine inlet temperatures at $H = 10\,000\,m$, compressor inlet temperature $T_{2t} = 251\,K$, and inflow air Mach number $M_1 = 0.8$.

12.3.1 Performance Maps

These characteristics, sometimes referred to as *carpet plots* (Mattingly et al., 2002), show the *sfc* as a function of the specific thrust f at constant temperature ratio lines T_{4t}/T_{2t} (or constant turbine entry temperature lines T_{4t}) and constant compressor pressure ratio lines r_c (Figure 12.10). If we replace the *sfc* with the propulsive efficiency, a performance map similar to the one shown in Figure 12.11 is obtained.

Figure 12.10 confirms the strong dependence of the thrust on turbine entry temperature observed from the general characteristics and shows the maximum possible thrust and the corresponding compressor pressure ratio for a given T_{4t} that will keep the engine as small as possible. For each value of T_{4t}, there is an initial increase in specific thrust with increasing r_c, which then starts to decrease after a certain value of r_c is reached. For the assumed flight

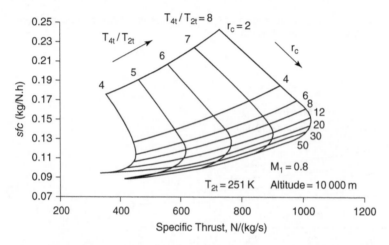

Figure 12.10 Performance map in terms of *sfc* vs. specific thrust for the turbojet engine at $H = 10\,000\,m$ and inflow air Mach number $M_1 = 0.8$.

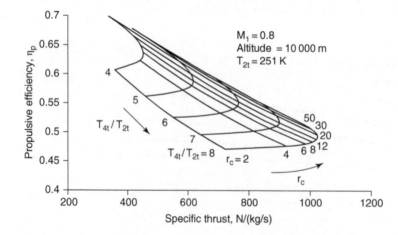

Figure 12.11 Performance map in terms of η_p vs. specific thrust for the turbojet engine at $H = 10\,000\,m$ and inflow air Mach number $M_1 = 0.8$.

conditions, the maximum thrust of 1024 N is obtained when $T_{4t} = 2008\ K$ and $r_c = 16$. The value of r_c at which this occurs increases with T_{4t}. At a constant r_c, both the specific thrust and the *sfc* increase with increasing T_{4t}. The effect of r_c on the *sfc* becomes less pronounced at higher values, as shown by the increase in density of the constant r_c lines at the bottom of the plot. At the lowest temperature ratio of 4 ($T_{4t} = 1004\ K$), operating the engine at pressure ratios above 12 causes the *sfc* to decrease slightly at the expense of sharply decreasing specific thrust.

For a given pressure ratio, the propulsive efficiency decreases as the temperature ratio increases. For a given temperature ratio, as the pressure ratio increases, both the specific thrust and the propulsive efficiency initially increase, and then the specific thrust starts decreasing while the propulsive efficiency continues to increase (Figure 12.11).

A three-dimensional surface plot of the performance characteristics in the form $sfc = \psi(r_c, f)$ can be obtained by fitting the results of the calculations to the correlation

$$sfc = a + \frac{b}{r_c} + c\ \ln(f) + \frac{d}{r_c^2} + e\ \ln(f^2) + \frac{g\ \ln(f)}{r_c}\ \ kg/N.h \tag{12.44}$$

where f in the specific thrust, $a = 2.013731$, $b = 0.675000$, $c = 0.641320$, $d = 0.168114$, $e = 0.053319$, and $g = 0.131156$.

Figure 12.12 shows the surface plot in which the compressor pressure ratio r_c and the specific thrust f are the independent variables and the *sfc* is the dependent variable. The plot also shows the constant turbine entry temperature lines (T_{4t}). For a compressor entry temperature of $T_{2t} = 251\ K$, the temperatures shown correspond to the temperature ratios T_{4t}/T_{2t} of 4, 5, 6, 7, and 8 respectively. The dependent variable axis is inverted to make the plot easier to read. The dark area on the top of the plot shows the lowest attainable *sfc* and corresponding values of the specific thrust and compressor pressure ratio. The specific thrust is defined as $f = F/(\dot{m}_a + \dot{m}_f)$ in $N/kg/s$. Equation (12.44) is valid for compressor

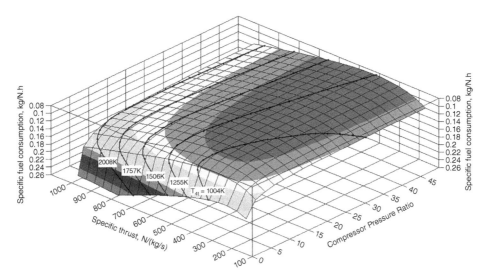

Figure 12.12 3-D surface plot of the performance of the simple turbojet engine at $H = 10\,000\ m$, $T_{2t} = 251\ K$, and inflow air Mach number $M_1 = 0.8$.

pressure ratio range 2–50 and specific thrust range 168–1024 N/kg (air)/s corresponding to a turbine entry temperature range of 1004–2008 K.

12.3.2 Effect of Flight Mach Number

The effect of flight Mach number on the performance of the turbojet engine is investigated at an altitude of $H = 10\,000\ m$ and constant turbine entry temperature of 1757 K. The Mach numbers, corresponding flight speed C_a, compressor inlet pressure p_{2t}, and temperature T_{2t} are given in Table 12.2. It should be noted that these calculations are for an uninstalled engine, and the calculations are design-point calculations at each Mach number for which the nozzle is choked over the entire compressor pressure ratio range.

Figures 12.13–12.15 show the specific thrust, sfc, and propulsive efficiency versus compressor pressure ratio at different Mach numbers. At a constant Mach number, the specific thrust initially increases rapidly with the compressor pressure ratio and then decreases slowly after reaching a peak (Figure 12.13). The specific thrust plotted versus compressor pressure ratio exhibits a peak at all Mach numbers.

The sfc, plotted versus the compressor pressure ratio (Figure 12.14), decreases continuously with r_c without reaching a minimum for all constant M_1 curves. For a given compressor pressure ratio, sfc increases with increasing M_1.

Figure 12.15 shows that the propulsive efficiency increases with both the compressor pressure ratio and the Mach number. The change with the Mach number is more pronounced than with the compressor pressure ratio. This trend can be explained by the approximate relationship $\eta_p = 2/(1 + C_c/C_1)$. At constant M_1 (M_a), C_1 is constant, and C_c decreases by about 14% over the range $r_c = 2$–50. Hence the moderate increase in the

Table 12.2 Operating conditions for variable flight Mach numbers at constant $T_{4t} = 1757\ K$ and $H = 10\,000\ m$, $p_{1t} = 26.4\ kPa$, $T_{1t} = 223.2\ K$.

M_1	0.2	0.6	0.7	0.75	0.8	0.85
C_a, m/s	59.94	179.72	209.72	224.72	239.58	254.58
p_{2t}, kPa	27.14	33.32	36.10	37.72	39.49	41.60
T_{2t}, K	225	239	245	248	252	256

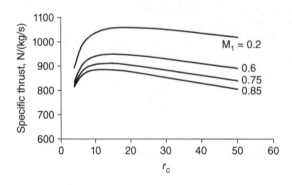

Figure 12.13 Specific thrust of the turbojet engine vs. compressor pressure ratio at different flight Mach numbers, $H = 10\,000\ m$, and $T_{4t} = 1757\ K$.

Figure 12.14 Specific fuel consumption of the turbojet engine vs. compressor pressure ratio at different flight Mach numbers, $H = 10\,000\,m$, and $T_{4t} = 1757\,K$.

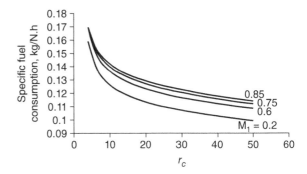

Figure 12.15 Propulsive efficiency of the turbojet engine vs. compressor pressure ratio at different flight Mach numbers, $H = 10\,000\,m$, and $T_{4t} = 1757\,K$.

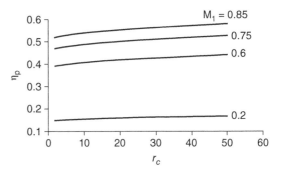

propulsive efficiency with increasing r_c. At constant r_c, increasing the Mach number from 0.2 to 0.85 results in a threefold increase in C_1 and a corresponding increase of more than twofold in propulsive efficiency.

The specific thrust and sfc plotted versus the Mach number at a fixed compressor pressure ratio is shown in Figures 12.16. The specific thrust decreases continuously with increasing flight Mach number, while the sfc increases. If we consider the specific thrust equation

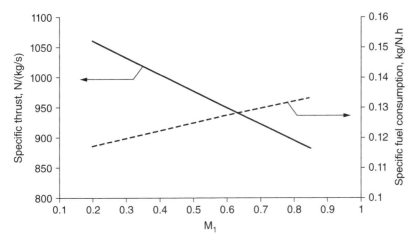

Figure 12.16 Specific thrust and specific fuel consumption vs. flight Mach number ($M_a = M_1$) at $H = 10\,000\,m$, $T_{4t} = 1757\,K$, and $r_c = 16$.

for the choked nozzle, $f = (C_c - C_1) + (p_c - p_1)A_8/\dot{m}_g$, calculations show that the rate of decrease of $(C_c - C_1)$ is greater than the rate of increase of $(p_c - p_1)$ (for constant ambient conditions); hence the reason for decreasing f with increasing M_1 (increasing C_1). The sfc increases with M_1 because the rate of decrease of f (16.9%) with M_1 is greater than the rate of decrease of the fuel flow rate (5.3%).

12.3.3 Effect of Flight Altitude

The performance characteristics considered are the same as previously with the calculations conducted at each flight altitude at constant turbine entry temperature $T_{4t} = 1757\ K$ and flight Mach number $M_1 = 0.85$. The other operating conditions are given in Table 12.3.

The variation of the specific thrust with the compressor pressure ratio at four altitudes is shown in Figure 12.17. The specific thrust exhibits a peak around $r_c = 11$ at all altitudes. This is due to the combined effect of the fuel mass flow rate $\dot{m}_f$, critical flow velocity C_c, and critical pressure p_c in the nozzle throat. The increase in specific thrust up to $r_c = 11$ is due to the rate of increase in p_c, with the compressor pressure ratio outweighing the rate of decrease in $\dot{m}_f$ and C_c. As the rate of increase of p_c drops significantly afterwards, the specific thrust decreases continuously for all altitudes for $r_c > 11$. For a given pressure ratio, the specific thrust increases with altitude due to the favourable effect of lower engine entry temperature (Saravanamuttoo et al., 2001). The rate of increase of the specific thrust is greater for higher compressor pressure ratios (Figure 12.17).

The sfc decreases continuously with the compressor pressure ratio, as seen in Figures 12.18 and 12.19. The rate of decrease is greater at the lower range of r_c.

Table 12.3 Operating conditions at $T_{4t} = 1757\ K$ for variable altitude at $M_1 = 0.85$.

H, m	4 000	8 000	10 000	12 000
C_a, m/s	276	262	254.6	247.1
p_{1t}, kPa	61.6	35.6	26.4	19.3
p_{2t}, kPa	96.7	55.8	41.5	30.3
T_{2t}, K	300	270	256	241

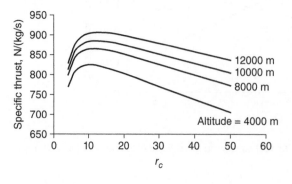

Figure 12.17 Specific thrust of the turbojet engine vs. compressor pressure ratio at different altitudes ($T_{4t} = 1757\ K, M_1 = 0.85$).

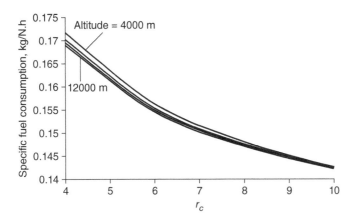

Figure 12.18 Specific fuel consumption of the turbojet engine vs. compressor pressure ratio at different altitudes ($r_c = 4$–10, $T_{4t} = 1757\,K$, $M_1 = 0.85$).

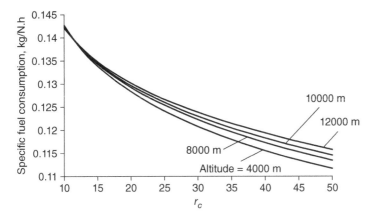

Figure 12.19 Specific fuel consumption of the turbojet engine vs. compressor pressure ratio at different altitudes ($r_c = 10$–50, $T_{4t} = 1757\,K$, $M_1 = 0.85$).

For compressor pressure ratios less than 11, the *sfc* decreases with altitude. For compressor pressure ratios greater than 11, the *sfc* increases with altitude. At $r_c = 11$, the altitude has no effect on the *sfc*.

The effect of altitude on the propulsive efficiency is shown in Figure 12.20. The efficiency increases with the compressor pressure ratio and decreases with increasing altitude at all pressure ratios. The rate of decrease of the propulsive efficiency with altitude increases with increasing compressor pressure ratio.

12.4 Two-Spool Unmixed-Flow Turbofan Engine

Modern aircraft engines are of a type known as *turbofan* engines. A turbofan engine can be the mixed-flow kind used mainly as high-bypass-ratio engines in civil aviation or the

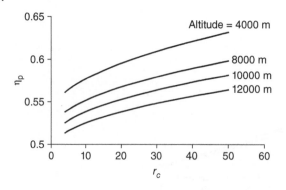

Figure 12.20 Propulsive efficiency of the turbojet engine vs. compressor pressure ratio at different altitudes ($T_{4t} = 1757\,K$, $M_1 = 0.85$).

mixed-flow type used predominantly as low-bypass-ratio engines in military aviation. A schematic diagram of an unmixed-flow simple turbofan engine with the cold and hot streams flowing through two separate nozzles is shown in Figure 12.21. This basic configuration comprises a diffuser, the core engine, and a bypass section. The components of the core engine are the root section of fan F, high-pressure compressor C, combustion chamber CC, high-pressure turbine HPT, low-pressure turbine LPT, and hot nozzle HN. The bypass section comprises the tip section of fan F, the concentric jet tube section, and a cold nozzle CN. The total mass flow rate of the air $\dot{m}_1$ flowing through the diffuser is split into 'hot' flow $\dot{m}_h$ in the core engine and 'cold' flow $\dot{m}_c$ in the bypass section. The HN in the core engine produces a jet of combustion products of mass flow rate $\dot{m}_g$ (hot flow rate plus fuel flow rate) at speed C_8, and the CN of the bypass section produces a jet of air of mass flow rate $\dot{m}_c$ at speed C_{18}. The fan is driven by the LPT while the compressor is driven by the HPT. The fan acts as a low-pressure compressor in the core engine and provides compressed air to the bypass section to be expanded in the CN. The total thrust is made up of the thrust produced by the core engine and the thrust produced by the bypass section.

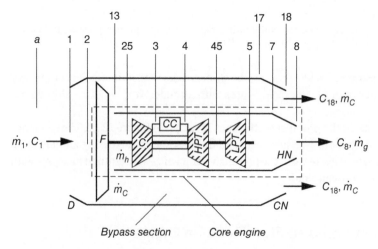

Figure 12.21 Configuration of a simple unmixed-flow turbofan engine with separate cold and hot streams: D – diffuser, F – fan, C – compressor, CC – combustion chamber, HPT – high-pressure turbine, LPT – low-pressure turbine, CN – cold nozzle, HN – hot nozzle.

In the schematic shown in Figure 12.21 the air is assumed to have uniform properties over the entire front face of the fan (station 2). There is no afterburner in the configuration, no power take-off to operate auxiliary equipment, and no air bleeding for blade-cooling purposes.

The station numbering shown in Figure 12.21 is based on the Society of Automotive Engineers (SAE) recommended practice (SAE, 1974; Mattingly et al., 2002), and an explanation is given in Table 12.4. Both the CN and HN are convergent nozzles.

Turbojets operated at high subsonic speeds have the disadvantage of relatively low propulsion efficiency. This is due to the high value of the nozzle exit speed term in Eq. (12.42) for the propulsive efficiency, rewritten as $\eta_p = 2/(1 + C_8/C_1)$ in this case. To increase this efficiency for turbojet engines, the jet velocity must be decreased significantly; a maximum value of 100% is reached when $C_8 = C_1$. However, decreasing C_8 reduces the thrust produced by the engine, rendering this measure ineffective in turbojet engines. The turbofan configuration allows both the reduction of C_8 and increase of the thrust by providing large amounts of air at high flow rates to compensate for the reduction of the jet velocity. All of the air that enters the engine passes through the fan, but only some of this air passes through the core engine; most of the air bypasses the core engine and flows through the annular bypass section, with the potential to produce the bulk of the total thrust of the engine. The turbofan engine provides high propulsive and overall efficiencies in the flight speed range 400–1000 km/h. This is the speed range in which the vast majority of commercial jet aircraft operate. Furthermore, the exhaust noise of turbofan engines is lower than that of jet engines due to the low mean jet velocity.

Figure 12.22 shows the two main configurations for commercial aircraft engines with unmixed hot and cold flows: the two-spool and the three-spool turbofan engines. In the

Table 12.4 Station numbering in the unmixed-flow turbofan engine.

Station	Location
a	Ambient conditions upstream from the diffuser
1	Entry into diffuser D
13	Exit from fan F tip part (entry into cold jet tube in the bypass section)
17	Entry into cold nozzle CN
18	Exit from cold nozzle CN (nozzle throat)
2	Entry into both root and tip parts of fan F (uniform properties over entire fan face)
25	Exit from root part of fan F (entry into compressor C)
3	Exit from compressor C (entry into combustion chamber CC)
4	Exit from combustion chamber CC (entry into high-pressure turbine HPT)
45	Exit from high-pressure turbine HPT (entry into low-pressure turbine LPT)
5	Exit from low-pressure turbine LPT (entry into hot jet tube in core engine)
7	Entry into hot nozzle HN
8	Exit from hot nozzle HN (nozzle throat)

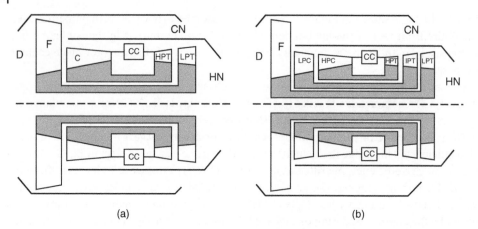

Figure 12.22 Two-spool (a) and three-spool (b) unmixed-flow turbofan engine configurations: D – diffuser, F – fan, C – compressor, LPC – low-pressure compressor, HPC – high-pressure compressor, CC – combustion chamber, HPT – high-pressure turbine, IPT – intermediate-pressure turbine, LPT – low-pressure turbine, CN – cold nozzle, HN – hot nozzle.

two-spool (two-shaft) engine, the intake air is compressed in two stages, with the fan forming the first stage; and the combustion products expand in two turbine stages, HPT and LPT (Figure 12.22a). The fan is driven by the LPT, and the compressor in the gas generator is driven by the HPT. The gases leaving the LPT expand in the HN, and the bypass air compressed by the fan expands in the CN. In the three-spool (three-shaft) turbofan engine, shown schematically in Figure 12.22b, the core engine has three compression stages (the fan F, LPC, and high-pressure compressor HPC) and three expansion stages (HPT, intermediate-pressure turbine [IPT], and LPT). The fan is driven by the LPT, the LPC is driven by the IPT, and the HPC is driven by the HPT. The gases leaving the LPT expand in the HN, and the bypass air expands in the CN. We will limit our study to the two-pool engine in this book.

12.4.1 Design-Point Calculations of the Core Engine

Figure 12.23 shows the T-s diagrams for the core engine of the two-spool turbofan engine with unmixed flow depicted in Figures 12.21 and 12.22a. The inset in Figure 12.23 is a magnification of the processes between stations 1 and 2. We are going to assume that the conditions at stations a (the ambient air) and 1 (entry into the diffuser) are the same and the conditions at stations 5 and 7 are also the same. In addition to the ram effect, the inlet air is compressed in two stages: in fan F and in compressor C. The combustion products expand in two turbine stages (HPT and LPT) before they are expanded in the hot converging nozzle of the core engine HN. The calculation methodology for the core engine is similar to the one used for the turbojet engine cycle.

12.4.1.1 Bypass Ratio

Of the total air intake into the engine $\dot{m}_1$, only a small amount $\dot{m}_h$ flows through the core engine; the rest $\dot{m}_c$, known as the *bypass air*, flows through the bypass section. The bypass

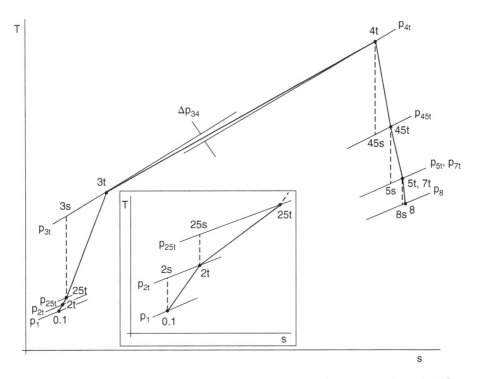

Figure 12.23 T-s diagrams of the processes in the core engine of the two-spool, unmixed-flow turbofan engine.

ratio B of a turbofan engine is defined as

$$B = \frac{\dot{m}_c}{\dot{m}_h} \tag{12.45}$$

Since $\dot{m}_c + \dot{m}_h = \dot{m}_1$,

$$\dot{m}_c = \frac{\dot{m}_1 B}{B+1} \tag{12.46}$$

$$\dot{m}_h = \frac{\dot{m}_1}{B+1} \tag{12.47}$$

The bypass ratio can be as high as 10 in some modern commercial aircraft engines.

12.4.1.2 Intake
The isentropic intake efficiency (see inset in Figure 12.23) is

$$\eta_i = \frac{T_{2s} - T_1}{T_{2t} - T_1}$$

The energy equation applied to the diffuser gives

$$T_{2t} - T_1 = \frac{C_1^{\,2}}{2c_{pa}}$$

Hence,

$$T_{2s} - T_1 = \eta_i \frac{C_1^{\,2}}{2c_{pa}}$$

or

$$\frac{T_{2s}}{T_1} = 1 + \eta_i \frac{C_1^{\,2}}{2c_{pa}T_1} \tag{12.48}$$

Therefore, the intake pressure ratio is

$$\frac{P_{2t}}{P_1} = \left(\frac{T_{2s}}{T_1}\right)^{\gamma_a/(\gamma_a - 1)} = \left[1 + \eta_i \frac{C_1^{\,2}}{2c_{pa}T_1}\right]^{\gamma_a/(\gamma_a - 1)} \tag{12.49}$$

Or, since $M_1 = C_1/\sqrt{\gamma_a R_a T_1}$ and $\gamma_a R_a = c_{pa}(\gamma_a - 1)$,

$$\frac{P_{2t}}{P_1} = \left[1 + \eta_i \frac{\gamma_a - 1}{2}M_1^{\,2}\right]^{\gamma_a/(\gamma_a - 1)} \tag{12.50}$$

$$\frac{T_{2t}}{T_1} = 1 + \frac{\gamma_a - 1}{2}M_1^{\,2} \tag{12.51}$$

These equations are valid only when the flow in the intake is subsonic ($M_1 < 1$) or sonic ($M_1 = 1$).

12.4.1.3 Fan

The temperature equivalent of the fan work, using the subscript f for fan, is

$$\Delta T_{2-25} = T_{25t} - T_{2t} = \frac{T_{2t}}{\eta_f}\,[r_f^{(\gamma_a-1)/\gamma_a} - 1]$$

$$\Delta T_{2-25} = T_{25t} - T_{2t} = T_{2t}[r_f^{(\gamma_a-1)/\gamma_a\eta_{pf}} - 1]$$

η_f, η_{fp}, and r_f are, respectively, the fan isentropic efficiency, polytropic efficiency, and pressure ratio.

If r_f is known, the fan exit pressure within the core engine is

$$P_{25t} = r_f P_{2t}$$

The power absorbed by the fan is

$$\dot{W}_f = \dot{m}_1 c_{pa} \Delta T_{2-25} \tag{12.52}$$

where c_{pa} is the mean specific heat at constant pressure across the fan, and $\dot{m}_1$ is the total flow rate of the air entering the diffuser. This flow is the sum of the cold mass flow rate in the bypass duct $\dot{m}_c$ and the mass flow rate of the air $\dot{m}_h$ entering the core engine.

12.4.1.4 Compressor

The temperature equivalent of the compressor work in the gas generator is

$$\Delta T_{25-3} = T_{3t} - T_{25t} = \frac{T_{25t}}{\eta_c}\,[r_c^{(\gamma_a-1)/\gamma_a} - 1]$$

$$\Delta T_{25-3} = T_{3t} - T_{25t} = T_{25t}[r_c^{(\gamma_a-1)/\gamma_a\eta_{pc}} - 1]$$

where η_c, η_{cp}, and r_c are, respectively, the compressor isentropic efficiency, polytropic efficiency, and pressure ratio.

The overall pressure ratio of the turbofan engine is

$$r_o = r_f r_c$$

The power absorbed by the compressor in the core engine (and provided by the HPT) is

$$\dot{W}_c = \dot{m}_h c_{pa} \Delta T_{25-3} \qquad (12.53)$$

where c_{pa} is the mean specific heat at constant pressure across the compressor.

12.4.1.5 Combustion Chamber

The rate of heat liberated during the combustion process is the heat added to the cycle and can be found by writing the energy equation without work transfer and no change in kinetic energy:

$$\dot{Q}_{in} = \dot{m}_g c_{pg} T_{4t} - \dot{m}_h c_{pa} T_{3t} \qquad (12.54)$$

where $\dot{m}_g = \dot{m}_h + \dot{m}_f$, c_{pa} is the specific heat of air at station 3, c_{pg} is the specific heat of the combustion products at station 4, and $\dot{m}_f$ is the mass flow rate of the fuel in the combustion chamber.

12.4.1.6 Compressor Turbine (High-Pressure Turbine)

The HPT generates just enough power to drive the compressor and overcome the frictional losses in the compressor-turbine unit. The temperature equivalent of the HPT work is

$$\Delta T_{4-45} = T_{4t} - T_{45t} = \eta_t T_{4t} \left[1 - \frac{1}{r_{HPT}^{(\gamma_g - 1)/\gamma_g}} \right]$$

$$\Delta T_{4-45} = T_{4t} - T_{45t} = T_{4t} \left[1 - \frac{1}{r_{HPT}^{\eta_{pt}(\gamma_g - 1)/\gamma_g}} \right]$$

where, η_t, η_{cp}, and r_{HPT} are, respectively, the isentropic efficiency, polytropic efficiency, and expansion ratio of the HPT.

The power produced by the HPT is

$$\dot{W}_{HPT} = \dot{m}_g c_{pg} \Delta T_{4-45} = \dot{m}_g c_{pg} (T_{4t} - T_{45t}) \qquad (12.55)$$

$$\dot{W}_{HPT} = \frac{\dot{W}_c}{\eta_{m2}}$$

where c_{pg} is the mean specific heat for the gases in the HPT, and η_{m2} is the mechanical efficiency of spool 2 (C-HPT drive).

Hence,

$$\dot{W}_{HPT} = \dot{m}_g c_{pg} (T_{4t} - T_{45t}) = \frac{\dot{m}_h c_{pa} (T_{3t} - T_{25t})}{\eta_{m2}} \qquad (12.56)$$

$$\Delta T_{4-45} = T_{4t} - T_{45t} = \frac{1}{\eta_{m2}} \frac{\dot{m}_h}{\dot{m}_g} \frac{c_{pa}}{c_{pg}} (T_{3t} - T_{25t}) \qquad (12.57)$$

Depending on the turbine efficiency used, Eq. (12.57) can be rewritten as

$$\Delta T_{4-45} = \frac{1}{\eta_{m2}} \frac{\dot{m}_h}{\dot{m}_g} \frac{c_{pa}}{c_{pg}} (T_{3t} - T_{25t}) = \eta_t T_{4t} \left[1 - \frac{1}{r_{HPT}^{(\gamma_g-1)/\gamma_g}} \right] \tag{12.58}$$

$$\Delta T_{4-45} = \frac{1}{\eta_{m2}} \frac{\dot{m}_h}{\dot{m}_g} \frac{c_{pa}}{c_{pg}} (T_{3t} - T_{25t}) = T_{4t} \left[1 - \frac{1}{r_{HPT}^{\eta_{pt}(\gamma_g-1)/\gamma_g}} \right] \tag{12.59}$$

Equation (12.57) can be used to determine the temperature drop ΔT_{4-45} in the HPT, and either Eq. (12.58) or (12.59) can be used to determine the pressure ratio in the HPT.

12.4.1.7 Low-Pressure Turbine

The LPT generates just enough power to drive the fan and overcome the frictional losses in the fan-LPT drive. The temperature equivalent of the LPT work is

$$\Delta T_{45-5} = T_{45t} - T_{5t} = \eta_t T_{45t} \left[1 - \frac{1}{r_{LPT}^{(\gamma_g-1)/\gamma_g}} \right]$$

$$\Delta T_{45-5} = T_{45t} - T_{5t} = T_{45t} \left[1 - \frac{1}{r_{LPT}^{\eta_{pt}(\gamma_g-1)/\gamma_g}} \right]$$

where η_t, η_{cp}, and r_{LPT} are, respectively, the isentropic efficiency, polytropic efficiency, and expansion ratio of the LPT.

The work produced by the HPT is

$$\dot{W}_{LPT} = \dot{m}_g c_{pg} \Delta T_{45-5} = \dot{m}_g c_{pg} (T_{45t} - T_{5t}) \tag{12.60}$$

$$\dot{W}_{LPT} = \frac{\dot{W}_f}{\eta_{m1}}$$

where c_{pg} is the mean specific heat in the LPT, and η_{m1} is the mechanical efficiency of spool 1 (F-LPT drive).

Combining the last two equations yields

$$\dot{W}_{LPT} = \dot{m}_g c_{pg} (T_{45t} - T_{5t}) = \frac{\dot{m}_1 c_{pa} (T_{25t} - T_{2t})}{\eta_{m1}} \tag{12.61}$$

or

$$\Delta T_{45-5} = T_{45t} - T_{5t} = \frac{1}{\eta_{m1}} \frac{\dot{m}_1}{\dot{m}_g} \frac{c_{pa}}{c_{pg}} (T_{25t} - T_{2t}) \tag{12.62}$$

Depending on the turbine efficiency used, Eq. (12.62) can be rewritten as

$$\Delta T_{45-5} = \frac{1}{\eta_{m1}} \frac{\dot{m}_1}{\dot{m}_g} \frac{c_{pa}}{c_{pg}} (T_{25t} - T_{2t}) = \eta_t T_{45t} \left[1 - \frac{1}{r_{LPT}^{(\gamma_g-1)/\gamma_g}} \right] \tag{12.63}$$

$$\Delta T_{45-5} = \frac{1}{\eta_{m1}} \frac{\dot{m}_1}{\dot{m}_g} \frac{c_{pa}}{c_{pg}} (T_{25t} - T_{2t}) = T_{45t} \left[1 - \frac{1}{r_{LPT}^{\eta_{pt}(\gamma_g-1)/\gamma_g}} \right] \tag{12.64}$$

Equation (12.62) can be used to determine the temperature drop ΔT_{45-5} in the compressor turbine, and either Eq. (12.63) or (12.64) can be used to determine the pressure ratio in the HPT.

12.4.1.8 Hot Nozzle

The isentropic efficiency of the adiabatic HN is defined as

$$\eta_{hn} = \frac{T_{7t} - T_8}{T_{7t} - T_{8s}} = \frac{T_{5t} - T_8}{T_{5t} - T_{8s}} \tag{12.65}$$

For the isentropic process $5t - 8s$,

$$\frac{T_{5t}}{T_{8s}} = \left(\frac{p_{5t}}{p_8}\right)^{(\gamma_g - 1)/\gamma_g} \tag{12.66}$$

Combining Eqs. (12.65) and (12.66) yields

$$\Delta T_{5-8} = T_{5t} - T_8 = \eta_{hn} T_{5t} \left[1 - \frac{1}{(p_{5t}/p_8)^{(\gamma_g - 1)/\gamma_g}}\right] \tag{12.67}$$

The temperature and pressure ratio in the nozzle can be written in terms of the Mach number, as was the case with the diffuser. The Mach number in the nozzle throat is

$$M_8 = \frac{C_8}{\sqrt{\gamma_g R_g T_8}}$$

where $\gamma_g R_g = c_{pg}(\gamma_g - 1)$

For a nozzle exit velocity of C_8, the energy equation for the flow between stations 5 and 8 is

$$h_{5t} = h_8 + \frac{C_8^2}{2}$$

For a perfect gas, the equation can be rewritten as

$$T_{5t} - T_8 = \frac{C_8^2}{2c_{pg}}$$

or

$$C_8 = [2c_{pg}(T_{5t} - T_8)]^{1/2}$$

or

$$\frac{T_{5t}}{T_8} = 1 + \frac{C_8^2}{2c_{pg}T_8} = 1 + \frac{\gamma_g - 1}{2}M_8^2 \tag{12.68}$$

From Eqs. (12.65) and (12.66),

$$\frac{T_{8s}}{T_{5t}} = 1 - \frac{1}{\eta_{hn}}\left(1 - \frac{T_8}{T_{5t}}\right)$$

and

$$\frac{p_8}{p_{5t}} = \left(\frac{T_{8s}}{T_{5t}}\right)^{\gamma_g/(\gamma_g - 1)} = \left[1 - \frac{1}{\eta_{hn}}\left(1 - \frac{T_8}{T_{5t}}\right)\right]^{\gamma_g/(\gamma_g - 1)} \tag{12.69}$$

Using Eq. (12.68), we finally obtain

$$\frac{p_8}{p_{5t}} = \left[1 - \frac{1}{\eta_{hn}}\left(1 - \frac{1}{1 + \frac{\gamma_g - 1}{2}M_8^2}\right)\right]^{\gamma_g/(\gamma_g - 1)} \tag{12.70}$$

12.4.1.9 Sonic Flow in the Hot Nozzle (Nozzle Choked)

When the flow in the nozzle throat is sonic, $M_8 = 1$, and the gases are at the critical conditions of p_c and T_c. These gases expand first from the pressure at station 5 to the critical pressure p_c ($p_c > p_a$) at station 8, and then to the ambient pressure p_a.

$$C_8 = C_c = \sqrt{\gamma_g R_g T_c} \tag{12.71}$$

The isentropic nozzle efficiency is

$$\eta_{hn} = \frac{T_{5t} - T_c}{T_{5t} - T_{cs}} \tag{12.72}$$

where T_c is the critical temperature and T_{cs} is the temperature of the gases at the end of the isentropic expansion with sonic flow. The temperature ratio from Eq. (12.68) becomes

$$\frac{T_{5t}}{T_c} = \frac{\gamma_g + 1}{2} \tag{12.73}$$

The pressure ratio from Eq. (12.70) can be rewritten as

$$\frac{p_c}{p_{5t}} = \left[1 - \frac{1}{\eta_{hn}} \left(\frac{\gamma_g - 1}{\gamma_g + 1} \right) \right]^{\gamma_g/(\gamma_g - 1)} \tag{12.74}$$

or

$$\frac{p_{5t}}{p_c} = \frac{1}{\left[1 - \dfrac{1}{\eta_{hn}} \left(\dfrac{\gamma_g - 1}{\gamma_g + 1} \right) \right]^{\gamma_g/(\gamma_g - 1)}} \tag{12.75}$$

If the calculated pressure ratio p_{5t}/p_a is greater than the critical pressure ratio p_{5t}/p_c, then the nozzle is choked, and the critical velocity, temperature, and pressure are found from Eqs. (12.71), (12.73), and (12.74), respectively.

The critical density is

$$\rho_8 = \rho_c = \frac{p_c}{R_g T_c} \tag{12.76}$$

The nozzle throat area is

$$A_8 = \frac{\dot{m}_g}{\rho_c C_c} \tag{12.77}$$

Since the gas in the nozzle expands first to the critical pressure p_c and then to the ambient pressure p_a, the thrust produced by the core engine is made up of two components: the momentum thrust and the pressure thrust:

$$F_h = (\dot{m}_g C_8 - \dot{m}_h C_1) + (p_{c8} - p_1) A_8 \quad N \tag{12.78}$$

where C_8 is the jet speed (equal to the critical speed C_c in the nozzle throat) of the gases in the HN when the gas pressure in the nozzle throat is equal to the critical pressure p_c.

The *sfc* is

$$sfc = \frac{\dot{m}_f}{F_h} \tag{12.79}$$

The units of *sfc* depend on the units of $\dot{m}_f$ and F_h, as stated previously.

12.4.1.10 Subsonic Flow in the Hot Nozzle (Nozzle Unchoked)

If the gases in the nozzle are expanded to the atmospheric pressure ($p_8 = p_a$) and the pressure ratio p_{5t}/p_a is less than the critical pressure ratio p_{5t}/p_c, the nozzle is unchoked with $M_8 < 1$, and the temperature and pressure ratios in the nozzle are given by Eqs. (12.68) and (12.70) as follows:

$$\frac{T_{5t}}{T_8} = 1 + \frac{C_8^2}{2c_{pg}T_8} = 1 + \frac{\gamma_g - 1}{2}M_8^2 \tag{12.80}$$

$$\frac{p_8}{p_{5t}} = \left[1 - \frac{1}{\eta_{hn}} \left(1 - \frac{1}{1 + \frac{\gamma_g - 1}{2}M_7^2} \right) \right]^{\gamma_g/(\gamma_g - 1)} \tag{12.81}$$

The gas density in the nozzle throat is

$$\rho_8 = \frac{p_a}{R_g T_8} \tag{12.82}$$

The nozzle throat area is

$$A_8 = \frac{\dot{m}_g}{\rho_8 C_8} \tag{12.83}$$

Since the pressure in the nozzle expands to p_1, the pressure component of the thrust is equal to zero, reducing the thrust equation to

$$F_h = (\dot{m}_g C_8 - \dot{m}_h C_1) \tag{12.84}$$

where C_8 is the jet speed (equal to the speed in the nozzle throat) of the gases in the HN when the gas pressure in the nozzle throat is equal to the ambient pressure p_a.

The *sfc* is

$$sfc = \frac{\dot{m}_f}{F_h}$$

12.4.2 Design-Point Calculations of the Engine Bypass Section

The T-s diagram in Figure 12.24 is used for the bypass section, and calculations are conducted for the compression process in the intake, the compression process in the fan, and the expansion process in the CN. The mass flow rate of the air flowing through the bypass section is $\dot{m}_c$, and it is assumed that the pressure loss in the annular duct is negligible.

12.4.2.1 Intake and Fan

The calculations are the same as for the fan in the core engine.

12.4.2.2 Cold Nozzle

The isentropic efficiency of the adiabatic CN is defined as

$$\eta_{cn} = \frac{T_{13t} - T_{18}}{T_{13t} - T_{18s}} \tag{12.85}$$

For the isentropic process $13t - 18s$,

$$\frac{T_{13t}}{T_{18s}} = \left(\frac{p_{13t}}{p_{18}} \right)^{(\gamma_a - 1)/\gamma_a}$$

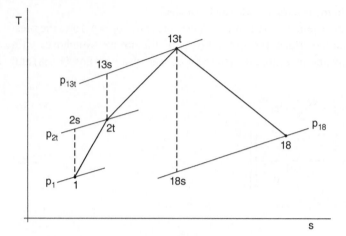

Figure 12.24 T-s diagrams of the processes in the bypass section of two-spool, unmixed-flow turbofan engine.

Combining this equation with Eq. (12.85) yields

$$\Delta T_{13-18} = T_{13t} - T_{18} = \eta_{cn} T_{13t} \left\{ 1 - \left[\frac{1}{(p_{13t}/p_{18})^{(\gamma_a - 1)/\gamma_a}} \right] \right\}$$

(12.86)

The temperature and pressure ratios in the nozzle throat can be written in terms of the Mach number as

$$M_{18} = \frac{C_{18}}{\sqrt{\gamma_a R_a T_{18}}}$$

where $\gamma_a R_a = c_{pa}(\gamma_a - 1)$

For a nozzle exit speed of C_{18}, the energy equation for the flow between stations 13 and 18 is

$$h_{13t} = h_{18} + \frac{C_{18}^2}{2}$$

For a perfect gas, the equation can be rewritten as

$$T_{13t} - T_{18} = \frac{C_{18}^2}{2c_{pa}}$$

from which

$$C_{18} = [2c_{pa}(T_{13t} - T_{18})]^{1/2}$$

or

$$\frac{T_{13t}}{T_{18}} = 1 + \frac{C_{18}^2}{2c_{pa} T_{18}} = 1 + \frac{\gamma_g - 1}{2} M_{18}^2$$

(12.87)

From Eq. (12.85),

$$\frac{T_{18s}}{T_{13t}} = 1 - \frac{1}{\eta_{cn}} \left(1 - \frac{T_{18}}{T_{13t}} \right)$$

$$\frac{P_{18}}{P_{13t}} = \left(\frac{T_{18s}}{T_{13t}}\right)^{\gamma_a/(\gamma_a-1)} = \left[1 - \frac{1}{\eta_{cn}}\left(1 - \frac{T_{18}}{T_{13t}}\right)\right]^{\gamma_a/(\gamma_a-1)} \tag{12.88}$$

$$\frac{P_{18}}{P_{13t}} = \left[1 - \frac{1}{\eta_{cn}}\left(1 - \frac{1}{1 + \frac{\gamma_a - 1}{2}M_{18}^2}\right)\right]^{\gamma_a/(\gamma_a-1)} \tag{12.89}$$

The CN can be operated choked or unchoked depending on the pressure at the CN throat (station 18). The calculations are the same as for the HN, with the gas having the properties of air at the mean temperature across the nozzle.

For the choked nozzle,

$$F_c = \dot{m}_c(C_{18} - C_1) + (p_{c18} - p_1)A_{18} \ \ N \tag{12.90}$$

For the unchoked nozzle,

$$F_c = \dot{m}_c(C_{18} - C_1) \ \ N \tag{12.91}$$

The total thrust is the sum of the thrusts produced by the cold and hot streams for the engine shown in Figure 12.21 in which the two flows are separate.

For both the CN and HN choked,

$$F_t = \dot{m}_g C_8 + \dot{m}_c C_{18} - C_1(\dot{m}_h + \dot{m}_c) + (p_{c8} - p_1)A_8 + (p_{c18} - p_1)A_{18} \tag{12.92}$$

For both the CN and HN unchoked,

$$F_t = \dot{m}_g C_8 + \dot{m}_c C_{18} - C_1(\dot{m}_h + \dot{m}_c) \tag{12.93}$$

The *sfc* for the turbofan engine, defined as the mass flow rate of the fuel in the core engine divided by the total thrust, is

$$sfc = \frac{\dot{m}_f}{F_t} \tag{12.94}$$

The overall efficiency is defined as the thrust power divided by the chemical energy rate:

$$\eta_o = \frac{C_1 F_t}{\dot{m}_f H_l} \tag{12.95}$$

The propulsive efficiency is defined as the ratio of the power provided to the aircraft to the power generated by the change of kinetic energy of the gases flowing through the engine:

$$\eta_p = \frac{C_1[(\dot{m}_g C_8 - \dot{m}_h C_1) + \dot{m}_c(C_{18} - C_1)]}{\frac{1}{2}(\dot{m}_g C_8^2 - \dot{m}_h C_1^2) + \dot{m}_c(C_{18}^2 - C_1^2)} \tag{12.96}$$

The thermal efficiency, defined the same way as for the turbojet engine, can be written in this case as

$$\eta_{th} = \frac{\eta_o}{\eta_p} = \frac{C_1 F_t/\dot{m}_f(Hl)}{\left\{\frac{2C_1[(\dot{m}_g C_8 - \dot{m}_h C_1) + \dot{m}_c(C_{18} - C_1)]}{(\dot{m}_g C_8^2 - \dot{m}_h C_1^2) + \dot{m}_c(C_{18}^2 - C_1^2)}\right\}}$$

$$\eta_{th} = \frac{0.5F_t}{\dot{m}_f H_l} \times \frac{(\dot{m}_g C_8^2 - \dot{m}_h C_1^2) + \dot{m}_c(C_{18}^2 - C_1^2)}{[(\dot{m}_g C_8 - \dot{m}_h C_1) + \dot{m}_c(C_{18} - C_1)]} \tag{12.97}$$

where C_{18} and C_8 are the speeds of the jets in the CN and HN throats, respectively. C_1 is the aircraft forward speed, and H_l is the lower heating value of the fuel (the water in the products remains in vapour form). $\dot{m}_f$, $\dot{m}_h$, $\dot{m}_g$, and $\dot{m}_c$ are the mass flow rates of the fuel, the air in the core engine, and the combustion products in the core engine air in the bypass section.

12.5 Performance of Two-Spool Unmixed-Flow Turbofan Engine – Case Study

The turbofan engine is complex, and the performance analysis involves more independent parameters than the analysis of the turbojet engine. In addition to the variables seen previously, such as the compressor pressure ratio and turbine entry temperature, two more parameters need to be considered: the fan pressure ratio and the bypass ratio. The analysis based on the data in Table 12.5 is conducted at a constant turbine inlet temperature $T_{4t} = 1757\ K$, constant fan pressure ratio $r_f = 1.6$, and different bypass ratios varying between 2 and 16. It is also assumed that the mass flow rate of the air through the fan (overall air flow rate) $\dot{m}_1 = 1\ kg/s$. The cold and hot flow rates are then given, respectively, by

$$\dot{m}_c = \frac{B\dot{m}_1}{B+1} = \frac{B}{B+1}, \dot{m}_h = \frac{\dot{m}_1}{B+1} = \frac{1}{B+1}$$

As with the case of the turbojet engine, the results of the analysis is shown in the form of general performance characteristics, performance maps and 3-D surface data.

The effect of the bypass ratio at a constant fan pressure ratio can be summarised as follows:

1. The power needed to run the fan is independent of B (see Eq. (12.52)).
2. Less air flows through the core engine as B increases.
3. The power needed to run the compressor decreases with increasing B, mainly due to a decrease in the rate of air flow through the core engine.
4. The condition of the CN is unaffected by B, and it remains choked at all compressor pressure ratios.
5. The HN is more sensitive to the value of B. For values of B up to 6, the HN is choked at all compressor pressure ratios. For value of B from 8–12, the HN is unchoked at $r_c = 2$ and 4 only. For $B = 14$, the engine cannot be operated at $r_c = 2$, and the HN is unchoked at values of $r_c = 4$–10, becoming choked at higher values. At $B = 16$, the engine can only be operated within the range $r_c = 6$–36 with the HN unchoked throughout.

Figure 12.25 shows the temperature and pressure in the core engine and bypass section of the turbofan engine operating at altitude 10 000 m with compressor pressure ratio $r_c = 10$, fan pressure ratio $r_f = 1.6$, flight Mach number $M_1 = 0.85$, turbine inlet temperature $T_{4t} = 1757\ K$, and varying bypass ratios B ranging from 4–12. The bypass ratio B has no effect on the magnitude of the pressure and temperature in the bypass duct. In the core engine, both the pressure and temperature vary with the bypass ratio in the LPT and HN. Both the pressure and temperature decrease with increasing B between stations 45 and 8.

Figure 12.26 shows the variation of the mean specific heat and corresponding ratio of the specific heats across the various components of the turbofan engine when the engine

Table 12.5 Data for the calculation of the turbofan engine performance characteristics.

	Fuel $C_{12}H_{24}$	
Turbine inlet temperature, K	$T_{max} = T_{4t}$	1757
Fan pressure ratio	r_f	1.6
Fan polytropic efficiency	η_{pf}	0.9
Cold nozzle isentropic efficiency	η_{cn}	0.9
Diffuser Efficiency	η_i	0.95
Compressor polytropic efficiency	η_{pc}	0.9
Turbine polytropic efficiency	η_{pt}	0.9
Mechanical efficiency	η_m	0.99
Hot nozzle isentropic efficiency	η_{hn}	0.95
Aircraft forward speed, km/h	C_1	916.8
Flight altitude, m	H	10 000
Ambient pressure, kPa	p_a	26.428
Ambient temperature, K	T_a	223.2
Flight Mach number	M_1	0.85
Fan inlet temperature, K	T_{2t}	255.55
Bypass ratio	B	2; 4; 6; 8; 10; 12; 14; 16
Combustion chamber Pressure losses, kPa	Δp_{34}	1.5
Overall air flow rate, kg/s	$\dot{m}_1$	1.0
Air flow rate in bypass duct, kg/s	$\dot{m}_c$	variable
Air flow rate in core engine, kg/s	$\dot{m}_h$	variable
Gas flow rate, kg/s	$\dot{m}_g$	variable
Relative air-fuel ratio	λ	variable
Fuel flow rate, kg/s	$\dot{m}_f$	variable
Lower heating value, kJ/kg	H_l	44 340

is operated at a fixed compressor pressure ratio of 10 and bypass ratio of 6. The highest specific-heat value is observed in the HPT. Both c_p and γ vary significantly along the flow path of the gases in the engine.

Figures 12.27–12.29 show the effect of the compressor pressure ratio on the specific thrust, *sfc*, and propulsive efficiency at five selected bypass ratios for a turbofan engine operated at constant flight conditions, fan pressure ratio, and turbine entry temperature. These figures show that, if we exclude $B = 16$, increasing the bypass ratio at any given compressor pressure ratio up to 50 causes the specific thrust and *sfc* to decrease and the propulsive efficiency to increase.

Figure 12.30 shows the effect of the bypass ratio on the specific thrust and *sfc* at three compressor pressure ratios for a turbofan engine operated at constant flight conditions and constant fan pressure ratio. At $r_c = 36$, the *sfc* starts increasing when B exceeds 12. This

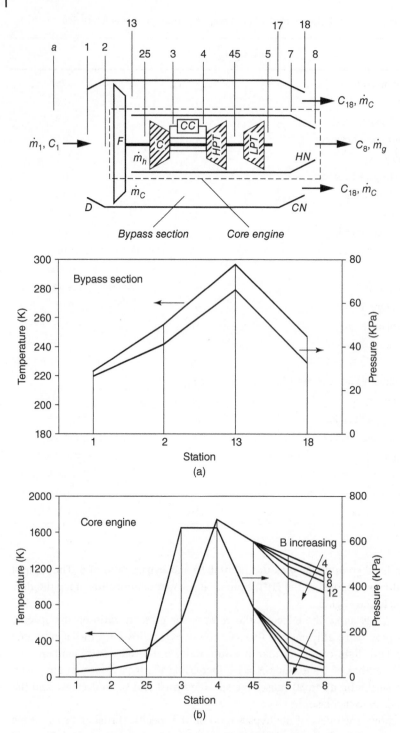

Figure 12.25 Temperature and pressure profiles in the turbofan engine ($H = 10\,000\,m$, $M_1 = 0.85$, $B = 4 - 12$, $r_f = 1.6$, $r_c = 10$, $T_{4t} = 1757\,K$): (a) bypass section; (b) core engine.

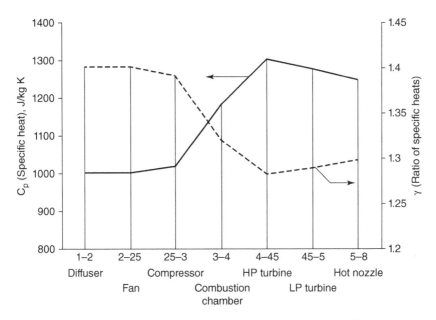

Figure 12.26 Mean specific heats and ratios of specific heats across different components in the turbofan engine ($H = 10\,000\ m$, $M_1 = 0.85$, $B = 6$, $r_f = 1.6$, $r_c = 10$, $T_{4t} = 1757\ K$).

Figure 12.27 Specific thrust vs. compressor pressure ratio of the turbofan engine at different bypass ratios ($H = 10\,000\ m$, $M_1 = 0.85$, $T_{4t} = 1757\ K$, $r_f = 1.6$).

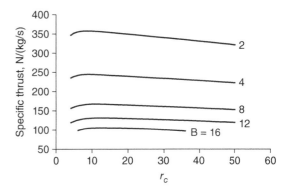

Figure 12.28 Specific fuel consumption vs. compressor pressure ratio of the turbofan engine at different bypass ratios ($H = 10\,000\ m$, $M_1 = 0.85$, $T_{4t} = 1757\ K$, $r_f = 1.6$).

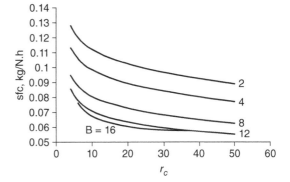

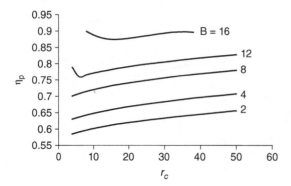

Figure 12.29 Propulsive efficiency vs. compressor pressure ratio of the turbofan engine at different bypass ratios ($H = 10\,000\,m$, $M_1 = 0.85$, $T_{4t} = 1757\,K$, $r_f = 1.6$).

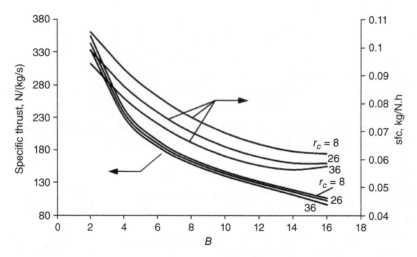

Figure 12.30 Specific thrust and specific fuel consumption versus the bypass ratio at different compressor pressure ratios for a turbofan engine ($H = 10\,000\,m$, $M_1 = 0.85$, $T_{4t} = 1757\,K$, $r_f = 1.6$).

figure also shows that at any given bypass ratio, the specific thrust is much less sensitive to the change of the compressor pressure ratio than the *sfc*.

Figure 12.31 shows the performance map of the turbofan engine under consideration. The map is plotted on thrust-*sfc* axes with constant bypass and compressor pressure ratio lines. The entire investigated ranges of the bypass ratio (2–16) and compressor pressure ratio (4–50) are shown in the figure.

When dealing with high-bypass-ratio turbofan engines, it is common practice to show the performance parameters for a range of bypass ratios starting at a higher minimum value: say, from 6 instead of 2 (Figure 12.32).

A 3-D surface representation of the performance characteristics in the form $sfc = \psi(r_c, f)$ can be obtained by fitting the results of the calculations to the correlation

$$sfc = a + \frac{b}{r_c} + \frac{c}{r_c^2} + \frac{d}{r_c^3} + \frac{e}{f} + \frac{g}{f^2} \ kg/N.h \tag{12.98}$$

where, $a = 0.134214$, $b = 0.260902$, $c = 0.66739$, $d = 0.79581$, $e = 17.8379$, and $g = 935.3016$.

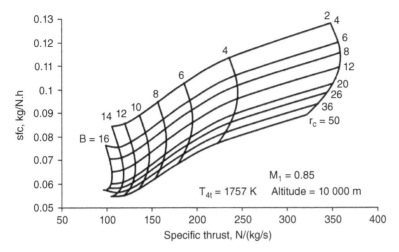

Figure 12.31 Performance map in terms of *sfc* vs. specific thrust for the turbofan engine within a wide range of bypass and compressor pressure ratios and $r_f = 1.6$.

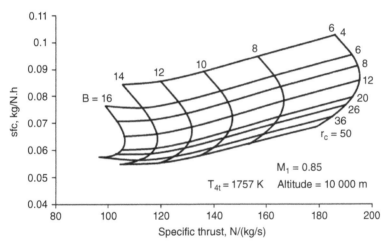

Figure 12.32 Performance map in terms of *sfc* vs. specific thrust for the high-bypass turbofan engine ($B = 6$–16, $r_f = 1.6$).

A plot of Eq. (12.98) is shown in Figure 12.33 in which the values of the *sfc* are shown in descending order for the purpose of clarity. Shown also on this plot are the constant bypass ratio lines from $B = 2$ to 16.

Inspection of the results presented in Figures 12.27–12.33 for an uninstalled engine with a fan pressure ratio of 1.6 for an aircraft flying at an altitude of 10 000 *m* and flight Mach number of 0.85 leads to the following conclusions:

1. The *sfc* decreases significantly with increasing bypass ratio and compressor pressure ratio.
2. For a given compressor pressure ratio, increasing the bypass ratio reduces both the specific thrust and *sfc* of the engine. The specific thrust continues to decrease up to

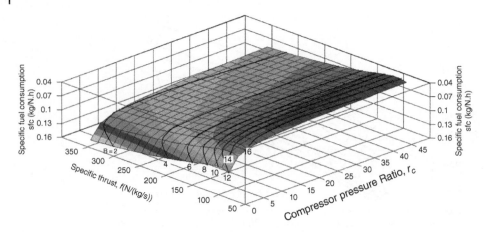

Figure 12.33 3-D surface plot of the performance of the turbofan engine ($H = 10\,000\,m$, $M_1 = 0.85$, $T_{4t} = 1757\,K$, $r_f = 1.6$).

the maximum bypass ratio examined ($B = 16$). The *sfc* exhibits minima at specific compressor pressure ratios and bypass ratios. For example, at $r_c = 36$, the minimum *sfc* is reached at $B = 14$; at $r_c = 26$, it shifts to about $B = 16$. It can be concluded that even if a significant reduction in specific thrust is tolerated, there is no fuel savings benefit in increasing B beyond 14.

3. For a given bypass ratio, the *sfc* is greatly reduced with increasing compressor pressure ratio, while the specific thrust changes within a relatively narrow range.

An alternative approach to analysing turbofan engine performance is to assume a constant core engine flow rate of hot air $\dot{m}_h = 1\,kg/s$, which gives an overall flow rate of $\dot{m}_1 = B + 1$ and cold stream flow rate of $\dot{m}_c = B$ (Cumpsty, 2003). The thrust per *kg* of air in the core engine appears to increase with the bypass ratio, while the *sfc* decreases (Figure 12.34). To convert back to the previous results, the specific thrust results should be divided by $B + 1$.

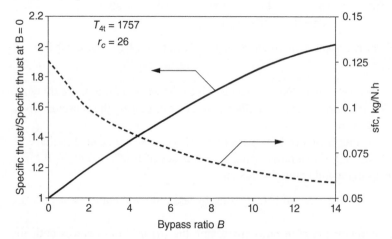

Figure 12.34 Relative thrust and specific fuel consumption versus bypass ratio with fixed airflow rate through the core engine ($\dot{m}_h = 1\,kg/s$, $H = 10\,000\,m$, $M_1 = 0.85$, $T_{4t} = 1757\,K$, $r_f = 1.6$, $r_c = 26$).

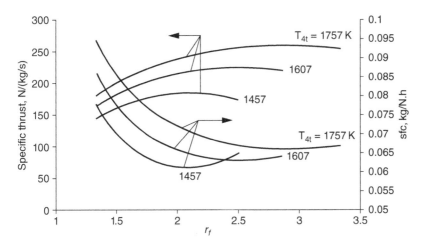

Figure 12.35 Specific thrust and specific fuel consumption vs. fan pressure ratio at different turbine entry temperatures ($H = 10\,000\,m$, $M_1 = 0.85$, $B = 5$, $r_o = 40$).

In the previous analysis, the fan pressure ratio was kept constant at 1.6. Strictly speaking, the fan pressure ratio should be selected depending on the turbine inlet temperature, overall pressure ratio, and bypass ratio. Figure 12.35 shows the specific thrust and the *sfc* plotted versus the fan pressure ratio r_f at three different turbine entry temperatures T_{4t} for a turbofan engine operated at constant flight conditions, overall pressure ratio $r_o \, (= r_f \times r_c)$ equal to 40, and bypass ratio $B = 5$.

It is apparent that there is an optimum fan pressure ratio at each value of T_{4t} at which the specific thrust is a maximum and the *sfc* is a minimum. The value of the optimum r_f increases with increasing T_{4t}. One more result, not shown here, is that increasing the bypass ratio at a constant T_{4t} decreases the optimum value of r_f. For example, at $T_{4t} = 1757\ K$, increasing B from 5 to 8, keeping everything else unchanged, causes the optimum value of r_f to decrease from 2.86 to 2.22.

Other factors, not discussed here, that affect the optimum value of the fan pressure ratio are the bypass ratio and overall engine pressure ratio. According to Walsh and Fletcher (2004), the optimum fan pressure ratio increases with decreasing bypass pressure ratio at constant overall pressure ratio and turbine entry temperature and decreases with increasing overall pressure ratio at constant bypass ratio and turbine entry temperature.

12.6 Two-Spool Mixed-Flow Turbofan Engine

In these engines the hot and cold streams are mixed before expanding in a single propelling nozzle. This arrangement is predominantly used in supersonic combat military aircraft engines with low bypass ratios. It is also used in some subsonic civil aircraft. In the analysis that follows, the inner (root) and outer (tip) sections of the fan are assumed to have different

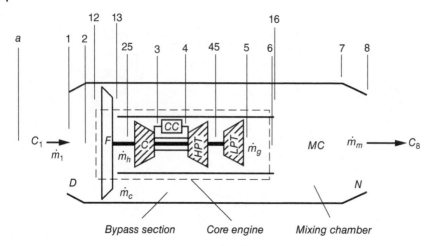

Figure 12.36 Configuration of a simple mixed-flow turbofan engine: D – diffuser, F – fan, C – compressor, CC – combustion chamber, HPT – high-pressure turbine, LPT – low-pressure turbine, MC – mixing chamber, N – nozzle.

pressure ratios, with the inner fan section serving as the low-pressure compression stage of the core engine. There is no afterburner in the configuration, no power take-off to operate auxiliary equipment, and no air bleeding for blade-cooling purposes.

A schematic diagram of the simple mixed-flow turbofan engine is shown in Figure 12.36 and the station numbering is detailed in Table 12.6. This basic configuration comprises a

Table 12.6 Station numbering in the simple mixed-flow turbofan engine.

Station	Location
0	Ambient conditions upstream from the diffuser
1	Entry into diffuser D
12	Tip front face of fan F
13	Exit from tip part of fan F (entry into cold jet tube in the bypass section)
16	Exit from the bypass section/cold mixer inlet
2	Entry into both root and tip parts of fan F (uniform properties over entire fan face)
25	Exit from root part of fan F (entry into compressor C)
3	Exit from compressor C (entry into combustion chamber CC)
4	Exit from combustion chamber CC (entry into high-pressure turbine HPT)
45	Exit from high-pressure turbine HPT (entry into low-pressure turbine LPT)
5	Exit from low-pressure turbine LPT (entry into hot jet tube in core engine)
6	Hot mixer inlet
7	Mixer outlet/entry into propelling nozzle N
8	Exit from propelling nozzle N/nozzle throat

Figure 12.37 Schematic T-s diagram for the mixed-flow turbofan engine.

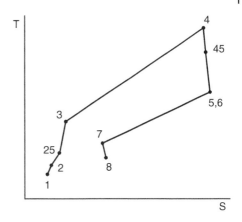

diffuser D, core engine, bypass section, mixing chamber MC, and single propelling nozzle N. The components of the core engine are the root section of fan F, high-pressure compressor C, combustion chamber CC, HPT, and LPT. The cold section comprises the tip section of fan F, the bypass concentric tube, and the exit from the bypass section (cold mixer inlet). The mixer is assumed to be ideal and to have a simple uniform cylindrical shape. The propelling nozzle is a convergent nozzle. Stations 13 to 16 and 5 to 6 represent the cold and hot jet tubes, respectively. The conditions are assumed constant along these sections.

The total mass flow rate of the air $\dot{m}_1$ flowing through the diffuser is split into 'hot' mass flow rate $\dot{m}_h$ in the core engine and 'cold' flow rate $\dot{m}_c$ in the bypass section. The core engine produces a jet of combustion products of mass flow rate $\dot{m}_g$ (hot flow rate plus fuel flow rate). The fan is driven by the LPT, while the compressor is driven by the HPT. The fan acts as a LPC in the core engine and provides compressed air to the bypass section. The cold and hot flows are mixed in the mixing chamber and then expelled from the propelling nozzle with a flow rate of $\dot{m}_m$ and jet velocity of C_8.

Figure 12.37 shows a simplified schematic of the T-s diagram and the processes in the diffuser, core engine, mixer, and propelling nozzle in a typical mixed-flow turbofan engine.

12.6.1 Design-Point Calculations of Engine Core

The T-s diagrams used in the calculations are shown in Figures 12.38 and 12.39. Figure 12.38 shows the isentropic and polytropic processes in the diffuser, core engine (including the root section of the fan), mixer, and propelling nozzle. Figure 12.39 shows the T-s diagram of the diffuser, tip section of the fan, and cold jet tube. In all of the calculations, it is assumed that conditions at the diffuser inlet are the same as the ambient conditions. There are no changes of conditions between stations 5 and 6 and between stations 13 and 16. Conditions at the front face of the fan (stations 2 and 12) are also the same, but the pressure ratios p_{25}/p_2 and p_{13}/p_{12} are different.

The temperatures and pressures at all stations are shown as total temperatures and total pressures, with the exception of the entrance into the diffuser and the nozzle throat. The pressure at 8 is equal to the ambient pressure if the nozzle is unchoked and equal to the critical pressure if the nozzle is choked. The compression, combustion, and expansion processes are similar to the processes in the simple gas turbine cycle.

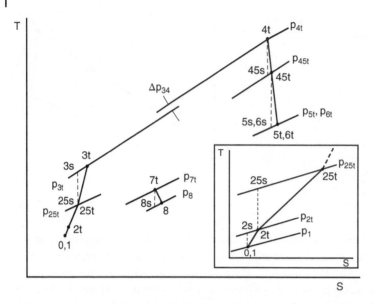

Figure 12.38 T-s diagram of the processes in the diffuser, core engine, mixer, and propelling nozzle (processes 1–2 and 2–25 are magnified in the inset).

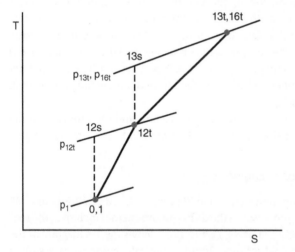

Figure 12.39 T-s diagram of the processes in the diffuser, fan, and cold jet tube

The computational procedure for the core engine and bypass duct in the mixed-flow turbofan engine is similar to the unmixed-flow turbofan engine described previously. The main differences between the two configurations are in the absence of a second nozzle (CN) and the presence of a mixing chamber upstream of nozzle N in which the cold air stream in the bypass duct and hot gas stream in the core engine are mixed before expanding in the nozzle. The mixing chamber (MC) is represented by the planes at stations 6 and 7, as shown in Figure 12.36 and the mixing process is accompanied by a decrease in temperature, pressure, and entropy.

12.6.1.1 Intake

The isentropic intake efficiency (see inset in Figure 12.38) is

$$\eta_i = \frac{T_{2s} - T_1}{T_{2t} - T_1}$$

The energy equation applied to the diffuser gives

$$T_{2t} - T_1 = \frac{C_1^2}{2c_{pa}}$$

Hence,

$$T_{2s} - T_1 = \eta_i \frac{C_1^2}{2c_{pa}}$$

$$\frac{T_{2s}}{T_1} = 1 + \eta_i \frac{C_1^2}{2c_{pa}T_1} \tag{12.99}$$

Therefore, the intake pressure ratio is

$$\frac{p_{2t}}{p_1} = \left(\frac{T_{2s}}{T_1}\right)^{\gamma_a/(\gamma_a-1)} = \left[1 + \eta_i \frac{C_1^2}{2c_{pa}T_1}\right]^{\gamma_a/(\gamma_a-1)} \tag{12.100}$$

or, since $M_1 = C_1/\sqrt{\gamma_a R_a T_1}$ and $\gamma_a R_a = c_{pa}(\gamma_a - 1)$,

$$\frac{p_{2t}}{p_1} = \left[1 + \eta_i \frac{\gamma_a - 1}{2} M_1^2\right]^{\gamma_a/(\gamma_a-1)} \tag{12.101}$$

and

$$\frac{T_{2t}}{T_1} = 1 + \frac{\gamma_a - 1}{2} M_1^2 \tag{12.102}$$

These equations are valid only when the flow in the intake is subsonic ($M_1 < 1$) or sonic ($M_1 = 1$).

12.6.1.2 Fan

Using the subscript fr (fan root) for the root section of the fan, the temperature equivalent of the fan work can be written as

$$\Delta T_{2-25} = T_{25t} - T_{2t} = \frac{T_{2t}}{\eta_f} [r_{fr}^{(\gamma_a-1)/\gamma_a} - 1]$$

$$\Delta T_{2-25} = T_{25t} - T_{2t} = T_{2t}[r_{fr}^{(\gamma_a-1)/\gamma_a \eta_{pf}} - 1]$$

η_f, η_{fp}, and r_{fr} are, respectively, the fan isentropic efficiency, polytropic efficiency, and pressure ratio.

If r_{fr} is known, the fan exit pressure within the core engine is

$$p_{25t} = r_{fr} p_{2t}$$

The power absorbed by the fan root section is

$$\dot{W}_{fr} = \dot{m}_h c_{pa} \Delta T_{2-25} \tag{12.103}$$

where c_{pa} is the mean specific heat at constant pressure of air across the fan, and $\dot{m}_h$ is the flow rate of the air entering the core engine.

12.6.1.3 Compressor

The temperature equivalent of the compressor work in the gas generator is

$$\Delta T_{25-3} = T_{3t} - T_{25t} = \frac{T_{25t}}{\eta_c} [r_c^{(\gamma_a-1)/\gamma_a} - 1]$$

$$\Delta T_{25-3} = T_{3t} - T_{25t} = T_{25t}[r_c^{(\gamma_a-1)/\gamma_a\eta_{pc}} - 1]$$

where, η_c, η_{cp}, and r_c are, respectively, the compressor isentropic efficiency, polytropic efficiency, and pressure ratio.

The overall pressure ratio of the turbofan engine is

$$r_o = r_{fr}r_c$$

The power absorbed by the compressor in the core engine (and provided by the HPT), is

$$\dot{W}_c = \dot{m}_h c_{pa} \Delta T_{25-3} \tag{12.104}$$

where c_{pa} is the mean specific heat at constant pressure of air across the compressor.

12.6.1.4 Combustion Chamber

The rate of heat liberated during the combustion process is the heat added to the cycle and can be found by writing the energy equation without work transfer and no change in kinetic energy:

$$\dot{Q}_{in} = \dot{m}_g c_{pg} T_{4t} - \dot{m}_h c_{pa} T_{3t} \tag{12.105}$$

where $\dot{m}_g = \dot{m}_h + \dot{m}_f$, c_{pa} is the specific heat of air at station 3, c_{pg} is the specific heat of the combustion products at station 4, and $\dot{m}_f$ is the mass flow rate of the fuel in the combustion chamber.

12.6.1.5 Compressor Turbine (High-Pressure Turbine)

The HPT generates just enough power to drive the compressor and overcome the frictional losses in the compressor-turbine unit.

The temperature equivalent of the HPT work is

$$\Delta T_{4-45} = T_{4t} - T_{45t} = \eta_t T_{4t} \left[1 - \frac{1}{r_{HPT}^{(\gamma_g-1)/\gamma_g}}\right]$$

$$\Delta T_{4-45} = T_{4t} - T_{45t} = T_{4t} \left[1 - \frac{1}{r_{HPT}^{\eta_{pt}(\gamma_g-1)/\gamma_g}}\right]$$

where η_t, η_{cp}, and r_{HPT} are, respectively, the isentropic efficiency, polytropic efficiency, and expansion ratio of the HPT.

The power produced by the HPT is

$$\dot{W}_{HPT} = \dot{m}_g c_{pg} \Delta T_{4-45} = \dot{m}_g c_{pg} (T_{4t} - T_{45t}) \tag{12.106}$$

or

$$\dot{W}_{HPT} = \frac{\dot{W}_c}{\eta_{m2}}$$

where c_{pg} is the mean specific heat for the gases in the HPT, and η_{m2} is the mechanical efficiency of spool 2 (C-HPT drive).

Hence,

$$\dot{W}_{HPT} = \dot{m}_g c_{pg}(T_{4t} - T_{45t}) = \frac{\dot{m}_h c_{pa}(T_{3t} - T_{25t})}{\eta_{m2}} \tag{12.107}$$

and

$$\Delta T_{4-45} = T_{4t} - T_{45t} = \frac{1}{\eta_{m2}} \frac{\dot{m}_h}{\dot{m}_g} \frac{c_{pa}}{c_{pg}}(T_{3t} - T_{25t}) \tag{12.108}$$

Depending on the turbine efficiency used, Eq. (12.108) can be rewritten as

$$\Delta T_{4-45} = \frac{1}{\eta_{m2}} \frac{\dot{m}_h}{\dot{m}_g} \frac{c_{pa}}{c_{pg}}(T_{3t} - T_{25t}) = \eta_t T_{4t} \left[1 - \frac{1}{r_{HPT}^{(\gamma_g-1)/\gamma_g}} \right] \tag{12.109}$$

or

$$\Delta T_{4-45} = \frac{1}{\eta_{m2}} \frac{\dot{m}_h}{\dot{m}_g} \frac{c_{pa}}{c_{pg}}(T_{3t} - T_{25t}) = T_{4t} \left[1 - \frac{1}{r_{HPT}^{\eta_{pt}(\gamma_g-1)/\gamma_g}} \right] \tag{12.110}$$

Equation (12.108) can be used to determine the temperature drop ΔT_{4-45} in the HPT, and either Eq. (12.109) or (12.110) can be used to determine the pressure ratio in the HPT.

12.6.1.6 Low-Pressure Turbine
The LPT generates just enough power to drive the fan and overcome the frictional losses in the fan-LPT drive. The temperature equivalent of the LPT work is

$$\Delta T_{45-5} = T_{45t} - T_{5t} = \eta_t T_{45t} \left[1 - \frac{1}{r_{LPT}^{(\gamma_g-1)/\gamma_g}} \right]$$

$$\Delta T_{45-5} = T_{45t} - T_{5t} = T_{45t} \left[1 - \frac{1}{r_{LPT}^{\eta_{pt}(\gamma_g-1)/\gamma_g}} \right]$$

where, η_t, η_{cp}, and r_{LPT} are, respectively, the isentropic efficiency, polytropic efficiency, and expansion ratio of the LPT.

The work produced by the LPT is

$$\dot{W}_{LPT} = \dot{m}_g c_{pg} \Delta T_{45-5} = \dot{m}_g c_{pg}(T_{45t} - T_{5t}) \tag{12.111}$$

or

$$\dot{W}_{LPT} = \frac{\dot{W}_f}{\eta_{m1}}$$

where $\dot{W}_f$ is the total work absorbed by the fan (sum of the power at the root and at the tip sections), c_{pg} is the mean specific heat in the LPT, and η_{m1} is the mechanical efficiency of spool 1 (F-LPT drive).

$$\dot{W}_f = \dot{W}_{fr} + \dot{W}_{ft} = \dot{m}_h c_{pa} \Delta T_{2-25} + \dot{m}_c c_{pa} \Delta T_{12-13}$$

Hence,

$$\dot{W}_{LPT} = \dot{m}_g c_{pg}(T_{45t} - T_{5t}) = \frac{\dot{m}_h c_{pa} \Delta T_{2-25} + \dot{m}_c c_{pa} \Delta T_{12-13}}{\eta_{m1}} \tag{12.112}$$

or

$$\Delta T_{45-5} = T_{45t} - T_{5t} = \frac{\dot{m}_h c_{pa} \Delta T_{2-25} + \dot{m}_c c_{pa} \Delta T_{12-13}}{\eta_{m1} \dot{m}_g c_{pg}} \tag{12.113}$$

Depending on the turbine efficiency used, Eq. (12.113) can be rewritten as

$$\Delta T_{45-5} = \frac{\dot{m}_h c_{pa} \Delta T_{2-25} + \dot{m}_c c_{pa} \Delta T_{12-13}}{\eta_{m1} \dot{m}_g c_{pg}} = \eta_t T_{45t} \left[1 - \frac{1}{r_{LPT}^{(\gamma_g-1)/\gamma_g}} \right] \tag{12.114}$$

$$\Delta T_{45-5} = \frac{\dot{m}_h c_{pa} \Delta T_{2-25} + \dot{m}_c c_{pa} \Delta T_{12-13}}{\eta_{m1} \dot{m}_g c_{pg}} = T_{45t} \left[1 - \frac{1}{r_{LPT}^{\eta_{pt}(\gamma_g-1)/\gamma_g}} \right] \tag{12.115}$$

Equation (12.112) can be used to determine the temperature drop ΔT_{45-5} in the compressor turbine, and either Eq. (12.114) or (12.115) can be used to determine the pressure ratio in the LPT.

12.6.2 Design-Point Calculations of Bypass Section

The bypass section includes the intake, fan, and bypass duct (cold jet tube). The mixer and propelling nozzle will be treated as separate components. The processes in the intake, fan, and cold jet tube are depicted in the T-s diagram in Figure 12.39.

12.6.2.1 Intake

The isentropic intake efficiency is

$$\eta_i = \frac{T_{12s} - T_1}{T_{12t} - T_1}$$

The energy equation applied to the diffuser gives

$$T_{12t} - T_1 = \frac{C_1^2}{2c_{pa}}$$

$$T_{12s} - T_1 = \eta_i \frac{C_1^2}{2c_{pa}}$$

$$\frac{T_{12s}}{T_1} = 1 + \eta_i \frac{C_1^2}{2c_{pa} T_1} \tag{12.116}$$

Therefore, the intake pressure ratio is

$$\frac{p_{12t}}{p_1} = \left(\frac{T_{12s}}{T_1} \right)^{\gamma_a/(\gamma_a-1)} = \left[1 + \eta_i \frac{C_1^2}{2c_{pa} T_1} \right]^{\gamma_a/(\gamma_a-1)} \tag{12.117}$$

or, since $M_1 = C_1/\sqrt{\gamma_a R_a T_1}$ and $\gamma_a R_a = c_{pa}(\gamma_a - 1)$,

$$\frac{p_{12t}}{p_1} = \left[1 + \eta_i \frac{\gamma_a - 1}{2} M_1^2 \right]^{\gamma_a/(\gamma_a-1)} \tag{12.118}$$

and

$$\frac{T_{12t}}{T_1} = 1 + \frac{\gamma_a - 1}{2} M_1^2 \tag{12.119}$$

These equations are valid only when the flow in the intake is subsonic ($M_1 < 1$) or sonic ($M_1 = 1$).

12.6.2.2 Fan

Using the subscript ft (fan tip) for the tip section of the fan, and assuming the isentropic and polytropic efficiencies are the same at the root and tip, the temperature equivalent of the fan work can be written as

$$\Delta T_{12-13} = T_{13t} - T_{12t} = \frac{T_{12t}}{\eta_f} [r_{ft}^{(\gamma_a-1)/\gamma_a} - 1]$$

$$\Delta T_{12-13} = T_{13t} - T_{12t} = T_{12t}[r_{ft}^{(\gamma_a-1)/\gamma_a\eta_{pf}} - 1]$$

η_f, η_{fp}, and r_{ft} are, respectively, the fan isentropic efficiency, fan polytropic efficiency, and fan tip pressure ratio.

If r_{ft} is given, the fan exit pressure in the bypass section is

$$p_{13t} = r_{ft}p_{12t}$$

The power absorbed by the fan tip section is

$$\dot{W}_{ft} = \dot{m}_c c_{pa}\Delta T_{12-13} \tag{12.120}$$

where c_{pa} is the mean specific heat at constant pressure of air across the fan, and $\dot{m}_c$ is the flow rate of the air entering the bypass section.

12.6.2.3 Bypass Duct (Cold Jet Tube)

The mass flow rate of the air flowing through this section is $\dot{m}_c$, and the pressure losses in the annular duct are assumed to be negligible.

12.6.3 Mixer

12.6.3.1 Assumptions

The assumptions made for the mixing scheme of the cold and hot gases are:

- The fan generates two different pressure ratios. The root (inner) fan pressure ratio is the LPC pressure ratio in the core engine, while the tip (outer) fan pressure ratio is the driving force for the secondary flow in the bypass section.
- The mixing chamber is a constant-area cylindrical duct between stations 6 and 7.
- No pressure or heat losses occur in the mixing chamber (100% mixing efficiency).
- The conditions at stations 5 and 6 are identical.
- The static and total pressures of the core and secondary flows are equal at the entry into the mixer.
- The Mach number M_6 of the hot core stream at station 6 is taken equal to 0.5.

12.6.3.2 Governing Equations

The governing equations used for the mixing process are the energy balance, momentum balance, and continuity equations together with the property equations of the mixed stream.

Energy equation:

$$\dot{m}_c c_{pa}T_{16t} + \dot{m}_g c_{pg}T_{6t} = \dot{m}_m c_{pm}T_{7t} \tag{12.121}$$

where, $\dot{m}_m = \dot{m}_c + \dot{m}_g = \dot{m}_c + (\dot{m}_h + \dot{m}_f) = \dot{m}_1 + \dot{m}_f$ ($\dot{m}_f$ – fuel flow rate).

Momentum equation:

$$(\dot{m}_c C_{16} + p_{16} A_{16}) + (\dot{m}_g C_6 + p_6 A_6) = (\dot{m}_m C_7 + p_7 A_7) \tag{12.122}$$

where A_{16} is the annular cross-sectional area of the bypass section, A_6 is the cross-sectional area of the core engine, and A_7 is the cross-sectional area of the mixing chamber.

Continuity equation:

$$\dot{m}_m = \rho_m C_7 A_7 = p_7 C_7 A_7 / R_m T_7 \tag{12.123}$$

Property equation (mixture specific heat):

$$c_{pm} = \frac{\dot{m}_c c_{pa} + \dot{m}_g c_{pg}}{\dot{m}_c + \dot{m}_g} \tag{12.124}$$

The gas constant is taken as constant throughout and is equal to 0.287 kJ/kg K.

12.6.3.3 Computational Procedure

The procedure to calculate the properties of the mixed gases at station 7 (particularly p_{7t}) is as follows:

- For a given fan root pressure, the value of the fan tip pressure ratio is adjusted until equality of the total pressures of the cold and hot air streams at the entry into the mixer is achieved ($p_{16t} \approx p_{6t}$).
- Knowing M_6 ($=0.5$), p_{6t}, and T_{6t} (calculated using primary flow conditions in the core engine), we can determine, T_6, C_6, p_6, ρ_6, and A_6, respectively, from

$$\frac{T_{6t}}{T_6} = 1 + \frac{\gamma_g - 1}{2} M_6{}^2 = 1 + \frac{C_6{}^2}{2 c_{pg} T_6}$$

$$\frac{p_{6t}}{p_6} = \left(\frac{T_{6t}}{T_6}\right)^{\gamma_g/(\gamma_g - 1)} = \left[1 + \frac{\gamma_g - 1}{2} M_6{}^2\right]^{\gamma_g/(\gamma_g - 1)}$$

$$\rho_6 = \frac{p_6}{R T_6}$$

$$A_6 = \frac{\dot{m}_h}{C_6 \rho_6}$$

The value of $(\dot{m}_g C_6 + p_6 A_6)$ in the momentum equation can now be determined.

- Knowing p_{16} ($= p_6$), p_{16t}, and T_{16t} (calculated using secondary flow conditions in the bypass section), M_{16} can be found from

$$\frac{p_{16t}}{p_{16}} = \left(\frac{T_{16t}}{T_{16}}\right)^{\gamma_g/(\gamma_g - 1)} = \left[1 + \frac{\gamma_g - 1}{2} M_{16}{}^2\right]^{\gamma_g/(\gamma_g - 1)}$$

- Knowing M_{16}, p_{16t} ($= p_{13t}$), and T_{16t} ($= T_{13t}$), we can determine T_{16}, C_{16}, ρ_{16}, and A_{16} from the following equations:

$$\frac{T_{16t}}{T_{16}} = 1 + \frac{\gamma_a - 1}{2} M_{16}{}^2$$

$$\frac{T_{16t}}{T_{16}} = 1 + \frac{C_{16}^2}{2c_{pg}T_{16}}$$

$$\rho_{16} = \frac{p_{16}}{RT_{16}}$$

$$A_{16} = \frac{\dot{m}_h}{C_{16}\rho_{16}}$$

- The value of $(\dot{m}_cC_{16} + p_7A_{16})$ in the momentum equation can now be determined.
- The momentum equation then yields $(\dot{m}_mC_7 + p_7A_7)$.
- $A_7 = A_6 + A_{16}$, and the continuity equation give $\dot{m}_m = \rho_mC_7A_7 = p_7C_7A_7/R_mT_7$.
- T_{7t} is known from the energy equation, but we need to determine p_7 and p_{7t}.
- Guess a value of M_7, and then find T_7 and C_7 from

$$\frac{T_{7t}}{T_7} = 1 + \frac{C_7^2}{2c_{pm}T_7} = 1 + \frac{\gamma_m - 1}{2}M_7^2$$

- The continuity equation gives p_7.
- Iterate by checking that $(\dot{m}_mC_7 + p_7A_7)$ is equal to the left-hand side of the momentum equation.
- Once the correct p_7 is found by the iterative method, p_{7t} can be determined from

$$\frac{p_{7t}}{p_7} = \left[1 + \frac{\gamma_m - 1}{2}M_7^2\right]^{\gamma_m/(\gamma_m-1)}$$

12.6.4 Propelling Nozzle

The calculation of the process in the nozzle is the same as before and depends on whether the nozzle is choked (sonic flow) or unchoked (subsonic flow).

12.6.4.1 Sonic Flow in the Propelling Nozzle (Nozzle Choked)
When the flow in the nozzle throat is sonic, $M_8 = 1$, and the gases are at the critical conditions of p_c and T_c. These gases expand first from the pressure at station 7 to the critical pressure p_c ($p_c > p_a$) at station 8 and then to the ambient pressure p_a (p_1).

$$C_8 = C_c = \sqrt{\gamma_m R_m T_c} \tag{12.125}$$

The isentropic nozzle efficiency is

$$\eta_n = \frac{T_{7t} - T_c}{T_{7t} - T_{cs}} \tag{12.126}$$

where T_c is the critical temperature and T_{cs} is the temperature of the gases at the end of the isentropic expansion with the nozzle choked. The temperature ratio from Eq. (12.68) becomes

$$\frac{T_{7t}}{T_c} = \frac{\gamma_m + 1}{2} \tag{12.127}$$

The pressure ratio from Eq. (12.74)

$$\frac{p_c}{p_{7t}} = \left[1 - \frac{1}{\eta_n}\left(\frac{\gamma_m - 1}{\gamma_m + 1}\right)\right]^{\gamma_m/(\gamma_m-1)} \tag{12.128}$$

For the conditions in the nozzle throat at $M_8 = 1$, Eq. (12.70) can be rewritten as

$$\frac{p_{7t}}{p_c} = \frac{1}{\left[1 - \frac{1}{\eta_n}\left(\frac{\gamma_m - 1}{\gamma_m + 1}\right)\right]^{\gamma_m/(\gamma_m - 1)}} \tag{12.129}$$

The critical density is

$$\rho_8 = \rho_c = \frac{p_c}{RT_c} \tag{12.130}$$

The nozzle throat area is

$$A_8 = \frac{\dot{m}_m}{\rho_c C_c} \tag{12.131}$$

Since the gas in the nozzle expands first to the critical pressure p_c and then to the ambient pressure p_1, the thrust produced by the core engine is made up of two components: the momentum thrust and the pressure thrust:

$$F = (\dot{m}_m C_c - \dot{m}_1 C_1) + (p_c - p_1)A_8 \ \ N \tag{12.132}$$

where C_c is the critical velocity of the gases in the nozzle throat when the gas pressure in the nozzle throat is equal to the critical pressure p_c.

The *sfc* is

$$sfc = \frac{\dot{m}_f}{F} \tag{12.133}$$

12.6.4.2 Subsonic Flow in the Hot Nozzle (Nozzle Unchoked)

If the gases in the nozzle are expanded to the atmospheric pressure ($p_8 = p_a$) and the pressure ratio p_{7t}/p_a is less than the critical pressure ratio p_{7t}/p_c, the nozzle is unchoked with $M_8 < 1$, and the temperature and pressure ratios in the nozzle are

$$\frac{T_{7t}}{T_8} = 1 + \frac{C_8^2}{2c_{pm}T_8} = 1 + \frac{\gamma_m - 1}{2}M_8^2 \tag{12.134}$$

$$\frac{p_8}{p_{7t}} = \left[1 - \frac{1}{\eta_n}\left(1 - \frac{1}{1 + \frac{\gamma_m - 1}{2}M_8^2}\right)\right]^{\gamma_m/(\gamma_m - 1)} \tag{12.135}$$

The gas density in the nozzle throat is

$$\rho_8 = \frac{p_1}{RT_8} \tag{12.136}$$

The nozzle throat area is

$$A_8 = \frac{\dot{m}_m}{\rho_8 C_8} \tag{12.137}$$

Since the pressure in the nozzle expands to the ambient pressure p_a, the pressure component of the thrust is equal to zero, reducing the thrust equation to

$$F = (\dot{m}_m C_8 - \dot{m}_1 C_1) \ \ N \tag{12.138}$$

where C_8 is the jet velocity (equal to the velocity in the nozzle throat) of the gases when the gas pressure in the nozzle throat is equal to the ambient pressure p_a.

The *sfc* is

$$sfc = \frac{\dot{m}_f}{F} \tag{12.139}$$

The overall efficiency is

$$\eta_o = \eta_p \eta_{th} = \frac{FC_1}{\dot{m}_f H_l} \tag{12.140}$$

The propulsive efficiency is

$$\eta_p = \frac{C_1[(\dot{m}_1 + \dot{m}_f)C_8 - \dot{m}_1 C_1]}{\frac{1}{2}[(\dot{m}_1 + \dot{m}_f)C_8^2 - \dot{m}_1 C_1^2]} = \frac{2C_1[\dot{m}_m C_8 - \dot{m}_1 C_1]}{[\dot{m}_m C_8^2 - \dot{m}_1 C_1^2]} \tag{12.141}$$

The thermal efficiency is

$$\eta_{th} = \frac{\eta_o}{\eta_p} = \frac{FC_1}{\dot{m}_f H_l} \times \frac{[\dot{m}_m C_8^2 - \dot{m}_1 C_1^2]}{2C_1[\dot{m}_m C_8 - \dot{m}_1 C_1]}$$

$$\eta_{th} = \frac{0.5F}{\dot{m}_f H_l} \times \frac{[\dot{m}_m C_8^2 - \dot{m}_1 C_1^2]}{[\dot{m}_m C_8 - \dot{m}_1 C_1]} \tag{12.142}$$

12.7 Performance of Two-Spool Mixed-Flow Turbofan Engine – Case Study

The performance parameters presented are for the engine and flight data given in Table 12.7. It should be noted that for the constant value of the fan root pressure ratio $r_{fr} = 3$, the value of the tip pressure ratio r_{ft} changes with the overall pressure ratio r_o to satisfy the condition of equality of static and total pressures of the core and secondary flows at the entry into the mixer: $p_6 = p_{16}$, $p_{6t} = p_{16t}$.

Figure 12.40 shows the variation of r_{ft} with the overall pressure ratio r_o for two values of r_{fr}. For a given fan-root pressure ratio r_{fr}, the fan-tip pressure ratio increases quickly with r_o up to $r_o = 20$ and then tapers off afterwards. The effect of r_{fr} on r_{ft} decreases with increasing r_o until it diminishes completely when $r_o \approx 45$. For a given overall pressure ratio, r_{ft} increases with increasing r_{fr} over most of the investigated range of r_o. It is observed that the variation of r_{ft} with r_{fr} has no effect on the specific thrust and *sfc* plotted versus r_o.

Figure 12.41 shows the temperature and pressure profiles in the mixed-flow turbofan engine operated at the conditions in Table 12.7 at an overall compression pressure ratio $r_o = 38$. Both the temperature and pressure vary significantly along the engine axis, the former impacting significantly on the mean specific heat of the air and combustion products within the different components of the engine, as seen in Figure 12.42. The largest value of the mean specific heat is observed in the HPT, followed by the LPT, and then the combustion chamber.

Figure 12.43 shows the specific thrust and *sfc* of the mixed-flow turbofan engine as functions of the overall compression pressure ratio. The specific thrust reaches a maximum at $r_o \approx 16$ and then decreases continuously with increasing r_o at a maximum rate of 2.5%. This small decrease can also be noticed by comparing the T-s diagrams at overall compression

Table 12.7 Data for the calculation of the mixed-flow turbofan engine performance.

Fuel $C_{12}H_{24}$		
Turbine inlet temperature, K	$T_{max} = T_{4t}$	1973
Inner fan (root) pressure ratio	r_{fr}	3.0
Outer fan (tip) pressure ratio	r_{ft}	variable
Fan polytropic efficiency	η_{pf}	0.9
Nozzle isentropic efficiency	η_n	0.9
Diffuser isentropic efficiency	η_i	0.95
Compressor polytropic efficiency	η_{pc}	0.9
Turbine polytropic efficiency	η_{pt}	0.9
Mechanical efficiency	η_m	0.99
Aircraft forward speed, km/h	C_1	855
Flight altitude, m	H	10 675
Ambient pressure, kPa	p_1	23.81
Ambient temperature, K	T_1	218.8
Flight Mach number	M_1	0.8
Fan inlet temperature, K	T_{2t}, T_{12t}	246.95
Bypass ratio	B	6.5
Pressure losses in combustion chamber, kPa	Δp_{34}	1.5
Overall air flow rate, kg/s	$\dot{m}_1$	1.0
Air flow rate in bypass duct, kg/s	$\dot{m}_c$	0.866667
Air flow rate in core engine, kg/s	$\dot{m}_h$	0.133333
Gas flow rate in core engine, kg/s	$\dot{m}_g$	variable
Flow rate in the mixer and nozzle, kg/s	$\dot{m}_m$	variable
Relative air-fuel ratio	λ	variable
Fuel flow rate, kg/s	$\dot{m}_f$	variable
Lower heating value, kJ/kg	H_l	44 340

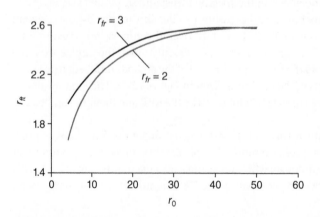

Figure 12.40 Effect of the fan-root pressure ratio and overall pressure ratio on the fan-tip pressure ratio in the mixed-flow turbofan engine.

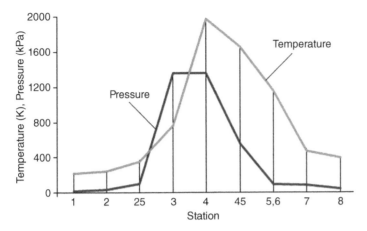

Figure 12.41 Temperature and pressure profiles in the engine ($H = 10\,675\,m$, $M_1 = 0.8$, $T_{4t} = 1973\,K$, $B = 6.5$, $r_{fr} = 3$, $r_o = 38$).

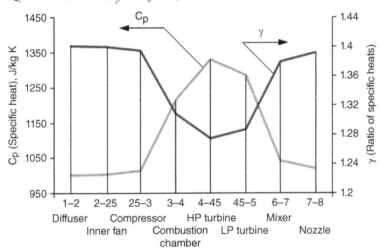

Figure 12.42 Profiles of the specific heat and ratio of specific heats ($H = 10\,675\,m$, $M_1 = 0.8$, $T_{4t} = 1973\,K$, $B = 6.5$, $r_{fr} = 3$, $r_o = 38$).

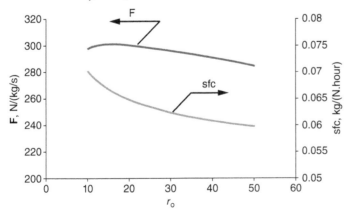

Figure 12.43 Specific thrust and specific fuel consumption of the mixed-flow turbofan engine ($H = 10\,675\,m$, $M_1 = 0.8$, $T_{4t} = 1973\,K$, $B = 6.5$).

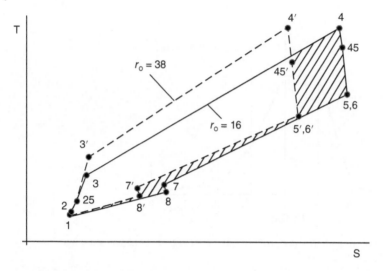

Figure 12.44 Effect of the overall compression pressure ratio on the T-s diagram ($H = 10\,675\,m$, $M_1 = 0.8$, $T_{4t} = 1973\,K$, $B = 6.5$, $r_{fr} = 3$).

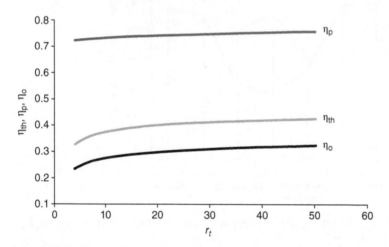

Figure 12.45 Thermal, propulsive, and overall efficiencies of the mixed-flow turbofan engine ($H = 10\,675\,m$, $M_1 = 0.8$, $T_{4t} = 1973\,K$, $B = 6.5$).

pressure ratios of 16 and 38 (Figure 12.44). The dashed area shows that there is a gain in cycle work as the overall pressure ratio is reduced.

Figure 12.45 shows the thermal, propulsive, and overall efficiencies of the mixed-flow turbofan engine as functions of the overall compression pressure ratio. All the efficiencies increase marginally with the overall pressure ratio, albeit at different rates.

Problems

12.1 A jet aircraft is flying at 815 km/h at an altitude of 10 000 m. Ignoring the mass flow of the fuel, determine the net thrust generated on the basis of the following data:

Compressor pressure ratio	10 : 1
Compressor isentropic efficiency	0.85
Turbine isentropic efficiency	0.88
Intake isentropic efficiency	0.9
Nozzle isentropic efficiency	0.9
Mass flow rate of air (gases)	50 kg/s
Lower heating value of the fuel	43 400 kJ/kg
Turbine entry temperature	1200 K
Specific heats of air and products, respectively	1.005, 1.148 kJ/kg K
Pressure losses in combustion chamber	6% of compressor delivery pressure

12.2 Repeat Problem 12.1, allowing for temperature-dependent specific heats.

12.3 A jet aircraft is flying at a speed of 280 m/s and altitude of 7000 m. The compressor pressure ratio is 8 : 1, the turbine inlet temperature is 1200 K, and the mass flow rate of air is 16.0 kg/s. If the mass flow rate of the fuel is 0.269 kg/s, calculate the propelling nozzle area required, the net thrust developed, and the specific fuel consumption, assuming the following component efficiencies:

Compressor polytropic efficiency	0.86
Turbine polytropic efficiency	0.88
Intake isentropic efficiency	0.95
Nozzle isentropic efficiency	0.95
Mechanical transmission efficiency	0.99
Lower heating value of the fuel	43 400 kJ/kg
Pressure losses in combustion chamber	6% of compressor delivery pressure

(Take the gas constant for both air and the combustion products as 0.287 kJ/kg K, and the ratios of specific heats for the air and products as 1.4 and 1.333, respectively.)

12.4 The following data apply to an aircraft turbofan engine at an altitude of 11 000 m and Mach number 0.82. The engine is twin-spool with the fan and the LP compressor driven by the LP turbine rotor and the HP compressor driven by the HP turbine rotor. Separate hot and cold exhaust nozzles are used.

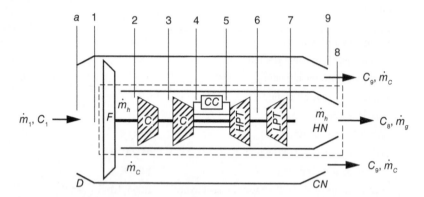

Determine the cold and hot nozzle areas and total thrust for the following data:

Fan pressure ratio	1.8
LP compressor pressure ratio	1.5
HP compressor pressure ratio	7.0
Bypass ratio	6.0
Turbine entry temperature, K	1388
Lower fuel heating value, kJ/kg	43 260
Fan, compressor, and turbine isentropic efficiencies	0.90
Isentropic efficiencies of the nozzles and inlet duct	0.95
Mechanical efficiency of each spool	0.99
Total air mass flow rate	200
Gas constant for air and combustion products, $J/kg\ K$	287
Ratio of specific heats of air and combustion products	1.4
Ratio of specific heats of combustion products	1.333

12.5 On the basis of the solution of Problem 12.4, determine further
(a) The rate and specific fuel consumptions
(b) The propulsion efficiency of the engine
(c) The thermal and overall efficiencies of the engine

12.6 Using the charts obtained from the design-point case study of the turbojet engine, estimate the design-point parameters, and justify your choice for
  (a) Turbine entry temperature fixed by metallurgical considerations
  (b) Turbine entry temperature dictated by a high thermal efficiency requirement.
    If assumptions need to be made, state them clearly.

12.7 Repeat Problem 12.6 for the two-spool turbofan engine case study.

13

Design-Point Calculations of Industrial Gas Turbines

The purpose of the design-point calculations for industrial gas turbine engines is to determine the airflow rate, specific fuel consumption, and efficiency for a given power demand. The input data include the ambient conditions, compressor pressure ratio, turbine entry temperature (TET) (or air-fuel ratio in the combustion chamber), and component efficiencies. Based on these calculations, engine components such as the compressor, combustion chamber, and turbine can be designed. Unlike aero engines, size and weight are of secondary importance in the design process. Two cases will be considered: single-shaft and two-shaft gas turbines for stationary applications.

13.1 Single-Shaft Gas Turbine Engine

Figure 13.1 shows a single-shaft stationary gas turbine cycle in its simplest configuration. The compressor and turbine are fixed to a common shaft, with part of the power generated by the turbine expended to run the compressor. The surplus power is used to drive different types of machines via a gearbox.

In most stationary applications, the pressure loses in the inlet (Δp_{1a}) are small and can be ignored, which results in $p_1 = p_a$ as shown on the T-s diagram in Figure 13.1b.

13.1.1 Design-Point Calculations

Some of the equations are the same as those used for other cycles and are repeated here to avoid continuous reference to previous chapters when setting up the calculation procedure.

13.1.1.1 Compressor
Depending on the efficiency term used, the temperature equivalents of the compressor work can be written as

$$\Delta T_{12} = T_2 - T_1 = \frac{T_1}{\eta_c}[r_c^{(\gamma_a-1)/\gamma_a} - 1]$$

$$\Delta T_{12} = T_2 - T_1 = T_1[r_c^{(\gamma_a-1)/\gamma_a\eta_{pc}} - 1]$$

Fundamentals of Heat Engines: Reciprocating and Gas Turbine Internal Combustion Engines, First Edition. Jamil Ghojel.
© 2020 John Wiley & Sons Ltd. This Work is a co-publication between John Wiley & Sons Ltd and ASME Press.
Companion website: www.wiley.com/go/JamilGhojel_Fundamentals of Heat Engines

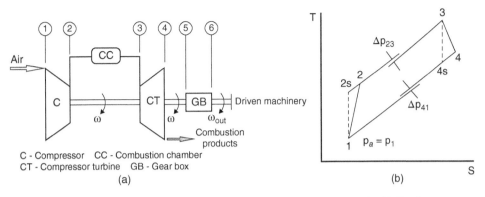

Figure 13.1 Single-shaft industrial gas turbine engine: (a) engine schematic; (b) T-s diagram.

where

η_c: isentropic efficiency of the compressor
η_{pc}: polytropic efficiency of the compressor

The compressor temperature ratio is given by

$$\frac{T_2}{T_1} = r_c^{(\gamma-1)/\gamma\eta_{pc}}$$

The turbine power required to drive the compressor, expressed as the power $\dot{W}_c$ absorbed by the compressor, is

$$\dot{W}_c = \frac{\dot{m}_a}{\eta_m} c_{pa} \Delta T_{12} \tag{13.1}$$

where

$\dot{m}_a$: mass flow rate of air
η_m: mechanical efficiency of the compressor-turbine unit
c_{pa}: specific heat at constant pressure of the air in the compressor

13.1.1.2 Combustion Chamber

The energy balance in the combustion chamber in terms of heat addition and corresponding temperature rise can be written as

$$\eta_{com} H_l \dot{m}_f = \dot{m}_a c_{pm}(T_3 - T_2)$$

where

η_{com}: combustion efficiency
H_l: lower heating value of the fuel
$\dot{m}_f$: mass flow rate of the fuel
$\dot{m}_a$: mass flow rate of the air
c_{pm}: mean specific heat of the air

13.1.1.3 Turbine

The temperature equivalent of the total turbine work (including work to drive the compressor and surplus output shaft work) can be written as

$$\Delta T_{34} = T_3 - T_4 = \eta_t T_3 \left[1 - \frac{1}{r_t^{(\gamma_g - 1)/\gamma_g}} \right]$$

$$\Delta T_{34} = T_3 - T_4 = T_3 \left[1 - \frac{1}{r_t^{\eta_{pt}(\gamma_g - 1)/\gamma_g}} \right]$$

η_t: isentropic efficiency of the turbine
η_{pt}: polytropic efficiency of the turbine

The turbine temperature ratio is

$$\frac{T_3}{T_4} = r_t^{\eta_{pt}(\gamma - 1)/\gamma}$$

If the pressure losses in the inlet duct are ignored, i.e. $p_1 = p_a$, the turbine pressure ratio (expansion ratio) is then

$$r_t = r_c \left(\frac{1 - x}{1 + y} \right)$$

where r_c is the compressor pressure ratio, $x = \Delta p_{23}/p_2$, and $y = \Delta p_{4a}/p_a = \Delta p_{41}/p_1$.

The total turbine power is

$$\dot{W}_t = \dot{m}_g c_{pg} \Delta T_{34} \tag{13.2}$$

where

$\dot{m}_g$: mass flow rate of combustion gases ($= \dot{m}_a + \dot{m}_f$)
$\dot{m}_f$: mass flow rate of fuel
c_{pg}: specific heat of the combustion gases at constant pressure

The turbine net shaft power is

$$\dot{W}_{net} = \dot{W}_t - \dot{W}_c = \dot{m}_g c_{pg} \Delta T_{34} - \frac{\dot{m}_a}{\eta_m} c_{pa} \Delta T_{12} \tag{13.3}$$

13.1.1.4 Specific Fuel Consumption

This is a measure of the mass of fuel used to produce a unit of power over a period of time:

$$sfc = \frac{\dot{m}_f}{\dot{W}_{net}} \tag{13.4}$$

The units of the specific fuel consumption (sfc) depend on the units of the mass flow rate and power. The most widely used unit is kg/kWh, where $\dot{m}_f$ is in kg/h and $\dot{W}_{net}$ is in kW.

13.1.1.5 Cycle Thermodynamic Efficiency

This is defined as the ratio of net power output to the rate of heat supply to the engine:

$$\eta_{th} = \frac{\dot{W}_{net}}{\dot{m}_f H_l} \tag{13.5}$$

where H_l is the lower heating value of the fuel in kJ/kg.

Table 13.1 Data for design point calculations for single-shaft gas turbine $\lambda = var$.

	Fuel: $C_{12}H_{24}$				
Relative air-fuel ratio	λ	2	4	6	8
Compressor polytropic efficiency	η_{pc}	0.85	0.85	0.85	0.85
Turbine polytropic efficiency	η_{pt}	0.88	0.88	0.88	0.88
Mechanical efficiency	η_m	0.98	0.98	0.98	0.98
Ambient pressure, kPa	p_a	100	100	100	100
Ambient temperature, K	T_a	298	298	298	298
Combustion chamber pressure losses, kPa	Δp_{23}	4.0	4.0	4.0	4.0
Exit pressure loss, kPa	Δp_{4a}	4.0	4.0	4.0	4.0
Air flow rate, kg/s	$\dot{m}_a$	1.0	1.0	1.0	1.0
Fuel flow rate, kg/s	$\dot{m}_f$	0.033981	0.01699	0.011327	0.008495
Gas flow rate, kg/s	$\dot{m}_g$	1.033981	1.01699	1.011327	1.008495
Lower heating value, kJ/kg	H_l	44 340	44 340	44 340	44 340

Combining Eqs. (13.4) and (13.5) gives

$$\eta_{th} = \frac{1}{(sfc)H_l} \tag{13.6}$$

which indicates that for a given fuel, the thermal efficiency is inversely proportional to the *sfc*.

13.2 Performance of Single-Shaft Gas Turbine Engine – Case Study

Performance of a single-shaft gas turbine engine can be characterised in terms of the relative air-fuel ratio λ or maximum temperature T_3. The cycle maximum temperature is usually the combustion temperature at turbine entry (TET). In the analysis to follow, total temperatures are not used, as was done in the case of aviation gas turbines, because the flow speeds are so low in industrial engines that static and total temperatures become almost equal.

13.2.1 In Terms of Relative Air-Fuel Ratio λ

Using the methodology presented for the practical cycles in Chapter 11, the design-point calculations can be conducted for the cycle in Figure 13.1b. It is assumed that the losses in the inlet duct are negligible and the pressure at the compressor inlet is equal to the ambient pressure ($p_1 = p_a$).

As a case study, the calculations are conducted for dodecene $C_{12}H_{24}$, assuming an air mass flow rate of 1 kg/s and compressor inlet pressure equal to the ambient pressure taken as 100 kPa. The data used for the calculations are presented in Table 13.1. The independent design parameters considered are the compressor pressure ratio r_c, relative air fuel ratio λ, and cycle maximum temperature T_3.

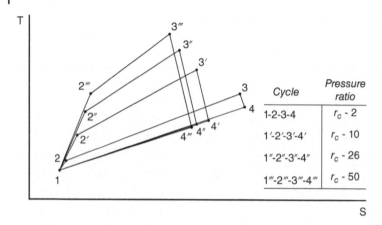

Figure 13.2 T-s diagrams of the simple cycle for $\lambda = 4$ at different pressure ratios.

Figure 13.2 shows the schematic T-s diagrams for a constant relative air-fuel ratio λ equal to 4 and four compressor pressure ratios. For a given air-fuel ratio (or input heat), as the pressure ratio in the compressor increases, the TET (maximum cycle temperature) increases and the turbine exit temperature decreases. The entropy at the exit from the turbine shifts to the left, indicating a decrease in the heat rejected from the cycle. This together with increasing area of the cycle point to increased output work and thermal efficiency as the pressure ratio is increased. These results are further illustrated in the plots in Figure 13.3.

For a constant air-fuel ratio, the TET increases steadily with the compressor pressure ratio. The lower the air-fuel ratio, the higher this temperature at a given pressure ratio. The maximum temperature attained when $\lambda = 2$ is 2091 K at $r_c = 50$. The power increases steadily for $\lambda = 2$, $\lambda = 4$, and $\lambda = 6$ without reaching a maximum in the selected compressor pressure ratio range. As the air-fuel mixture becomes leaner, the power decreases; at $\lambda = 8$, the power reaches a maximum at a pressure ratio of around 38 and then starts dropping slowly. The thermal efficiency follows the change in specific output power, reaching a maximum value of just under 46% at $\lambda = 2$ and $r_c = 50$. The *sfc* is inversely proportional to the thermal efficiency: it initially drops sharply with increasing pressure ratio and then drops very slowly in the pressure ratio range 10–50 for all air-fuel ratios. A minimum *sfc* is observed when $\lambda = 8$ at $r_c = 38$. For any fixed compressor pressure ratio, the *sfc* decreases steadily with decreasing relative air-fuel ratio.

Another way of representing the results of the design-point calculations is by means of performance maps (carpet plots), as seen in Figures 13.4 and 13.5. These plots show the *sfc* (or thermal efficiency) versus the specific net output power at constant lines of relative air-fuel ratio λ and constant lines of compressor pressure ratio r_c. These plots are convenient qualitative indicators of the performance of the engine in terms of two important parameters, λ and r_c, simultaneously.

At a constant relative air-fuel ratio λ, increasing the specific output power (and corresponding compressor pressure ratio) causes the *sfc* to decrease, with the rate of decrease rising with increasing λ. At a constant pressure ratio, the *sfc* decreases continuously as the output power increases. If the *sfc* is replaced by the thermal efficiency, the carpet plot shows an inverse trend, as can be seen in Figure 13.5.

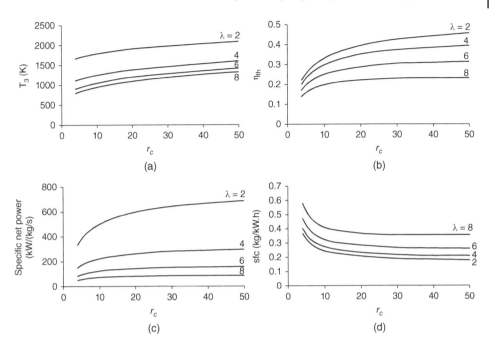

Figure 13.3 Performance characteristics of single-shaft gas turbine versus compressor pressure ratio r_c and relative air-fuel ratio λ: (a) turbine entry temperature T_3; (b) thermal efficiency η_{th}; (c) specific net power; (d) specific fuel consumption.

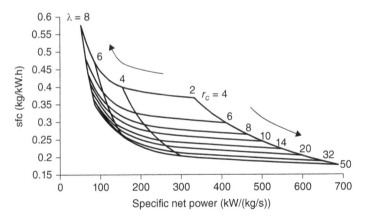

Figure 13.4 Performance map for the single-shaft gas turbine: *sfc* vs. specific power at various constant values of λ and r_c.

For lean air-fuel mixtures, the efficiency increases sharply with increasing r_c accompanied by a small increase in output power. With richer mixtures, the increase in efficiency is accompanied by a greater rate of output power increase. For the data shown in Figure 13.5, maximum output power and thermal efficiency can be achieved by operating the engine at $\lambda = 2$ and $r_c = 50$, provided that the engine can operate safely at the ensuing TET (2091 K) and emissions do not exceed the regulated levels.

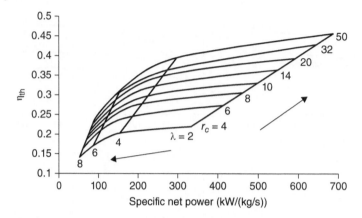

Figure 13.5 Performance map for the single-shaft gas turbine: η_{th} vs. specific net power at various constant values of λ and r_c.

Table 13.2 Data for design point calculations for a single-shaft gas turbine $T_3 = var.$

	Fuel: $C_{12}H_{24}$				
Turbine entry temperature, K	$T_{max} = T_3$	1192	1490	1788	2086
Compressor polytropic efficiency	η_{pc}	0.85	0.85	0.85	0.85
Turbine polytropic efficiency	η_{pt}	0.88	0.88	0.88	0.88
Mechanical efficiency	η_m	0.98	0.98	0.98	0.98
Ambient pressure, kPa	p_a	100	100	100	100
Ambient temperature, K	$T_a = T_1$	298	298	298	298
Combustion chamber pressure losses, kPa	Δp_{23}	4	4	4	4
Exit pressure loss, kPa	Δp_{4a}	4	4	4	4
Air flow rate, kg/s	$\dot{m}_a$	1.0	1.0	1.0	1.0
Relative air-fuel ratio	λ	variable	variable	variable	variable
Fuel flow rate, kg/s	$\dot{m}_f$	variable	variable	variable	variable
Gas flow rate, kg/s	$\dot{m}_g$	variable	variable	variable	variable
Lower heating value, kJ/kg	H_l	44 340	44 340	44 340	44 340

13.2.2 In Terms of Cycle Maximum Temperature T_3

In previous discussion, T_3 increases with r_c, and direct comparison with the theoretical cycles discussed earlier cannot be made. To make such a comparison possible, the performance parameters are reproduced as functions of r_c and T_3 (T_{max}). For a fixed inlet temperature T_1 equal to 298 K, T_3 can be kept constant as r_c varies by changing the relative air-fuel ratio λ. Computationally, this can be done using Eq. (11.18) for $C_{12}H_{24}$ (from Chapter 11) for any given compressor delivery temperature T_2. Table 13.2 shows the data for

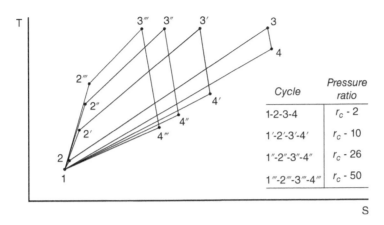

Figure 13.6 T-s diagrams of the simple cycle for $T_3 = 1490\,K$ at different pressure ratios.

four cases where the constant turbine entry temperatures are 1192, 1490, 1788, and 2086 K, corresponding to temperature ratios of 4, 5, 6, and 7 when T_1 is taken as equal to 298 K.

The T-s diagrams for a constant $T_3 = 1490\,K$ at four compressor pressure ratios are shown in Figure 13.6. The pressure losses are accounted for in theses diagrams, but the heat addition and rejection processes are shown as continuous lines. The T-s diagrams indicate that for the given maximum cycle temperature of 1490 K, as the pressure ratio in the compressor increases, the turbine exit temperature decreases and the entropy values shift significantly to the left. This is a sign of decreasing heat rejection. The trend of cycle work is more difficult to predict from the T-s diagrams; however, the performance characteristics in Figure 13.7 show that specific output power at $T_3 = 1490\,K$ exhibits a peak at a specific pressure ratio. Similar results are shown for temperatures of 1192, 1788, and 2086 K.

For the TETs selected (1192–2086 K), the relative air-fuel ratio λ changes within the range 1.336–14.845. For a constant T_3, λ increases with r_c; the increase becomes steeper at lower values of T_3 (Figure 13.7a). For a given pressure ratio, the specific output (net) power increases with increasing T_3 (decreasing λ). For each constant T_3 value used in the analysis, the specific power exhibits a maximum at a specific pressure ratio with maximum point shifting towards higher r_c as T_3 is increased (Figure 13.7c). The thermal efficiency plots in Figure 13.7b exhibit peaks at temperatures of 1490 and 1192 K. At 1192 K, the efficiency reaches its peak much earlier and then drops sharply as the compressor pressure ratio increases. The maximum efficiency observed is 45.68% at $r_c = 50$ when $T_3 = 2086\,K$. For a given compressor pressure ratio, the thermal efficiency increases with peak temperature T_3. The behaviour of the thermal efficiency at different temperature ratios is due to the net effect of the changes in cycle heat rejection and output work when the compressor pressure ratio increases. At higher maximum cycle temperatures, the net effect is towards increasing the thermal efficiency despite the decrease in output cycle work at the higher end of the pressure ratio. The plots of the *sfc* are shown in Figure 13.7d. As stated earlier, the *sfc* is inversely proportional to the thermal efficiency, and that is clearly seen in the plots. For a given compressor pressure ratio, the *sfc* increases with decreasing T_3.

The performance map in terms of the temperature ratios and compressor pressure ratios is shown in Figure 13.8. If the lowest temperature ratio is kept constant at 4 $(T_3/T_1 = 4)$

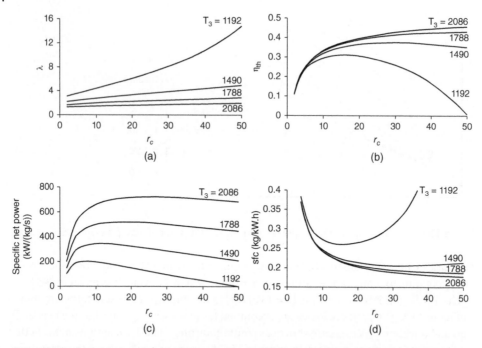

Figure 13.7 Performance characteristics of a single-shaft gas turbine vs. compressor pressure ratio r_c and TET T_3: (a) relative air-fuel ratio λ; (b) thermal efficiency η_{th}; (c) specific net power; (d) specific fuel consumption.

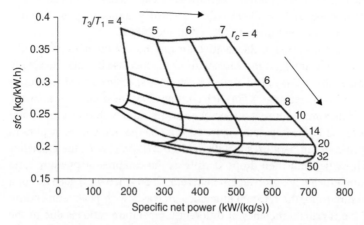

Figure 13.8 Performance map for the single-shaft gas turbine: specific fuel consumption vs. specific net power at various T_3/T_1 and r_c ($T_1 = 298$ K).

while increasing the pressure ratio, the *sfc* initially drops with specific power and then starts increasing at $r_c = 10$ with the specific power almost unchanged. Similar tendencies are observed for the other temperature ratios, with the points of maximum specific power occurring at $r_c = 14$, $r_c = 20$, and $r_c = 30$ for temperature ratios 5, 6, and 7, respectively. The lower the temperature ratio, the steeper the decrease in specific power with increasing

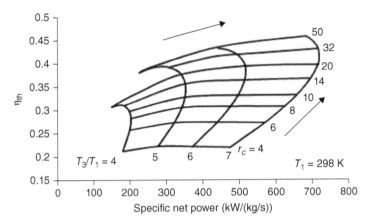

Figure 13.9 Performance map for the single-shaft gas turbine: η_{th} vs. specific net power at various T_3/T_1 and r_c ($T_1 = 298\ K$).

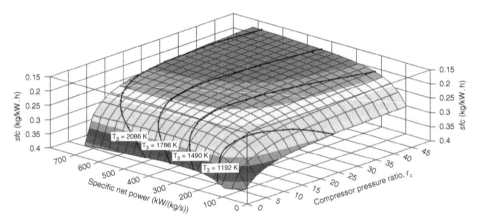

Figure 13.10 3-D surface plot of the specific fuel consumption as a function of the compressor pressure ratio, specific net power, and TET for the single-shaft gas

compressor pressure ratio (after reaching a peak). To obtain maximum specific power (at $r_c = 30$) or minimum *sfc* (at $r_c = 50$), a temperature ratio of $T_3/T_1 = 7$ must be maintained. Such elevated operating temperatures (2086 K) are not feasible in industrial gas turbine engines due to metallurgical limitations, and the normal practice is to find a trade-off between fuel consumption, power, and metallurgical limits.

Figure 13.9 shows a performance map in which the *sfc* is replaced by the thermal efficiency. Conclusions similar to those made regarding Figure 13.8 can also be made here.

In addition to the 2D plots of the performance characteristics, calculated data for the *sfc*, compressor pressure ratio r_c, and specific net power (w) were fitted to a 3D surface function $sfc = \psi(r_c, w)$ and plotted as shown in Figure 13.10. The plot also shows lines of constant turbine-entry temperatures T_3. The *sfc* axis is inverted to make the plot easier to read. The plot is a smooth surface that shows the region of optimum operation as a dark grey area at the top. The mathematical function describing this surface to a good degree of accuracy can be written as

$$sfc = a + \frac{b}{r_c} + \frac{c}{w} + \frac{d}{r_c^2} + \frac{e}{w^2} + \frac{f}{r_c w} \quad kg/kW.h \tag{13.7}$$

where $a = 0.151\,529$, $b = 0.791\,964$, $c = 8.282\,785$, $d = 0.233\,021$, $e = 617.468\,601$, and $f = -35.390\,325$.

The function is valid for the case study under consideration only and for the compressor pressure ratio range 4–50 and specific output power range 97–720 $kW/kg\,(air)/s$ corresponding to TET range of 1192–2086 K.

13.2.3 Comparison with Practical Cycles

Comparison will be made between the practical cycle of the single-shaft gas turbine with the irreversible air-standard cycle of the simple gas turbine discussed in Chapter 10, with the polytropic efficiencies and pressure losses accounted for as shown in Table 13.2. The comparison made for the maximum-to-minimum cycle temperature ratios assumes $T_1 = T_{min} = 298\ K$. The specific net power for the practical cycle is converted to a non-dimensional work term by dividing it by the product $(c_p T_1)$. The results are shown in Figures 13.11 and 13.12. As expected, the characteristics are similar in shape but different in magnitude. Compared with the irreversible air-standard cycle, the thermal efficiency of the practical cycle could be lower or higher depending on the compressor pressure ratio and temperature ratio. The net work is higher for the irreversible air-standard cycle up to a certain value of r_c for both temperature ratios followed by trend reversal as r_c is increased further. However, the point of intersection for the lower temperature ratio is reached much earlier than for the higher temperature ratio. Additionally, the maxima for the practical cycles occur at higher pressure ratios.

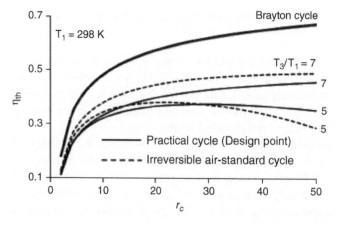

Figure 13.11 Thermal efficiencies of the irreversible air-standard and practical (design-point calculations) cycles of the simple gas turbine.

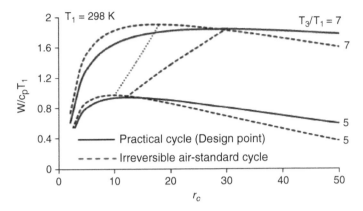

Figure 13.12 Output work of the irreversible air-standard and practical (design-point) calculations) cycles of the simple gas turbine.

13.3 Two-Shaft Gas Turbine Engine

Two-shaft gas turbines are predominately used in power generation, marine propulsion, and oil and gas pumping stations. The arrangement of the components and the schematic T-s diagram of this turbine are shown in Figure 13.13. The compressor C and compressor turbine CT form the gas generator, with the CT generating just enough power to run the compressor and overcome the frictional losses in the gas generator.

The gases leaving the CT expand further in the power turbine PT, which is mounted on a separate shaft, to generate the output shaft power of the engine.

13.3.1 Design-Point Calculations

The calculation procedure is similar to that for the single-shaft turbine, with some modification to the subscripts when dealing with the CT and PT.

13.3.1.1 Mechanical Efficiency of the Gas Generator
The mechanical efficiency of the gas generator is given by

$$\eta_m = \frac{\dot{W}_c}{\dot{W}_{ct}} = \frac{\dot{m}_a c_{pa}(T_2 - T_1)}{\dot{W}_{ct}} \tag{13.8}$$

where $\dot{W}_{ct}$ is the power generated by the CT.

13.3.1.2 Temperature Equivalents of the Compressor Work
ΔT_{12} can be determined from either of the following two equations:

$$\Delta T_{12} = T_2 - T_1 = \frac{T_1}{\eta_c}[r_c^{(\gamma_a-1)/\gamma_a} - 1]$$

$$\Delta T_{12} = T_2 - T_1 = T_1[r_c^{(\gamma_a-1)/\gamma_a \eta_{pc}} - 1]$$

where $r_c = p_2/p_1, p_1 = p_a$.

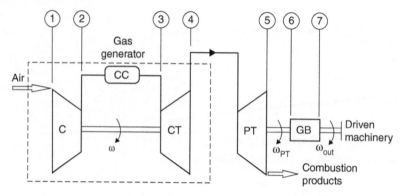

C - Compressor CC - Combustion chamber CT - Compressor turbine
PT - Power turbine GB - Gear box

(a)

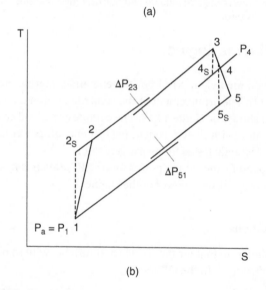

(b)

Figure 13.13 Two-shaft gas turbine cycle: (a) engine schematic; (b) T-s diagram.

13.3.1.3 Pressure Losses

Pressure losses usually occur in the intake, exhaust ducts, and combustion chamber. These are given by Δp_{1a}, Δp_{51} (or Δp_{5a}), and Δp_{23}, respectively, where subscript a indicates *ambient*. In most stationary applications Δp_{1a} and Δp_{51} can be neglected, which results in $p_1 = p_a$ and $p_5 = p_a$. The pressure loss in the duct between the CT and PT is ignored.

13.3.1.4 Temperature Drop in the Compressor Turbine

Depending on the adopted turbine efficiency,

$$\Delta T_{34} = T_3 - T_4 = \eta_t T_3 \left[1 - \frac{1}{r_{ct}^{(\gamma_g-1)/\gamma_g}} \right] \tag{13.9}$$

$$\Delta T_{34} = T_3 - T_4 = T_3 \left[1 - \frac{1}{r_{ct}^{\eta_{pt}(\gamma_g-1)/\gamma_g}} \right] \tag{13.10}$$

where $r_{ct} = p_3/p_4$, $p_3 = p_2 - \Delta p_{23}$.

The CT has only one function – to drive the compressor – and nothing else. As for the single shaft gas turbine, the power required to drive the compressor $\dot{W}_{ct}$ is

$$\dot{W}_{ct} = \frac{\dot{m}_a}{\eta_m} c_{pa} \Delta T_{12} = \dot{m}_g c_{pg} \Delta T_{34} \tag{13.11}$$

from which

$$\Delta T_{34} = T_3 - T_4 = \frac{1}{\eta_m} \frac{\dot{m}_a}{\dot{m}_g} \frac{c_{pa}}{c_{pg}} \Delta T_{12} \tag{13.12}$$

TET T_3 is calculated from Eq. (11.18), the entry temperature into the PT T_4 can be determined from Eq. (13.12) and the pressure ratio in the CT can be calculated from

$$r_{ct} = \left[\frac{1}{1 - \Delta T_{34}/T_3} \right]^{\gamma_g/\eta_{pt}(\gamma_g-1)} \tag{13.13}$$

or

$$r_{ct} = \left(\frac{T_3}{T_4} \right)^{\gamma_g/\eta_{pt}(\gamma_g-1)} \tag{13.14}$$

13.3.1.5 Temperature Equivalent of the Power-Turbine Work
Depending on the adopted turbine efficiency, the power-turbine work can be calculated from either of the following two equations:

$$\Delta T_{45} = T_4 - T_5 = \eta_t T_4 \left[1 - \frac{1}{r_{pt}^{(\gamma_g-1)/\gamma_g}} \right] \tag{13.15}$$

$$\Delta T_{45} = T_4 - T_5 = T_4 \left[1 - \frac{1}{r_{pt}^{\eta_{pt}(\gamma_g-1)/\gamma_g}} \right] \tag{13.16}$$

$r_{pt} = p_4/p_5$, $p_5 = p_a + \Delta p_{51}$ (with losses in the exit duct), or $p_5 = p_1$ (no losses in the exit duct).

$$r_{pt} = \left(\frac{T_4}{T_5} \right)^{\gamma_g/\eta_{pt}(\gamma_g-1)} \tag{13.17}$$

The PT provides power to the driven machinery via the output shaft:

$$\dot{W}_{pt} = \dot{m}_g c_{pg} (T_4 - T_5) \tag{13.18}$$

13.3.1.6 Specific Fuel Consumption
The sfc is defined as for the simple turbine, with the net power being replaced by the power of the PT:

$$sfc = \frac{\dot{m}_f}{\dot{W}_{pt}} \tag{13.19}$$

13.3.1.7 Cycle Thermal Efficiency
The thermal efficiency is calculated on the basis of the output power generated by the PT and the total energy input through the fuel burned in the combustor:

$$\eta_{th} = \frac{\dot{W}_{pt}}{\dot{m}_f H_l} \tag{13.20}$$

where $\dot{m}_f$ is the mass flow rate of the fuel and H_l is the lower heating value of the fuel.

13.4 Performance of Two-Shaft Gas Turbine Engine – Case Study

As a case study, the calculations are conducted for dodecene $C_{12}H_{24}$ (as a surrogate fuel for kerosene) assuming an air mass flow rate of $1\,kg/s$ and a compressor inlet pressure equal to the ambient pressure taken as $100\,kPa$. The analysis is conducted in terms of the relative air-fuel ratio λ and in terms of the maximum cycle temperature $T_{max} = T_3$.

13.4.1 In Terms of Relative Air-Fuel Ratio λ

The performance characteristics obtained for the two-shaft gas turbine are based on the data in Table 13.1 and the methodology explained previously. The T-s diagrams constructed for $\lambda = 4$ at four compressor pressure ratios are shown in Figure 13.14. For a given air-fuel ratio, as the pressure ratio in the compressor r_c is increased, the maximum cycle tempera-ture T_3 (combustion temperature) increases, the compressor-turbine exit temperature T_4 remains unchanged, and the power-turbine temperature decreases. Simultaneously, the entropy at point 5 shifts to the left, indicating reduced heat rejection from the cycle. Inspec-tion of the T-s diagrams also indicates an increase in cycle work. As a consequence of these variations, it can be concluded that the output work and thermal efficiency of the cycle increase continuously with increasing pressure ratio. Similar diagrams can be constructed for different values of λ.

A more comprehensive picture of the performance of the two-shaft gas turbine cycle emerges from the performance characteristics shown in Figure 13.15 obtained using the data in Table 13.1. The effect of the compressor pressure ratio and λ shown in Figure 13.15a is a confirmation of the results from the T-s diagrams. As expected, the combustion temperature at any given r_c increases with decreasing λ. The specific power developed

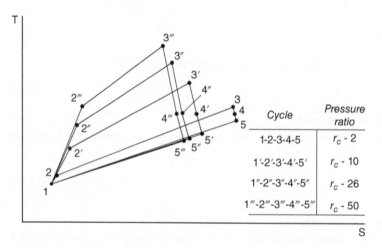

Figure 13.14 T-s diagrams of the two-shaft practical gas turbine cycle for $\lambda = 4$ and different compressor pressure ratios.

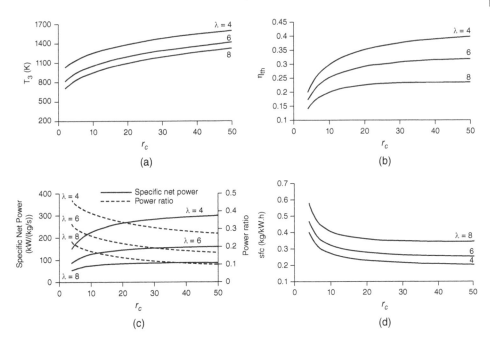

Figure 13.15 Calculated performance characteristics of the two-shaft gas turbine in terms of λ.

by the power-turbine and the power ratio (PR) are shown in Figure 13.15c. The PR is defined as

$$PR = \frac{\dot{W}_{pt}}{\dot{W}_{pt} + \dot{W}_{ct}} \tag{13.21}$$

The compressor-turbine power increases sharply with the compressor pressure ratio, as can be gauged by the increase in temperature drop ΔT_{34} with increasing r_c in the T-s diagrams in Figure 13.14. On the other hand, the power generated by the PT tends to increase slowly when r_c exceeds 10; hence the significant drop in power ratio as r_c increases.

The thermal efficiency and *sfc* are shown in Figures 13.15b,d. The efficiency initially increases sharply and then tapers off, reaching a maximum of about 39.8%. The *sfc* initially drops quickly to about $r_c = 10$ and then decreases very slowly afterwards.

The characteristics indicate that, for the given relative air-fuel ratio of $\lambda = 4$, the engine is best operated at around $r_c = 30$, which results in relatively low compressor work and an *sfc* at a peak power-turbine entry temperature of $T_3 = 1490\,K$. T_3 can be further lowered by reducing r_c or increasing the relative air-fuel ratio λ; however, both measures lead to a decreased net power output and thermal efficiency.

The performance map for three relative air-fuel ratios is shown in Figure 13.16, which allows the prediction of engine performance in terms of specific output power and *sfc* when r_c and λ are varied.

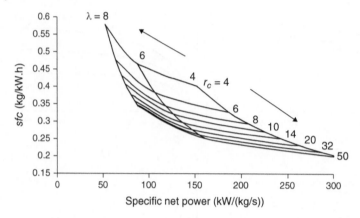

Figure 13.16 Performance map for the two-shaft gas turbine: specific fuel consumption vs. specific net power at constant λ and r_c lines.

13.4.2 In Terms of Cycle Maximum Temperature T_3

The performance characteristics can also be plotted for constant T_3 as functions of the compressor pressure ratio (T_3 can be kept constant as r_c is varied by changing the relative air-fuel ratio λ). The T-s diagrams for the two-shaft gas turbine cycle at $T_3 = 1490\,K$ and four pressure ratios are shown in Figure 13.17. The data in Table 13.2 are used in the calculations for this case study.

For a given maximum cycle temperature, increasing the pressure ratio leads to decreased PT entry and exit temperatures. The entropy at point 5 (exit from the PT) shifts to the left,

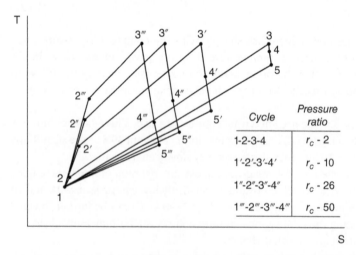

Cycle	Pressure ratio
1-2-3-4	r_c - 2
1'-2'-3'-4'	r_c - 10
1''-2''-3''-4''	r_c - 26
1'''-2'''-3'''-4'''	r_c - 50

Figure 13.17 T-s diagrams of the practical two-shaft gas turbine cycle for $T_3 = 1490\,K$ and different compressor pressure ratios r_c.

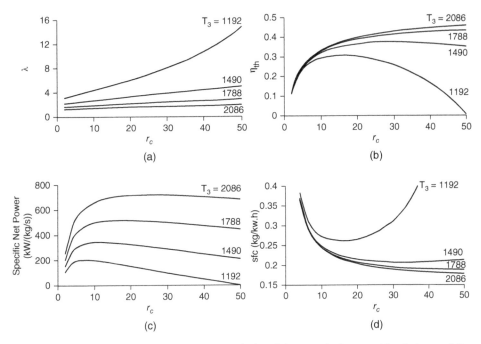

Figure 13.18 Calculated performance characteristics of the two-shaft gas turbine in terms of T_3.

indicating reduced heat rejection from the cycle with increased pressure ratio. The temperature drop in the PT appears to increase and then decrease, indicating that the power generated by the PT exhibits a peak value at around $r_c = 10$. The temperature drop in the CT increases continuously with the compressor pressure ratio, indicating increased levels of power required to run the compressor. These observations are confirmed by the plots of the characteristics shown in Figure 13.18. The specific net power curves display peaks at different compressor pressure ratios, with the peak shifting towards higher values as the maximum temperature increases. The curves for the thermal efficiency exhibit peaks only at lower values of T_3. The location of the peak of the thermal efficiency curve is determined by the combined effect of reduced heat rejection from the cycle with increasing pressure ratio and reduced specific net power at compressor pressure ratios above 10. The *sfc* curves exhibit minima at the pressure ratios corresponding to maximum thermal efficiencies. The curve at the lowest temperature ratio is U-shaped, with the *sfc* increasing sharply at very low and very high pressure ratios. At a constant temperature T_3, the power ratio curve (not shown here) drops as the compressor pressure ratio increases; and the higher T_3, the higher the power ratio at a given pressure ratio.

The performance maps in term of the *sfc* and the efficiency for the two-shaft gas turbine cycle are shown in Figures 13.19 and 13.20. The maps are very similar to the ones for the single-shaft gas turbine discussed earlier.

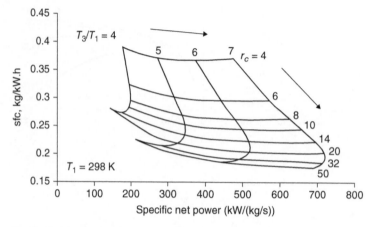

Figure 13.19 Performance map for the two-shaft gas turbine: specific fuel consumption vs. specific power at various T_3/T_1 and r_c ($T_1 = 298\ K$).

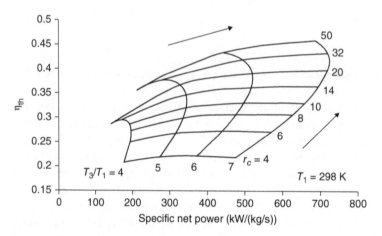

Figure 13.20 Performance map for the two-shaft gas turbine: efficiency vs. specific net power at various T_3/T_1 and r_c ($T_1 = 298\ K$).

Problems

13.1 A gas turbine engine for power production is working on the simple cycle at a maximum gas temperature of $800\,°C$ and pressure ratio of $6:1$, and the flow rate of the air is $20\ kg/s$. The polytropic efficiency is 0.87 for both the compression and expansion processes. The gas constant for both air and products is $0.287\ kJ/kg\ K$. The ambient conditions are $0.101\ MPa$ and $15\,°C$. If the combustion pressure loss is $13.8\ kPa$ and the ratios of specific heats for air and gases are 1.4 and 1.33, respectively, calculate
  (a) The overall thermal efficiency of the engine
  (b) The specific fuel consumption
      (Take the mass flow rate of the fuel as $0.283\ kg/s$ and the H_l as $43\,150\ kJ/kg$.)

13.2 The following data without pressure losses are provided for a single-shaft gas turbine power plant:

Maximum cycle temperature	1200 K
Compressor pressure ratio	5:1
Isentropic turbine efficiency	0.85
Isentropic compressor efficiency	0.80
Specific heat for both air and products	1100 J/kg K
Gas constant for both air and products	287 J/kg K
Compressor inlet temperature	289 K

Calculate:
(a) The compressor delivery temperature
(b) The turbine discharge temperature

13.3 A single-shaft gas turbine unit generating 400 MW of net mechanical power at 3000 rpm is operating under the following conditions:

Air mass flow rate	845 kg/s
Compressor pressure ratio	19.2
Isentropic efficiency of the compressor	0.87
Isentropic efficiency of the turbine	0.87
Specific heat at constant pressure for air	1.005 kJ/kg K
Specific heat for the products	1.148 kJ/kg K
Lower heating value of the fuel	43 150 kJ/kg
Turbine exhaust temperature	900 K
Ambient conditions	293 K, 0.1 MPa
Pressure losses in the combustion chamber	6% of compressor delivery pressure

Calculate:
(a) Power required to run the compressor
(b) Power generated by the turbine
(c) Turbine entry temperature
(d) Mass flow rate of the fuel
(e) Thermal efficiency of the engine
   Assume that the mass flow rate of air bleed taken from the compressor to cool the turbine blades is equal to the mass flow rate of the fuel.

13.4 For Problem 13.3, what is the overall air-fuel ratio in the combustor if the stoichiometric air-fuel ratio for the fuel is 14.714? Also, determine the pressure ratio in the turbine.

13.5 A two-shaft gas turbine with air bleed has a compressor with a pressure ratio of 19 : 1. The compressor inlet temperature is 300 K at 1 bar pressure. 6% of the air supplied by the compressor is used for blade cooling since the turbine entry temperature is 2000 K. The compressor and turbine isentropic efficiencies are 0.85 and 0.89, respectively. Assuming the ratio of specific heat $\gamma = 1.35$ and molecular weight 28 for the combustion products, calculate the temperature of the products at the outlet from the power turbine. Losses at the entry and exit can be ignored.

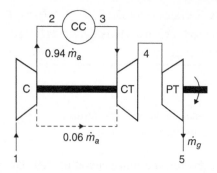

13.6 In a two-shaft gas turbine engine, the compressor turbine drives the compressor, and the power turbine is coupled with a generator. The pressure ratio in the compressor is 5 : 1, and the flow rate of the air is 50 kg/s. The ambient conditions are 0.101 MPa and 20 °C. Additionally, we know the following:

Isentropic efficiency of the compressor	0.85
Isentropic efficiency of the turbine	0.87
Maximum cycle temperature	850 °C
Pressure losses in the combustion chamber	20 kPa
Specific heat ratio for air	1.4
Specific heat ratio for combustion products	1.33
Gas constant for both air and products	0.287 kJ/kg K
Mechanical efficiency of the gas generator	0.98

Calculate:
(a) Output power
(b) Specific fuel consumption
(c) Thermal efficiency of the power plant
    (Take the fuel flow rate as 0.879 kg/s and H_l as 43 150 kJ/kg.)

13.7 A large air-cooled two-shaft gas turbine engine designed for mechanical drive applications is to run at a speed of 3000 rpm. The pressure ratio in the compressor is 17 : 1, and the flow rate of the exhaust gas is 296 kg/s. The ambient conditions are 0.1 MPa and 20 °C. Additionally, we know the following:

Isentropic efficiency of the compressor	0.85
Isentropic efficiency of the turbine	0.87
Maximum cycle temperature	$1300\,^\circ C$
Pressure losses in the combustion chamber	$20\,kPa$
Specific heat ratio for air	1.4
Specific heat ratio for the combustion products	1.333
The gas constant for both air and products	$0.287\,kJ/kg\,K$
Combustion efficiency	0.98
Lower heating value of the fuel	$43\,150\,kJ/kg$

Calculate:
(a) Power needed to run the compressor
(b) Output power
(c) Mass flow rate of the fuel
(d) The thermal efficiency of the engine
  The mass flow rate of air bleed taken from the compressor to cool the turbine blades is equal to the mass flow rate of the fuel.

13.8 Using the charts obtained from the design-point case study of the single-shaft industrial engine, estimate the design-point parameters, and justify your choice for
(a) Turbine entry temperature fixed by metallurgical considerations
(b) Turbine entry temperature dictated by a high-thermal-efficiency requirement
  If assumptions need to be made, state them clearly.

14

Work-Transfer System in Gas Turbines

For engines intended to drive mechanical systems in stationary and mobile applications or midrange and peak load electric power generation, the output is rotary mechanical power. For engines used in aviation, both civil and military, the output is propulsive energy, but rotary mechanical energy is produced internally to drive compressors and fans depending on the engine configuration (turbojet, two-spool, three-spool). *Work transfer* (the conversion of thermal energy to mechanical or propulsive energy) in gas turbines occurs without the need for a complicated and highly unbalanced system like the reciprocating mechanism in piston engines. The flow of combustion products through the turbine causes the rotation of the various shafts in the engine, producing the required torque for a driven machinery or thrust for propelling an aircraft. Work transfer takes place through the two main components of a gas turbine engine: the compressor and the turbine. The function of the compressor in a gas turbine engine is to provide high-temperature, high-pressure air to the combustion chamber to generate a high-energy gas flow into the turbine, where it is converted to mechanical energy. The compressor and turbine are broadly classified as axial or radial turbomachines.

14.1   Axial-Flow Compressors

Axial-flow compressors are widely used in applications where large volumes of air need to be compressed to a high level, such as in aircraft engines and large gas turbine power plants. They are rotary machines in which the flow of the fluid is in the direction of the axis of rotation of the machine through a series of stages, each comprising one row of rotating blades followed by one row of stationary blades.

Axial compressors absorb energy as the angular momentum in the moving blades changes, which results in a rise in pressure. Figures 14.1 shows two- and three-dimensional representations of rotor and stator blades and variations of pressure, absolute velocity, and enthalpy in the axial compressor stage. The compressor stage comprises one row of rotor blades RB mounted on a rotor turning at angular velocity ω and a row of stationary or stator blades SB mounted in the casing of the compressor (Figure 14.1a). The velocity diagram in Figure 14.2 shows that air enters the rotor-blade passage at angle α_1 with absolute velocity C_1 and leaves at angle α_2 with absolute velocity C_2. The work imparted

Fundamentals of Heat Engines: Reciprocating and Gas Turbine Internal Combustion Engines, First Edition. Jamil Ghojel.
© 2020 John Wiley & Sons Ltd. This Work is a co-publication between John Wiley & Sons Ltd and ASME Press.
Companion website: www.wiley.com/go/JamilGhojel_Fundamentals of Heat Engines

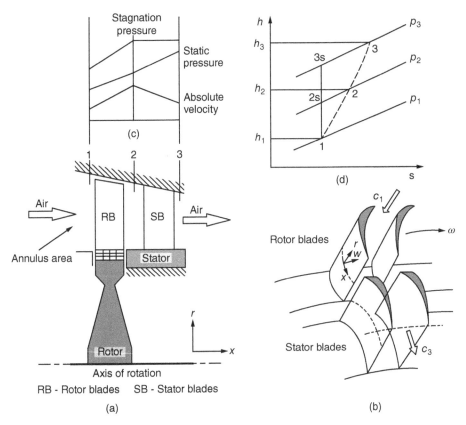

Figure 14.1 2-D and 3-D representation of rotor and stator blades and variation of pressure, absolute velocity, and enthalpy in an axial compressor stage.

to the fluid by rotor rotation causes the absolute flow velocity across the rotor blades to increase from C_1 to C_2 and the relative velocity to decrease from V_1 to V_2 as the air flowing in the diverging rotor-blade passage is diffused (decelerated), causing both the static and stagnation enthalpy (or temperature) and pressure to increase (Figures 14.1c,d). Air enters the stator-blade passage at angle α_2 with absolute velocity C_2 and leaves at angle α_3 with reduced absolute velocity C_3, causing further diffusion (deceleration) of the air and an increase in static enthalpy (temperature) and pressure. The stagnation enthalpy (temperature) and pressure across the stator remain constant because there is zero input work. Figure 14.1b shows a 3-D representation of the rotor and stator blades together with the Cartesian coordinates $x - r - w$ used throughout the book in the analysis of flows in both compressors and turbines, where x is the axial direction, r is radial, and w is rotational.

The blade ahead of the rotor blades shown in Figure 14.2 is part of an additional row of fixed blades known as *guide vanes,* mounted at the inlet into the compressor with the purpose of ensuring that air enters the rotor blades of the first stage at the correct angle. In a multistage compressor, the air leaving a stage enters the next similar stage at an absolute velocity equal to the entry velocity C_1, i.e. $C_3 = C_1$, and $\alpha_3 = \alpha_1$. The circled numbers in Figure 14.2 are the three station numbers shown in Figure 14.1a.

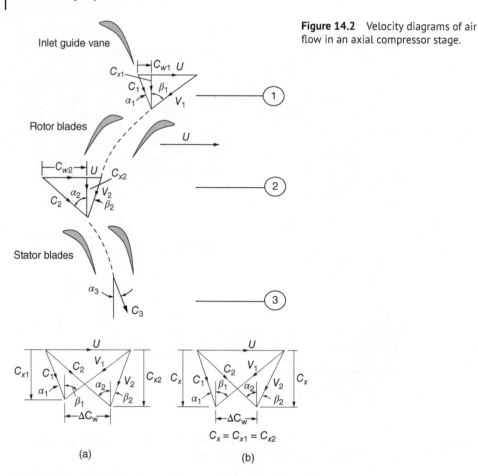

Figure 14.2 Velocity diagrams of air flow in an axial compressor stage.

Figure 14.3 Combined velocity diagrams of flow in and out of rotor blades: (a) varying axial velocity C_{x1} to C_{x2}; (b) constant axial velocity C_x.

In multistage compressors, to maintain flow continuity and a constant axial velocity along the axis of the machine, the height of the blades (and hence the annulus area) must gradually decrease as the density of air increases. This is shown schematically for the single stage in the 2-D representation in Figure 14.1a. The superimposed velocity diagrams for the flow of air at the inlet and outlet of the rotor blades for the case of varying axial velocity is shown in Figure 14.3a and for the case of constant axial-flow velocity in Figure 14.3b.

14.1.1 Input Power

The power absorbed in the compressor stage can be written in terms of the change in angular momentum of the air passing through the rotor blade passages:

$$\dot{W}_{cs} = \dot{m}_a U (C_{w2} - C_{w1}) \tag{14.1}$$

where $\dot{m}_a$ is the mass flow rate of air, U is the blade linear speed at the mean diameter, and C_{w1} and C_{w2} are the components of air velocity before and after the rotor in the w-axis direction (Figures 14.2 and 14.3). The velocities in the w-axis direction are known as *whirl velocities*.

From Figure 14.2,

$$C_{w2} - C_{w1} = C_2 \sin \alpha_2 - C_1 \sin \alpha_1 = \Delta C_w$$

$$C_{w2} - C_{w1} = V_1 \sin \beta_1 - V_2 \sin \beta_2 = \Delta V_w$$

Hence, the power absorbed in a axial compressor stage is

$$\dot{W}_{cs} = \dot{m}_a U \Delta C_w = \dot{m}_g U \Delta V_w \tag{14.2}$$

$\dot{W}_{cs}$ is in W when $\dot{m}_a$ is in kg/s and U is in m/s.

From the velocity diagrams in Figure 14.2 at stations 1 and 2,

$$C_{w1} = U - C_{x1} \tan \beta_1 = C_{x1} \tan \alpha_1$$

$$C_{w2} = U - C_{x2} \tan \beta_2 = C_{x2} \tan \alpha_2$$

Equation (14.1) can be written in terms of the rotor blade angles as

$$\dot{W}_{cs} = \dot{m}_a U (C_{x1} \tan \beta_1 - C_{x2} \tan \beta_2) \tag{14.3}$$

or, in terms of the air angles in the rotor, as

$$\dot{W}_{cs} = \dot{m}_a U (C_{x2} \tan \alpha_2 - C_{x1} \tan \alpha_1) \tag{14.4}$$

If we assume that the axial-flow velocity through the rotor is constant ($C_x = C_{x1} = C_{x2}$), as shown in Figure 14.3b, Eqs. (14.3) and (14.4) become

$$\dot{W}_{cs} = \dot{m}_a U C_x (\tan \beta_1 - \tan \beta_2) \tag{14.5}$$

$$\dot{W}_{cs} = \dot{m}_a U C_x (\tan \alpha_2 - \tan \alpha_1) \tag{14.6}$$

14.1.2 Degree of Reaction D

The *degree of reaction* in the compressor is defined as the ratio of the enthalpy increase in the rotor to the enthalpy increase in the stage. Referring to Figure 14.1d, the degree of reaction is

$$D = \frac{enthalpy\ increase\ in\ the\ rotor\ blades}{enthalpy\ increase\ in\ the\ stage} = \frac{h_2 - h_1}{h_3 - h_1} \tag{14.7}$$

The optimum value of D is found to be 50%, i.e. when the rise in pressure in the stage is equally divided between the rotor and stator. A symmetrical velocity diagram is obtained when the inlet and outlet absolute velocities are equal and the axial velocity is constant, as shown in Figure 14.4. From this figure, it can be established that $V_1 = C_2, V_2 = C_1, \beta_1 = \alpha_2, \beta_2 = \alpha_1$, and $\alpha_1 + \beta_1 = \beta_2 + \alpha_2$.

The steady energy equation for a flow of perfect gas applied to the stage in Figure 14.1 is

$$\dot{W}_{cs} = \dot{m}_a (h_3 - h_1) = \dot{m}_a c_p (T_3 - T_1)$$

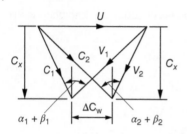

Figure 14.4 Symmetrical velocity diagram of an axial compressor stage with 50% reaction.

Combining this equation with Eq. (14.2), we obtain

$$U \Delta C_w = c_p(T_3 - T_1) \tag{14.8}$$

The isentropic efficiency of the stage from Figure 14.1d is

$$\eta_s = \frac{h_{3s} - h_1}{h_3 - h_1} = \frac{T_{3s} - T_1}{T_3 - T_1}$$

$$\frac{T_{3s}}{T_1} = \frac{T_1 + T_{3s} - T_1}{T_1} = 1 + \eta_s \frac{T_3 - T_1}{T_1} = 1 + \eta_s \frac{U \Delta C_w}{c_p T_1}$$

Also,

$$\frac{p_3}{p_1} = \left(\frac{T_{3s}}{T_1}\right)^{\gamma/(\gamma-1)}$$

Hence,

$$\frac{p_3}{p_1} = \left(1 + \eta_s \frac{U \Delta C_w}{c_p T_1}\right)^{\gamma/(\gamma-1)} \tag{14.9}$$

For a multistage axial compressor with inlet point 1 and exit point 2 and overall isentropic efficiency η_c, the overall pressure ratio is

$$\frac{p_2}{p_1} = \left(1 + \eta_c \frac{U \Delta C_w}{c_p T_1}\right)^{\gamma/(\gamma-1)} \tag{14.10}$$

Having calculated $r_c = p_2/p_1$ from Eq. (14.10), the temperature at the exit from the multistage compressor can be calculated from

$$T_2 = T_1 \left(\frac{p_2}{p_1}\right)^{(\gamma-1)/\gamma \eta_{pc}} \tag{14.11}$$

where η_{pc} is the compression polytropic efficiency, which can be determined from the following equation (see Eq. (10.6) in Chapter 10):

$$\eta_c = \frac{r_c^{(\gamma-1)/\gamma} - 1}{r_c^{(\gamma-1)/\gamma \eta_{pc}} - 1}$$

14.1.3 Compressor Performance Characteristics

Characteristics here mean the flowspeed-pressure ratio relationships for the compressor. The following equations can be written for the compressor delivery pressure and isentropic efficiency in terms of the compressor inlet conditions p_{1t} and T_{1t}, characteristic linear dimension D, shaft speed N, mass flow rate $\dot{m}_a$, molecular mass M_a, ratio of specific heats γ_a, and dynamic viscosity μ_a:

$$p_{2t} = f_1(p_{1t}, T_{1t}, D, N, \dot{m}_a, M_a, \gamma_a, \mu_a) \tag{14.12a}$$

$$\eta_c = f_2(p_{1t}, T_{1t}, D, N, \dot{m}_a, M_a, \gamma_a, \mu_a) \tag{14.12b}$$

which can be transformed to

$$\frac{p_{2t}}{p_{1t}} = f_1\left(\frac{ND}{\sqrt{T_{1t}R}}, \frac{\dot{m}_a\sqrt{T_{1t}}}{p_{1t}D^2}, \frac{p_{1t}D}{\mu_a\sqrt{T_{1t}}}, M_a, \gamma_a\right) \tag{14.13a}$$

$$\eta_c = f_2\left(\frac{ND}{\sqrt{T_{1t}R}}, \frac{\dot{m}_a\sqrt{T_{1t}}}{p_{1t}D^2}, \frac{p_{1t}D}{\mu_a\sqrt{T_{1t}}}, M_a, \gamma_a\right) \tag{14.13b}$$

For a particular compressor working with air, M_a, γ_a, R, and D can be taken as constants, which simplifies Eqs. (14.13a and 14.13b) to

$$\frac{p_{2t}}{p_{1t}} = f_1\left(\frac{N}{\sqrt{T_{1t}}}, \frac{\dot{m}_a\sqrt{T_{1t}}}{p_{1t}}, \frac{p_{1t}}{\mu_a\sqrt{T_{1t}}}\right) \tag{14.14a}$$

$$\eta_c = f_2\left(\frac{N}{\sqrt{T_{1t}}}, \frac{\dot{m}_a\sqrt{T_{1t}}}{p_{1t}}, \frac{p_{1t}}{\mu_a\sqrt{T_{1t}}}\right) \tag{14.14b}$$

The term $p_{1t}/\mu_a\sqrt{T_{1t}}$ is usually of minor importance and can be ignored, yielding the final equations for the pressure ratio and isentropic efficiency as functions of the loosely termed *non-dimensional speed* and *non-dimensional mass flow* parameters:

$$\frac{p_{2t}}{p_{1t}} = f_1\left(\frac{N}{\sqrt{T_{1t}}}, \frac{\dot{m}_a\sqrt{T_{1t}}}{p_{1t}}\right) \tag{14.15a}$$

$$\eta_c = f_2\left(\frac{N}{\sqrt{T_{1t}}}, \frac{\dot{m}_a\sqrt{T_{1t}}}{p_{1t}}\right) \tag{14.15b}$$

It is common practice to present the compressor performance characteristic as a single function by combining the two equations in (14.15a) and (14.15b) into

$$\frac{p_{2t}}{p_{1t}} = f_1\left(\frac{N}{\sqrt{T_{1t}}}, \frac{\dot{m}_a\sqrt{T_{1t}}}{p_{1t}}, \eta_c\right) \tag{14.15c}$$

A single plot of function (14.15c) features the compressor pressure ratio versus the mass flow parameter with constant lines of the compressor speeds and compressor isentropic efficiencies, as shown in Figure 14.5. The lines of constant efficiency increase in value towards the centre of the concentric ovals.

Another way of representing the compressor characteristics is to use Eqs. (14.15a) and (14.15b) to plot the compressor pressure ratio and efficiency versus the mass flow parameter at constant compressor speeds as two separate characteristics, as shown in Figure 14.6 for example. Actual compressor characteristics differ widely from one machine to another and can only be accurately determined for a specific compressor by laboratory testing. Quite often, the mass flow parameters and compressor speeds are presented as fractions or percentages of the design values instead of absolute values. The surge line, representing the onset of unstable compressor operation caused by the presence of positive pressure gradient

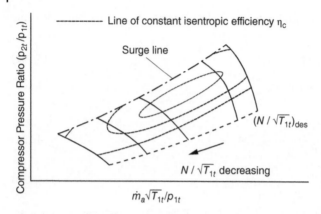

Figure 14.5 Compressor performance map on a single pair of coordinates.

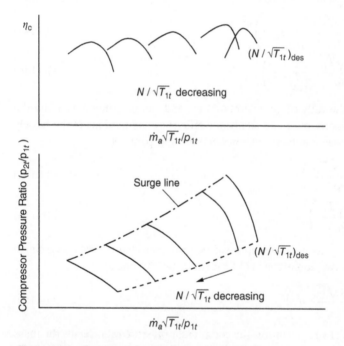

Figure 14.6 Typical axial-compressor characteristics.

in the direction of flow, is the line connecting the lower limits of the mass flow rates for the constant-speed lines. The choking line represents the maximum (and constant) mass flow rates that are achievable at all speeds as the local sonic velocity is reached.

14.2 Radial-Flow Compressors

Radial compressors, also known as *centrifugal* compressors, are used in turbochargers and superchargers and in small gas turbine plants in the process industry and for power generation. The main advantages of radial compressors are smooth operation and high

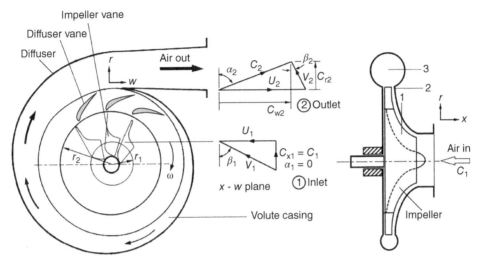

Figure 14.7 Radial-flow compressor with a single-sided impeller and diverging diffuser blades (velocity diagram with subscript 1 for inlet and subscript 2 for outlet).

Figure 14.8 Radial compressor diffuser with tangential diverging passages.

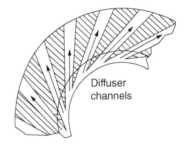

reliability, and they are particularly suited for high-pressure, low-flowrate applications. A simplified diagram of a radial-flow compressor with a single-sided impeller is shown in Figure 14.7. The main components of the compressor are the impeller with diverging vanes (blades), diffuser, and volute casing. The inlet diffuser occupies the annular space surrounding the impeller and can be of the form of a row of fixed nozzle vanes mounted tangent to the impeller rim (Figure 14.7) or in the form of diverging passages in the volute body (Figure 14.8). The high-speed impeller motion increases the angular momentum of the air entering the compressor axially (along the x-axis); the air gains flow speed as the direction changes to radial (along the r-axis). Subsequent diffusion of the flow in the diverging impeller vanes causes the flow to decelerate and the static pressure to increase. Further increase in static pressure occurs as the flow decelerates in the diffuser space. Radial compressors are usually designed in such a way that half the rise in pressure takes place within the impeller vanes and half in the diffuser.

Velocity diagram 1 at entry into the compressor is drawn in the $x - w$ plane at radius r_1, where air enters the compressor along the x-axis at absolute velocity C_1, equal to the axial velocity C_{x1}, and then is forced by the high-speed impeller vanes into the vane passages at high relative velocity V_1. The flow changes direction, and the kinetic energy is converted to pressure as the flow decelerates in the diverging impeller vanes. Velocity diagram 2 drawn in the $w - r$ plane shows that the air leaves the impeller vanes at absolute velocity C_2, which is higher than the entry velocity C_1. C_2 is reduced in the diffuser, resulting in a further increase in static pressure.

According to the Euler equation, the power input when the absolute inlet velocity is in the axial direction (see velocity diagram 1 in Figure 14.7) is

$$\dot{W}_c = \dot{m}_a U_2 C_{w2} \tag{14.16}$$

In an ideal case, the air would leave the impeller at a radial-flow velocity so that $V_2 = C_{r2}$ and the whirl velocity is equal to the impeller blade tip velocity ($C_{w2} = U_2$). The deviation from the ideal case is explained by the slip effect, which can be accounted for by the slip factor σ defined for radial blades as $\sigma = C_{w2}/U_2$. The slip factor changes between 0.83 and 0.95; the higher value is associated with a large number of blades Z, as shown by Stanitz slip-factor equation (Stanitz, 1952):

$$\sigma = 1 - \frac{0.63\pi}{Z}$$

From Eq. (14.16) and the definition of the slip factor, the power input required by the compressor becomes

$$\dot{W}_c = \dot{m}_a \sigma U_2^2 \tag{14.17}$$

The energy equation between points 1 and 2 can be written as

$$\dot{W}_c = \dot{m}_a (h_2 - h_1) = \dot{m}_a c_p (T_{2t} - T_1)$$

The isentropic efficiency of the compressor is

$$\eta_c = \frac{(T_{2s} - T_{1t})}{(T_{2t} - T_1)}$$

From the previous two equations, we obtain

$$\dot{W}_c = \dot{m}_a \frac{c_p}{\eta_c}(T_{2t} - T_{1t}) \tag{14.18}$$

The pressure ratio across the impeller is

$$\frac{p_{2t}}{p_{1t}} = \left(\frac{T_{2s}}{T_{1t}}\right)^{\gamma/(\gamma-1)} = \left(1 + \frac{T_{2s} - T_{1t}}{T_{1t}}\right)^{\gamma/(\gamma-1)}$$

Combining last equation and the efficiency equation with equations (14.17) and (14.18) yields

$$\frac{p_{2t}}{p_{1t}} = \left(1 + \frac{\eta_c \sigma U_2^2}{c_p T_{1t}}\right)^{\gamma/(\gamma-1)} \tag{14.19}$$

14.2.1 Radial-Flow Compressor Characteristics

These characteristics, schematically drawn in Figure 14.9, resemble the characteristics in the lower part of the axial-flow compressor characteristics shown in Figure 14.6, with some important differences: (i) pressure ratio curves at constant speed lines are not as steep as in axial compressors, and they all exhibit maxima; (ii) all characteristics for constant speed

Figure 14.9 Schematic of radial compressor characteristics.

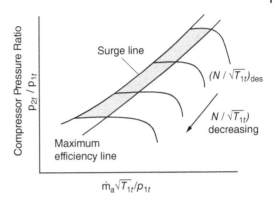

values cover a much wider range of mass flow than in axial compressors; and (iii) the radial compressor is more stable: the maximum efficiency line is located a good distance to the right of the surge line, as shown by the grey area in Figure 14.9.

14.3 Axial-Flow Turbines

The function of a turbine is to convert the high-temperature, high-pressure products of combustion into mechanical and/or propulsive power, depending on the intended application. In a single-shaft engine, a single turbine drives the compressor and provides external output power. In a two-shaft engine, the compressor turbine drives the compressor and the power turbine provides external output power. In a turbojet engine, a single turbine powers the compressor and a nozzle generates the thrust. In more complicated aircraft engines, two or three turbines are used to run different stages of the compression process in the engine.

   Gas turbines are predominantly of the axial-flow type. In these turbines, power extraction from the gases at high temperature and pressure is effected as the gases flow between aerodynamically shaped rows of blades mounted circumferentially on a stationary disk and rows of blades mounted on a rotating disk (Figure 14.10). The stationary blades are called *stator blades* or *nozzle blades*, and the moving blades are called *rotor blades*. The direction of flow changes within the blades, as shown by the thick arrows, causing the rotor to turn at the required speed. The combination of a stator and a rotor is known as a *turbine stage*. The pressure drops continuously across the stage as the fluid accelerates in the nozzle blades and then expands in the rotor blades. As the pressure drops, the enthalpy of the fluid also drop, as can be seen in the h-s diagram. The enthalpy drop is divided between the stator and rotor blades, and the proportions of this division are given by the degree of reaction D, which is defined as

$$D = \frac{enthalpy\ drop\ in\ the\ rotor\ blades}{enthalpy\ drop\ in\ the\ stage} = \frac{h_2 - h_3}{h_1 - h_3} \tag{14.20}$$

14.3.1 Velocity Diagrams

Figure 14.10 shows the turbine stage in 2-D and 3-D with a Cartesian coordinate system comprising the axial, radial, and tangential coordinates (x, r, w). For the development of

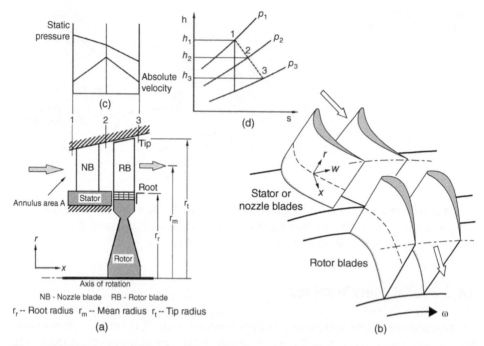

Figure 14.10 Schematic diagrams of an axial-flow reaction turbine stage.

the velocity diagram (triangle), consider the 2-D diagram of the turbine stage as shown in Figure 14.10a. The fluid enters the stator blades, which are arranged like a nozzle, at station 1 and then expands in the turbine blades from station 2 to station 3. To maintain flow continuity with the decrease in density as the gases expand and the pressure drops, the annulus area A increases gradually in the x direction.

Figure 14.11 is an illustration of the construction of velocity diagrams of the axial-flow reaction stage shown in Figure 14.10. The labelling in the velocity diagrams, taking into account the coordinate system used and station numbering, is as follows:

U: Tangential velocity of the rotor at the mean radius
C_1: Fluid absolute velocity at the nozzle blade inlet (station 1)
α_1: Stator blade entry angle
C_{x1}: Axial fluid velocity at the nozzle blade inlet
C_2: Fluid absolute velocity at the nozzle blade outlet (station 2)
α_2: Stator blade outlet angle
V_2: Velocity of fluid relative to the rotor blade at the inlet
β_2: Rotor blade inlet angle
C_{w2}: Whirl (tangential) velocity at the nozzle blade outlet
C_3: Fluid velocity at the rotor blade outlet (station 3)
α_3: Fluid absolute velocity angle at the rotor blade outlet
V_3: Velocity of fluid relative to the rotor blade at outlet
β_3: Rotor blade outlet angle
C_{w2}: Whirl velocity at the rotor blade outlet

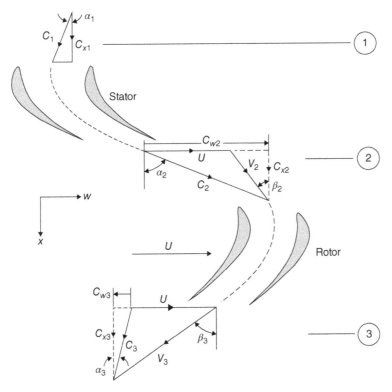

Figure 14.11 Velocity diagrams of an axial reaction turbine stage.

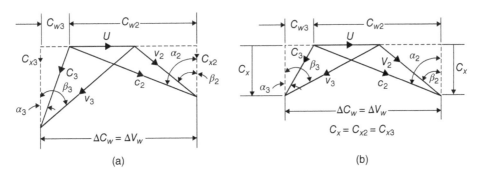

Figure 14.12 Superimposed velocity diagrams of the flow through the rotor: (a) variable axial-flow velocity; (b) axial-flow velocity constant.

A more convenient way to draw the velocity diagrams is by superimposing the velocity diagrams at the inlet and outlet of the rotor blades, as seen in Figure 14.12.

In a stand-alone single-stage turbine, $\alpha_1 = 0$ and $C_1 = C_{x1}$. In a multistage axial turbine with identical blades for all stages, $\alpha_1 = \alpha_3$ and $C_1 = C_3$; i.e. the fluid entry velocity into successive stages remains constant and equal to C_1. From Figure 14.12, it can be seen that there are two distinct changes in fluid velocity as a result of the flow through the stage:

- Change in magnitude of the whirl-velocity component $C_{w2} - (-C_{w3}) = (C_{w2} + C_{w3})$.
- Change in magnitude of the axial-velocity component $(C_{x2} - C_{x3})$.

14.3.2 Stage Output Power

From the Euler turbomachinery equation for flow at the mean radius r_m, the stage specific-work output (work per unit mass flow) in the turbine can be written as

$$w_{ts} = \omega r_m (C_{w2} + C_{w3}) \ N, kN \ or \ MN$$

And the stage output power for the gas flow rate of $\dot{m}_g$ is

$$\dot{W}_{ts} = \dot{m}_g \omega r_m (C_{w2} + C_{w3}) \tag{14.21}$$

From Figure 14.12,

$$C_{w2} + C_{w3} = C_2 \sin \alpha_2 + C_3 \sin \alpha_3 = \Delta C_w$$

$$C_{w2} + C_{w3} = V_2 \sin \beta_2 + V_3 \sin \beta_3 = \Delta V_w$$

Hence,

$$\dot{W}_{ts} = \dot{m}_g U \Delta C_w = \dot{m}_g U \Delta V_w \tag{14.22}$$

$\dot{W}_s$ is in W when $\dot{m}_g$ is in kg/s, and ΔV_w is in m/s.

Also from Figure 14.12,

$$C_{w2} = U + C_{x2} \tan \beta_2$$

$$C_{w3} = C_{x3} \tan \beta_3 - U$$

Equation (14.22) can now be rewritten as

$$\dot{W}_{ts} = \dot{m}_g U (C_{x2} \tan \beta_2 + C_{x3} \tan \beta_3) \tag{14.23}$$

Gas turbine engines are often designed with constant axial-flow velocity; $C_x = C_{x2} = C_{x3}$. The velocity diagrams for this special case are shown in Figure 14.12b, from which it can be shown that

$$\tan \beta_2 - \tan \alpha_3 = \tan \alpha_2 - \tan \beta_3$$

or

$$\tan \beta_2 + \tan \beta_3 = \tan \alpha_2 + \tan \alpha_3$$

As a result, Eq. (14.23) becomes

$$\dot{W}_{ts} = \dot{m}_g U C_x (\tan \beta_2 + \tan \beta_3) = \dot{m}_g U C_x (\tan \alpha_2 + \tan \alpha_3) \tag{14.24}$$

For a single-stage turbine engine with fixed blade geometries and axial-flow velocity, the power output from the turbine will be directly proportional to the mass flow rate of the gases and blade speed. If the engine is multistage, the power output will be a function of number of stages in addition to the abovementioned parameters. However, it should be noted that changing the parameters affecting turbine engine power output may have negative consequences:

- For a given turbine size, increasing the mass flow rate leads to increased flow velocities in the blades, causing an increase in frictional losses.
- Increased flow velocities will increase the fraction of kinetic energy lost in the fluid leaving rotor blades.

- Maximum blade speed is limited by the allowable centrifugal stresses in rotating parts in the engine.
- Increasing the number of stages will cause an increase in engine size and cost. The engine design objective should be to use the smallest number of stages that maximises the utilisation of the available enthalpy drop across the engine with the least frictional losses and minimum stresses on the rotating parts.

If N is shaft speed in revolutions per minute (rpm), the blade speed at mean blade radius r_m is

$$U = r_m \omega = \frac{\pi N r_m}{30}$$

And the power can be written as

$$\dot{W}_{ts} = \frac{\pi}{30} N r_m \dot{m}_g C_x (\tan \beta_2 + \tan \beta_3) \tag{14.25}$$

The torque generated by the rotor blades using Eq. (14.24) is

$$T = \frac{\dot{W}_{ts}}{\omega} = \dot{m}_g r_m C_x (\tan \beta_2 + \tan \beta_3) \tag{14.26}$$

The power generated by a single-shaft gas turbine based on the energy equation given by Eq. (13.2) in Chapter 13 is

$$\dot{W}_t = \dot{m}_g c_{pg} \Delta T_{34}$$

If we assume that the turbine is single-stage, then $\Delta T_{34} = \Delta T_{13} = \Delta T_s$, where ΔT_s is the stagnation temperature drop in the stage.

Equating the previous equation with Eq. (14.24), we obtain

$c_{pg} \Delta T_s = (h_1 - h_3) = U C_x (\tan \beta_2 + \tan \beta_3)$, from which

$$\Delta T_s = \frac{U C_x}{c_{pg}} (\tan \beta_2 + \tan \beta_3) \tag{14.27}$$

Equation (14.27) shows that the stagnation temperature drop in the turbine stage is dependent on a combination of the blade speed, gas thermodynamic properties, axial-flow velocity, and rotor blade geometry.

14.3.3 Multistage Turbine Output Power

A multistage turbine comprises a series of stages with identical blades and constant inlet fluid velocity into each stage. Figure 14.13 shows a schematic diagram of two-stage reaction turbine, stator and rotor blade arrangements, and variation of absolute velocity and static pressure across the turbine.

Consider the general steady-state energy equation:

$$q - w = \Delta(h + KE)$$

If it is assumed that the process in each stages is adiabatic and the steady-state energy equation is applied between the inlet at point 1 and outlet at point 2 of each stage relative to the casing (Figure 14.13), the total power for the multistage compressor with n stages can be written, according to Rogers and Mayhew (1992), as the sum of all stage powers from the first stage $\dot{W}_{ts(1)}$ to the nth stage $\dot{W}_{ts(n)}$:

$$\dot{W}_{ts(1)} = \dot{W}_{1-2} = \dot{m}_g \left[(h_1 - h_2) + \frac{1}{2}(C_1^2 - C_2^2) \right]$$

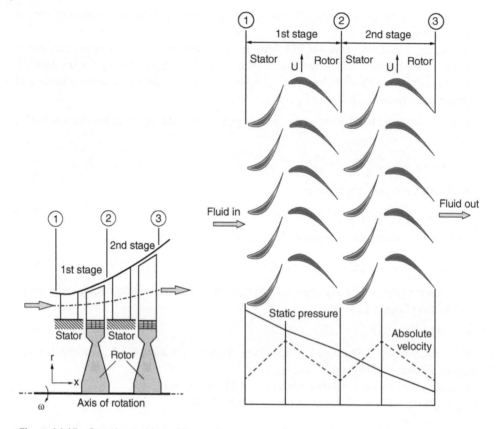

Figure 14.13 Reaction turbine with two stages.

$$\dot{W}_{ts(2)} = \dot{W}_{2-3} = \dot{m}_g \left[(h_2 - h_3) + \frac{1}{2}(C_2^2 - C_3^2) \right]$$

$$\vdots$$

$$\dot{W}_{ts(n)} = \dot{W}_{n-(n+1)} = \dot{m}_g \left[(h_n - h_{n+1}) + \frac{1}{2}(C_n^2 - C_{n+1}^2) \right]$$

Summing up, we obtain the power output for an entire multistage turbine with n stages:

$$\dot{W}_t = \sum_1^n \dot{W}_{ts} = \dot{m}_g \left[(h_1 - h_{2n+1}) + \frac{1}{2}(C_1^2 - C_{n+1}^2) \right] \tag{14.28}$$

The power output of a multistage turbine can also be written, according to Korpela (2011), as

$$\dot{W}_t = \dot{m}_g c_{pg} T_{1t}[1 - (1 - \eta_{ts}x)^n] \tag{14.29}$$

where the turbine stage efficiency η_{ts} is assumed to be the same for all stages and x for a single stage is given by

$$x = \left[1 - \left(\frac{p_{2t}}{p_{1t}} \right)^{(\gamma_g-1)/\gamma_g} \right]$$

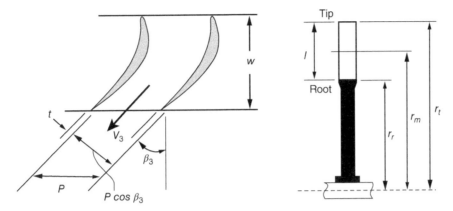

Figure 14.14 Fluid flow passage in moving blades at exit w – blade width, p – blade pitch, t – blade edge thickness, l – blade height, r_r – blade root radius, r_m – blade mean radius, r_t – blade tip radius.

14.3.4 Blade Profile

Fluid flow characteristics in a turbine stage are dependent on the entry conditions in the flow channels between blades as well as the shape and height of the blades. The geometrical characteristics of both the nozzle and moving blades have a significant influence on the losses in a turbine stage. Additionally, the shape of the blades is important with respect to the ease of their design, manufacture, and attachment to the turbine wheels. A certain blade thickness needs to be maintained at the edges of blades in order to maintain structural rigidity and strength in order to resist centrifugal and dynamic stresses and vibration and withstand machining operations during manufacturing. This edge thickness will tend to reduce the flow cross-section and disturb the flow. Figure 14.14 shows a schematic diagram of the rotor blades of a reaction turbine stage. Since the blade pitch increases with blade height from the root to the tip, the effective pitch (p_m) is taken at the mean radius r_m of the blade, and the effective area at the exit from the blades is

$$A = \frac{2\pi r_m}{p_m}(p_m \cos \beta_3 - t)l \tag{14.30}$$

The volume rate of the fluid flowing through the blades is the product of the relative velocity of flow V_3 and the total effective area of the passages perpendicular to the velocity

$$\frac{\dot{m}}{\rho} = AV_3 = \frac{2\pi r_m}{p_m}(p_m \cos \beta_3 - t)lV_3 \tag{14.31}$$

where

ρ : local fluid density
V_3: relative velocity of fluid at the outlet from moving blades

In gas turbine blades, frictional losses are dominant, compared with steam turbines, in which leakage losses are dominant. In multistage reaction turbines, fluid friction losses in the blades cause the relative pressure and exit velocity from each stage V_3 to continuously decrease. To maintain continuity of flow, the volume flow rate $\dot{m}/\rho$ must be constant; i.e. the

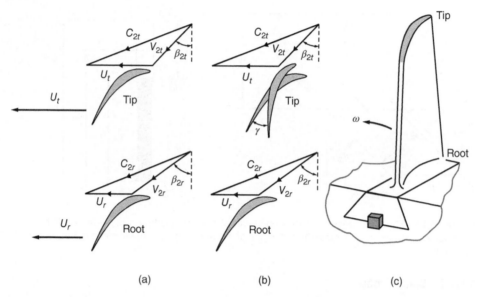

Figure 14.15 Blade profiles: (a) velocity diagrams at the root and tip of a uniform blade; (b) velocity diagrams for a twisted blade at the tip; (c) fully twisted blade designed to accommodate a decreasing angle of incidence β_2 with blade height.

total annulus area A must be increased, and hence the blade height must be continuously increased along the rotor axis. As a result, the conditions of flow will vary significantly from the root to the tip due to the fact that the pressure of the fluid must be adjusted to satisfy the requirement of radial equilibrium. According to vortex theory, for constant axial velocity across the rotor blades, the following condition must be observed to maintain radial equilibrium (Bayley, 1958):

$$C_w r^\eta = const \tag{14.32}$$

where C_w is the whirl, or tangential velocity, of the fluid and η is the isentropic efficiency of the flow in the blade. This means there could be a significant decrease in whirl velocity (and hence a decrease in rotor blade angle of incidence β_{2t} at the tip of the blade) and increased separation losses with increased blade height (Figure 14.15a). To maintain radial equilibrium of the free-vortex flow and prevent separation of flow at the tip, the leading-edge angle of the blade can be increased with the blade height by twisting the blade by angle γ, as shown schematically in Figure 14.15b and 14.15c.

14.3.5 Degree of Reaction

The definition of the degree of reaction with reference to Figure 14.10 and station numbering between 1 and 3 was presented earlier in the form of the following equation:

$$D = \frac{h_2 - h_3}{h_1 - h_3} \tag{14.33}$$

$D = 0\%$ for the pure impulse stage and $D = 100\%$ for the pure reaction stage. Most reaction turbines are designed with 50% degree of reaction in which the pressure ratios across the stator and rotor blades are equally divided.

14.3.5.1 Degree of Reaction in Terms of Fluid Velocities

Applying the steady-flow energy equation to the flow in the rotor relative to the casing, assuming an adiabatic process, yields the following equation for specific work:

$$w = (h_2 - h_3) + \frac{1}{2}(C_2^2 - C_3^2) \tag{14.34}$$

If we apply the same equation to the flow in the rotor blades relative to the rotor ($w = 0$), we obtain

$$h_2 - h_3 = \frac{1}{2}(V_3^2 - V_2^2) \tag{14.35}$$

Eliminating the enthalpy terms from Eqs. (14.34) and (14.35) gives

$$w = \frac{1}{2}(V_3^2 - V_2^2) + \frac{1}{2}(C_2^2 - C_3^2) \tag{14.36}$$

where w is the specific work (work per mass fluid or power per fluid mass flow rate in J/kg).

If we apply the energy equation without heat transfer to the whole stage, we obtain

$$-w = (h_3 - h_1) + \frac{1}{2}(C_3^2 - C_1^2)$$

or

$$(h_1 - h_3) = w + \frac{1}{2}(C_3^2 - C_1^2) \tag{14.37}$$

Equation (14.34) can be rewritten as

$$(h_2 - h_3) = w + \frac{1}{2}(C_3^2 - C_2^2) \tag{14.38}$$

Substituting Eqs. (14.37) and (14.38) into Eq. (14.33) results in

$$D = \frac{w + \frac{1}{2}(C_3^2 - C_2^2)}{w + \frac{1}{2}(C_3^2 - C_1^2)} \tag{14.39}$$

Substituting Eq. (14.36) into Eq. (14.39) results in

$$D = \frac{(V_3^2 - V_2^2)}{(C_2^2 - C_1^2) + (V_3^2 - V_2^2)} \tag{14.40}$$

For a multistage turbine ($C_1 = C_3$, $C_{x2} = C_{x3} = C_x$), and if $D = 0.5$, Eq. (14.40) yields

$$C_2^2 - C_3^2 = V_3^2 - V_2^2 \tag{14.41}$$

Equality (14.41) can be true if $C_2 = V_3$, $C_3 = V_2$. As a result, the velocity diagrams in Figure 14.12 are transformed to the symmetrical velocity diagrams in Figure 14.16.

14.3.5.2 Degree of Reaction in Terms of Blade Characteristics

Consider a 50% reaction turbine and the velocity diagrams in Figure 14.16. The steady-state flow energy equation for the entire stage per unit mass flow rate is

$$w = h_1 - h_3 \tag{14.42}$$

Rewriting Eq. (14.24) in terms of specific work and combining with Eq. (14.42) results in

$$w = \frac{W_{ts}}{\dot{m}_g} = UC_x(\tan \beta_2 + \tan \beta_3)$$

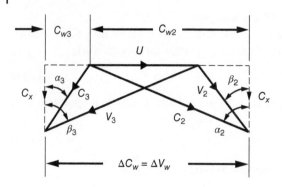

or

$$h_1 - h_3 = UC_x(\tan \beta_2 + \tan \beta_3) \tag{14.43}$$

From Eq. (14.35) and the velocity diagram in Figure 14.16,

$$(h_2 - h_3) = \frac{C_x^2}{2}\left(\frac{1}{\cos^2 \beta_3} - \frac{1}{\cos^2 \beta_2}\right) \tag{14.44}$$

Substituting Eqs. (14.43) and (14.44) into Eq. (14.33) yields

$$D = \frac{h_2 - h_3}{h_1 - h_3} = \frac{\dfrac{C_x^2}{2}\left(\dfrac{1}{\cos^2 \beta_3} - \dfrac{1}{\cos^2 \beta_2}\right)}{UC_x(\tan \beta_2 + \tan \beta_3)}$$

$$D = \frac{C_x\left(\dfrac{1}{\cos^2 \beta_3} - \dfrac{1}{\cos^2 \beta_2}\right)}{2U(\tan \beta_2 + \tan \beta_3)} \tag{14.45}$$

Since

$$\frac{1}{\cos \beta_3} = \pm\sqrt{\frac{\sin^2 \beta_3 + \cos^2 \beta_3}{\cos^2 \beta_3}} = \pm\sqrt{1 + \tan^2 \beta_3}$$

and

$$\frac{1}{\cos \beta_2} = \pm\sqrt{1 + \tan^2 \beta_2},$$

$$D = \frac{C_x}{2U}\frac{(\tan^2 \beta_3 - \tan^2 \beta_2)}{(\tan \beta_2 + \tan \beta_3)} = \frac{C_x}{2U}\frac{(\tan \beta_3 + \tan \beta_2)(\tan \beta_3 - \tan \beta_2)}{(\tan \beta_2 + \tan \beta_3)}$$

Finally,

$$D = \frac{C_x}{2U}(\tan \beta_3 - \tan \beta_2) \tag{14.46}$$

From Figure 14.16, it is apparent that $\alpha_2 = \beta_3$, $\alpha_3 = \beta_2$ as well as $C_2 = V_3$, $C_3 = V_2$, as stated previously.

14.3.6 Utilisation Factor (Diagram Efficiency)

The utilisation factor or diagram efficiency E is defined as

$$E = \frac{ideal\ stage\ work}{energy\ supplied} = \frac{w}{w + \dfrac{C_3^2}{2}} \tag{14.47}$$

Substituting for w from Eq. (14.36) in Eq. (14.47) yields

$$E = \frac{(V_3^2 - V_2^2) + (C_2^2 - C_3^2)}{C_2^2 + (V_3^2 - V_2^2)} \tag{14.48}$$

Equation (14.47) indicates that, for a given stage work w, the utilisation factor is at a maximum when the exit velocity from stage C_3 is at a minimum. The symmetrical velocity diagram for a 50% reaction shown in Figure 14.16 indicates that C_3 is at a minimum when it is axial in direction; i.e., $C_3 = C_x = V_2$, $\beta_3 = \alpha_2$, $\beta_2 = \alpha_3 = 0$. The velocity diagrams for these conditions are shown in Figure 14.17.

The utilisation factor can also be expressed in terms of blade characteristics. The numerator in Eq. (14.48) is written as

$$(V_3^2 - V_2^2) + (C_2^2 - C_3^2) = V_3^2 - V_2^2 + V_3^2 - V_2^2 = 2(V_3^2 - V_2^2)$$

Referring to Figure 14.16, the last term on the right-hand side can be written as

$$2(V_3^2 - V_2^2) = 2[C_2^2 - (U^2 + C_2^2 - 2UC_2 \sin \alpha_2)] = 2U(2C_2 \sin \alpha_2 - U)$$

The denominator can be written as

$$C_2^2 + (V_3^2 - V_2^2) = C_2^2 + (C_2^2 - V_2^2) = 2C_2^2 - V_2^2$$

where

$$2C_2^2 - V_2^2 = 2C_2^2 - (U^2 + C_2^2 - 2UC_2 \sin \alpha_2) = C_2^2 - U^2 + 2UC_2 \sin \alpha_2$$

Hence,

$$E = \frac{2U(2C_2 \sin \alpha_2 - U)}{C_2^2 - U^2 + 2UC_2 \sin \alpha_2}$$

Figure 14.17 Velocity diagram for maximum utilisation.

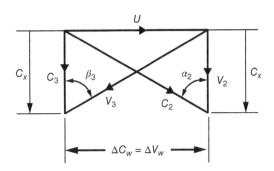

Finally,

$$E = \frac{\frac{2U}{C_2}\left(2\sin\alpha_2 - \frac{U}{C_2}\right)}{1 - \left(\frac{U}{C_2}\right)^2 + 2\left(\frac{U}{C_2}\right)\sin\alpha_2} \tag{14.49}$$

Differentiating Eq. (14.49) with respect to the ratio U/C_2 and equating to zero yields the optimum value of the blade velocity ratio for maximum E, i.e.,

$$\left(\frac{U}{C_2}\right)_{E_{max}} = \sin\alpha_2 \tag{14.50}$$

$$E_{max} = \frac{2\sin^2\alpha_2}{1 + \sin^2\alpha_2} \tag{14.51}$$

Equation (14.50) confirms the conclusion made earlier about the shape of the velocity diagram shown in Figure 14.17. The variation of E with U/C_2 is shown in Figure 14.18. The maximum value E_{max} is reached at $U/C_2 = \sin\alpha_2$. E is zero when $U/C_2 = 0$ and $U/C_2 = 2\sin\alpha_2$. Increasing the fluid inlet angle α_2 causes the maximum utilisation factor to increase with E, as shown in Figure 14.19, which depicts a series of curves $E = f(U/C_2)$

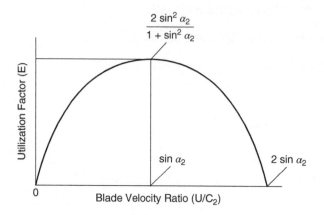

Figure 14.18 Effect of the blade velocity ratio on utilisation factor for a given α_2.

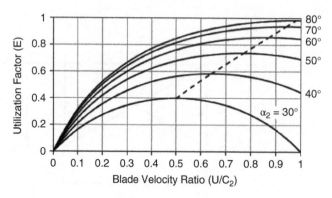

Figure 14.19 Effect of the blade velocity ratio on utilisation factor for various values of α_2.

for different values of α_2 ranging from 30° to 80°. As α_2 increases, E_{max} increases and the value of (U/C_2) at which it occurs shifts closer to 1 (see the dashed line connecting the maxima). The reduced gap between consecutive curves as α_2 increases is an indication of a reduced rate of increase of E_{max}.

Since maximum utilisation occurs when $\beta_2 = 0$ (Figure 14.17), Eq. (14.24) for the stage output power for a 50% reaction turbine can be rewritten as

$$\dot{W}_{ts} = \dot{m}_g U C_x (\tan \beta_2 + \tan \beta_3) = \dot{m}_g U C_x \tan \beta_3 \qquad (14.52)$$

From the velocity diagram in Figure 14.17,

$$\frac{U}{C_3} = \frac{U}{V_2} = \frac{U}{C_x} = \tan \beta_3$$

The equation for the power output in a 50% reaction turbine stage is then

$$\dot{W}_{ts} = \dot{m}_g U^2 \qquad (14.53)$$

14.3.7 Axial Turbine Coefficients

Four dimensional parameters can be identified as useful tools in designing gas turbines. These are the degree of reaction D, utilisation factor E, stage loading coefficient ψ, and flow coefficient ϕ. We have already defined D and E and analysed their significance in detail, and we will now discuss the other two coefficients.

The stage loading coefficient ψ is a measure of the work done in the stage and is defined as

$$\psi = \frac{w_{ts}}{U^2} \qquad (14.54)$$

where w_{ts} is the specific stage work and U is the mean blade velocity.

The flow coefficient ϕ is defined as the ratio of the axial velocity of the fluid C_x and the blade mean speed U:

$$\phi = \frac{C_x}{U} \qquad (14.55)$$

The degree of reaction in Eq. (14.46) can be rewritten as

$$D = \frac{\phi}{2}(\tan \beta_3 - \tan \beta_2) \qquad (14.56)$$

The specific work from Eq. (14.24) can be rewritten as

$$\frac{\dot{W}_{ts}}{\dot{m}_g} = w_{ts} = \psi U^2 = U C_x (\tan \beta_2 + \tan \beta_3), \text{ or}$$

$$\psi = \phi(\tan \beta_2 + \tan \beta_3) \qquad (14.57)$$

From Eqs. (14.56) and (14.57),

$$\tan \beta_2 = \frac{1}{2\phi}(\psi - 2D) \qquad (14.58)$$

$$\tan \beta_3 = \frac{1}{2\phi}(\psi + 2D) \qquad (14.59)$$

From Eq. (14.24) and Figure 14.12b,

$$\tan \alpha_2 - \tan \beta_2 = \tan \beta_3 - \tan \alpha_3 = \frac{U}{C_x} \qquad (14.60)$$

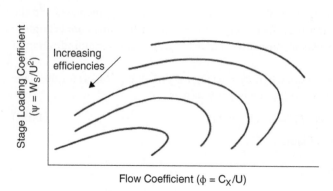

Figure 14.20 caption area:

Stage Loading Coefficient ($\psi = W_S/U^2$)

Increasing efficiencies

Flow Coefficient ($\phi = C_X/U$)

Figure 14.20 Variation of the measured stage efficiency for axial-flow turbines plotted on a $\psi - \phi$ coordinate system.

Combining Eqs. (14.55), (14.58), and (14.60) results in

$$\tan \alpha_2 = \tan \beta_2 + \frac{1}{\phi} = \frac{1}{\phi}\left(\frac{\psi}{2} - D + 1\right) \tag{14.61}$$

$$\tan \alpha_3 = \tan \beta_3 - \frac{1}{\phi} = \frac{1}{\phi}\left(\frac{\psi}{2} + D - 1\right) \tag{14.62}$$

Knowing the gas turbine coefficients D, ϕ, and ψ, Eqs. (14.58), (14.59), (14.61), and (14.62) fully determine the velocity angles in the stage.

For a 50% degree of reaction, from Eqs. (14.56), (14.58), and (14.59),

$$(\tan \beta_3 - \tan \beta_2) = \frac{1}{\phi} \tag{14.63}$$

$$\tan \beta_2 = \frac{1}{2\phi}(\psi - 1) \tag{14.64}$$

$$\tan \beta_3 = \frac{1}{2\phi}(\psi + 1) \tag{14.65}$$

The combination of Eqs. (14.60, 14.63, 14.65) allows all blade and gas angles in the stage to be determined in terms of ϕ and ψ. On the basis of this and estimated losses in the blades of varying geometries, the contours of isentropic stage efficiency of 50% reaction turbines can be plotted on a $\psi - \phi$ coordinate system as shown in Figure 14.20. A plot of such contours and superimposed collection of various actual turbine performance characteristics is known as the *Smith plot* (Smith, 1965) and is used by designers as the first step in estimating the efficiency that can be obtained for a specific design. A comprehensive Smith chart can be found in Korpela (2011).

14.3.8 Axial-Flow Turbine Performance Characteristics

The suffixes for the gas turbine characteristics vary depending on the engine type and the station numbering adopted. Using the station numbering used earlier for the single-shaft, simple turbojet and twin-shaft engines, the non-dimensional expansion ratios, mass flow rates, and speeds can be written as in Table 14.1.

Two variations of generic axial turbine characteristics are shown in Figures 14.21 and 14.22. In both cases, the isentropic efficiencies vary within a very narrow range with turbine

Table 14.1 Designation of non-dimensional parameters for different gas-turbine engines.

Non-dimensional parameter	Single-shaft engine	Simple turbojet engine	Two-shaft engine	
			Compressor turbine	Power turbine
Expansion ratio	p_{3t}/p_{4t}	p_{4t}/p_{5t}	p_{3t}/p_{4t}	p_{5t}/p_{6t}
Mass flow rate	$\dot{m}_g\sqrt{T_{3t}}/p_{3t}$	$\dot{m}_g\sqrt{T_{4t}}/p_{4t}$	$\dot{m}_g\sqrt{T_{3t}}/p_{3t}$	$\dot{m}_g\sqrt{T_{5t}}/p_{5t}$
Rotational speed	$N/\sqrt{T_{3t}}$	$N/\sqrt{T_{4t}}$	$N/\sqrt{T_{3t}}$	$N/\sqrt{T_{5t}}$

Figure 14.21 Generic axial turbine characteristics with choking in the stator: (a) isentropic efficiency; (b) mass flow parameter.

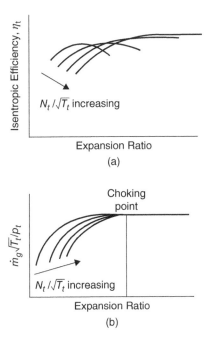

pressure ratio, particularly in the high-ratio range close to the design speed, as a result of which the efficiency is often assumed constant in off-design calculations. The variation of the mass flow parameter $\dot{m}_a\sqrt{T_t}/p_t$ with the expansion ratio, shown in Figure 14.21, indicates the incidence of choking at a certain expansion ratio and the merging of the constant speed lines to form a single line as the flow reaches its maximum. In Figure 14.22 there is no merging of constant speed lines at any value of the expansion ratio, despite the fact that choking may be occurring in the turbine. The mass flow parameter in the former remains constant due to choking at the inlet of the turbine; in the latter, choking is occurring in the turbine rotor, accompanied by small variations of the mass flow with turbine speed (Razak, 2007).

The turbine performance parameters in Figures 14.21 and 14.22 are given without specification of station numbering. The numbers to precede the subscript t for different engine configurations are given in Table 14.1.

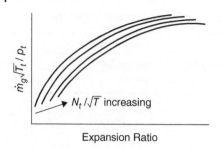

Figure 14.22 Generic axial turbine characteristics with choking in the rotor.

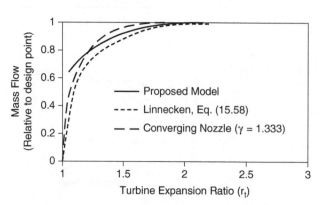

Figure 14.23 Single-curve models for axial turbine mass flow characteristic.

To simplify the calculations, turbine normalised flow characteristics are sometimes replaced with a single-curve model that ignores the effect of turbine speed at the lower range of the expansion ratios. Figure 14.23 depicts three such models; the one labelled "Proposed Model" will be used in this book when considering off-design calculations. The proposed model is a fourth-order inverse polynomial of the following form:

$$(\dot{m}\sqrt{T_t}/p_t)/(\dot{m}\sqrt{T_t}/p_t)_{des} = a + \frac{b}{r_t} + \frac{c}{r_t^2} + \frac{d}{r_t^3} + \frac{e}{r_t^4} \tag{14.66}$$

where $a = 1.0889$, $b = -1.2261$, $c = 4.8277$, $d = -6.9116$, and $e = 2.8067$.

The converging nozzle characteristic has been used in the past to approximate turbine mass flow characteristic in off-design calculations (Mirza-Baig and Saravanamuttoo, 1991; Saravanamuttoo et al., 2001; Suraweera, 2011). The Linnecken model was used by Wittenberg in his off-design calculation models (Wittenberg, 1981).

14.4 Radial-Flow Turbines

The radial-flow turbine is essentially a radial compressor with reversed flow. It is used mainly in the chemical industries, small power-generation plants, and transportation. Its use in transportation is limited to turbochargers whose function is to boost the power output of reciprocating engines – both spark ignition and compressor ignition – using the exhaust gases to run a radial turbine-compressor unit (Figure 14.24).

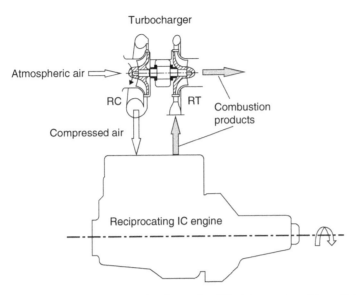

Figure 14.24 Turbocharged reciprocating (IC) piston engine: RT – radial turbine; RC – radial compressor.

14.4.1 Turbine Design

The radial turbine comprises a row of fixed converging vanes tangent to the impeller acting like nozzles, an impeller with radial vanes, a volute casing, and a diffuser (Figure 14.25). In the $r-w$ plane, the fluid (combustion products) from the combustion chamber accelerates to velocity C_2 in the nozzle vanes and then enters the impeller vanes at radius r_2 with velocity V_2 relative to the impeller tip moving at speed U_2 (see velocity diagram 2 in Figure 14.25). The outlet velocity diagram 3 is drawn in the $x-w$ plane at radius r_3 with the

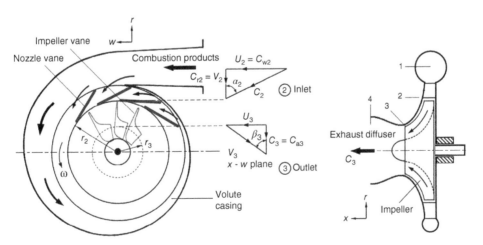

Figure 14.25 Radial-flow turbine with a single-sided impeller (velocity diagram with subscript 2 for inlet conditions and subscript 3 for outlet conditions).

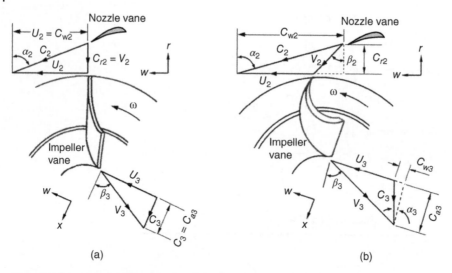

Figure 14.26 Velocity diagrams of radial inflow turbines: (a) radial impeller vane; (b) forward-curved impeller vane.

assumption that the fluid leaves the impeller blades at an absolute velocity C_3 in the axial direction (along the x-axis). The outlet velocity relative to the impeller vane at r_3 moving at speed U_3 is V_3. The planes of the velocity diagrams may be easier to visualise if drawn in 3-D, as in Figure 14.26, which shows the velocity diagrams for radial and forward-curved impeller vanes.

The Euler equation for the forward-curved impeller vanes shown in Figure 14.26b is

$$\dot{W}_t = \dot{m}_g(U_2 C_{w2} + U_3 C_{w3}) \tag{14.67}$$

We will limit the following discussion to the radial-flow turbine with radial impeller vanes shown in Figures 14.25 and 14.26a, for which the angular momentum at radius r_3 is zero because of the orientation of C_3 in the axial direction; therefore, the power output from the turbine is

$$\dot{W}_t = \dot{m}_g U_2 C_{w2}$$

Since we have assumed that the fluid enters the impeller vanes radially, as shown in velocity diagram 2, $V_2 = C_{r2}$, and $C_{w2} = U_2$, the power output is reduced to

$$\dot{W}_t = \dot{m}_g U_2^2 \tag{14.68}$$

Applying the steady-flow energy equation between stations 1 and 3, we can write the power output in terms of total (stagnation) temperatures as follows:

$$\dot{W}_t = \dot{m}_g(h_1 - h_3) = \dot{m}_g c_p(T_{1t} - T_{3t})$$

The isentropic efficiency of the turbine is defined as

$$\eta_t = \frac{(T_{1t} - T_{3t})}{(T_{1t} - T_3)}$$

From the last two equations, we obtain

$$\dot{W}_t = \dot{m}_g c_p \eta_t(T_{1t} - T_3) \tag{14.69}$$

The pressure ratio across the impeller is

$$\frac{p_3}{p_{1t}} = \left(\frac{T_3}{T_{1t}}\right)^{\gamma/(\gamma-1)} = \left(\frac{T_{1t} + T_3 - T_{1t}}{T_{1t}}\right)^{\gamma/(\gamma-1)} = \left(1 - \frac{T_{1t} - T_3}{T_{1t}}\right)^{\gamma/(\gamma-1)} \tag{14.70}$$

Equating (14.68) and (14.69) and substituting for $(T_{1t} - T_{3t})$ in (14.70) results in

$$\frac{p_{1t}}{p_3} = \left(1 - \frac{U_2^{\,2}}{c_p \eta_t T_{1t}}\right)^{-\gamma/(\gamma-1)} \tag{14.71}$$

The turbine efficiency defined in terms of the maximum enthalpy drop is known as the *total-to-static* isentropic efficiency (η_{ts}), which is defined as follows from Figure 14.27:

$$\eta_{ts} = \frac{h_{1t} - h_{3t}}{h_{1t} - h_{3s}} \tag{14.72}$$

For a perfect gas,

$$\eta_{ts} = \frac{T_{1t} - T_{3t}}{T_{1t} - T_{3s}} \tag{14.73}$$

The total-to-static efficiency can be rewritten in terms of the nozzle loss coefficient λ_N and rotor loss coefficient λ_R using the following equation (Saravanamuttoo et al., 2001 and Korpela, 2011) for the turbine shown in Figure 14.25:

$$\eta_{ts} = \frac{1}{1 + \dfrac{1}{2}\left[\left(\dfrac{r_3}{r_2}\right)^2\left(\dfrac{1}{\tan^2\beta_3} + \dfrac{\lambda_R}{\sin^2\beta_3}\right) + \dfrac{\lambda_N}{\sin^2\alpha_3}\left(\dfrac{T_{3s}}{T_{2s}}\right)\right]} \tag{14.74}$$

where

$$\frac{T_{3s}}{T_{2s}} = 1 - \frac{U_2^{\,2}}{c_p T_2}\left\{1 + \left(\frac{r_3}{r_2}\right)^2\left[\frac{(1 + \lambda_R)}{\sin^2\beta_3} - 1\right] - \frac{1}{\tan^2\alpha_3}\right\} \tag{14.75}$$

Figure 14.27 *T − s* diagram for the radial-flow turbine in Figure 14.25.

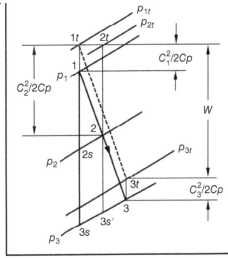

The loss coefficients in Eq. (14.74) can be written in terms of enthalpy changes as follows:

$$h_3 - h_{3s'} = c_p(T_3 - T_{3s'}) = \frac{1}{2}\lambda_R V_3^2$$

$$h_2 - h_{2s} = c_p(T_2 - T_{2s}) = \frac{1}{2}\lambda_N C_2^2$$

The temperature drop between $1t$ and 2 in Figure 14.27 is

$$T_{1t} - T_2 = \frac{C_2^2}{2c_p} = \frac{U_2^2}{2c_p \sin^2 \alpha_2}$$

from which

$$T_2 = T_{1t} - \frac{U_2^2}{2c_p \sin^2 \alpha_2} \tag{14.76}$$

The temperature drop across the entire stage between stations 1 and 3 can be determined by equating Eqs. (14.68) and (14.69):

$$T_{1t} - T_3 = \frac{U_2{}^2}{c_p \eta_t}$$

from which

$$T_3 = T_{1t} - \frac{U_2{}^2}{c_p \eta_t} \tag{14.77}$$

14.4.2 Turbine Characteristics

Radial turbine characteristics are generally similar in shape to the axial turbine characteristic discussed earlier. Figure 14.28 shows the characteristics of a radial-flow turbine of a turbocharger. The design-point speed in such a turbine can be as high as $160\,000\,rpm$ or more.

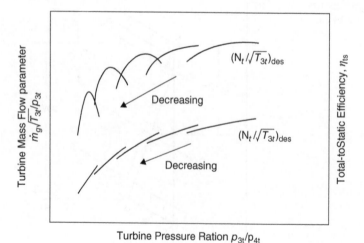

Figure 14.28   Radial-flow turbine characteristics.

Problems

14.1 The pressure ratio and temperature rise in a single-stage axial compressor are 1.5 and 42 K, respectively. The inlet temperature into the stage is 298 K, and the specific heat for air is 1.005 $kJ/kg.K$. If the rotor rotational speed is 6000 rpm and the mean blade speed is 400 m/s, determine:
   (a) Exit stage temperature
   (b) Stage efficiency
   (c) Mean diameter of the rotor
   (d) Specific power required to run the compressor
       (Take the gas constant as 0.287 $kJ/kg.K$.)

14.2 For the compressor in Problem 14.1, the degree of reaction is 50% and the axial velocity is constant through the stage and equal to 120 m/s. If the relative velocity outlet angle is 26°, determine the absolute velocity outlet angle.

14.3 Determine the input power, the isentropic efficiency, and the stage outlet temperature of a radial compressor, given the following information:

Mass flow rate of air	20 kg/s
Compressor pressure ratio	1.8
Inlet temperature	288 K
Inlet pressure	0.1 MPa
Ratio of specific heats for air	1.38
Impeller tip diameter	0.9 m
Impeller rotational speed	6000 rpm
Slip factor	0.9

14.4 The following information is provided for a reaction turbine stage with constant axial velocity throughout the stage:

Mass flow of the gases	25 kg/s
Rotor entry temperature	1079 K
Polytropic turbine efficiency	0.86
Stator blade exit angle α_2	60
Gas velocity at rotor outlet	720 m/s
Rotor blade outlet angle β_3	55°
Turbine expansion ratio	5
Specific heat of gases	1.148 $kJ/kg.K$
Gas constant	0.287 $kJ/kg.K$

Determine:

(a) Turbine exit temperature
(b) Whirl velocities at rotor inlet and outlet
(c) Stage power
(d) Absolute velocity of gases at rotor exit

14.5 The information in the following table is for a gas turbine stage in which the inlet and outlet velocities are equal and the axial velocity is constant throughout the stage:

Mass flow of the gases	20 kg/s
Axial velocity	220 m/s
Mean rotor diameter	0.6 m
Rotational speed	10 000 rpm
Stator blade exit angle α_2	60°
Absolute velocity angle at rotor blade outlet α_3	12°

Determine:

(a) Rotor blade outlet angle β_3
(b) Output power
(c) Degree of reaction

15

Off-Design Performance of Gas Turbines

Design-point calculations fix the operating conditions under which the engine will normally operate. However, gas turbines, like other internal combustion engines, operate outside their design operating conditions for long periods of time, and the performance of the engine under these conditions, known as *off-design performance*, is of vital importance. Off-design performance could involve changes such as load, thrust, compressor speed, or ambient conditions, and its prediction requires special computational techniques known as *off-design calculations*. The need for such a computational approach is dictated by the fact that operating conditions such as maximum cycle temperature, shaft speed, output power, and mass flow rates cannot be selected arbitrarily and must be determined by observing the compatibility conditions and limitations of the different components making up the gas turbine engine. Compatibility is governed by a number of factors such as flow dynamics; component characteristics; and laws of thermodynamic, aerodynamic, and gas dynamics.

This chapter discusses the prediction of the off-design steady-state performance of simple industrial and aviation gas turbine engines using two methods:

1. The component-matching method, in which calculations are based on actual performance data for gas turbine engine components: compressor and turbine(s) characteristics and, if applicable, propelling nozzle characteristics.
2. The thermo-gas-dynamic matching method, in which calculations are based on thermodynamic relationships and generalised mass flow characteristics of the turbine or nozzle (if applicable), and no prior knowledge of actual compressor and turbine characteristics is required.

Information on off-design performance of more complex gas turbine engine configurations can be found in, to name a few, Walsh and Fletcher (2004), Wittenberg (1976, 1981), and Saravanamuttoo et al. (2001).

15.1 Component-Matching Method

This method, which is also known as *serial nested loops* (Walsh and Fletcher, 2004), is the simplest and most widely used for the determination of off-design performance of gas turbine engines. Figure 15.1 is a schematic representation of the required data for the component-matching scheme. An essential part of all gas turbine configurations is the gas

Fundamentals of Heat Engines: Reciprocating and Gas Turbine Internal Combustion Engines, First Edition. Jamil Ghojel.
© 2020 John Wiley & Sons Ltd. This Work is a co-publication between John Wiley & Sons Ltd and ASME Press.
Companion website: www.wiley.com/go/JamilGhojel_Fundamentals of Heat Engines

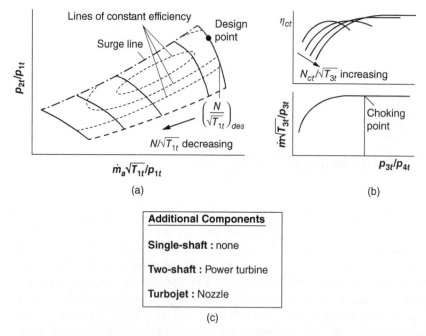

Figure 15.1 Component-matching scheme requirements: (a) compressor characteristics; (b) turbine characteristics; (c) characteristics determined by engine type.

generator, which comprises a compressor, a combustor, and a compressor turbine (CT). More components can be added further downstream of the gas generator, depending on the specific engine application.

The methodology used for off-design performance is loosely based on the method outlined by Saravanamuttoo et al. (2001), and three simple gas turbine engine configurations will be considered:

1. Single-shaft gas turbine engine that can be used for power generation or marine propulsion
2. Two-shaft gas turbine that can be used for power generation, marine propulsion, or automotive propulsion
3. Simple turbojet engine used for aircraft propulsion

The characteristics of arbitrary compressors and turbines are shown schematically in Figure 15.1a and b (the combustor does not figure in off-design calculations as a separate component). The box in Figure 15.1c shows what type of additional component will be needed, if any, for the engines under consideration. For the off-design performance calculation of two-shaft and turbojet engines, the performance characteristics of the power turbine (PT) and exhaust nozzle must be provided.

All off-design calculations must satisfy the compatibility conditions between the components making up the engine:

1. Compatibility of mass flow
2. Compatibility of rotational speeds
3. Compatibility of power

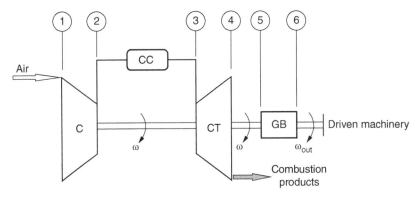

C - Compressor   CC - Combustion chamber
CT - Compressor turbine   GB - Gear box

Figure 15.2   Schematic diagram of a single-shaft gas turbine.

Additionally, the following assumptions are made to simplify the calculations:

1. No change in mass flow throughout the engine (fuel mass flow is ignored).
2. No air bleeds at any stage in the engine.
3. Inlet and exhaust pressure losses are ignored.
4. The pressure ratio in the combustion chamber (CC) is constant.
5. The mass flow function for any turbine is represented by a single curve.

15.1.1   Off-Design Performance of Single-Shaft Gas Turbine

Figure 15.2 shows a schematic representation of the single-shaft gas turbine engine, and the performance characteristics of the gas-generator components are shown in Figure 15.1a,b.

15.1.1.1   Calculation Procedure

The design point is shown on the compressor characteristics as a black dot on the design speed line in Figure 15.1a. To find the performance at any point on the design speed line other than the design point, or at different compressor speeds and efficiencies, the following steps can be followed:

1. A constant-speed line is selected on the compressor characteristics, and a point is selected on this line. The values of $\dot{m}_a\sqrt{T_{1t}}/p_{1t}$, p_{2t}/p_{1t}, η_c, and $N/\sqrt{T_{1t}}$ are determined for the selected point.
2. The corresponding points on the turbine characteristics are obtained from consideration of the compatibility of speed and continuity of flow:

$$\frac{N}{\sqrt{T_{3t}}} = \frac{N}{\sqrt{T_{1t}}} \cdot \sqrt{\frac{T_{1t}}{T_{3t}}} \qquad (15.1)$$

$$\frac{\dot{m}_g\sqrt{T_{3t}}}{p_{3t}} = \frac{\dot{m}_a\sqrt{T_{1t}}}{p_{1t}} \cdot \frac{p_{1t}}{p_{2t}} \cdot \frac{p_{2t}}{p_{3t}} \cdot \sqrt{\frac{T_{3t}}{T_{1t}}} \cdot \frac{\dot{m}_g}{\dot{m}_a} \qquad (15.2)$$

Since we have already assumed $\dot{m}_a = \dot{m}_g = \dot{m}$, the relationship for flow continuity is reduced to

$$\frac{\dot{m}\sqrt{T_{3t}}}{p_{3t}} = \frac{\dot{m}\sqrt{T_{1t}}}{p_{1t}} \cdot \frac{p_{1t}}{p_{2t}} \cdot \frac{p_{2t}}{p_{3t}} \cdot \sqrt{\frac{T_{3t}}{T_{1t}}} \tag{15.3}$$

3. The values of $\dot{m}\sqrt{T_{1t}}/p_{1t}$ and p_{2t}/p_{1t} are defined when the operating point is selected on the compressor characteristics. The pressure ratio p_{3t}/p_{2t} is assumed to be constant for all turbine settings. Since we have neglected the pressure losses in the inlet and outlet ducts, $p_a = p_{1t} = p_{4t}$, and the turbine pressure ratio $(p_{3t}/p_{4t}) = (p_{3t}/p_{1t}) = (p_{3t}/p_{2t})(p_{2t}/p_{1t})$, which allows the flow parameter $\dot{m}\sqrt{T_{3t}}/p_{3t}$ to be determined from turbine characteristic similar to Figure 15.1b.

4. $\sqrt{T_{3t}/T_{1t}}$ is determined from Eq. (15.3); and, knowing the ambient temperature T_{1t}, the turbine inlet temperature T_{3t} can now be determined.

5. The turbine speed $N/\sqrt{T_{3t}}$ is calculated using Eq. (15.1); and, knowing $N/\sqrt{T_{3t}}$ and p_{3t}/p_{4t}, the turbine efficiency η_{ct} can be determined from the turbine characteristics (Figure 15.1b).

6. The compressor and turbine temperature gradients can now be calculated, respectively, from

$$\Delta T_{12t} = (T_{2t} - T_{1t}) = \frac{T_{1t}}{\eta_c} \left[\left(\frac{p_{2t}}{p_{1t}} \right)^{(\gamma-1)/\gamma} - 1 \right] \tag{15.4}$$

$$\Delta T_{34t} = (T_{3t} - T_{4t}) = \eta_{ct} T_{3t} \left[1 - \left(\frac{p_{4t}}{p_{3t}} \right)^{(\gamma-1)/\gamma} \right] \tag{15.5}$$

7. The net engine output power is

$$\dot{W}_{net} = \dot{m}c_{pg}\Delta T_{34t} - \frac{\dot{m}}{\eta_m}c_{pa}\Delta T_{12t} \tag{15.6}$$

where η_m is the mechanical efficiency of the CT assembly, and the mass flow rate can be written as

$$\dot{m} = \frac{\dot{m}\sqrt{T_{1t}}}{p_{1t}} \cdot \frac{p_1}{\sqrt{T_1}} \tag{15.7}$$

where p_1 and T_1 are the ambient conditions at the compressor inlet.

8. The turbine inlet temperature T_{3t} is known, and $T_{2t} = T_{1t} + \Delta T_{12t}$. This information can be used to determine the air-fuel ratio from temperature charts or computed as discussed in Chapter 11. The fuel flow rate can now be determined from

$$\dot{m}_f = \frac{\dot{m}/\lambda}{(A/F)_{stoich}} \tag{15.8}$$

where λ is the relative air-fuel ratio $(\lambda = 1/\phi)$, and $(A/F)_{stoich}$ is the stoichiometric air-fuel ratio for the fuel used.

9. The compressor operating point now needs to be matched with the characteristics of the load imposed, as described next.

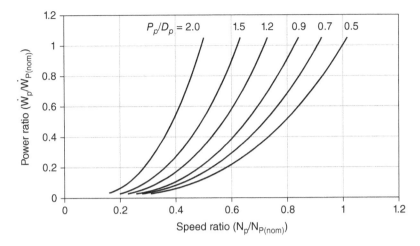

Figure 15.3 Variable-pitch propeller load characteristics.

15.1.1.2 Propeller Load

The power characteristic for a constant-pitch propeller, be it in marine or aircraft applications, is usually written as

$$\dot{W}_p = K_1 N_p^{\,3} \tag{15.9}$$

where K_1 is a constant of proportionality and N_p is the propeller speed.

If the propeller is coupled to the gas turbine engine with a constant gear-ratio transmission, the propeller power is proportional to the cube of the speed of the gas turbine output shaft N_t:

$$\dot{W}_p = K_2 N_t^{\,3} \tag{15.10}$$

where K_2 is another constant of proportionality.

The power for a variable-pitch ship propeller, based on graphical data provided by Orlin et al. (1971), can be written as

$$\left(\frac{\dot{W}_p}{\dot{W}_{p(nom)}} \right) = a b^{(P_p/D_p)} \left(\frac{N_p}{N_{p(nom)}} \right)^3 \tag{15.11}$$

where $a = 0.5$, $b = 4.1$, and P_p and D_p are the propeller pitch and diameter, respectively.

Note that the propeller power and speed are presented relative to the nominal (design) values. Figure 15.3 is a plot of correlation (15.11) for a number of pitch-to-diameter ratios.

The design point shown on the constant-speed line (100% of the design speed) on the compressor characteristics in Figure 15.1a will correspond to the design point on the propeller load characteristics. For a propeller with constant pitch, the power is reduced by reducing the rotational speed, and only one point on a constant-speed line on the compressor characteristics will satisfy the net power requirement of the propeller at the reduced propeller speed. This point (operating point) can be found by trial and error by taking several points on the compressor characteristics at the selected constant speed and repeating steps 1–9 of the earlier calculation procedure. The calculations are repeated for

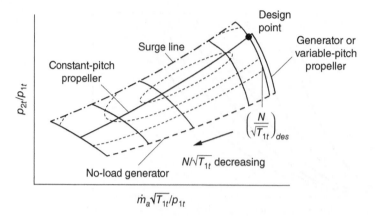

Figure 15.4 Operating lines on the compressor characteristic of a constant-pitch propeller, a variable-pitch propeller, and a generator.

other constant-speed lines to obtain the equilibrium operating line for the constant-pitch propeller, as shown in Figure 15.4.

For the variable-pitch propeller, the load is reduced by reducing the propeller pitch-to-diameter ratio at a constant propeller speed, and the corresponding operating line on the compressor characteristics is along a constant-speed line as shown in Figure 15.4. A comparison of the operating lines of constant- and variable-pitch propellers shows that a decrease in propeller power is accompanied in the latter by a greater reduction in compressor isentropic efficiency (note that the isentropic efficiency increases towards the centre of the oval-shaped constant-efficiency lines).

15.1.1.3 Electric Generator

In this configuration, a single-shaft gas turbine engine drives an electrical generator directly, mostly through a gearbox, with the speed being determined by the A/C frequency to be generated (50 or 60 Hz). Hence, generators run at constant speed, and the turbine power output varies with the electrical load imposed. The resulting operating line usually coincides with the design speed line on the compressor characteristics, as shown in Figure 15.4 (this operating line is identical to the line for a variable-pitch propeller). Each point on this line corresponds to a different turbine inlet temperature and output power. When the turbine output power becomes equal to the compressor power requirement at a certain pressure ratio for each speed line, there is no net shaft output power, and this *no-load* condition is represented by the lower operating broken line.

15.1.2 Off-Design Performance of Two-Shaft Gas Turbine (Free-Turbine Engine)

Figure 15.5 shows a schematic diagram of a two-shaft gas turbine comprising a compressor C, combustion chamber CC, compressor turbine CT, and power turbine PT. The CT generates just enough power to operate the compressor and overcome any friction in the gas generator; the remaining energy in the combustion products is utilised to generate shaft power by the PT.

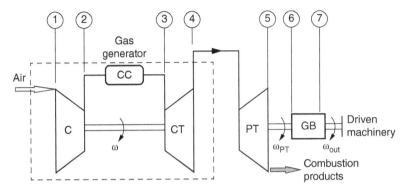

C - Compressor CC - Combustion chamber CT - Compressor turbine
PT - Power turbine GB - Gear box

Figure 15.5 Schematic diagram of a two-shaft gas turbine.

For the calculation of off-design performance of the engine by component-matching, the PT characteristics must be included in the computational scheme in Figure 15.1. The PT characteristics are schematically depicted in Figure 15.6. The calculations for the determination of the operating conditions of a gas generator in a two-shaft gas turbine are generally identical to the calculations for the single-shaft gas turbine. The main difference is that the pressure ratio of the CT cannot be determined directly from gas generator parameters and need to be found by iteration. The procedure used is presented here in full for the sake of completeness:

1. A constant-speed line is selected on the compressor characteristics, and a point is selected on this line. The values of $\dot{m}_a \sqrt{T_{1t}}/p_{1t}$, p_{2t}/p_{1t}, η_c, and $N/\sqrt{T_{1t}}$ are determined for the selected point.

2. The corresponding points on the CT characteristics are obtained from consideration of compatibility of speed and continuity of flow:

$$\frac{\dot{m}\sqrt{T_{3t}}}{p_{3t}} = \frac{\dot{m}\sqrt{T_{1t}}}{p_{1t}} \cdot \frac{p_{1t}}{p_{2t}} \cdot \frac{p_{2t}}{p_{3t}} \cdot \sqrt{\frac{T_{3t}}{T_{1t}}} \tag{15.12}$$

$$\frac{N}{\sqrt{T_{3t}}} = \frac{N}{\sqrt{T_{1t}}} \cdot \sqrt{\frac{T_{1t}}{T_{3t}}} \tag{15.13}$$

3. The pressure ratio (p_{3t}/p_{2t}) across the CC can be found directly from the assumed pressure loss in the chamber:

$$\frac{p_{3t}}{p_{2t}} = \frac{p_{2t} - \Delta p_{23}}{p_{2t}} = 1 - \frac{\Delta p_{23}}{p_{2t}} \tag{15.14}$$

4. Equating the compressor and CT powers yields

$$c_{pa}\Delta T_{12t} = \eta_m c_{pg}\Delta T_{34t} \tag{15.15}$$

Rewriting Eq. (15.15) as an identity results in

$$\frac{\Delta T_{34t}}{T_{3t}} = \frac{1}{\eta_m} \cdot \frac{\Delta T_{12t}}{T_{1t}} \cdot \frac{T_{1t}}{T_{3t}} \cdot \frac{c_{pa}}{c_{pg}} \tag{15.16}$$

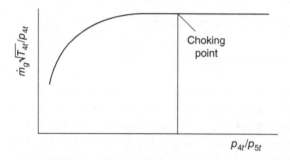

Figure 15.6 Power-turbine characteristics.

From Eq. (15.4),

$$\frac{\Delta T_{12t}}{T_{1t}} = \frac{1}{\eta_c}\left[\left(\frac{p_{2t}}{p_{1t}}\right)^{(\gamma-1)/\gamma} - 1\right] \qquad (15.17)$$

5. The pressure ratio p_{3t}/p_{4t} is guessed, and the mass flow parameter $\dot{m}\sqrt{T_{3t}}/p_{3t}$ is obtained from the CT characteristics. The temperature ratio T_{3t}/T_{1t} can now be calculated from Eq. (15.12).

6. Knowing T_{3t}/T_{1t}, the speed parameter of the CT $N/\sqrt{T_{3t}}$ can be determined from Eq. (15.13).

7. With p_{3t}/p_{4t} and $N/\sqrt{T_{3t}}$ known, the isentropic efficiency η_{ct} can be determined from the CT characteristics (Figure 15.1b).

8. The ratio $\Delta T_{34t}/T_{1t}$ from Eq. (15.5) is

$$\frac{\Delta T_{34t}}{T_{3t}} = \eta_{ct}\left[1 - \left(\frac{p_{4t}}{p_{3t}}\right)^{(\gamma-1)/\gamma}\right]$$

This equation can be used with Eq. (15.16) to calculate another value of T_{3t}/T_{1t}.

9. If the temperature ratio T_{3t}/T_{1t} calculated in step 8 does not agree with the value calculated from Eq. (15.12), a second value of the pressure ratio p_{3t}/p_{4t} is guessed, and the calculations are repeated from step 5 until the two values of T_{3t}/T_{1t} converge to the required accuracy.

10. The mass flow parameter into the PT can be written as

$$\frac{\dot{m}_g\sqrt{T_{4t}}}{p_{4t}} = \frac{\dot{m}_a\sqrt{T_{3t}}}{p_{3t}} \cdot \frac{p_{3t}}{p_{4t}} \cdot \sqrt{\frac{T_{4t}}{T_{3t}}} \cdot \frac{\dot{m}_g}{\dot{m}_a} \qquad (15.18)$$

or, in terms of constant mass flow throughout the engine $\dot{m}_a = \dot{m}_g = \dot{m}$,

$$\frac{\dot{m}\sqrt{T_{4t}}}{p_{4t}} = \frac{\dot{m}\sqrt{T_{3t}}}{p_{3t}} \cdot \frac{p_{3t}}{p_{4t}} \cdot \sqrt{\frac{T_{4t}}{T_{3t}}} \tag{15.19}$$

In this equation,

$$\sqrt{\frac{T_{4t}}{T_{3t}}} = \sqrt{\left(1 - \frac{T_{3t} - T_{4t}}{T_{3t}}\right)} = \sqrt{\left(1 - \frac{\Delta T_{34t}}{T_{3t}}\right)}$$

where, from Eq. (15.5),

$$\frac{\Delta T_{34t}}{T_{3t}} = \eta_{pt}\left[1 - \left(\frac{p_{4t}}{p_{3t}}\right)^{(\gamma-1)/\gamma}\right]$$

11. The pressure ratio across the PT can be written as

$$\frac{p_{4t}}{p_{5t}} = \frac{p_{4t}}{p_{3t}} \cdot \frac{p_{3t}}{p_{2t}} \cdot \frac{p_{2t}}{p_{5t}}$$

Since $p_{1t} = p_{5t} = p_a$, the pressure ratio can be rewritten as

$$\frac{p_{4t}}{p_a} = \frac{p_{4t}}{p_{3t}} \cdot \frac{p_{3t}}{p_{2t}} \cdot \frac{p_{2t}}{p_a} \tag{15.20}$$

12. Knowing the pressure ratio across the PT p_{4t}/p_a, the mass flow parameter $\dot{m}\sqrt{T_{4t}}/p_{4t}$ can be found from the characteristics of the PT in Figure 15.6.

13. If agreement is not attained between the mass flow rate parameters determined in step 12 and from Eq. (15.19), another point needs to be selected on the same speed line on the compressor characteristic, and the previous procedure is repeated until the condition of mass flow continuity between the two turbines is satisfied. For each constant-speed line on the compressor characteristics, there is only one point that will satisfy both the power compatibility of the gas generator and mass flow continuity with PT. Joining these points forms the operating line for the load imposed by the particular machinery driven by the PT. Figure 15.7 shows the operating lines for an electrical generator and a propeller. If needed, constant lines of ratios of the CT entry temperature T_{3t} to the compressor inlet temperature T_{1t} can be plotted on the compressor characteristics.

15.1.2.1 Power Turbine Output

The output of the PT is given by

$$\dot{W}_{pt} = \dot{m}c_{pg}(T_{4t} - T_{5t}) \tag{15.21}$$

where

$$T_{4t} - T_{5t} = \eta_{pt}T_{4t}\left[1 - \left(\frac{p_{5t}}{p_{4t}}\right)^{(\gamma-1)/\gamma}\right]$$

Compressor-turbine exit temperature (power-turbine entry temperature) is

$$T_{4t} = T_{3t} - \Delta T_{34t} = T_{3t} - \eta_{ct}T_{3t}\left[1 - \left(\frac{p_{4t}}{p_{3t}}\right)^{(\gamma-1)/\gamma}\right]$$

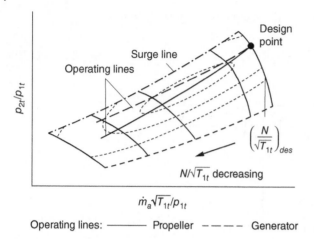

Operating lines: —————— Propeller — — — — Generator

Figure 15.7 Power-turbine operating lines on compressor characteristics.

And the mass flow rate at ambient conditions of pressure and temperature (p_1, T_2) is

$$\dot{m} = \frac{\dot{m}\sqrt{T_{1t}}}{p_{1t}} \cdot \frac{p_1}{\sqrt{T_1}}$$

Unlike the single-shaft gas turbine, the gas generator speed and running lines for the two-shaft gas turbine shown in Figure 15.7 are independent of the applied load on the PT. Power-turbine performance maps are plots of power and/or torque as functions of the output shaft speed at constant compressor speed lines, and specific fuel consumption (*sfc*) as a function of power at constant compressor speed lines. An example of a power-speed map is shown in Figure 15.8 together with three different superimposed load curves representing a constant-pitch propeller, vehicle traction at a constant gear ratio, and road gradient and electrical power generation at synchronous speed. Each line of the gas-generator speed parameter on the PT map is a fixed operating point on the gas-generator running line, shown in Figure 15.7.

The power generator line is a vertical line at the required synchronous speed, and the applied load is varied at constant speed by changing the gas-generator speed. For the propeller line, also known as the *cube law*, if a proper propeller pitch is selected, the running line roughly coincides with the peaks of the gas-generator speed lines (Figure 15.8) and troughs of the *sfc* lines (Walsh and Fletcher, 2004).

15.1.3 Off-Design Performance of Turbojet Engine

The gas generator in a turbojet engine performs the same function as in a two-shaft gas turbine: the generation of a continuous flow of high-temperature, high-pressure gas. This gas is expanded in the expanding nozzle to produce a high-velocity propulsive flow of gas. Figure 15.9 shows a schematic diagram of a simple jet engine comprising a diffuser D, compressor C, combustion chamber CC, turbine T, and converging nozzle N.

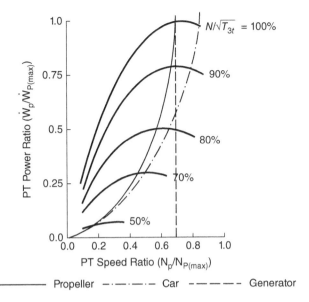

Figure 15.8 Operating lines for a propeller, a car, and a generator superimposed on power-turbine output characteristics.

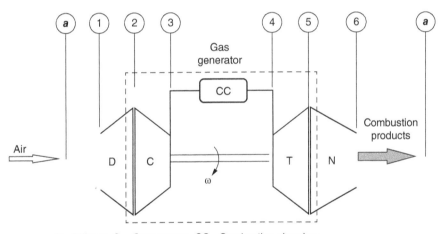

D - Diffuser C - Compressor CC - Combustion chamber
T - Compressor turbine N - Nozzle

Figure 15.9 Schematic diagram of a simple turbojet engine.

Two more components, in addition to the gas generator, must be considered for the calculation of off-design performance of the turbojet engine: the nozzle and diffuser (intake duct).

15.1.3.1 Converging Nozzle

Consider the converging nozzle shown in Figure 15.10, and assume that the flow starts from a big plenum chamber with zero velocity and that gas properties are in stagnation.

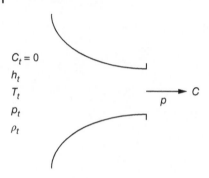

Figure 15.10 Compressible flow through nozzle.

$C_t = 0$
h_t
T_t
p_t
ρ_t

The steady flow energy equation with no work and an adiabatic flow condition applied to the nozzle is

$$h_t + 0 = h + \frac{1}{2}c^2$$

$$c^2 = 2(h_t - h) = 2c_{pg}(T_t - T) \tag{15.22}$$

The velocity at any point in the nozzle is

$$C = \sqrt{2(h_t - h)} = \sqrt{2c_{pg}(T_t - T)} = \sqrt{\frac{2\gamma}{\gamma - 1}RT_t\left(1 - \frac{T}{T_t}\right)}$$

If we replace the temperature ratio with the pressure ratio, we obtain

$$C = \sqrt{\frac{2\gamma}{\gamma - 1}RT_t\left[1 - \left(\frac{p}{p_t}\right)^{(\gamma-1)/\gamma}\right]} \tag{15.23}$$

The gas mass flow rate is

$$\dot{m}_g = \rho AC = AC\left(\frac{P}{RT}\right) \tag{15.24}$$

$$\frac{P}{RT} = \left(\frac{P_t}{RT_t} \cdot \frac{T_t}{T} \cdot \frac{p}{P_t}\right) = \left(\frac{P_t}{RT_t}\right) \cdot \left(\frac{P_t}{p}\right)^{(\gamma-1)/\gamma} \cdot \left(\frac{p}{P_t}\right) = \left(\frac{P_t}{RT_t}\right) \cdot \left(\frac{p}{P_t}\right)^{1/\gamma} \tag{15.25}$$

where ρ is the gas density and A is the cross-sectional area of the nozzle at the point under consideration.

Substituting Eqs. (15.23) and (15.25) into Eq. (15.24) yields

$$\dot{m}_g = A\sqrt{\frac{2\gamma}{R(\gamma - 1)} \frac{P_t^2}{T_t}\left[\left(\frac{p}{P_t}\right)^{2/\gamma} - \left(\frac{p}{P_t}\right)^{(\gamma+1)/\gamma}\right]} \tag{15.26}$$

In this equation, A is the nozzle throat area (minimum area), R is the gas constant, and γ is the ratio of specific heats of the gas. For a given nozzle inlet pressure p_t and temperature T_t, the mass flow rate is a function mainly of the discharge pressure p and, to much lesser extent, the ratio of specific heats. For fixed stagnation conditions and nozzle throat area,

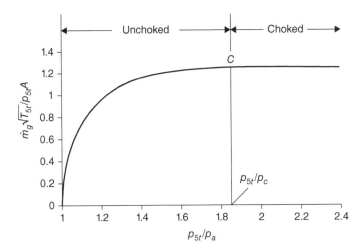

Figure 15.11 Nozzle characteristics for the flow of combustion products ($\gamma = 1.333$)

there is a maximum flow rate at a specific pressure known as the critical pressure p_c that can be found by differentiating Eq. (15.26) and equating the result to zero:

$$\frac{p_c}{p_t} = \left(\frac{2}{\gamma + 1}\right)^{\gamma/(\gamma-1)} \tag{15.27}$$

The critical temperature ratio is therefore

$$\frac{T_c}{T_t} = \left(\frac{2}{\gamma + 1}\right) \tag{15.28}$$

This pressure ratio p_c/p_t is approximately equal to 0.528 ($p_t/p_c = 1.93$) for air with $\gamma = 1.4$, and to 0.54 for combustion products with $\gamma = 1.333$ ($p_t/p_c = 1.85$).

If we replace the mass flow rate $\dot{m}_g$ with the semi-dimensional mass flow parameter, Eq. (15.26) is transformed to

$$\frac{\dot{m}_g \sqrt{T_t}}{p_t A} = \sqrt{\frac{2\gamma}{R(\gamma - 1)} \left[\left(\frac{p}{p_t}\right)^{2/\gamma} - \left(\frac{p}{p_t}\right)^{(\gamma+1)/\gamma}\right]} \tag{15.29}$$

The gas mass flow parameter for nozzle N in Figure 15.9, $\dot{m}_g \sqrt{T_{5t}}/p_{5t} A$, plotted versus the overall expansion pressure ratio down to atmospheric pressure $p_t/p = p_{5t}/p_a$, is shown in Figure 15.11.

Two operating conditions can be identified from the nozzle characteristics in Figure 15.11:

- Initially, the mass flow parameter $\dot{m}_g \sqrt{T_{5t}}/p_{5t} A$ increases steadily with the pressure ratio until the critical pressure ratio p_{4t}/p_c is reached at point C. During this period, the pressure in the nozzle throat remains constant and equal to the ambient pressure p_a, and the nozzle is referred to as an *unchoked nozzle*.
- Thereafter, the maximum mass flow parameter, reached at $p_{5t}/p_a = p_{5t}/p_c$, remains constant and independent of the pressure ratio, and the pressure at the nozzle throat remains constant and equal to the critical pressure p_c ($p_c > p_a$). The maximum mass flow

parameter is known as the *choking mass*, and a nozzle operating under these conditions is called a *choked nozzle*.

Using the station numbering in Figure 15.9 and the continuity Eq. (15.24) at the nozzle throat, the mass flow parameter can be rewritten as

$$\frac{\dot{m}_g \sqrt{T_{5t}}}{p_{5t}} = C_6 A_6 \frac{p_6}{T_6 R_g} \frac{\sqrt{T_{5t}}}{p_{5t}}$$

or as the identity

$$\frac{\dot{m}_g \sqrt{T_{5t}}}{p_{5t}} = \frac{C_6}{\sqrt{T_{5t}}} \cdot \frac{T_{5t}}{T_6} \cdot \frac{p_6}{p_{5t}} \cdot \frac{A_6}{R_g} \tag{15.30}$$

where

A_6: effective nozzle throat area (determined from design-point calculations)
R_g: gas constant
C_6: gas flow velocity at the nozzle outlet (throat)
T_6: gas temperature in the nozzle throat

From the steady flow energy equation for the adiabatic expansion process in the nozzle without work transfer using Eq. (15.22),

$$C_6^2 = 2(h_{5t} - h_{6t}) = 2c_{pg}(T_{5t} - T_6) = 2c_{pg}\Delta T_{56t}$$

From the definition of the isentropic efficiency of the nozzle,

$$\eta_n = \frac{T_{5t} - T_6}{T_{5t} - T_{6s}}$$

The temperature difference $(T_{5t} - T_6)$ can be written as

$$\Delta T_{56t} = T_{5t} - T_6 = T_{5t}\eta_n \left[1 - \left(\frac{p_6}{p_{5t}} \right)^{(\gamma-1)/\gamma} \right]$$

Hence, the ratio $C_6/\sqrt{T_{5t}}$ can be determined from

$$\frac{C_6^2}{T_{5t}} = 2c_{pg}\eta_n \left[1 - \left(\frac{p_6}{p_{5t}} \right)^{(\gamma-1)/\gamma} \right] \tag{15.31}$$

And the temperature ratio T_6/T_{5t} can be written as

$$\frac{T_6}{T_{5t}} = 1 - \frac{T_{5t} - T_6}{T_{5t}} = 1 - \frac{\Delta T_{56t}}{T_{5t}} = 1 - \eta_n \left[1 - \left(\frac{p_6}{p_{5t}} \right)^{(\gamma-1)/\gamma} \right] \tag{15.32}$$

The temperature and pressure ratios in a converging nozzle in terms of the ratio of specific heats and Mach number were determined in Chapter 12 as

$$\frac{T_{5t}}{T_6} = \left[1 + \frac{1}{2}(\gamma - 1)M_6^2 \right] \tag{15.33}$$

$$\frac{p_{5t}}{p_6} = \frac{1}{\left[1 - \frac{1}{\eta_n} \left(1 - \frac{1}{1 + \frac{\gamma - 1}{2}M_6^2} \right) \right]^{\gamma/(\gamma-1)}} \tag{15.34}$$

The Mach number at the nozzle throat (station 6 in Figure 15.9) $M_6 = C_6/\sqrt{\gamma R T_6}$. Substituting T_6 from this equation into Eq. (15.33) yields

$$\frac{C_6}{\sqrt{T_{5t}}} = \frac{M_6\sqrt{\gamma R}}{\sqrt{\left(1 + \frac{\gamma - 1}{2}M_6^2\right)}} \tag{15.35}$$

Equations (15.33)–(15.35) are valid only for pressure ratios up to the critical point, i.e. while the nozzle is unchoked. If we combine Eqs. (15.28) and (15.32), we can write the pressure ratio when the nozzle becomes choked as

$$\frac{p_{5t}}{p_c} = \frac{1}{\left[1 - \frac{1}{\eta_n}\left(\frac{\gamma - 1}{\gamma + 1}\right)\right]^{\gamma/(\gamma-1)}} \tag{15.36}$$

Equations (15.34) and (15.36) become equal when $M_6 = 1$; i.e. the nozzle becomes choked when the flow in the throat becomes sonic. To summarise, when the nozzle starts choking, $M_6 = 1$, and Eqs. (15.33), (15.34), and (15.35) are transformed into

$$\frac{p_{5t}}{p_c} = \frac{1}{\left[1 - \frac{1}{\eta_n}\left(\frac{\gamma - 1}{\gamma + 1}\right)\right]^{\gamma/(\gamma-1)}}, \frac{T_{5t}}{T_c} = \left(\frac{\gamma + 1}{2}\right), \text{and} \frac{C_6}{\sqrt{T_{5t}}} = \sqrt{\frac{2\gamma R}{\gamma + 1}}$$

15.1.3.2 Intake Duct

Consider the inlet duct (diffuser) shown in Figure 15.12 with inlet ambient conditions at 1 $(p_1 = p_a, T_1 = T_a)$ and exit stagnation conditions at 2 (compressor inlet point) in accordance with the station numbering in Figure 15.9.

The forward speed of the aircraft is C_a $(=C_1)$, and the exit speed from the diffuser is C_2. Knowing the following relationships for the stagnation temperature, isentropic pressure ratio, intake isentropic efficiency, and flight Mach number, respectively,

$$T_{2t} = T_a + \frac{C_a^2}{2c_{pa}}, \quad \frac{p_{2t}}{p_a} = \left(\frac{T_{2s}}{T_a}\right)^{\gamma/(\gamma-1)}, \quad \eta_i = \frac{T_{2s} - T_a}{T_{2t} - T_a}, \quad M_a = \frac{C_a}{\sqrt{\gamma R T_a}},$$

Figure 15.12 Schematic diagram of the intake duct in the turbojet engine.

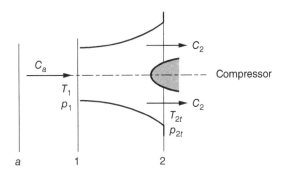

the ram pressure ratio p_{2t}/p_a and temperature ratio T_{2t}/T_a can be calculated from

$$\frac{p_{2t}}{p_a} = \left[1 + \eta_i \frac{\gamma - 1}{2} M_a^2\right]^{\gamma/(\gamma-1)}$$

(15.37)

$$\frac{T_{2t}}{T_a} = \left[1 + \frac{\gamma - 1}{2} M_a^2\right]$$

(15.38)

15.1.3.3 Matching the Gas Generator and Nozzle

The forward motion of the aircraft generates a ram pressure ratio p_{2t}/p_a that is a function of both the forward speed (flight Mach number) and the efficiency of the intake duct (diffuser). The nozzle pressure ratio p_{5t}/p_a can be expressed as an identity comprising four ratios: the turbine expansion ratio, combustion-chamber pressure ratio, compressor pressure ratio, and ram pressure ratio:

$$\frac{p_{5t}}{p_a} = \frac{p_{5t}}{p_{4t}} \cdot \frac{p_{4t}}{p_{3t}} \cdot \frac{p_{3t}}{p_{2t}} \cdot \frac{p_{2t}}{p_a}$$

(15.39)

All jet engines operate with the nozzle choked during take-off and cruising, and unchoked when preparing to land or when manoeuvring on the ground. Equation (15.37) shows that the nozzle pressure ratio is a function of the intake efficiency, gas thermodynamic properties, and flight Mach number. The same procedure used for component matching of the two-shaft gas turbine is used here, with identity (15.20) replaced by identity (15.39), which includes an additional term: the ram pressure ratio. For each selected compressor speed line, the calculations are repeated at several values of the ram pressure ratio p_{2t}/p_a or Mach number M_a over the required range of flight speeds. The result is a number of running lines that can be plotted on the compressor characteristics, each representing a unique line for a fixed flight Mach number (Figure 15.13). Once the propelling nozzle becomes choked, the running lines coincide to form a single line for all Mach numbers.

15.1.3.4 Thrust Calculation

The thrust generated by the propelling nozzle is equal to the change of momentum of the combustion products and can be written in the following general form:

$$F = \dot{m}_g(C_6 - C_a) + (p_6 - p_a)A_6$$

(15.40)

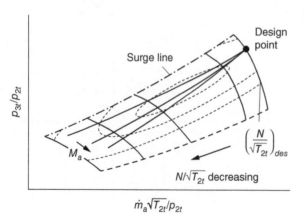

Figure 15.13 Compressor characteristics with running lines for a turbojet engine at different Mach numbers.

If the pressure p_6 at station 6 is equal to the ambient pressure p_a (nozzle is unchoked), Eq. (15.40) is reduced to

$$F = \dot{m}_g(C_6 - C_a) \tag{15.41}$$

If the pressure p_6 is equal to the critical pressure p_c (nozzle is choked), Eq. (15.40) is rewritten as

$$F = \dot{m}_g(C_c - C_a) + (p_c - p_a)A_6 \tag{15.42}$$

Equation (15.42) shows that there is further expansion of the gases outside the nozzle from pressure p_c to the ambient pressure p_a, generating what is known as *pressure thrust*. The thrust can be expressed in terms of various semi-dimensional parameters as follows:

$$\frac{F}{p_a} = \frac{\dot{m}_g\sqrt{T_{2t}}}{p_{2t}} \cdot \frac{p_{2t}}{p_a} \left[\frac{C_6}{\sqrt{T_{5t}}} \sqrt{\left(\frac{T_{5t}}{T_{4t}} \cdot \frac{T_{4t}}{T_{2t}}\right)} - \frac{C_a}{\sqrt{T_{2t}}} \right] + \left(\frac{p_6}{p_a} - 1\right) A_6 \tag{15.43}$$

All but three of the parameters in Eq. (15.43) can be determined directly from plots similar to Figure 15.13. The three parameters are $C_a/\sqrt{T_{2t}}$, $C_6/\sqrt{T_{5t}}$ and p_6/p_a. The first can be found by analogy with Eq. (15.35) as follows:

$$\frac{C_a}{\sqrt{T_{2t}}} = \frac{M_a\sqrt{\gamma R}}{\sqrt{\left(1 + \frac{\gamma-1}{2}M_a^2\right)}} \tag{15.44}$$

When the nozzle is unchoked, $p_6/p_a = 1$ and $C_6/\sqrt{T_{5t}}$ can be determined from Eq. (15.31) by substituting p_a for p_6:

$$\frac{C_6^2}{T_{5t}} = 2c_{pg}\eta_n \left[1 - \left(\frac{p_a}{p_{5t}}\right)^{(\gamma-1)/\gamma} \right]$$

When the nozzle is choked, $C_6/\sqrt{T_{5t}}$ is given by

$$\frac{C_6}{\sqrt{T_{45t}}} = \sqrt{\frac{2\gamma R}{\gamma + 1}}$$

Also, bearing in mind that $p_6 = p_c$, the pressure ratio p_6/p_a can be determined from

$$\frac{p_6}{p_a} = \frac{p_c}{p_a} = \frac{p_c}{p_{5t}} \cdot \frac{p_{5t}}{p_a}$$

Turbojet engine performance can be characterised by plots of the following forms:

$$\frac{F}{p_a} = f_1\left(\frac{N}{\sqrt{T_{2t}}}, M_a\right) \tag{15.45}$$

$$\frac{\dot{m}_f}{F/\sqrt{T_{2t}}} = f_2\left(\frac{F}{p_a}, M_a\right) \tag{15.46}$$

Such plots are shown in Figure 15.14 for two speeds: zero and cruise speed.

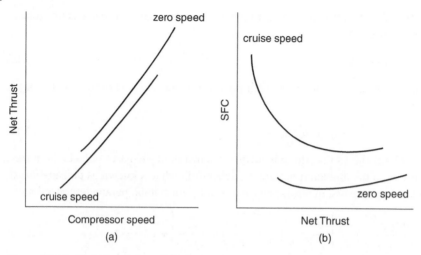

Figure 15.14 Turbojet engine off-design performance characteristics for two Mach numbers: (a) net thrust vs. compressor speed; (b) specific fuel consumption vs. net thrust.

15.2 Thermo-Gas-Dynamic Matching Method

It was mentioned in the previous section that the off-design component-matching method requires prior knowledge of detailed performance characteristics of the compressor and turbine(s). But such characteristics are not usually available for the design of new engines, and the usual practice in such cases is to rely on existing scaled characteristics of comparable components or accumulated experience with similar engine designs. Another approach that does not require prior knowledge of component characteristics and that is based mainly on thermodynamic and gas dynamic relationships and design-point calculation results is the methodology developed by Wittenberg (1976, 1981). This methodology is referred to in this book as the *thermo-gas-dynamic matching method,* and three examples will be given: a single-shaft gas turbine, a two-shaft gas turbine, and a turbojet engine. Examples of applications to more complicated configurations can be found in Wittenberg (1976, 1981), Mirza-Baig and Saravanamuttoo (1991, 1997), and Suraweera (2011).

15.2.1 Single-Shaft Gas Turbine

The calculations are based on the schematic diagram and station numbering shown in Figure 15.2. Since component characteristics are not to be used and isentropic efficiencies data will not be available, the governing equations of compressor and turbine performance will be based on their polytropic efficiencies, which will be assumed constant and independent of the pressure ratio. Properties such as specific heat and ratio of specific heats for air and combustion products will be assumed constant in the analysis, but temperature-dependent properties can be used without significantly complicating the calculations.

15.2.1.1 Compressor

The compressor pressure and temperature ratios are

$$r_c = \frac{p_{2t}}{p_{1t}}$$

$$\frac{p_{2t}}{p_{1t}} = \left(\frac{T_{2t}}{T_{1t}}\right)^{\gamma_a \eta_c/(\gamma_a - 1)} \tag{15.47}$$

$$\frac{T_{2t}}{T_{1t}} = \left(\frac{p_{2t}}{p_{1t}}\right)^{(\gamma_a - 1)/\gamma_a \eta_c} = r_c^{(\gamma_a - 1)/\gamma_a \eta_c} \tag{15.48}$$

where

γ_a: specific heat ratio for air

η_{cp}: compressor polytropic efficiency

15.2.1.2 Combustion Chamber

The energy balance in the CC in terms of heat addition and corresponding temperature rise can be written as

$$\eta_{com} H_l \dot{m}_f = \dot{m}_a c_{pm}(T_{3t} - T_{2t}) \tag{15.49}$$

where

η_{com}: combustion efficiency

H_l: lower heating value of the fuel

$\dot{m}_f$: mass flow rate of the fuel

$\dot{m}_a$: mass flow rate of the air

c_{pm}: mean specific heat of the gas during the combustion process

The pressure drop in the CC is assumed constant and is usually given as a fraction of the compressor delivery pressure p_{2t}; hence, the ratio p_{3t}/p_{2t} is a constant value. Additionally, the turbine entry temperature T_{3t} is fixed unless it is investigated as an off-design variable.

15.2.1.3 Compressor Turbine

The pressure ratio in the turbine is

$$r_{ct} = \frac{p_{3t}}{p_{4t}}$$

$$\frac{p_{4t}}{p_{3t}} = \left(\frac{T_{4t}}{T_{3t}}\right)^{\gamma_g/\eta_{ct}(\gamma_g - 1)} \tag{15.50}$$

$$\frac{T_{4t}}{T_{3t}} = \left(\frac{p_{4t}}{p_{3t}}\right)^{\eta_{ct}(\gamma_g - 1)/\gamma_g} = \frac{1}{r_{ct}^{\eta_{ct}(\gamma_g - 1)/\gamma_g}} \tag{15.51}$$

where

γ_g: specific heat ratio for the combustion products

η_{ct}: polytropic efficiency of the CT

The net output power is

$$\dot{W}_{net} = \dot{m}_g c_g(T_{3t} - T_{4t}) - \frac{1}{\eta_m} \dot{m}_a c_a(T_{2t} - T_{1t}) \tag{15.52}$$

15.2.1.4 Flow Compatibility

Compatibility of flow between the compressor and the CT, using the symbol r_c for the compressor pressure ratio, is

$$\frac{\dot{m}\sqrt{T_{3t}}}{p_{3t}} = \frac{\dot{m}\sqrt{T_{1t}}}{p_{1t}} \cdot \frac{1}{r_c} \cdot \frac{p_{2t}}{p_{3t}} \cdot \sqrt{\frac{T_{3t}}{T_{1t}}} \tag{15.53}$$

Since $p_{2t}/p_{3t} = $ const., Eq. (15.53) can be rewritten as

$$\sqrt{\frac{T_{3t}}{T_{1t}}} \cdot \frac{1}{r_c} \cdot \frac{\dot{m}\sqrt{T_{1t}}}{p_{1t}} \cdot \frac{1}{\dot{m}\sqrt{T_{3t}}/p_{3t}} = \text{const.}$$

Hence, we obtain

$$\sqrt{\frac{T_{3t}}{T_{1t}}} \cdot \frac{1}{r_c} \cdot \frac{\dot{m}\sqrt{T_{1t}}}{p_{1t}} \cdot \frac{1}{\dot{m}\sqrt{T_{3t}}/p_{3t}} = \left(\sqrt{\frac{T_{3t}}{T_{1t}}}\right)_{des} \cdot \frac{1}{(r_c)_{des}} \cdot \left(\frac{\dot{m}\sqrt{T_{1t}}}{p_{1t}}\right)_{des} \cdot \frac{1}{(\dot{m}\sqrt{T_{3t}}/p_{3t})_{des}}$$

from which

$$\frac{\sqrt{T_{3t}/T_{1t}}}{(\sqrt{T_{3t}/T_{1t}})_{des}} = \frac{r_c}{(r_c)_{des}} \cdot \frac{(\dot{m}\sqrt{T_{1t}}/p_{1t})_{des}}{\dot{m}\sqrt{T_{1t}}/p_{1t}} \cdot \frac{\dot{m}\sqrt{T_{3t}}/p_{3t}}{(\dot{m}\sqrt{T_{3t}}/p_{3t})_{des}} \tag{15.54}$$

The subscript *des* indicates that the parameter is a result of design-point calculations.

Let

$$\varphi = \frac{(T_{3t}/T_{1t})}{(T_{3t}/T_{1t})_{des}} \tag{15.55}$$

Equation (15.54) becomes

$$\sqrt{\varphi} = \frac{r_c}{(r_c)_{des}} \cdot \frac{(\dot{m}\sqrt{T_{1t}}/p_{1t})_{des}}{\dot{m}\sqrt{T_{1t}}/p_{1t}} \cdot \frac{\dot{m}\sqrt{T_{3t}}/p_{3t}}{(\dot{m}\sqrt{T_{3t}}/p_{3t})_{des}} \tag{15.56}$$

To simplify the calculations in the absence of actual component characteristics, the mass flow parameter of turbines (CT and PT) are usually approximated by a single curve with constant turbine polytropic efficiency. Any of the following correlations can be used in the calculations:

1. Nozzle equation (15.29), used by Mirza-Baig and Saravanamuttoo (1991), Suraweera (2011), can be written for any turbine as

$$\frac{\dot{m}_g\sqrt{T_t}}{p_t} = \text{const.}\sqrt{\left[\left(\frac{1}{r_t}\right)^{2/\gamma} - \left(\frac{1}{r_t}\right)^{(\gamma+1)/\gamma}\right]} \tag{15.57}$$

2. The Linnecken correlation used by Wittenberg (1981):

$$\frac{\dot{m}_g\sqrt{T_t}}{p_t} = \text{const.}\sqrt{\left[\left(1 - \frac{1}{(r_t)_{crit}}\right)^2 - \left(\frac{1}{r_t} - \frac{1}{(r_t)_{crit}}\right)^2\right]} \tag{15.58}$$

where $(r_t)_{crit}$ is the critical pressure ratio of the CT, taken by Linnecken as 2.5.

3. Ellipse law:

$$\frac{\dot{m}_g\sqrt{T_t}}{p_t} = \text{const.}\sqrt{\left[1 - \left(\frac{1}{r_t}\right)^2\right]}$$ (15.59)

4. The correlation proposed by this author on the basis of analysis of a large number of turbine characteristics in the open literature:

$$\frac{\dot{m}\sqrt{T_t}/p_t}{(\dot{m}\sqrt{T_t}/p_t)_{des}} = \left[a + \frac{b}{r_t} + \frac{c}{r_t^2} + \frac{d}{r_t^3} + \frac{e}{r_t^4}\right]$$ (15.60)

$$(a = 1.0889, b = -1.2261, c = 4.8277, d = -6.9116, e = 2.8067)$$

To use these correlations in CT and PT flow approximations, the subscript t is replaced with ct for the former and pt for the latter.

Unless specifically stated, all turbines will be assumed to choke at the inlet; as a result, the conditions at the end of the expansion process will be calculated using the normal thermodynamic relations when calculating temperatures and pressures.

The compressor flow parameter $\dot{m}\sqrt{T_{1t}}/p_{1t}$ does not change much with the compressor pressure ratio at a fixed compressor speed, particularly at or close to the design speed (see Figure 15.13). In some compressors, the mass flow parameter shows no change at all with the compressor pressure ratio at or near the design speed. Hence, Eq. (15.56) is reduced to

$$\sqrt{\varphi} = \frac{r_c}{(r_c)_{des}} \cdot \frac{\dot{m}\sqrt{T_{3t}}/p_{3t}}{(\dot{m}\sqrt{T_{3t}}/p_{3t})_{des}}$$ (15.61)

Neglecting the pressure losses in the inlet and exhaust, $p_a = p_{1t} = p_{4t}$. The pressure ratio in the turbine is then

$$\frac{p_{3t}}{p_{4t}} = \frac{p_{3t}}{p_{2t}} \cdot \frac{p_{2t}}{p_{4t}} = \frac{p_{3t}}{p_{2t}} \cdot \frac{p_{2t}}{p_{1t}}$$ (15.62)

15.2.1.5 Solution Procedure

The aim is to find the operating line for a generator at constant compressor speed line (design-point line) given the design-point performance parameters. The following data are specified for the task:

- Ambient conditions: p_{1t}, T_{1t}
- Thermodynamic data for air and combustion products: γ_a, c_a, γ_g, c_{pg}
- Combustion, mechanical, and component polytropic efficiencies: η_{com}, η_m, η_c, η_{ct}
- Combustion-chamber pressure ratio: p_{3t}/p_{2t}

15.2.1.6 Procedure

1. Assume a value of the compressor pressure ratio r_c other than the design-point value.
2. T_{2t} is calculated from Eq. (15.48)
3. Knowing the pressure drop p_{3t}/p_{2t} in the CC, the turbine pressure ratio $r_t(=p_{3t}/p_{4t})$ is calculated using Eq. (15.62)
4. Knowing r_t, the turbine mass flow parameter $\dot{m}\sqrt{T_{3t}}/p_{3t}$ can be calculated using any of the turbine mass flow characteristics given by Eqs. (15.57–15.60).

5. Calculate φ from Eq. (15.61) and T_{3t} from Eq. (15.55).
6. T_{4t} is calculated from (15.51).
7. With $\dot{m}\sqrt{T_{1t}}/p_{1t} =$ const. and unchanged ambient conditions, the fluid mass flow rate in the engine $\dot{m}$ will remain constant as the compressor pressure ratio is varied along the constant compressor speed line: $\dot{m} = (\dot{m})_{des}$.
8. Power absorbed by the compressor is calculated from $\dot{W}_c = \dot{m}c_{pa}(T_{2t} - T_{1t})$.
9. Power generated by the turbine is calculated from $\dot{W}_t = \dot{m}c_{pg}(T_{3t} - T_{4t})$.
10. Net power produced by the engine is calculated from $\dot{W}_{net} = \dot{W}_t - \dot{W}_c$.
11. For a given fuel type, the fuel mass flow rate $\dot{m}_f$ can be determined from Eq. (15.49).
12. The sfc is then

$$sfc = \frac{\dot{m}_f}{\dot{W}_{net}}$$

Figure 15.15 shows the effect on the net output power and turbine inlet temperature of increasing or decreasing the compressor pressure ratio relative to the design-point value. To obtain the operating line for a generator, the calculations can be repeated for more compressor pressure ratios.

15.2.2 Two-Shaft Gas Turbine

The method described uses the schematic diagram and station numbering shown in Figure 15.5. Again, polytropic efficiencies will be used, and the governing equations for the different engine components will be the same as in the case of the single-shaft engine. The main difference is in referring to the CT unit as a gas generator, and the presence of an additional component: the PT. The method can be applied to predict off-design turbine performance at one constant PT speed only (usually the design speed).

The station numbering for the gas generator in the two-shaft engine is the same as in the single-shaft engine; therefore, the governing equations for the compressor, CC, and CT are the same and will not be repeated here. The main difference between the two engines is the function of the gas generator.

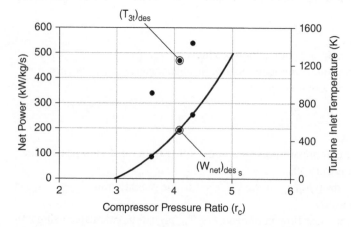

Figure 15.15 Off-design performance of a single-shaft gas turbine.

15.2.2.1 Gas Generator

The energy produced by the gas generator comprises mechanical power generated by the CT to run the compressor and overcome mechanical losses, and a high-temperature, high-pressure gas stream to run the PT. The power balance in the gas generator can be written as

$$\eta_m \dot{W}_{ct} = \dot{W}_c$$

or

$$\eta_m \dot{m}_g c_{pg}(T_{3t} - T_{4t}) = \dot{m}_a c_{pa}(T_{2t} - T_{1t}) \tag{15.63}$$

15.2.2.2 Matching the Compressor and Compressor Turbine

Combining Eqs. (15.48) and (15.63) and assuming that $\dot{m} = \dot{m}_a = \dot{m}_g$ results in

$$r_c^{(\gamma_a-1)/\gamma_a\eta_c} - 1 = \eta_m \frac{c_{pg}}{c_{pa}} \frac{T_{3t}}{T_{1t}} \left(1 - \frac{T_{4t}}{T_{3t}}\right) \tag{15.64}$$

15.2.2.3 Power Turbine

The expansion ratio in the PT is

$$r_{pt} = \frac{p_{4t}}{p_{5t}}$$

$$\frac{p_{5t}}{p_{4t}} = \left(\frac{T_{5t}}{T_{4t}}\right)^{\gamma_g/\eta_{pt}(\gamma_g-1)} \tag{15.65}$$

$$\frac{T_{5t}}{T_{4t}} = \left(\frac{p_{5t}}{p_{4t}}\right)^{\eta_{pt}(\gamma_g-1)/\gamma_g} = \frac{1}{r_{pt}^{\eta_{pt}(\gamma_g-1)/\gamma_g}} \tag{15.66}$$

γ_g: specific heat ratio for the combustion products
η_{pt}: polytropic efficiency of the PT

The PT output is

$$\dot{W}_{pt} = \eta_m \dot{m}_g c_{pg}(T_{4t} - T_{5t}) \tag{15.67}$$

15.2.2.4 Matching Mass Flow of the Compressor Turbine and Power Turbine

The continuity equation for the mass flow leaving the CT in the gas generator and entering the PT can be represented by the following identity:

$$\frac{\dot{m}\sqrt{T_{4t}}}{p_{4t}} = \frac{\dot{m}\sqrt{T_{3t}}}{p_{3t}} \cdot \frac{p_{3t}}{p_{4t}} \cdot \sqrt{\frac{T_{4t}}{T_{3t}}} \tag{15.68}$$

If the mass flow characteristic $\dot{m}\sqrt{T_{3t}}/p_{3t}$ of the CT is known, the mass flow entering the PT can be calculated for a series of CT pressure ratios p_{3t}/p_{4t} (and corresponding temperature ratios T_{3t}/T_{4t}) to obtain the dashed curve in Figure 15.16a. For matching the two components, the mass flow characteristic of the PT is plotted to the right, as shown in Figure 15.16b. If the PT is choked (point A_1 in Figure 15.16b), $\dot{m}\sqrt{T_{4t}}/p_{4t}$ = const., and the CT will operate at point A_2 at a fixed pressure ratio r_{ct1} (and fixed temperature ratio as a consequence). If the PT is unchoked, a point such as B_1 on its characteristic will correspond

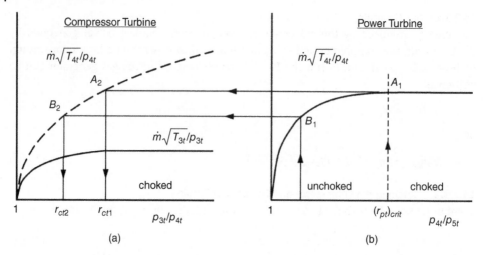

Figure 15.16 Matching the compressor turbine and power turbine in the two-shaft engine.

to point B_2 at which the CT will operate at a specific pressure ratio r_{ct2}. Multiple PT pressure ratios r_{pt} within the range 1 to $(r_{pt})_{crit}$ can be assumed, for each of which a fixed CT pressure ratio r_{ct} can be found within the range 1 to r_{ct1}.

Substituting T_{4t}/T_{3t} from Eq. (15.51) into Eq. (15.68) results in

$$\frac{\dot{m}\sqrt{T_{4t}}}{P_{4t}} = \frac{\dot{m}\sqrt{T_{3t}}}{P_{3t}} \cdot \left(\frac{P_{3t}}{P_{4t}}\right)^{[1-\eta_{ct}(\gamma_g-1)/2\gamma_g]} \tag{15.69}$$

Equation (15.69) can be written in terms of parameter ratios relative to the design point as

$$\frac{\dot{m}\sqrt{T_{4t}}/P_{4t}}{(\dot{m}\sqrt{T_{4t}}/P_{4t})_{des}} = \frac{\dot{m}\sqrt{T_{3t}}/P_{3t}}{(\dot{m}\sqrt{T_{3t}}/P_{3t})_{des}} \left[\frac{P_{3t}/P_{4t}}{(P_{3t}/P_{4t})_{des}}\right]^{[1-\eta_{ct}(\gamma_g-1)/2\gamma_g]} \tag{15.70}$$

In most cases, the CT is choked at the inlet (in the stator) so that $\dot{m}\sqrt{T_{3t}}/P_{3t} = $ const., and the matching Eq. (15.70) then becomes

$$\frac{\dot{m}\sqrt{T_{4t}}/P_{4t}}{(\dot{m}\sqrt{T_{4t}}/P_{4t})_{des}} = \left[\frac{P_{3t}/P_{4t}}{(P_{3t}/P_{4t})_{des}}\right]^{[1-\eta_{ct}(\gamma_g-1)/2\gamma_g]} \tag{15.71}$$

If the PT is choked, the operating point of the CT is constant and independent of the PT pressure ratio and equal to the design pressure ratio.

The pressure ratio over the engine is

$$\frac{P_{5t}}{P_{1t}} = \frac{P_{5t}}{P_{4t}} \cdot \frac{P_{4t}}{P_{3t}} \cdot \frac{P_{3t}}{P_{2t}} \cdot \frac{P_{2t}}{P_{1t}} \tag{15.72}$$

If $P_{5t}/P_{1t} = 1$ and $P_{3t}/P_{2t} = $ const, the compressor pressure ratio can be written as

$$\frac{r_c}{(r_c)_{des}} = \frac{P_{4t}/P_{5t}}{(P_{4t}/P_{5t})_{des}} \cdot \frac{P_{3t}/P_{4t}}{(P_{3t}/P_{4t})_{des}} \tag{15.73}$$

If $\eta_m c_{pg}/c_{pa}$ is kept constant as engine operating conditions change, Eq. (15.64) can be written as

$$\frac{r_c^{(\gamma_a-1)/\gamma_a\eta_c} - 1}{\dfrac{T_{3t}}{T_{1t}}\left(1 - \dfrac{T_{4t}}{T_{3t}}\right)} = \frac{[r_c^{(\gamma_a-1)/\gamma_a\eta_c} - 1]_{des}}{\left(\dfrac{T_{3t}}{T_{1t}}\right)_{des}\left(1 - \dfrac{T_{4t}}{T_{3t}}\right)_{des}}$$

from which

$$\frac{r_c^{(\gamma_a-1)/\gamma_a\eta_c} - 1}{(r_c^{(\gamma_a-1)/\gamma_a\eta_c} - 1)_{des}} = \varphi \frac{1 - \dfrac{T_{4t}}{T_{3t}}}{\left(1 - \dfrac{T_{4t}}{T_{3t}}\right)_{des}} \tag{15.74}$$

where

$$\varphi = \frac{(T_{3t}/T_{1t})}{(T_{3t}/T_{1t})_{des}}$$

If the CT is unchoked, the gas-generator flow compatibility is the same as Eq. (15.56)

$$\sqrt{\varphi} = \frac{r_c}{(r_c)_{des}} \cdot \frac{(\dot{m}\sqrt{T_{1t}}/p_{1t})_{des}}{\dot{m}\sqrt{T_{1t}}/p_{1t}} \cdot \frac{\dot{m}\sqrt{T_{3t}}/p_{3t}}{(\dot{m}\sqrt{T_{3t}}/p_{3t})_{des}} \tag{15.75}$$

and if the CT is choked, Eq. (15.75) is modified as

$$\sqrt{\varphi} = \frac{r_c}{(r_c)_{des}} \cdot \frac{(\dot{m}\sqrt{T_{1t}}/p_{1t})_{des}}{\dot{m}\sqrt{T_{1t}}/p_{1t}} \tag{15.76}$$

15.2.2.5 Off-Design Calculation Procedure

The independent variable for the off-design calculations for the two-shaft gas turbine is the PT pressure ratio $r_{pt} = p_{4t}/p_{5t}$. The calculation procedure is as follows:

1. Select correlations for the CT and PT mass flow parameters.
2. Calculate the pressure ratio in the CT $r_{ct} = p_{3t}/p_{4t}$ for a series of values of the assumed pressure ratios in the PT r_{pt} using Eq. (15.70) or (15.71), depending on the operating mode of the CT (unchoked or choked).
3. Calculate the compressor pressure ratio r_c from Eq. (15.73)
4. Calculate T_{2t} from Eq. (15.48).
5. Calculate φ by combining Eqs. (15.51) and (15.74).
6. Determine T_{3t} from $\varphi = (T_{3t}/T_{1t})/(T_{3t}/T_{1t})_{des}$.
7. Calculate the compressor mass flow $\dot{m}\sqrt{T_{1t}}/p_{1t}$ from Eq. (15.75) (or from Eq. (15.76) if the CT is choked).
8. Calculate the mass flow rate through the engine from $\dot{m} = (\dot{m}\sqrt{T_{1t}}/p_{1t}) \times p_{1t}/\sqrt{T_{1t}}$.
9. Calculate T_{4t} from Eq. (15.51).
10. Knowing T_{4t}, T_{5t} can be determined from Eq. (15.66).
11. The power generated by the PT output is calculated from

$$\dot{W}_{pt} = \dot{m}c_{pg}(T_{4t} - T_{5t})$$

12. The fuel flow rate is calculated from $\dot{m}_f = \dot{m}c_{pm}(T_{3t} - T_{2t})/\eta_{com}H_l$.
13. The sfc of the engine is calculated from $sfc = \dot{m}_f/\dot{W}_{pt}$.

15.2.2.6 Off-Design Prediction Results

Some results for off-design operation at 100% design speed of a two-shaft gas turbine engine are presented in Figures 15.17–15.20. Figures 15.17 and 15.18 indicate that increasing the PT pressure ratio above the design point (by increasing the CT entry temperature) causes the compressor pressure ratio and output power to increase significantly, while the CT pressure ratio remains almost constant. Decreasing the PT pressure ratio below the design point has the opposite effect, with the CT pressure ratio decreasing appreciably.

Figure 15.19 shows the predicted compressor pressure ratio r_c and CT entry temperature T_{3t} as functions of the compressor mass flow parameter relative to the design value for a typical two-shaft gas turbine engine. These two graphs represent compressor working lines

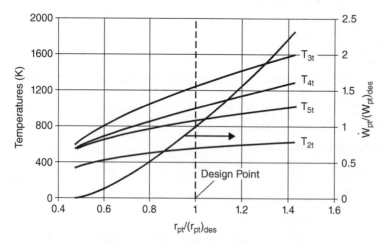

Figure 15.17 Off-design power and characteristic cycle temperatures as functions of the power-turbine pressure ratio (100% design speed).

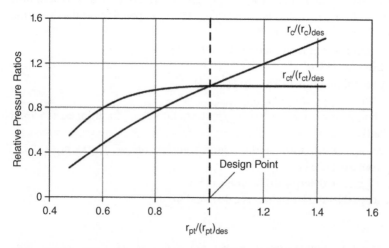

Figure 15.18 Off-design compressor and compressor-turbine pressure ratios as functions of the power-turbine pressure ratio (100% design speed).

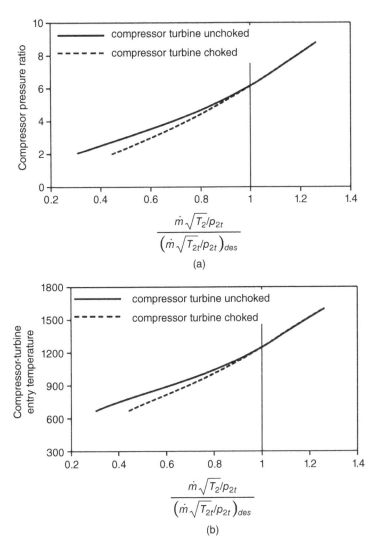

Figure 15.19 Predicted compressor pressure ratio (a) and compressor-turbine entry temperature (b) as functions of the relative compressor mass flow parameter.

for off-design performance of this engine. It is common to assume that the CT is choked over most of the operating range of the engine (Wittenberg, 1981); however, this assumption causes the predicted values of r_c and T_{3t} to be underestimated in the region below the design point, as indicated by the broken lines in Figure 15.19.

Figure 15.20 shows the predicted *sfc* plotted versus engine output power. It is apparent that the *sfc* increases significantly at very low engine output powers. The state of the CT has little effect on the *sfc*.

Quite often, the external load on the gas turbine engine is represented as power relative to the design value vs. PT speed ratio (see Figure 15.8). Therefore, it may be useful to represent

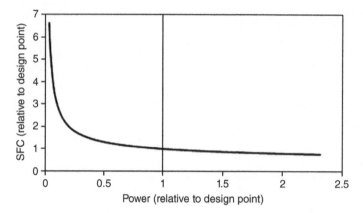

Figure 15.20 Predicted specific fuel consumption tendency with engine output load.

the power generated by the turbine the same way. Combining Eqs. (15.66) and (15.67), we get

$$\dot{W}_{pt} = \eta_m \dot{m}_g c_{pg} T_{4t} \left[1 - \left(\frac{p_{5t}}{p_{4t}} \right)^{\eta_{pt}(\gamma_g - 1)/\gamma_g} \right]$$

or

$$\frac{\dot{W}_{pt}}{p_{1t} \sqrt{T_{1t}}} = \eta_m \dot{m}_g c_{pg} \frac{\dot{m} \sqrt{T_{3t}}}{p_{3t}} \frac{T_{4t}}{T_{3t}} \frac{p_{3t}}{p_{2t}} \frac{p_{2t}}{p_{1t} c} \sqrt{\frac{T_{3t}}{T_{1t}}} \left[1 - \left(\frac{p_{5t}}{p_{4t}} \right)^{\eta_{pt}(\gamma_g - 1)/\gamma_g} \right] \qquad (15.77)$$

For a fixed operating point of a choked CT and constant pressure ratio in the CC, this equation can also be written relative to the design point parameters as

$$\frac{\dot{W}_{pt}/p_{1t} \sqrt{T_{1t}}}{(\dot{W}_{pt}/p_{1t} \sqrt{T_{1t}})_{des}} = \frac{r_c}{(r_c)_{des}} \sqrt{\varphi} \frac{1 - \left(\dfrac{p_{5t}}{p_{4t}} \right)^{\eta_{pt}(\gamma_g - 1)/\gamma_g}}{1 - \left(\dfrac{p_{5t}}{p_{4t}} \right)^{\eta_{pt}(\gamma_g - 1)/\gamma_g}_{des}} \qquad (15.78)$$

15.2.3 Turbojet Engine

The computational scheme of the turbojet engine, shown schematically in Figure 15.9, is the same as for the two-shaft scheme, with a nozzle replacing the PT in the calculations. The current task is to predict the off-design performance of a turbojet engine at a fixed compressor speed line and variable Mach numbers. Due to the differences in station numbering of the two engines, the governing equations for the compressor, CC, and CT are repeated here with the modified subscripts.

15.2.3.1 Compressor
The compressor pressure and temperature ratios are

$$r_c = \frac{p_{3t}}{p_{2t}}$$

$$\frac{p_{3t}}{p_{2t}} = \left(\frac{T_{3t}}{T_{2t}} \right)^{\gamma_a \eta_{cp}/(\gamma_a - 1)} \qquad (15.79)$$

$$\frac{T_{3t}}{T_{2t}} = \left(\frac{p_{3t}}{p_{2t}}\right)^{(\gamma_a-1)/\gamma_a\eta_{cp}} = r_c^{(\gamma_a-1)/\gamma_a\eta_{cp}} \tag{15.80}$$

where

γ_a: specific heat ratio for air

η_{cp}: compressor polytropic efficiency

15.2.3.2 Combustion Chamber

The energy balance in the CC in terms of heat addition and corresponding temperature rise can be written as

$$\eta_{com}H_l\dot{m}_f = \dot{m}_a c_{pm}(T_{4t} - T_{3t}) \tag{15.81}$$

where

η_{com}: combustion efficiency

H_l: lower heating value of the fuel

$\dot{m}_f$: mass flow rate of the fuel

$\dot{m}_a$: mass flow rate of the air

c_{pm}: mean specific heat of the gas during the combustion process

The pressure drop in the CC is assumed constant and is given as a fraction of the compressor outlet pressure p_{3t}; hence, the ratio p_{4t}/p_{3t} is constant.

15.2.3.3 Compressor Turbine

The pressure and temperature ratios in the CT are

$$r_{ct} = \frac{p_{4t}}{p_{5t}}$$

$$\frac{p_{5t}}{p_{4t}} = \left(\frac{T_{5t}}{T_{4t}}\right)^{\gamma_g/\eta_{ct}(\gamma_g-1)} \tag{15.82}$$

$$\frac{T_{5t}}{T_{4t}} = \left(\frac{p_{5t}}{p_{4t}}\right)^{\eta_{ct}(\gamma_g-1)/\gamma_g} = \frac{1}{r_{ct}^{\eta_{ct}(\gamma_g-1)/\gamma_g}} \tag{15.83}$$

where

γ_g: specific heat ratio for the combustion products

η_{ct}: polytropic efficiency of the CT

15.2.3.4 Gas Generator

The energy produced by the gas generator comprises mechanical power generated by the CT to run the compressor and overcome mechanical losses, and kinetic energy at the throat of the nozzle to generate the thrust required to propel the aircraft forward.

The power balance in the gas generator can be written as

$$\eta_m \dot{W}_{ct} = \dot{W}_c$$

or, taking into account the station numbering in Figure 15.9,

$$\eta_m \dot{m}_g c_{pg}(T_{4t} - T_{5t}) = \dot{m}_a c_{pa}(T_{3t} - T_{2t}) \tag{15.84}$$

15.2.3.5 Matching the Compressor and Compressor Turbine

Combining Eqs. (15.80) and (15.84) and assuming that $\dot{m} = \dot{m}_a = \dot{m}_g$ results in

$$r_c^{(\gamma_a-1)/\gamma_a\eta_c} - 1 = \eta_m \frac{c_{pg}}{c_{pa}} \frac{T_{4t}}{T_{2t}} \left(1 - \frac{T_{5t}}{T_{4t}}\right) \tag{15.85}$$

where

γ_g: specific heat ratio for the combustion products

η_{pt}: polytropic efficiency of the PT

15.2.3.6 Matching Mass Flow of the Compressor Turbine and Nozzle

The scheme for matching the CT in the gas generator and the propelling nozzle is similar to the scheme for the two-shaft gas turbine engine and is shown in Figure 15.21. Two operating conditions are noted: choked with constant flow parameter starting at A_1, resulting in compressor-turbine operation at a fixed pressure ratio r_{ct1} and corresponding fixed temperature ratio; and unchoked with variable flow parameter in which each nozzle pressure ratio is matched by a single compressor-turbine pressure ratio (points B_1, B_2, and r_{ct2}, as an example).

The mass flow of the CT can be expressed in terms of the exit conditions (station 5 in Figure 15.9) instead of the entry conditions, as in the following identity:

$$\frac{\dot{m}\sqrt{T_{5t}}}{p_{5t}} = \frac{\dot{m}\sqrt{T_{4t}}}{p_{4t}} \cdot \frac{p_{4t}}{p_{5t}} \cdot \sqrt{\frac{T_{5t}}{T_{4t}}} \tag{15.86}$$

Combining Eqs. (15.82) and (15.86),

$$\frac{\dot{m}\sqrt{T_{5t}}}{p_{5t}} = \frac{\dot{m}\sqrt{T_{4t}}}{p_{4t}} \cdot \left(\frac{p_{4t}}{p_{5t}}\right)^{[1-\eta_{ct}(\gamma_g-1)/2\gamma_g]} \tag{15.87}$$

This equation can be written relative to the parameters at the design point as follows

$$\frac{\dot{m}\sqrt{T_{5t}}/p_{5t}}{(\dot{m}\sqrt{T_{5t}}/p_{5t})_{des}} = \frac{\dot{m}\sqrt{T_{4t}}/p_{4t}}{(\dot{m}\sqrt{T_{4t}}/p_{4t})_{des}} \left[\frac{p_{4t}/p_{5t}}{(p_{4t}/p_{5t})_{des}}\right]^{[1-\eta_{ct}(\gamma_g-1)/2\gamma_g]} \tag{15.88}$$

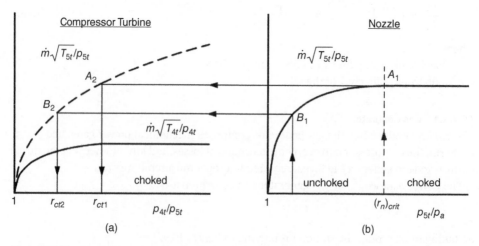

Figure 15.21 Matching the compressor turbine and propelling nozzle in the turbojet engine.

If the CT is choked, as it is the case over most of the operating range of turbojet engines, then $\dot{m}\sqrt{T_{4t}}/p_{4t} = $ const. and Eq. (15.88) is reduced to

$$\frac{\dot{m}\sqrt{T_{5t}}/p_{5t}}{(\dot{m}\sqrt{T_{5t}}/p_{5t})_{des}} = \left[\frac{p_{4t}/p_{5t}}{(p_{4t}/p_{5t})_{des}}\right]^{[1-\eta_{ct}(\gamma_{g-1})/2\gamma_g]} \tag{15.89}$$

The mass flow parameter for the nozzle can be determined from the nozzle Eq. (15.29), which can be rewritten as

$$\frac{\dot{m}_g\sqrt{T_{5t}}}{p_{5t}} = A\sqrt{\frac{2\gamma}{R(\gamma-1)}}\sqrt{\left[\left(\frac{p_6}{p_{5t}}\right)^{2/\gamma} - \left(\frac{p_6}{p_{5t}}\right)^{(\gamma+1)/\gamma}\right]}$$

or, for known nozzle throat area and gas properties,

$$\frac{\dot{m}_g\sqrt{T_{5t}}}{p_{5t}} = \text{const.}\sqrt{\left[\left(\frac{p_6}{p_{5t}}\right)^{2/\gamma} - \left(\frac{p_6}{p_{5t}}\right)^{(\gamma+1)/\gamma}\right]} \tag{15.90}$$

where
p_{5t}: stagnation pressure at the inlet into the nozzle
p_6: pressure at the nozzle throat (station 6)
Hence,

$$\frac{\dot{m}\sqrt{T_{5t}}/p_{5t}}{(\dot{m}\sqrt{T_{5t}}/p_{5t})_{des}} = \frac{\sqrt{\left[\left(\frac{p_6}{p_{5t}}\right)^{2/\gamma} - \left(\frac{p_6}{p_{5t}}\right)^{(\gamma+1)/\gamma}\right]}}{\sqrt{\left[\left(\frac{p_6}{p_{5t}}\right)^{2/\gamma}_{des} - \left(\frac{p_6}{p_{5t}}\right)^{(\gamma+1)/\gamma}_{des}\right]}} \tag{15.91}$$

Equation (15.91) shows that using the thermo-gas-dynamic method eliminates the need for knowledge of exact mass flow correlations, since any constant of proportionality is eliminated when divided by the design-point flow parameter. This is applicable to the nozzle equation (15.57), Linnecken correlation (15.58), and ellipse law (15.59). Correlation (15.60) can be used as is, since it is derived from data based on relative mass flow parameters.

Any of these equations can be used to approximate the relative mass flow parameter $\dot{m}\sqrt{T_{4t}}/p_{4t}/(\dot{m}\sqrt{T_{4t}}/p_{4t})_{des}$ in the CT.

Either Eq. (15.88) or (15.89) can be used to determine the nozzle pressure ratio for any given CT pressure ratio. The pressure in the nozzle throat at station 6 stays constant and equal to the ambient pressure p_a until the throat pressure reaches the critical value $p_6 = p_c > p_a$, whereupon the nozzle becomes choked; as a result, the left-hand sides of these equations tend to unity and the operating point of the CT becomes independent of the nozzle pressure ratio (Figure 15.21, points A_1, A_2, r_{ct1}).

The pressure ratio in the nozzle in terms of the ram pressure ratio can be written as

$$\frac{p_{5t}}{p_a} = \frac{p_{5t}}{p_{4t}} \cdot \frac{p_{4t}}{p_{3t}} \cdot \frac{p_{3t}}{p_{2t}} \cdot \frac{p_{2t}}{p_{1t}} \cdot \frac{p_{1t}}{p_a} \tag{15.92}$$

- Assuming no intake entry losses, $p_{1t} = p_a$.
- The ram pressure $p_{2t}/p_{1t} = p_{2t}/p_a$ is given by

$$\frac{p_{2t}}{p_{1t}} = \left[1 + \eta_i \left(\frac{\gamma - 1}{2}\right) M_a^2\right]^{\gamma/(\gamma-1)} \tag{15.93}$$

where η_i is the isentropic intake efficiency.

- For zero pressure drop in the CC, $p_{3t}/p_{4t} = 1.0$.

Equation (15.92) can now be written as

$$r_c = \frac{p_{5t}}{p_a} \cdot \frac{p_{4t}}{p_{5t}} \cdot \frac{p_{1t}}{p_{2t}}$$

or, it can be written relative to the design-point parameters as

$$\frac{r_c}{(r_c)_{des}} = \frac{p_{5t}/p_a}{(p_{5t}/p_a)_{des}} \cdot \frac{p_{4t}/p_{5t}}{(p_{4t}/p_{5t})_{des}} \cdot \frac{(p_{2t}/p_{1t})_{des}}{p_{2t}/p_{1t}} \tag{15.94}$$

Substituting the ram pressure ratios from Eq. (15.93) into Eq. (15.94) yields

$$\frac{r_c}{(r_c)_{des}} = \frac{p_{5t}/p_a}{(p_{5t}/p_a)_{des}} \cdot \frac{p_{4t}/p_{5t}}{(p_{4t}/p_{5t})_{des}} \cdot \frac{\left[1 + \eta_i \left(\frac{\gamma - 1}{2}\right) M_a^2\right]_{des}}{\left[1 + \eta_i \left(\frac{\gamma - 1}{2}\right) M_a^2\right]} \tag{15.95}$$

If $\eta_m c_{pg}/c_{pa} = const.$, Eq. (15.85), used to match the compressor and CT, can be rewritten as

$$\frac{r_c^{(\gamma_a-1)/\gamma_a \eta_c} - 1}{(r_c^{(\gamma_a-1)/\gamma_a \eta_c} - 1)_{des}} = \varphi \cdot \frac{1 - \dfrac{T_{5t}}{T_{4t}}}{\left(1 - \dfrac{T_{5t}}{T_{4t}}\right)_{des}} \tag{15.96}$$

The gas-generator flow compatibility equation in terms of the compressor flow parameter is

$$\frac{\dot{m}\sqrt{T_{2t}}}{p_{2t}} = \frac{\dot{m}\sqrt{T_{4t}}}{p_{4t}} \cdot \frac{p_{4t}}{p_{3t}} \cdot \frac{p_{3t}}{p_{2t}} \sqrt{\frac{T_{2t}}{T_{4t}}} \tag{15.97}$$

Since it was assumed that $p_{4t}/p_{3t} = 1.0$, Eq. (15.97) can be written as

$$\sqrt{\frac{T_{4t}/T_{2t}}{(T_{4t}/T_{2t})_{des}}} = \frac{r_c}{(r_c)_{des}} \cdot \frac{(\dot{m}\sqrt{T_{2t}}/p_{2t})_{des}}{\dot{m}\sqrt{T_{2t}}/p_{2t}} \cdot \frac{\dot{m}\sqrt{T_{4t}}/p_{4t}}{(\dot{m}\sqrt{T_{4t}}/p_{4t})_{des}} \tag{15.98}$$

Let $\varphi = \dfrac{(T_{4t}/T_{2t})}{(T_{4t}/T_{2t})_{des}}$

Equation (15.98) can now be written as

$$\sqrt{\varphi} = \frac{r_c}{(r_c)_{des}} \cdot \frac{(\dot{m}\sqrt{T_{2t}}/p_{2t})_{des}}{\dot{m}\sqrt{T_{2t}}/p_{2t}} \cdot \frac{\dot{m}\sqrt{T_{4t}}/p_{4t}}{(\dot{m}\sqrt{T_{4t}}/p_{4t})_{des}} \tag{15.99}$$

For a choked CT, this equation is reduced to

$$\sqrt{\varphi} = \frac{r_c}{(r_c)_{des}} \cdot \frac{(\dot{m}\sqrt{T_{2t}}/p_{2t})_{des}}{\dot{m}\sqrt{T_{2t}}/p_{2t}} \tag{15.100}$$

15.2.3.7 Off-Design Calculation Procedure

Since the operating conditions of the propelling nozzle determine the behaviour of the other engine components upstream, the nozzle expansion ratio p_{5t}/p_6 is taken as the independent variable for the off-design calculations for the turbojet engine. The calculation procedure is as follows:

1. Select a correlation to simulate the operation of the CT.
2. For a series of values of the expansion ratio in the nozzle p_{5t}/p_6, calculate the corresponding values of the pressure ratio in the CT $r_{ct} = p_{4t}/p_{5t}$ using Eq. (15.88) or (15.89), depending on the operating condition of the CT (unchoked or choked).
3. Assume a value for the flight Mach number M_a, and determine the compressor pressure ratio r_c from Eq. (15.95)
4. Calculate T_{3t} from Eq. (15.80)
5. Calculate φ by combining Eqs. (15.83) and (15.96).
6. Determine T_{4t} from $\varphi = (T_{4t}/T_{2t})/(T_{4t}/T_{2t})_{des}$.
7. Calculate the compressor mass flow $\dot{m}\sqrt{T_{2t}}/p_{2t}$ from Eq. (15.99) or (15.100), depending on whether the CT is unchoked or choked.
8. Calculate the mass flow rate through the engine from $\dot{m} = (\dot{m}\sqrt{T_{2t}}/p_{2t}) \times p_{2t}/\sqrt{T_{2t}}$.
9. Calculate T_{5t} from Eq. (15.83).
10. If the nozzle in unchoked, the gases expand to the ambient pressure in the throat at station 6 ($p_6 = p_a$), and T_6 can be determined from

$$\frac{T_6}{T_{5t}} = 1 - \eta_n \left[1 - \left(\frac{1}{p_{5t}/p_6} \right)^{(\gamma-1)/\gamma} \right]$$

If the nozzle is choked, the gases expand to the critical pressure at the throat ($p_6 = p_c > p_a$), and the critical pressure T_c can be determined from

$$\frac{T_c}{T_{5t}} = \frac{2}{\gamma + 1}$$

11. Knowing p_6 and T_6, the gas density in the throat can be determined from $\rho = p_6/RT_6$.
12. The flow speed in the nozzle throat can be calculated from $C_6 = \dot{m}/\rho A$, where A is the nozzle throat area determined during the design-point calculations, which remains constant over the off-design calculations.
13. The formula for the engine thrust F when the nozzle is choked is

$$F = \dot{m}(C_6 - C_a) + (p_6 - p_a)A_6$$

If the nozzle is unchoked, the thrust is given by

$$F = \dot{m}(C_5 - C_a)$$

14. The fuel flow rate is calculated from

$$\dot{m}_f = \frac{\dot{m}c_{pm}(T_{3t} - T_{2t})}{\eta_{com}H_l}$$

15. The *sfc* of the engine is calculated from $sfc = \dot{m}_f/F$.

15.2.3.8 Off-Design Prediction Results

Figures 15.22 and 15.23 show the predicted compressor pressure ratio r_c and CT entry temperature T_{4t} as functions of the compressor mass flow parameter relative to the design value for a typical turbojet engine operating at sea level at $M_a = 0$. The operating condition of the CT (unchoked or choked) has almost no effect over the entire compressor mass flow range.

Figure 15.24 shows the effect on the compressor pressure ratio of varying the ratio of the CT entry temperature to the compressor entry temperature. Figures 15.22 and 15.24 can be

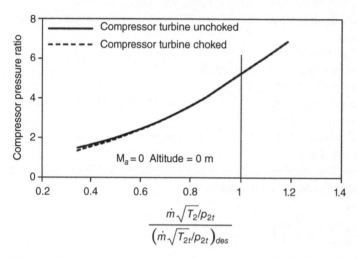

Figure 15.22 Predicted compressor operating line with the compressor turbine choked and unchoked.

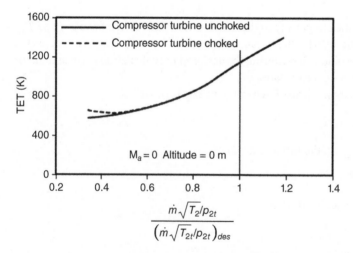

Figure 15.23 Predicted turbine entry temperature as a function of the compressor mass flow parameter with the compressor turbine choked and unchoked.

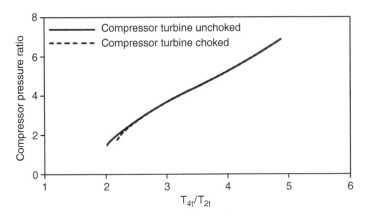

Figure 15.24 Effect of temperature ratio T_{4t}/T_{2t} on the compressor pressure ratio.

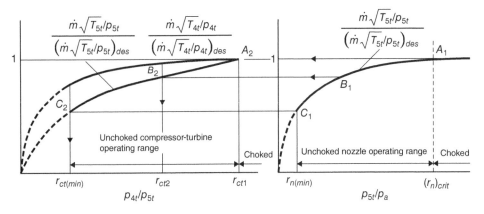

Figure 15.25 Modified scheme for matching the compressor turbine and the nozzle in a turbojet engine.

regarded as the off-design compressor operating lines of the turbojet engine operating along the compressor design speed line.

Figures 15.16 and 15.21, which were used to illustrate the matching schemes for off-design operation of the two-shaft and turbojet gas turbine engines, are generally the standard approach encountered in gas turbine practice. However, in the thermo-gas-dynamic calculation method, the governing equations are written in terms of parameter ratios (relative to the design-point values); hence, it may be more appropriate to modify the matching scheme by expressing the relative mass flow of the compressor turbine in terms of the exit conditions (Eq. (15.88)) and plotting it as a function of compressor-turbine pressure ratio as shown on the left-hand side of Figure 15.25.

The mass flow plots in this scheme are relative values with a maximum value of one. Points A_1 and C_1 on the nozzle characteristic and A_2 and C_2 on the CT characteristic define the operating range when the nozzle is unchoked.

Problems

15.1 The performance data for a single-shaft gas turbine at the design speed are given in the following table:

Compressor data			Turbine data	
$\dfrac{(\dot{m}\sqrt{T_{1t}}/p_{1t})}{(\dot{m}\sqrt{T_{1t}}/p_{1t})_{des}}$	p_{2t}/p_{1t}	η_c	p_{3t}/p_{4t}	$\dfrac{(\dot{m}\sqrt{T_{3t}}/p_{3t})}{(\dot{m}\sqrt{T_{3t}}/p_{3t})_{des}}$
0.968	4.309	88.1	4.309	1.0
0.97	4.312	88.05	4.312	1.0
0.979	4.262	87.8	4.262	1.0
0.99	4.168	87.26	4.168	1.0
0.995	4.129	86.8	4.129	1.0
1.0	4.089	86.1	4.089	1.0
1.006	4.027	84.85	4.027	1.0
1.01	3.971	84.0	3.971	1.0
1.017	3.788	81.8	3.788	1.0
1.022	3.598	79.69	3.598	1.0
1.024	3.469	77.59	3.469	0.996502

Design-point calculations (shadowed area in the table) showed that the turbine inlet temperature is 1254 K. Use the component-matching method to calculate the turbine inlet temperatures for the compressor pressure ratios 4.312 and 3.598. The ambient conditions are $p_{1t} = 101.3\ kPa$, $T_{1t} = 298\ K$. Ignore all pressure losses in the engine.

15.2 Use the thermo-gas-dynamic method to solve Problem 15.1

15.3 The following tabulated data and figure summarise the design conditions and results of design-point calculations for a two-shaft gas turbine:

Mass flow rate of air	1.5 kg/s
Compressor-turbine pressure ratio	6
Power turbine pressure ratio	2.084
Engine power output	252 kW
Compressor polytropic efficiency	0.8
Turbine polytropic efficiency	0.84
Pressure loss in the CC	30 kPa
Pressure loss in exit duct	6 kPa
Specific heat of air	1.005 kJ/kg. K
Specific heat of products	1.147 kJ/kg. K

Ignore the mass flow of the fuel and pressure losses in the inlet manifold.

The engine is to operate at the off-design ambient temperature of $37\,^\circ C$ while maintaining the design value of the compressor speed. What is the effect of this change in operating conditions on the following performance parameters: compressor-turbine pressure ratio, compressor pressure ratio, T_{2t}, T_{3t}, T_{4t}, T_{5t}, and output power $\dot{W}_{pt}$ at the power-turbine pressure ratios (1.8, 2.0, 2.2, 2.4, 2.6)? Use the thermo-gas-dynamic off-design calculation method, assuming that both the compressor turbine and power turbine are choked.

15.4 Repeat Problem 15.3 for a drop of ambient temperature by $25\,^\circ C$ relative to the design point.

15.5 Use the data from the solutions of Problems 15.3 and 15.4 to determine the performance parameters of the engine if the turbine entry temperature T_{3t} is to be kept equal to the design value as the ambient temperature changes from one extreme to another.

Bibliography

Arkhangelsky, V., Khovakh, M., Stepanov, Y. et al. (1971). *Motor Vehicle Engines*. Moscow, Russia: Mir Publishers.

Arkhangelsky, V.M., Vikhert, M.M., Voinov, A.N. et al. (1977). *Motor Vehicle Engines*. Moscow, Russia: Mashinostroyenie (In Russian).

Bayley, F.J. (1958). *An Introduction to Fluid Dynamics*. George Allen and Unwin.

Borganakke, K. and Sonntag, R.E. (2009). *Fundamentals of Thermodynamics*. Wiley.

Bosch, R. (2004). *Bosch Automotive Handbook*, 6e. SAE International.

Cencel, Y.A. and Boles, M.A. (2006). *Thermodynamics: An Engineering Approach*, 5e. McGraw Hill.

Cumpsty, N. (2003). *Jet Propulsion*, 2e. Cambridge University Press.

Dyechenko, N.K., Kostin, A.K., Pugachiuv, B.P. et al. (1974). *Theory of Internal Combustion Engines*. Mashinestroieniye (in Russian) (Теория двигателей внутреннего сгорания/ Н.Х.Дьяченко и др. - Л.: Машиностроение).

Gavrilov, A.A., Ugnatov, M.C., and Efros, V.V. (2003). *Computation of Piston Engine Cycles*. Vladimir: Vladimir State University Publications (in Russian).

Ghojel, J. (1994). Computer time reduction in first law analysis of combustion systems. In: *International Conference on Fluid and Thermal Energy Conversion* , Dec 12-15, 1994, Bali, Indonesia, vol. 1 (eds. A. Suwono, G.A. Mansoori and T. Hardiano), 335–342.

Ghojel, J.I. (1974). Some results of the investigation of direct injection diesel engine with cylindrical combustion chamber in the piston. PhD thesis, Moscow (in Russian).

Ghojel, J.I. (1982). A study of combustion chamber arrangements and heat release in DI diesel engines. Technical paper 821034 (SP-525). SAE International.

Ghojel, J.I. (1991). Effect of engine parameters on truck performance. Paper presented at the SAE-A Automotive Engine Conference, March 13 and 14, Melbourne.

Ghojel, J.I. (1992). Development of a computer program for the prediction of haul truck performance, SAE Paper 921745. *SAE Transactions, Journal of Commercial Vehicles* 101, Section 2: 630–642.

Ghojel, J.I. (1993). Haul truck performance prediction in open mining operations. In: *Proceedings of the Bulk Material Handling Conference*, 99–104. IEAust.

Ghojel, J.I. (2010). Review of the development and applications of the Wiebe function: a tribute to the contribution of Ivan Wiebe to engine research. *International Journal of Engine Research* 11 (4): 297–312.

Fundamentals of Heat Engines: Reciprocating and Gas Turbine Internal Combustion Engines, First Edition. Jamil Ghojel.
© 2020 John Wiley & Sons Ltd. This Work is a co-publication between John Wiley & Sons Ltd and ASME Press.
Companion website: www.wiley.com/go/JamilGhojel_Fundamentals of Heat Engines

Ghojel, J.I. and Watson, H.C. (1995). Relationship between road track cost and heavy vehicle fuel consumption. In: *Proceedings of the Fourth International Symposium on Heavy Vehicle Weights and Dimensions*, June 25-29, 1995, Ann Arbor, Michigan, USA (ed. C.B. Winkler), 31–38.

Ghojel, J.I., Watson, H.C., Dixon, P., et al. (1990). The impact that proper truck specification can have on a national economy. Paper presented at the International Conference on Auto Technology, November 12-14, Bangkok, Thailand.

Haywood, J.B. (1988). *Internal Combustion Engine Fundamentals*. McGraw Hill.

Heywood, J.B., Higgins, J.M., Watts, P.A., and Tabaczynski, R.J. (1979). Development and use of a cycle simulation to predict SI engine efficiency and NOx emissions. Technical paper 790291. SAE International.

Hill, P.G. and Peterson, C.R. (1992). *Mechanics and Thermodynamics of Propulsion*, 2e. Addison Wesly.

Jones, J.B. and Dugan, R.E. (1996). *Engineering Thermodynamics*. Prentice Hall.

Kolchin, A. and Demidov, V. (1984). *Design of Automotive Engines*. Moscow: Mir Publishers.

Korpela, S.A. (2011). *Principles of Turbomachinery*. Wiley.

Lemmon, E.W., Jacobsen, R.T., Penocello, S., and Friend, D.G. (2000). Thermodynamic properties of air and mixtures of nitrogen, argon, and oxygen from 60 to 2000 K at pressures to 2000 MPa. *Journal of Physical and Chemical Reference Data* 29 (3): 331–385.

Martel, C.R. (2000). Molecular weight and average composition of JP4, JP5 and Jet A. In: *Air-breathing Propulsion Manual*. Columbia, MD: Chemical Propulsion Information Agency.

Mattingly, J.D., Heiser, W.H., and Pratt, D.T. (2002). *Aircraft Engine Design*, 2e. AIAA Education Series.

Mirza-Baig, F. and Saravanamuttoo, H.I.H. (1997). Off-design performance prediction of single-spool turbojet engine using gasdynamics. *Journal of Propulsion* 13 (6): 808–810: Technical Notes.

Mirza-Baig, F.S. and Saravanamuttoo, H.I.H. (1991). Off-design performance prediction of turbofans using gasdynamics. Paper 91-GT-389 presented at the International Gas Turbine and Aeroengine Congress and Exposition.

Miyamoto, N., Chikahisa, T., Murayama, T., and Sawyer, R. (1985). Description and analysis of diesel engine rate of combustion and performance using Wiebe's functions. Technical paper 850107. SAE International.

Moran, M.J. and Shapiro, H.N. (2008). *Engineering Thermodynamics*, 6e. Wiley.

NIST (1971). *JANAF Thermochemical Tables*, 2e, NSRDS-NBS 37.

Obert, E.F. (1973). *Internal Combustion Engines and Air Pollution*. Harper & Row.

Orlin, A.C., Vierubov, D.N., Evin, V.E. et al. (1971). *Internal Combustion Engines: Theory of Working Processes of Piston and Combined Engines*. Moscow, Russia: Mochinostroyeniye (in Russian).

Razak, A.M.Y. (2007). *Industrial Gas Turbines – Performance and Operability*. CRC Press.

Rivkin, S.L. (1987). *Thermodynamic Properties of Gases*. Moscow: Energoatomizdat (in Russian) Ривкин С. Л. Термодиамические свойста газов, Енергоатомиздат, Москва, 1987.

Rogers, G. and Mayhew, Y. (1992). *Engineering Thermodynamics Work and Heat Transfer*, 4e. Longmans Scientific & Technical.

SAE. (1974). Gas turbine engine performance station identification and nomenclature. Aerospace Recommended Practice 755A. SAE International.

Sandfort, J.F. (1964). *Heat Engines – Thermodynamics in Theory and Practice*. London: Heinemann.

Saravanamuttoo, H.I.H., Rogers, G.F.C., and Cohen, H. (2001). *Gas Turbine Theory*, 5e. Pearson.

Sharaglazov, B.A., Farafontov, M.F., and Klementov, V.V. (2004). *Internal Combustion Engines: Theory, Modelling and Computational Processes*. Cheliabinsk: South Ural State University Publications (in Russian).

Smith, F.A. (1965). A simple correlation of turbine efficiency. *Journal of Royal Aeronautical Society* 69.

Stanitz, J.D. (1952). Some theoretical aerodynamic investigations of impellers in radial and mixed-flow centrifugal compressors. *ASME Transactions, Series A* 74: 473–497, 1A: A12I-A91.

Suraweera, J.K. (2011). Off-design performance of gas turbines without the use of compressor or turbine characteristics. Masters thesis, Carleton University, Ottawa, Ontario, Canada.

Taylor, C.F. (1980). *The Internal Combustion Engine in Theory and Practice*, vol. 1. MIT Press.

van Basshuysen, R. and Schafer, F. (2007). *Modern Engine Technology from A to Z*. SAE International.

Walsh, P.P. and Fletcher, P. (2004). *Gas Turbine Performance*, 2e. Blackwell Publishing, ASME Press.

Watson, N., Pilley, A.D., and Marzouk, M.A. (1980). Combustion correlation for diesel engine simulation. Technical paper 800029. SAE International.

Wiebe, I.I. (1956). Semi-empirical expression for combustion rate in engines. In: *Proceedings of Conference on Piston Engines*, 185–191. Moscow: USSR Academy of Sciences (in Russian) (Полуэмпирическое Уравнение Скорости Сгорания В Двигателях).

Wiebe, I.I. (1962). *Progress in Engine Cycle Analysis: Combustion Rate and Cycle Processes*. Mashgiz, Ural-Siberia Branch (in Russian) (Вибе И. И. Новое о рабочем цикле двигателя. М. – Свердловск: Машгиз, 1962).

Wittenberg, H. (1976). Prediction of off-design performance of turbojet and turbofan engines. AGARD CP-242-76, Proceedings of Performance Prediction Methods.

Wittenberg, H. (1981). Prediction of off-design performance of turbo-shaft engines. a simplified method. Paper 15 presented at the Seventh European Rotorcraft and Power Lift Aircraft Forum, Sepember 8-11, Germany.

Woschni, G. and Anisits, F. (1974). Experimental investigation of and mathematical presentation of rate of heat release in diesel engines dependent upon engine operating conditions. Technical paper 740086. SAE International.

Yasar, H., Soyhan, H.S., Walmsley, H. et al. (2008). Double-Wiebe function: an approach for single-zone HCCI engine modelling. *Applied Thermal Engineering* 28: 1284–1290.

Appendix A

Thermodynamic Tables

Table A.1 Specific heat at constant pressure C_p as per the correlations in Table 2.6 (enthalpy reference temperature $T_{ref} = 298.15\ K$).

T (K)	CO_2	CO	H_2O	H_2	O_2	N_2	Air
			C_p, kJ/kmole K				
298	37.143 69	29.145 15	33.579 19	28.868 26	29.338 14	29.144 75	29.185 36
300	37.226 96	29.144 23	33.590 16	28.875 44	29.350 79	29.142 28	29.186 07
400	41.290 41	29.319 67	34.304 35	29.140 47	30.162 74	29.216 7	29.415 37
500	44.611 39	29.807 95	35.222 11	29.264 08	31.105	29.591 73	29.909 52
600	47.337 12	30.460 39	36.294 24	29.352 76	32.049 82	30.141 5	30.542 25
700	49.584 97	31.175 93	37.475 06	29.468 97	32.923 82	30.776 47	31.227 42
800	51.448 3	31.889 21	38.723 64	29.644 05	33.691 71	31.435 2	31.909 07
900	53.001 27	32.560 94	40.004 5	29.888 27	34.343 66	32.077 52	32.553 41
1000	54.302 76	33.170 41	41.287 84	30.198 57	34.885 79	32.679 09	33.142 5
1100	55.399 55	33.709 6	42.549 53	30.564 21	35.333 1	33.227 01	33.669 29
1200	56.328 92	34.178 69	43.770 83	30.970 91	35.704 46	33.716 35	34.133 85
1300	57.120 67	34.582 81	44.937 91	31.403 67	36.019 17	34.147 52	34.540 56
1400	57.798 77	34.929 58	46.041 34	31.848 56	36.294 82	34.524 21	34.896 04
1500	58.382 61	35.227 52	47.075 49	32.293 82	36.546 09	34.851 96	35.207 73
1600	58.888	35.484 91	48.037 83	32.730 29	36.784 23	35.137 02	35.482 93
1700	59.327 94	35.709 22	48.928 42	33.151 52	37.017 12	35.385 68	35.728 28
1800	59.713 15	35.906 76	49.749 21	33.553 53	37.249 52	35.603 8	35.949 4
1900	60.052 57	36.082 58	50.503 63	33.934 49	37.483 67	35.796 56	36.150 85
2000	60.353 65	36.240 57	51.196 01	34.294 22	37.719 83	35.968 32	36.336 14
2100	60.622 63	36.383 6	51.831 25	34.633 74	37.956 96	36.122 64	36.507 85
2200	60.864 72	36.513 76	52.414 48	34.954 8	38.193 22	36.262 35	36.667 83
2300	61.084 25	36.632 5	52.950 76	35.259 56	38.426 5	36.389 61	36.817 36

(Continued)

Fundamentals of Heat Engines: Reciprocating and Gas Turbine Internal Combustion Engines, First Edition. Jamil Ghojel.
© 2020 John Wiley & Sons Ltd. This Work is a co-publication between John Wiley & Sons Ltd and ASME Press.
Companion website: www.wiley.com/go/JamilGhojel_Fundamentals of Heat Engines

Table A.1 (Continued)

T (K)	CO₂	CO	H₂O	H₂	O₂	N₂	Air
				C_p, kJ/kmole K			
2400	61.284 82	36.740 9	53.444 94	35.550 19	38.654 81	36.506 08	36.957 31
2500	61.469 37	36.839 83	53.901 48	35.828 71	38.876 47	36.613 01	37.088 34
2600	61.640 31	36.930 06	54.324 39	36.096 82	39.090 37	36.711 38	37.210 97
2700	61.799 59	37.012 34	54.717 22	36.355 82	39.295 93	36.801 97	37.325 7
2800	61.948 76	37.087 49	55.083 02	36.606 64	39.493 14	36.885 46	37.433 07
2900	62.089 06	37.156 39	55.424 38	36.849 84	39.682 43	36.962 46	37.533 66
3000	62.221 5	37.219 94	55.743 51	37.085 74	39.864 58	37.033 6	37.628 11
3100	62.346 89	37.279 1	56.042 27	37.314 51	40.040 5	37.099 49	37.717 1
3200	62.465 91	37.334 74	56.322 27	37.536 25	40.211 16	37.160 73	37.801 32
3300	62.579 14	37.387 69	56.584 88	37.751 11	40.377 39	37.217 94	37.881 43
3400	62.687 13	37.438 62	56.831 37	37.959 39	40.539 8	37.271 71	37.958 01
3500	62.790 36	37.488 04	57.062 89	38.161 54	40.698 7	37.322 59	38.031 57
3600	62.889 34	37.536 27	57.280 54	38.358 2	40.854 11	37.371 05	38.102 49
3700	62.984 55	37.583 47	57.485 39	38.550 2	41.005 79	37.417 51	38.171 05
3800	63.076 47	37.629 6	57.678 48	38.738 46	41.153 27	37.462 29	38.237 4
3900	63.165 57	37.674 51	57.860 86	38.923 97	41.296 01	37.505 64	38.301 62
4000	63.252 27	37.717 96	58.033 51	39.107 63	41.433 53	37.547 69	38.363 72
4100	63.337	37.759 7	58.197 4	39.290 19	41.565 52	37.588 55	38.423 71
4200	63.420 11	37.799 52	58.353 42	39.472 11	41.692 04	37.628 25	38.481 64
4300	63.501 91	37.837 31	58.502 38	39.653 51	41.813 56	37.666 79	38.537 61
4400	63.582 66	37.873 11	58.645 02	39.834 08	41.931 08	37.704 21	38.591 85
4500	63.662 57	37.907 14	58.781 98	40.013 11	42.046 14	37.740 53	38.644 71
4600	63.741 84	37.939 8	58.913 82	40.189 48	42.160 69	37.775 84	38.696 66
4700	63.820 66	37.971 61	59.041 07	40.361 81	42.277 05	37.810 29	38.748 31
4800	63.899 27	38.003 2	59.164 25	40.528 6	42.397 57	37.844 08	38.800 31
4900	63.978 02	38.035 17	59.283 91	40.688 41	42.524 45	37.877 44	38.853 31
5000	64.057 34	38.067 98	59.400 72	40.840 13	42.659 34	37.910 67	38.907 89
5100	64.137 82	38.101 85	59.515 47	40.983 23	42.803 09	37.944 03	38.964 43
5200	64.220 19	38.136 62	59.629 15	41.117 87	42.955 44	37.977 75	39.023 07
5300	64.305 28	38.171 69	59.742 87	41.245 1	43.114 97	38.011 99	39.083 62
5400	64.393 88	38.206 07	59.857 83	41.366 64	43.279 26	38.046 82	39.145 63
5500	64.486 53	38.238 53	59.975 12	41.484 55	43.445 5	38.082 2	39.208 49
5600	64.583 14	38.268 02	60.095 35	41.600 35	43.611 66	38.118 1	39.271 74
5700	64.682 49	38.294 34	60.218 17	41.713 61	43.778 49	38.154 59	39.335 61
5800	64.781 41	38.319 34	60.341 48	41.819 76	43.952 64	38.192 15	39.401 86
5900	64.873 73	38.348 53	60.460 26	41.906 81	44.151 07	38.232 04	39.475 04
6000	64.948 81	38.393 51	60.565 12	41.950 91	44.407 37	38.276 96	39.564 35

Table A.2 Internal energy U_T as per the correlations in Table 2.7 (enthalpy reference temperature $T_{ref} = 298.15$ K).

T (K)				U_T, MJ/kmole				MJ/kg
	CO_2	CO	H_2O	H_2	O_2	N_2	Air	Air
0	0	0	0	0	0	0	0	0
100	1.965 429	2.105 676	2.508 513	1.931 058	2.089 925	2.105 285	2.102 059	0.072 56
200	4.290 788	4.171 758	4.999 965	3.915 743	4.149 83	4.175 919	4.170 44	0.143 957
298	6.969 767	6.192 29	7.464 655	5.908 843	6.201 356	6.196 009	6.197 132	0.213 916
300	7.023 843	6.230 528	7.511 515	5.946 708	6.240 552	6.234 165	6.235 507	0.215 24
400	10.131 57	8.315 011	10.076 79	8.005 313	8.396 661	8.309 39	8.327 717	0.287 46
500	13.575 08	10.446 68	12.721 81	10.082 03	10.635 65	10.421 94	10.466 82	0.361 299
600	17.313 3	12.638 2	15.466	12.173 63	12.963 2	12.585 01	12.664 43	0.437 157
700	21.305 7	14.895 7	18.323 07	14.281 02	15.377 26	14.806 19	14.926 12	0.515 227
800	25.514 07	17.220 6	21.301 9	16.407 53	17.871 15	17.088 85	17.253 13	0.595 552
900	29.903 73	19.611 05	24.407 35	18.557 71	20.435 82	19.433 17	19.643 73	0.678 071
1000	34.444 11	22.063 07	27.640 95	20.736 41	23.061 53	21.837 11	22.094 24	0.762 659
1100	39.108 99	24.571 46	31.001 61	22.948 19	25.738 87	24.297 09	24.599 87	0.849 15
1200	43.876 45	27.130 4	34.486 14	25.196 9	28.459 47	26.808 65	27.155 32	0.937 36
1300	48.728 56	29.733 99	38.089 83	27.485 52	31.216 38	29.366 81	29.755 22	1.027 105
1400	53.650 99	32.376 54	41.806 87	29.816 06	34.004 13	31.966 52	32.394 42	1.118 206
1500	58.632 48	35.052 81	45.630 7	32.189 58	36.818 77	34.602 82	35.068 17	1.210 499
1600	63.664 38	37.758 11	49.554 39	34.606 27	39.657 64	37.271 09	37.772 27	1.303 841
1700	68.740 08	40.488 36	53.570 86	37.065 59	42.519 19	39.967 14	40.503 07	1.398 104
1800	73.854 6	43.240 1	57.673 1	39.566 42	45.402 71	42.687 24	43.257 49	1.493 182
1900	79.004 12	46.010 48	61.854 32	42.107 18	48.308 12	45.428 19	46.032 97	1.588 988
2000	84.185 62	48.797 12	66.108 1	44.686 03	51.235 65	48.187 27	48.827 43	1.685 448
2100	89.396 65	51.598 15	70.428 44	47.300 97	54.185 75	50.962 25	51.639 19	1.782 506
2200	94.635 03	54.412 06	74.809 8	49.949 96	57.158 82	53.751 3	54.466 88	1.880 113
2300	99.898 76	57.237 63	79.247 16	52.631 02	60.155 16	56.552 97	57.309 43	1.978 234
2400	105.185 9	60.073 91	83.735 98	55.342 3	63.174 83	59.366 08	60.165 92	2.076 835
2500	110.494 4	62.920 09	88.272 21	58.082 14	66.217 68	62.189 72	63.035 59	2.175 892
2600	115.822 5	65.775 51	92.852 23	60.849 07	69.283 27	65.023 15	65.917 78	2.275 381
2700	121.168 2	68.639 59	97.472 84	63.641 82	72.370 96	67.865 78	68.811 87	2.375 28
2800	126.529 8	71.511 79	102.131 2	66.459 35	75.479 91	70.717 08	71.717 28	2.475 57
2900	131.905 5	74.391 61	106.824 7	69.300 78	78.609 13	73.576 59	74.633 42	2.576 231
3000	137.293 8	77.278 58	111.551 2	72.165 41	81.757 59	76.443 87	77.559 75	2.677 244

(Continued)

Table A.2 (Continued)

T (K)	CO₂	CO	H₂O	H₂	O₂	N₂	Air	Air
			U_T, MJ/kmole					MJ/kg
3100	142.693 5	80.172 22	116.308 5	75.052 65	84.924 27	79.318 47	80.495 69	2.778 588
3200	148.103 7	83.072 1	121.094 8	77.962	88.108 17	82.199 96	83.440 68	2.880 245
3300	153.523 7	85.977 79	125.908 3	80.893 03	91.308 43	85.087 89	86.394 2	2.982 196
3400	158.953	88.888 9	130.747 5	83.845 35	94.524 32	87.981 81	89.355 73	3.084 423
3500	164.391 5	91.805 06	135.610 7	86.818 57	97.755 26	90.881 26	92.324 8	3.186 911
3600	169.839 3	94.725 96	140.496 7	89.812 29	101.000 8	93.785 82	95.300 97	3.289 644
3700	175.296 7	97.651 33	145.403 9	92.826 12	104.260 8	96.695 09	98.283 88	3.392 609
3800	180.763 9	100.581	150.331 1	95.859 65	107.534 9	99.608 71	101.273 2	3.495 796
3900	186.241 2	103.514 7	155.277 2	98.912 46	110.823 1	102.526 4	104.268 7	3.599 195
4000	191.728 9	106.452 4	160.241 1	101.984 1	114.125 3	105.447 9	107.270 1	3.702 801
4100	197.227	109.394	165.221 6	105.074 3	117.441 5	108.373	110.277 4	3.806 607
4200	202.735 3	112.339 5	170.217 9	108.182 6	120.771 4	111.301 8	113.290 4	3.910 61
4300	208.253 5	115.288 8	175.229 3	111.308 6	124.114 9	114.234	116.309	4.014 809
4400	213.781	118.241 9	180.254 9	114.452 1	127.471 6	117.169 9	119.333 3	4.119 202
4500	219.316 9	121.198 8	185.294 4	117.612 9	130.841 1	120.109 5	122.363 1	4.223 788
4600	224.860 4	124.159 5	190.347 1	120.790 8	134.222 9	123.052 9	125.398 6	4.328 566
4700	230.410 6	127.123 9	195.412 7	123.985 5	137.616 6	126.000 1	128.439 5	4.433 536
4800	235.966 6	130.091 8	200.491	127.196 9	141.021 8	128.951 2	131.486	4.538 697
4900	241.528	133.063 2	205.581 6	130.424 7	144.438 4	131.906 3	134.538	4.644 047
5000	247.094 8	136.038	210.684 4	133.668 5	147.866 4	134.865 2	137.595 5	4.749 585
5100	252.667 5	139.015 9	215.799 2	136.927 9	151.306 5	137.827 9	140.658 4	4.855 312
5200	258.247 5	141.996 9	220.925 8	140.202 1	154.759 7	140.793 9	143.726 7	4.961 227
5300	263.836 4	144.980 9	226.063 9	143.490 3	158.227 6	143.763 1	146.800 6	5.067 333
5400	269.436 6	147.967 8	231.213 3	146.791 4	161.711 8	146.735 1	149.880 2	5.173 635
5500	275.050 3	150.957 5	236.373 8	150.104 3	165.214 6	149.709 6	152.965 7	5.280 141
5600	280.678 8	153.950 4	241.545 5	153.427 6	168.737 6	152.686 7	156.057 4	5.386 862
5700	286.321 4	156.946 4	246.728 3	156.760 7	172.281 3	155.666 7	159.155 8	5.493 813
5800	291.973 6	159.945 8	251.923 1	160.103 1	175.844 3	158.650 7	162.261 4	5.601 013
5900	297.624 6	162.948 9	257.131 2	163.455 8	179.421	161.640 8	165.374 7	5.708 48
6000	303.254 1	165.955 7	262.355 2	166.821 8	183.000	164.640 7	168.496 1	5.816 228

Table A.3 Enthalpy change $\Delta H_T = H - H_0(T_{ref})$ as per the correlations in Table 2.9 (enthalpy reference temperature $T_{ref} = 298.15$ K).

	ΔH_T, MJ/kmole						
T (K)	CO_2	CO	H_2O	H_2	O_2	N_2	Air
298	0	0	0	0	0	0	0
300	0.068 089	0.056 128	0.060 593	0.049 347	0.059 696	0.055 512	0.056 267
400	4.006 382	2.969 628	3.455 42	2.964 984	3.016 379	2.969 033	2.979 562
500	8.306 034	5.926 356	6.928 975	5.883 3	6.083 341	5.908 128	5.946 024
600	12.905 02	8.942 989	10.501 87	8.807 705	9.250 579	8.894 479	8.970 63
700	17.752 16	12.027 55	14.188 4	11.743 55	12.506 91	11.940 18	12.060 59
800	22.805 33	15.182 02	17.997 62	14.697	15.840 99	15.050 43	15.217 67
900	28.029 95	18.404 29	21.934 31	17.674 27	19.242 03	18.225 67	18.439 98
1000	33.397 69	21.689 78	25.999 83	20.681 04	22.700 25	21.463 18	21.723 41
1100	38.885 39	25.032 52	30.192 83	23.722 16	26.207 14	24.758 41	25.062 59
1200	44.474 09	28.426 02	34.509 91	26.801 46	29.755 55	28.105 79	28.451 69
1300	50.148 34	31.863 82	38.946 18	29.921 67	33.339 73	31.499 5	31.884 93
1400	55.895 46	35.339 88	43.495 68	33.084 51	36.955 21	34.933 82	35.356 88
1500	61.705 09	38.848 77	48.151 79	36.290 7	40.598 67	38.403 48	38.862 68
1600	67.568 69	42.385 8	52.907 55	39.540 17	44.267 77	41.903 77	42.398 13
1700	73.479 22	45.947 03	57.755 87	42.832 17	47.960 99	45.430 62	45.959 7
1800	79.430 86	49.529 19	62.689 8	46.165 41	51.677 4	48.980 6	49.544 47
1900	85.418 73	53.129 67	67.702 61	49.538 25	55.416 53	52.550 89	53.150 1
2000	91.438 76	56.746 37	72.787 93	52.948 85	59.178 22	56.139 16	56.774 71
2100	97.487 52	60.377 62	77.939 8	56.395 21	62.962 45	59.743 51	60.416 78
2200	103.562 1	64.022 07	83.152 75	59.875 38	66.769 25	63.362 43	64.075 11
2300	109.659 9	67.678 64	88.421 78	63.387 46	70.598 62	66.994 62	67.748 68
2400	115.779	71.346 39	93.742 36	66.929 7	74.450 49	70.639 02	71.436 61
2500	121.917 4	75.024 5	99.110 43	70.500 5	78.324 66	74.294 68	75.138 11
2600	128.073 6	78.712 22	104.522 4	74.098 46	82.220 78	77.960 76	78.852 43
2700	134.246 3	82.408 85	109.974 9	77.722 38	86.138 37	81.636 46	82.578 87
2800	140.434 2	86.113 7	115.465 2	81.371 23	90.076 84	85.321 04	86.316 7
2900	146.636 4	89.826 12	120.990 7	85.044 15	94.035 49	89.013 8	90.065 22
3000	152.852	93.545 51	126.549	88.740 41	98.013 56	92.714 08	93.823 77
3100	159.080 3	97.271 27	132.138 1	92.459 4	102.010 2	96.421 25	97.591 67
3200	165.320 7	101.002 9	137.756	96.200 58	106.024 8	100.134 7	101.368 3
3300	171.572 5	104.739 9	143.401	99.963 48	110.056 3	103.854 1	105.153 2

(Continued)

Table A.3   (Continued)

T (K)	CO_2	CO	H_2O	H_2	O_2	N_2	Air
			ΔH_T, MJ/kmole				
3400	177.835 3	108.482	149.071 6	103.747 7	114.104 2	107.578 8	108.945 7
3500	184.108 8	112.228 9	154.766 1	107.552 7	118.167 8	111.308 5	112.745 6
3600	190.392 4	115.980 3	160.483 3	111.378 1	122.246 5	115.043	116.552 3
3700	196.685 9	119.736	166.221 8	115.223 5	126.339 7	118.782 1	120.365 8
3800	202.988 9	123.496 2	171.980 3	119.088 5	130.447 2	122.525 6	124.185 8
3900	209.301 1	127.260 6	177.757 6	122.972 7	134.568 4	126.273 5	128.012
4000	215.622 3	131.029 3	183.552 8	126.875 6	138.703 2	130.025 6	131.844 6
4100	221.952 2	134.802 3	189.364 8	130.797	142.851 2	133.782 1	135.683 2
4200	228.290 6	138.579 6	195.192 8	134.736 4	147.012 3	137.542 7	139.528
4300	234.637 1	142.361 1	201.035 8	138.693 7	151.186 2	141.307 5	143.378 6
4400	240.991 7	146.146 7	206.893 2	142.668 5	155.372 8	145.076 4	147.235 2
4500	247.354 2	149.936 2	212.764 4	146.660 6	159.571 8	148.849 1	151.097 4
4600	253.724 4	153.729 3	218.648 9	150.669 9	163.783 1	152.625 6	154.965 2
4700	260.102 3	157.526	224.546 2	154.696 2	168.006 6	156.405 6	158.838 2
4800	266.487 9	161.325 7	230.456 1	158.739 2	172.242 3	160.188 9	162.716 5
4900	272.881 3	165.128 5	236.378 1	162.798 8	176.490 2	163.975 3	166.599 8
5000	279.282 6	168.933 9	242.312 2	166.874 4	180.750 7	167.764 7	170.488 1
5100	285.691 9	172.742 1	248.258 2	170.965 5	185.024 2	171.557	174.381 5
5200	292.109 6	176.553 1	254.215 8	175.071 4	189.311 3	175.352 3	178.280 1
5300	298.536	180.367 1	260.185	179.191 1	193.612 9	179.150 8	182.184 3
5400	304.971 2	184.184 6	266.165 7	183.323 7	197.930 2	182.952 9	186.094 6
5500	311.415 7	188.005 9	272.157 8	187.467 8	202.264 4	186.759	190.011 6
5600	317.869 7	191.831 4	278.161 2	191.622 4	206.616 5	190.569 4	193.935 9
5700	324.333 2	195.661 2	284.176 2	195.786 7	210.987 5	194.384 4	197.867 8
5800	330.806 2	199.494 4	290.203 2	199.960 6	215.377 5	198.203 5	201.806 8
5900	337.288 4	203.329	296.242 7	204.145 1	219.785 2	202.025 1	205.751 6
6000	343.778 9	207.160 5	302.296 2	208.343 5	224.207 3	205.846	209.698 9

Table A.4 Absolute entropy s^o as per the correlations in Table 1.3 (reference pressure $p_{ref} = 0.1$ *MPa*).

T (K)	CO_2	CO	H_2O	H_2	O_2	N_2	T (K)	Air
			s^o_T, kJ/kmole K				s^o_T, kJ/kmole K	
0	0	0	0	0	0	0	220	185.153
100	179.009	165.85	152.388	100.3577	173.6279	160.3664	245	188.2783
200	199.975	186.025	175.485	119.8264	193.1668	179.7103	270	191.1002
298.15	213.795	197.653	188.834	130.9244	204.929	191.2404	295	193.6737
300	214.025	197.833	189.042	131.0969	205.1147	191.4217	320	196.0406
400	225.314	206.238	198.788	139.1642	213.8786	199.95	345	198.2331
500	234.901	212.831	206.534	145.5088	220.861	206.706	370	200.2766
600	243.283	218.319	213.052	150.7722	226.6958	212.3252	395	202.1916
700	250.75	223.066	218.739	155.291	231.726	217.1502	420	203.9947
800	257.494	227.277	223.825	159.2643	236.1586	221.3872	445	205.6996
900	263.645	231.074	228.459	162.8198	240.1288	225.1706	470	207.3176
1000	269.299	234.538	232.738	166.0444	243.7299	228.5929	495	208.8584
1100	274.528	237.726	236.731	169	247.0291	231.7205	520	210.3299
1200	279.39	240.679	240.485	171.7323	250.0764	234.6028	545	211.7392
1300	283.932	243.431	244.035	174.276	252.9101	237.2775	570	213.0922
1400	288.191	246.006	247.407	176.6583	255.5604	239.7743	595	214.394
1500	292.199	248.426	250.62	178.9004	258.0512	242.1166	620	215.6489
1600	295.983	250.707	253.69	181.02	260.402	244.3236	645	216.8609
1700	299.566	252.865	256.63	183.0311	262.6289	246.4109	670	218.0334
1800	302.968	254.912	259.451	184.9456	264.7452	248.3918	695	219.1693
1900	306.205	256.859	262.161	186.7734	266.7622	250.277	720	220.2713
2000	309.293	258.714	264.769	188.5231	268.6895	252.0761	745	221.3417
2100	312.244	260.486	267.282	190.2018	270.5355	253.7971	770	222.3826
2200	315.07	262.182	269.706	191.8157	272.3071	255.4468	795	223.3958
2300	317.781	263.808	272.048	193.3704	274.0107	257.0314	820	224.383
2400	320.385	265.369	274.312	194.8705	275.6516	258.556	845	225.3456
2500	322.89	266.871	276.503	196.3202	277.2347	260.0254	870	226.2851
2600	325.305	268.318	278.625	197.7232	278.7643	261.4437	895	227.2027
2700	327.634	269.713	280.683	199.0829	280.244	262.8145	920	228.0993
2800	329.885	271.06	282.68	200.4022	281.6774	264.1412	945	228.9762
2900	332.061	272.362	284.619	201.6837	283.0675	265.4266	970	229.8342
3000	334.169	273.623	286.504	202.9298	284.4171	266.6735	995	230.6741

(Continued)

Table A.4 (Continued)

T (K)	s_T^o, kJ/kmole K						T (K)	s_T^o, kJ/kmole K
	CO_2	CO	H_2O	H_2	O_2	N_2		Air
3100	336.211	274.844	288.337	204.1427	285.7285	267.8842	1020	231.4968
3200	338.192	276.029	290.12	205.3244	287.0042	269.0609	1045	232.303
3300	340.116	277.178	291.858	206.4765	288.2461	270.2056	1070	233.0933
3400	341.986	278.295	293.55	207.6009	289.4561	271.3201	1095	233.8685
3500	343.804	279.382	295.201	208.6988	290.636	272.4061	1120	234.6291
3600	345.574	280.438	296.812	209.7719	291.7874	273.4651	1145	235.3757
3700	347.299	281.468	298.384	210.8212	292.9117	274.4984	1170	236.1088
3800	348.979	282.471	299.919	211.8479	294.0103	275.5075	1195	236.8289
3900	350.619	283.449	301.42	212.8532	295.0844	276.4935	1220	237.5366
4000	352.219	284.403	302.887	213.8381	296.1351	277.4574	1245	238.2322
4100	353.782	285.335	304.322	214.8034	297.1637	278.4004	1270	238.9162
4200	355.31	286.245	305.726	215.75	298.171	279.3233	1295	239.589
4300	356.803	287.135	307.101	216.6788	299.1579	280.2271	1320	240.251
4400	358.264	288.005	308.448	217.5905	300.1255	281.1127	1345	240.9027
4500	359.694	288.856	309.767	218.4859	301.0744	281.9807	1370	241.5443
4600	361.094	289.69	311.061	219.3654	302.0054	282.8319	1395	242.1762
4700	362.466	290.506	312.329	220.2299	302.9193	283.667	1420	242.7988
4800	363.81	291.306	313.574	221.0798	303.8168	284.4866	1445	243.4124
4900	365.128	292.09	314.795	221.9158	304.6984	285.2914	1470	244.0173
5000	366.422	292.859	315.993	222.7383	305.5647	286.0819	1495	244.6137
5100	367.691	293.613	317.171	223.5478	306.4164	286.8586	1520	245.2021
5200	368.937	294.354	318.327	224.3448	307.254	287.6221	1545	245.7826
5300	370.161	295.08	319.464	225.1297	308.0779	288.3728	1570	246.3555
5400	371.364	295.794	320.582	225.903	308.8886	289.1112	1595	246.9211
5500	372.547	296.495	321.682	226.6649	309.6866	289.8376	1620	247.4796
5600	373.709	297.184	322.764	227.416	310.4723	290.5526	1645	248.0312
5700	374.853	297.862	323.828	228.1565	311.2462	291.2564	1670	248.5762
5800	375.979	298.528	324.877	228.8868	312.0085	291.9496	1695	249.1148
5900	377.087	299.184	325.909	229.6072	312.7597	292.6323	1720	249.6472
6000	378.178	299.829	326.926	230.318	313.5001	293.3049	1745	250.1737

Table A.5 Equilibrium constant $K_p = p_C^c p_D^d / p_A^a p_B^b$ for the reaction $aA + bB \rightleftarrows cC + dD$ (p_a, p_b, p_c, p_d are the partial pressures of the species A, B, C, and D; reference pressure = 1 *atm*).

T (K)	$H \rightleftarrows \frac{1}{2}H_2$	$O \rightleftarrows \frac{1}{2}O_2$	$N \rightleftarrows \frac{1}{2}N_2$	$H_2 + \frac{1}{2}O_2 \rightleftarrows H_2O$	$OH + \frac{1}{2}H_2 \rightleftarrows H_2O$	$\frac{1}{2}N_2 + \frac{1}{2}O_2 \rightleftarrows NO$
600	2.17E+16	3.75E+18	1.21E+38	4.30E+18	1.75E+21	6.17E−08
700	3.97E+13	2.81E+15	1.50E+32	3.83E+15	5.05E+17	8.20E−07
800	3.46E+11	1.26E+13	5.55E+27	1.95E+13	1.11E+15	5.72E−06
900	8.59E+09	1.87E+11	1.96E+24	3.15E+11	9.33E+12	2.59E−05
1000	4.43E+08	6.41E+09	3.37E+21	1.15E+10	2.04E+11	8.67E−05
1100	3.88E+07	4.04E+08	1.84E+19	7.64E+08	8.85E+09	2.33E−04
1200	5.09E+06	4.02E+07	2.38E+17	7.93E+07	6.47E+08	5.31E−04
1300	9.08E+05	5.69E+06	6.00E+15	1.16E+07	7.05E+07	1.07E−03
1400	2.07E+05	1.06E+06	2.55E+14	2.22E+06	1.05E+07	1.94E−03
1500	5.70E+04	2.48E+05	1.65E+13	5.31E+05	2.02E+06	3.26E−03
1600	1.85E+04	6.95E+04	1.50E+12	1.51E+05	4.75E+05	5.13E−03
1700	6.81E+03	2.25E+04	1.80E+11	5.00E+04	1.33E+05	7.66E−03
1800	2.81E+03	8.28E+03	2.74E+10	1.86E+04	4.28E+04	1.09E−02
1900	1.27E+03	3.38E+03	5.07E+09	7.69E+03	1.55E+04	1.50E−02
2000	6.17E+02	1.51E+03	1.11E+09	3.47E+03	6.21E+03	2.00E−02
2100	3.22E+02	7.24E+02	2.81E+08	1.69E+03	2.72E+03	2.59E−02
2200	1.78E+02	3.72E+02	8.04E+07	8.75E+02	1.28E+03	3.28E−02
2300	1.04E+02	2.03E+02	2.56E+07	4.81E+02	6.44E+02	4.06E−02
2400	6.31E+01	1.16E+02	9.00E+06	2.77E+02	3.43E+02	4.96E−02
2500	3.99E+01	6.95E+01	3.43E+06	1.68E+02	1.92E+02	5.93E−02
2600	2.61E+01	4.33E+01	1.41E+06	1.05E+02	1.13E+02	7.02E−02
2700	1.77E+01	2.79E+01	6.17E+05	6.81E+01	6.87E+01	8.19E−02
2800	1.23E+01	1.85E+01	2.86E+05	4.55E+01	4.34E+01	9.44E−02
2900	8.73E+00	1.27E+01	1.40E+05	3.13E+01	2.83E+01	1.08E−01
3000	6.35E+00	8.89E+00	7.21E+04	2.20E+01	1.90E+01	1.22E−01
3100	4.72E+00	6.38E+00	3.86E+04	1.59E+01	1.31E+01	1.37E−01
3200	3.57E+00	4.68E+00	2.15E+04	1.17E+01	9.18E+00	1.53E−01
3300	2.75E+00	3.49E+00	1.24E+04	8.75E+00	6.62E+00	1.69E−01
3400	2.15E+00	2.65E+00	7.38E+03	6.67E+00	4.86E+00	1.87E−01
3500	1.70E+00	2.04E+00	4.53E+03	5.15E+00	3.62E+00	2.04E−01

(Continued)

Table A.5 (Continued)

T (K)	$H \rightleftarrows \frac{1}{2}H_2$	$O \rightleftarrows \frac{1}{2}O_2$	$N \rightleftarrows \frac{1}{2}N_2$	$H_2 + \frac{1}{2}O_2 \rightleftarrows H_2O$	$OH + \frac{1}{2}H_2 \rightleftarrows H_2O$	$\frac{1}{2}N_2 + \frac{1}{2}O_2 \rightleftarrows NO$
3600	1.37E+00	1.60E+00	2.85E+03	4.05E+00	2.75E+00	2.22E−01
3700	1.11E+00	1.27E+00	1.84E+03	3.21E+00	2.12E+00	2.41E−01
3800	9.08E−01	1.02E+00	1.22E+03	2.59E+00	1.66E+00	2.60E−01
3900	7.53E−01	8.24E−01	8.22E+02	2.10E+00	1.31E+00	2.79E−01
4000	6.30E−01	6.76E−01	5.65E+02	1.73E+00	1.05E+00	2.99E−01
4100	5.32E−01	5.60E−01	3.96E+02	1.44E+00	8.51E−01	3.19E−01
4200	4.52E−01	4.68E−01	2.82E+02	1.20E+00	6.95E−01	3.39E−01
4300	3.87E−01	3.95E−01	2.04E+02	1.01E+00	5.73E−01	3.60E−01
4400	3.34E−01	3.35E−01	1.50E+02	8.61E−01	4.79E−01	3.80E−01
4500	2.90E−01	2.86E−01	1.11E+02	7.36E−01	4.01E−01	4.01E−01
4600	2.54E−01	2.47E−01	8.40E+01	6.35E−01	3.40E−01	4.22E−01
4700	2.23E−01	2.13E−01	6.41E+01	5.51E−01	2.89E−01	4.43E−01
4800	1.97E−01	1.86E−01	4.94E+01	4.80E−01	2.48E−01	4.65E−01
4900	1.75E−01	1.63E−01	3.85E+01	4.21E−01	2.14E−01	4.85E−01
5000	1.56E−01	1.44E−01	3.03E+01	3.72E−01	1.86E−01	5.06E−01
5100	1.40E−01	1.27E−01	2.40E+01	3.29E−01	1.62E−01	5.27E−01
5200	1.26E−01	1.13E−01	1.92E+01	2.92E−01	1.42E−01	5.48E−01
5300	1.14E−01	1.01E−01	1.55E+01	2.61E−01	1.26E−01	5.69E−01
5400	1.03E−01	9.08E−02	1.27E+01	2.34E−01	1.11E−01	5.90E−01
5500	9.40E−02	8.17E−02	1.04E+01	2.11E−01	9.91E−02	6.11E−01
5600	8.59E−02	7.38E−02	8.55E+00	1.91E−01	8.85E−02	6.32E−01
5700	7.87E−02	6.70E−02	7.11E+00	1.73E−01	7.93E−02	6.53E−01
5800	7.23E−02	6.10E−02	5.94E+00	1.57E−01	7.15E−02	6.73E−01
5900	6.65E−02	5.57E−02	4.99E+00	1.44E−01	6.46E−02	6.93E−01
6000	6.15E−02	5.11E−02	4.22E+00	1.32E−01	5.86E−02	7.13E−01

Table A.5   (Continued)

T (K)	$H_2O + CO$ $\rightleftarrows H_2 + CO_2$	$C + O_2$ $\rightleftarrows CO_2$	$C + 2H_2$ $\rightleftarrows CH_4$	$NO + \frac{1}{2}O_2$ $\rightleftarrows NO_2$	$\frac{1}{2}N_2 + \frac{3}{2}H_2$ $\rightleftarrows NH_3$	$3NO_2 + H_2O$ $\rightleftarrows 2HNO_3 + NO$
600	2.84E+01	2.54E+34	1.00E+02	1.26E+01	4.17E−02	3.08E−06
700	9.48E+00	3.21E+29	8.93E+00	2.36E+00	9.40E−03	1.16E−06
800	4.24E+00	6.76E+25	1.40E+00	6.70E−01	3.00E−03	5.74E−07
900	2.31E+00	9.33E+22	3.21E−01	2.52E−01	1.22E−03	3.39E−07
1000	1.45E+00	4.79E+20	9.75E−02	1.15E−01	5.85E−04	2.29E−07
1100	9.93E−01	6.40E+18	3.63E−02	6.10E−02	3.20E−04	1.69E−07
1200	7.33E−01	1.75E+17	1.58E−02	3.58E−02	1.92E−04	1.33E−07
1300	5.72E−01	8.32E+15	7.82E−03	2.29E−02	1.25E−04	1.10E−07
1400	4.65E−01	6.10E+14	4.25E−03	1.56E−02	8.63E−05	9.46E−08
1500	3.90E−01	6.32E+13	2.50E−03	1.12E−02	6.25E−05	8.40E−08
1600	3.36E−01	8.71E+12	1.57E−03	8.38E−03	4.73E−05	7.67E−08
1700	2.95E−01	1.51E+12	1.05E−03	6.49E−03	3.69E−05	7.10E−08
1800	2.65E−01	3.19E+11	7.26E−04	5.18E−03	2.97E−05	6.70E−08
1900	2.40E−01	7.91E+10	5.24E−04	4.22E−03	2.44E−05	6.44E−08
2000	2.21E−01	2.25E+10	3.91E−04	3.52E−03	2.05E−05	6.21E−08
2100	2.05E−01	7.24E+09	3.00E−04	2.99E−03	1.75E−05	6.05E−08
2200	1.92E−01	2.58E+09	2.36E−04	2.57E−03	1.52E−05	5.97E−08
2300	1.81E−01	1.00E+09	1.90E−04	2.24E−03	1.33E−05	5.90E−08
2400	1.72E−01	4.22E+08	1.55E−04	1.98E−03	1.18E−05	5.83E−08
2500	1.64E−01	1.91E+08	1.29E−04	1.77E−03	1.06E−05	5.81E−08
2600	1.58E−01	9.12E+07	1.09E−04	1.59E−03	9.64E−06	5.86E−08
2700	1.52E−01	4.61E+07	9.33E−05	1.44E−03	8.79E−06	5.86E−08
2800	1.47E−01	2.44E+07	8.07E−05	1.32E−03	8.09E−06	5.92E−08
2900	1.43E−01	1.36E+07	7.05E−05	1.21E−03	7.50E−06	5.97E−08
3000	1.39E−01	7.80E+06	6.22E−05	1.12E−03	6.98E−06	6.05E−08
3100	1.35E−01	4.66E+06	5.53E−05	1.04E−03	6.53E−06	6.14E−08
3200	1.32E−01	2.87E+06	4.97E−05	9.71E−04	6.15E−06	6.22E−08
3300	1.29E−01	1.82E+06	4.48E−05	9.10E−04	5.81E−06	6.32E−08
3400	1.27E−01	1.19E+06	4.06E−05	8.55E−04	5.51E−06	6.43E−08
3500	1.25E−01	7.91E+05	3.72E−05	8.07E−04	5.25E−06	6.59E−08

(Continued)

Table A.5   (Continued)

T (K)	$H_2O + CO$ $\rightleftarrows H_2 + CO_2$	$C + O_2$ $\rightleftarrows CO_2$	$C + 2H_2$ $\rightleftarrows CH_4$	$NO + \frac{1}{2}O_2$ $\rightleftarrows NO_2$	$\frac{1}{2}N_2 + \frac{3}{2}H_2$ $\rightleftarrows NH_3$	$3NO_2 + H_2O$ $\rightleftarrows 2HNO_3 + NO$
3600	1.23E−01	5.40E+05	3.41E−05	7.64E−04	5.01E−06	6.70E−08
3700	1.22E−01	3.75E+05	3.14E−05	7.26E−04	4.80E−06	6.82E−08
3800	1.20E−01	2.66E+05	2.91E−05	6.90E−04	4.61E−06	6.98E−08
3900	1.19E−01	1.92E+05	2.70E−05	6.61E−04	4.44E−06	7.10E−08
4000	1.18E−01	1.41E+05	2.52E−05	6.31E−04	4.29E−06	7.28E−08
4100	1.16E−01	1.05E+05	2.37E−05	6.05E−04	4.15E−06	7.43E−08
4200	1.16E−01	7.91E+04	2.22E−05	5.82E−04	4.02E−06	7.62E−08
4300	1.15E−01	6.04E+04	2.09E−05	5.60E−04	3.91E−06	7.78E−08
4400	1.14E−01	4.68E+04	1.98E−05	5.40E−04	3.80E−06	7.96E−08
4500	1.13E−01	3.66E+04	1.88E−05	5.21E−04	3.71E−06	8.17E−08
4600	1.12E−01	2.88E+04	1.78E−05	5.04E−04	3.62E−06	8.34E−08
4700	1.12E−01	2.31E+04	1.69E−05	4.88E−04	3.54E−06	8.53E−08
4800	1.11E−01	1.85E+04	1.61E−05	4.72E−04	3.47E−06	8.73E−08
4900	1.11E−01	1.51E+04	1.54E−05	4.59E−04	3.40E−06	8.91E−08
5000	1.11E−01	1.23E+04	1.47E−05	4.47E−04	3.33E−06	9.12E−08
5100	1.10E−01	1.02E+04	1.41E−05	4.34E−04	3.27E−06	9.38E−08
5200	1.10E−01	8.45E+03	1.35E−05	4.23E−04	3.22E−06	9.55E−08
5300	1.10E−01	7.08E+03	1.30E−05	4.12E−04	3.17E−06	9.77E−08
5400	1.10E−01	5.96E+03	1.25E−05	4.02E−04	3.13E−06	1.00E−07
5500	1.10E−01	5.05E+03	1.21E−05	3.92E−04	3.08E−06	1.02E−07
5600	1.10E−01	4.30E+03	1.16E−05	3.82E−04	3.04E−06	1.05E−07
5700	1.10E−01	3.68E+03	1.12E−05	3.74E−04	3.01E−06	1.07E−07
5800	1.10E−01	3.17E+03	1.08E−05	3.66E−04	2.97E−06	1.09E−07
5900	1.10E−01	2.74E+03	1.05E−05	3.58E−04	2.94E−06	1.13E−07
6000	1.09E−01	2.38E+03	1.02E−05	3.52E−04	2.91E−06	1.15E−07

Table A.5 (Continued)

T (K)	$C + \frac{1}{2}N_2 + \frac{1}{2}H_2 \rightleftarrows HCN$	$O_3 \rightleftarrows \frac{3}{2}O_2$	$\frac{1}{2}O_2 + \frac{1}{2}H_2 \rightleftarrows OH$	$C + \frac{1}{2}O_2 \rightleftarrows CO$	$CO + \frac{1}{2}O_2 \rightleftarrows CO_2$
600	2.81E−10	1.08E+16	2.46E−03	2.08E+14	1.22E+20
700	1.15E−08	1.82E+14	7.59E−03	8.83E+12	3.63E+16
800	1.87E−07	8.45E+12	1.76E−02	8.20E+11	8.24E+13
900	1.62E−06	7.75E+11	3.37E−02	1.28E+11	7.28E+11
1000	9.10E−06	1.14E+11	5.66E−02	2.86E+10	1.68E+10
1100	2.69E+04	2.38E+10	8.63E−02	8.43E+09	7.59E+08
1200	1.20E−04	6.41E+09	1.23E−01	3.01E+09	5.81E+07
1300	3.24E−04	2.11E+09	1.64E−01	1.26E+09	6.62E+06
1400	7.57E−04	8.13E+08	2.12E−01	5.90E+08	1.03E+06
1500	1.58E−03	3.55E+08	2.63E−01	3.06E+08	2.07E+05
1600	3.01E−03	1.72E+08	3.18E−01	1.71E+08	5.08E+04
1700	5.30E−03	9.06E+07	3.76E−01	1.03E+08	1.48E+04
1800	8.77E−03	5.12E+07	4.36E−01	6.47E+07	4.93E+03
1900	1.37E−02	3.06E+07	4.97E−01	4.28E+07	1.85E+03
2000	2.06E−02	1.93E+07	5.59E−01	2.94E+07	7.66E+02
2100	2.97E−02	1.27E+07	6.21E−01	2.09E+07	3.46E+02
2200	4.15E−02	8.71E+06	6.84E−01	1.53E+07	1.68E+02
2300	1.78E+01	6.15E+06	7.46E−01	1.15E+07	8.71E+01
2400	7.43E−02	4.48E+06	8.09E−01	8.83E+06	4.78E+01
2500	9.62E−02	3.33E+06	8.71E−01	6.92E+06	2.75E+01
2600	1.22E−01	2.55E+06	9.31E−01	5.51E+06	1.66E+01
2700	1.52E−01	1.98E+06	9.91E−01	4.46E+06	1.04E+01
2800	1.86E−01	1.57E+06	1.05E+00	3.66E+06	6.68E+00
2900	2.25E−01	1.27E+06	1.11E+00	3.04E+06	4.46E+00
3000	2.69E−01	1.03E+06	1.16E+00	2.55E+06	3.06E+00
3100	3.18E−01	8.55E+05	1.22E+00	2.17E+06	2.15E+00
3200	3.72E−01	7.16E+05	1.27E+00	1.86E+06	1.55E+00
3300	4.30E−01	6.07E+05	1.32E+00	1.61E+06	1.13E+00
3400	4.93E−01	5.19E+05	1.37E+00	1.40E+06	8.49E−01
3500	5.62E−01	4.48E+05	1.42E+00	1.23E+06	6.46E−01
3600	6.35E−01	3.89E+05	1.47E+00	1.08E+06	4.99E−01
3700	7.15E−01	3.41E+05	1.51E+00	9.59E+05	3.91E−01
3800	7.98E−01	3.01E+05	1.56E+00	8.57E+05	3.11E−01
3900	8.85E−01	2.68E+05	1.60E+00	7.69E+05	2.50E−01
4000	9.77E−01	2.40E+05	1.64E+00	6.93E+05	2.03E−01

(Continued)

Table A.5 (Continued)

T (K)	$C + \frac{1}{2}N_2 + \frac{1}{2}H_2 \rightleftarrows HCN$	$O_3 \rightleftarrows \frac{3}{2}O_2$	$\frac{1}{2}O_2 + \frac{1}{2}H_2 \rightleftarrows OH$	$C + \frac{1}{2}O_2 \rightleftarrows CO$	$CO + \frac{1}{2}O_2 \rightleftarrows CO_2$
4100	1.08E+00	2.16E+05	1.69E+00	6.28E+05	1.67E−01
4200	1.18E+00	1.95E+05	1.73E+00	5.70E+05	1.39E−01
4300	1.28E+00	1.77E+05	1.77E+00	5.21E+05	1.16E−01
4400	1.39E+00	1.61E+05	1.80E+00	4.78E+05	9.80E−02
4500	1.51E+00	1.48E+05	1.84E+00	4.39E+05	8.34E−02
4600	1.63E+00	1.37E+05	1.87E+00	4.05E+05	7.13E−02
4700	1.75E+00	1.26E+05	1.91E+00	3.74E+05	6.17E−02
4800	1.87E+00	1.17E+05	1.94E+00	3.47E+05	5.35E−02
4900	2.00E+00	1.08E+05	1.97E+00	3.22E+05	4.68E−02
5000	2.13E+00	1.01E+05	2.00E+00	3.00E+05	4.11E−02
5100	2.27E+00	9.46E+04	2.03E+00	2.81E+05	3.63E−02
5200	2.40E+00	8.87E+04	2.06E+00	2.62E+05	3.22E−02
5300	2.54E+00	8.34E+04	2.08E+00	2.46E+05	2.88E−02
5400	2.69E+00	7.85E+04	2.11E+00	2.31E+05	2.58E−02
5500	2.83E+00	7.43E+04	2.13E+00	2.18E+05	2.32E−02
5600	2.98E+00	7.03E+04	2.16E+00	2.05E+05	2.09E−02
5700	3.13E+00	6.67E+04	2.18E+00	1.94E+05	1.90E−02
5800	3.28E+00	6.34E+04	2.20E+00	1.84E+05	1.73E−02
5900	3.43E+00	6.04E+04	2.23E+00	1.74E+05	1.58E−02
6000	3.59E+00	5.75E+04	2.25E+00	1.65E+05	1.44E−02

Table A.6 Coefficients of correlations for enthalpies of reactants $H_R = a + bT_2 + c\lambda + dT_2^2 + e\lambda^2 + fT_2\lambda + gT_2^3 + h\lambda^3 + iT_2\lambda^2 + jT_2^2\lambda$ (kJ/kmole).

Coefficient	H_2	CH_4	C_3H_8	C_8H_{18}	$C_{12}H_{26}$	$C_{12}H_{25}$
a	$8.79502E+03$	$-2.82053E+02$	$1.63637E+00$	$1.66356E+03$	$1.76430E+03$	$2.42286E+03$
b	$-1.90392E+01$	$-9.51961E+00$	$-8.65419E-03$	$-8.35054E+00$	$-8.28766E+00$	$-8.15967E+00$
c	$-1.23981E+04$	$-6.19904E+03$	$-5.63549E+00$	$-5.43775E+03$	$-5.39681E+03$	$-5.31346E+03$
d	$1.04466E-02$	$5.22329E-03$	$4.74844E-06$	$4.58183E-03$	$4.54733E-03$	$4.47710E-03$
e	$-7.71553E-05$	$-1.36147E-07$	$-4.89257E-08$	$-4.87238E-07$	$2.91551E-07$	$-4.15375E-05$
f	$3.74633E+01$	$1.87316E+01$	$1.70288E-02$	$1.64313E+01$	$1.63075E+01$	$1.60557E+01$
g	$-1.62335E-06$	$-8.11673E-07$	$-7.37884E-10$	$-7.11993E-07$	$-7.06633E-07$	$-6.95719E-07$
h	$4.93565E-06$	$9.12031E-09$	$3.11616E-09$	$3.22756E-08$	$-2.49848E-08$	$2.65266E-06$
i	$4.03590E-09$	$-8.12101E-12$	$2.73203E-12$	$4.96818E-11$	$9.33819E-12$	$2.28862E-09$
j	$1.27325E-03$	$6.36625E-04$	$5.78750E-07$	$5.58443E-04$	$5.54238E-04$	$5.45679E-04$

Table A.7 Coefficients of correlations for enthalpies of products $H_P = a + bT_3 + c\lambda + dT_3^2 + e\lambda^2 + fT_3\lambda + gT_3^3 + h\lambda^3 + iT_3\lambda^2 + jT_3^2\lambda$ (kJ/kmole).

Coeffi-cient	H_2	CH_4	C_3H_8	C_8H_{18}	$C_{12}H_{26}$	$C_{12}H_{25}$
a	$-1.14180E+05$	$-5.12421E+04$	$-4.55204E+01$	$-4.35128E+04$	$-4.30971E+04$	$-4.22509E+04$
b	$-1.27141E+01$	$-7.18553E+00$	$-6.68294E-03$	$-6.50659E+00$	$-6.47008E+00$	$-6.39574E+00$
c	$-1.23981E+04$	$-6.19904E+03$	$-5.63549E+00$	$-5.43775E+03$	$-5.39681E+03$	$-5.31346E+03$
d	$1.38230E-02$	$6.33678E-03$	$5.65621E-06$	$5.41742E-03$	$5.36797E-03$	$5.26732E-03$
e	$-3.70066E-06$	$-2.41697E-06$	$-2.37280E-09$	$-2.46470E-06$	$-2.16633E-06$	$-1.98725E-06$
f	$3.74633E+01$	$1.87316E+01$	$1.70288E-02$	$1.64313E+01$	$1.63075E+01$	$1.60557E+01$
g	$-1.94235E-06$	$-9.26434E-07$	$-8.34078E-10$	$-8.01672E-07$	$-7.94962E-07$	$-7.81303E-07$
h	$2.68168E-07$	$1.57560E-07$	$1.56273E-10$	$1.66690E-07$	$1.40473E-07$	$1.26494E-07$
i	$-1.55419E-10$	$1.13088E-10$	$9.93343E-14$	$8.04214E-11$	$1.20620E-10$	$9.71552E-11$
j	$1.27325E-03$	$6.36625E-04$	$5.78750E-07$	$5.58443E-04$	$5.54238E-04$	$5.45679E-04$

Table A.8 Coefficients of adiabatic flame temperature correlations for some hydrocarbon fuel reactions without dissociation: $T_3 = a + \dfrac{b}{\lambda} + cT_2 + \dfrac{d}{\lambda^2} + eT_2^2 + f\dfrac{T_2}{\lambda} + \dfrac{g}{\lambda^3} + hT_2^3 + i\dfrac{T_2^2}{\lambda} + j\dfrac{T_2}{\lambda^2}$ (K).

Coeffi-cient	H_2	CH_4	C_3H_8	C_8H_{18}	$C_{12}H_{26}$	$C_{12}H_{25}$
a	$-8.98199E+00$	$-2.22600E+00$	$-1.14600E+01$	$-9.65541E+00$	$-2.40678E+01$	$-1.26729E+01$
b	$3.78227E+03$	$3.13811E+03$	$3.24152E+03$	$3.22234E+03$	$3.24076E+03$	$3.25940E+03$
c	$9.26981E-01$	$9.05532E-01$	$9.32138E-01$	$9.22352E-01$	$9.96977E-01$	$9.37338E-01$
d	$-1.67409E+03$	$-9.94348E+02$	$-1.07719E+03$	$-1.04883E+03$	$-1.05241E+03$	$-1.06247E+03$
e	$1.61958E-04$	$2.01217E-04$	$1.55335E-04$	$1.71724E-04$	$6.09055E-05$	$1.47381E-04$
f	$-1.03766E+00$	$-8.99630E-01$	$-9.13546E-01$	$-9.06293E-01$	$-9.94669E-01$	$-9.14949E-01$
g	$3.74522E+02$	$1.40576E+02$	$1.85327E+02$	$1.79136E+02$	$1.98246E+02$	$1.83875E+02$
h	$-8.20195E-08$	$-1.03345E-07$	$-7.81163E-08$	$-8.64585E-08$	$-3.85981E-08$	$-7.40568E-08$
i	$1.48830E-04$	$1.60817E-04$	$1.62041E-04$	$1.62731E-04$	$2.57445E-04$	$1.67612E-04$
j	$3.10173E-01$	$2.01934E-01$	$2.25758E-01$	$2.22798E-01$	$1.75793E-01$	$2.27784E-01$

Table A.9   Enthalpy of formation of selected chemical substances ($T_{ref} = 298.15\ K$, $p_{ref} = 0.1\ MPa$).

Fuel (state)	Formula	Molecular mass	Enthalpy of formation kJ/kmole
Acetylene (g)	C_2H_2	26	226 730
Benzene (l)	C_6H_6	78	49 000
Benzene (g)	C_6H_6	78	82 930
Butane (g)	C_4H_{10}	58	−126 150
Carbon dioxide (g)	CO_2	44	−393 520
Carbon monoxide (g)	CO	28	−110 530
Decane (l)	$C_{10}H_{22}$	142	−300 900
Dodecane (l)	$C_{12}H_{26}$	170	−350 900
Dodecane (g)	$C_{12}H_{26}$	170	−291 000
Dodecene (l)	$C_{12}H_{24}$	168	−165 460
Ethane (g)	C_2H_6	30	−84 680
Ethanol (l)	C_2H_5OH	46	−277 400
Ethanol (g)	C_2H_5OH	46	−235 000
Ethylene (g)	C_2H_4	28	52 288
Heptane (l)	C_7H_{16}	100	−224 200
Heptane (g)	C_7H_{16}	100	−187 900
Hydroxyl (g)	OH	17	39 000
Methane (g)	CH_4	16	−74 400
Methanol (l)	CH_3OH	32	−239 200
Methanol (g)	CH_3OH	32	−201 300
Octane (l)	C_8H_{18}	114	−249 910
Octane (g)	C_8H_{18}	114	−208 600
Propane (g)	C_3H_8	44	−103 850
Water (l)	H_2O	18	−285 230
Water (g)	H_2O	18	−241 820

Appendix B

Dynamics of the Reciprocating Mechanism

Dynamic forces acting on the crank mechanism in a spark ignition (SI) engine are shown in Figure B.1. Force F_p, the algebraic sum of the gas force F_g and inertia force F_i acting on the piston pin, is resolved into forces K acting along the connecting rod and Q acting perpendicular to the piston axis (cylinder wall). Force K, acting on the crank pin, generates the turning moment of the crank.

Tables B.1–B.4 show the unbalanced inertial forces and their moments for a selection of two- and four-stroke inline and V-engines. In referring to these tables, note the following:

1. Opposed-cylinder engines are treated as V-engines with $\gamma = 180°$.
2. Increasing the number of cylinders reduces the number of unbalanced forces and moments.
3. Six-, 8-, and 12-cylinder four-stroke inline engines are fully balanced.
4. Twelve- and 16-cylinder four-stroke V-engines are fully balanced.
5. Both two-stroke inline and V-engines are less balanced than equivalent four-stroke engines.
6. The values of the forces and moments shown in the tables are the maxima that can be reached.
7. Forces and moments that are complicated to balance are left unbalanced in real engines.
8. Cylinders in inline engines are numbered in ascending order when viewed from the front.

Fundamentals of Heat Engines: Reciprocating and Gas Turbine Internal Combustion Engines, First Edition. Jamil Ghojel.
© 2020 John Wiley & Sons Ltd. This Work is a co-publication between John Wiley & Sons Ltd and ASME Press.
Companion website: www.wiley.com/go/JamilGhojel_Fundamentals of Heat Engines

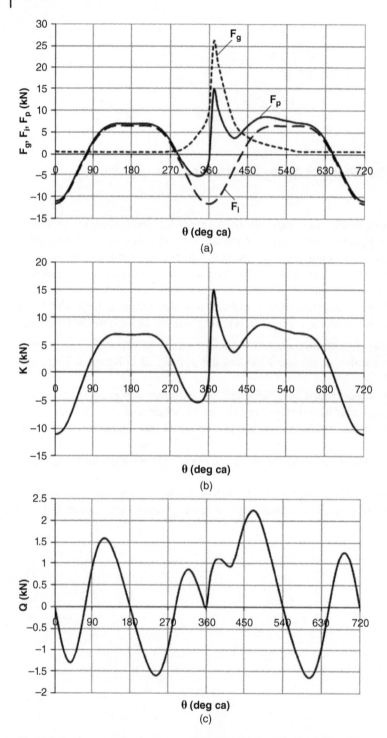

Figure B.1 Forces acting on the crank mechanism in a SI engine ($D = 78$ mm, $p_{max} = 64.7$ bar, $R = 39$ mm, $\tau = 0.285$, $\omega = 586$ rad/s): (a) gas, inertia, and resultant forces; (b) and (c) components K and Q of force F_p.

9. V-engine cylinders are numbered as shown in the example of an eight-cylinder V-engine in Figure B.2, as viewed from the front. The crank throws shared by cylinder pairs are numbered in ascending order.

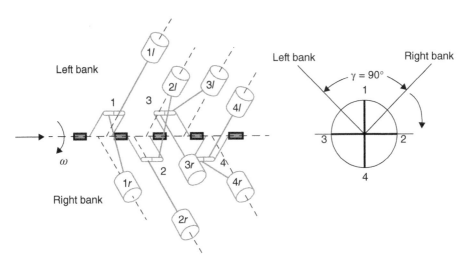

Figure B.2 V-engine cylinder and crank-throw numbering method used in Tables B.1–B.4, viewed from the front of the engine: *l* for left, *r* for right.

10. Equations of the forces and moments:

		Force	Moment
Reciprocating masses	First order	$\displaystyle\sum_{1}^{j} F_{il} = Z_i \cos\theta$	$\displaystyle\sum_{1}^{j} M_{il} = Z_i \cos\theta\, a$
	Second order	$\displaystyle\sum_{1}^{j} F_{il} = Z_i \tau \cos 2\theta$	$\displaystyle\sum_{1}^{j} M_{il} = Z_i \tau \cos 2\theta\, a$
Rotating masses	–	$\displaystyle\sum_{1}^{j} K_R = Z_r$	$\displaystyle\sum_{1}^{j} M_R = Z_r\, a$

$$Z_i = m_i R\omega^2,\ Z_r = m_R R\omega^2$$

Table B.1 Unbalanced inertial forces and moments in four-stroke inline engines.

				Unbalanced forces			Unbalanced moments		
No of cyl	Crank configuration	Angle between cranks	Firing order	$\sum F_{iI}$	$\sum F_{iII}$	$\sum K_R$	$\sum M_{iI}$	$\sum M_{iII}$	$\sum M_R$
3		120	1-2-3	0	0	0	$\sqrt{3}Z_i a$	$\sqrt{3}Z_i \tau a$	$\sqrt{3}Z_r a$
4		180	1-2-4-3	0	$4Z_i \tau$	0	0	0	0
5		72	1-2-4-5-3	0	0	0	$0.45Z_i a$	$4.98Z_i \tau a$	$0.45Z_r a$
6		120	1-5-3-6-2-4	0	0	0	0	0	0
8		90	1-6-2-5-8-3-7-4	0	0	0	0	0	0
12		60	1-6-9-2-8-3-12-7-4-11-5-10	0	0	0	0	0	0

Table B.2 Unbalanced inertial forces and moments in two-stroke inline engines.

No of cyl	Crank configuration	Angle between cranks	Firing order	Unbalanced forces			Unbalanced moments		
				$\sum F_{iI}$	$\sum F_{iII}$	$\sum K_R$	$\sum M_{iI}$	$\sum M_{iII}$	$\sum M_R$
3		120	1-2-3	0	0	0	$1.732 Z_i a$	$1.732 Z_i \tau a$	$1.732 Z_r a$
4		90	1-3-4-2	0	0	0	$3.16 Z_i a$	0	$3.16 Z_r a$
5		72	1-5-2-3-4	0	0	0	$0.45 Z_i a$	$4.98 Z_i \tau a$	$0.45 Z_r a$
6		60	1-6-2-4-3-5	0	0	0	0	$3.46 Z_i \tau a$	0
8		45	1-8-2-6-4-5-3-7	0	0	0	$0.45 Z_i a$	0	$0.45 Z_r a$
12		30	1-6-8-10-3-5-7-12-2-4-9-11	0	0	0	0	0	0

Table B.3 Unbalanced inertial forces and moments in four-stroke V-engines.

				Unbalanced forces			Unbalanced moments		
No of cyl	Crank configuration	Angle between cranks	Firing order	$\sum F_{II}$	$\sum F_{III}$	$\sum K_R$	$\sum M_{II}$	$\sum M_{III}$	$\sum M_R$
2	$\gamma = 180°$ (2 1)	0	1l-1r	$2Z_i$	0	$2Z_R$	0	0	0
4	$\gamma = 180°$ (1 2)	180	1l-2r-1r-2l	0	0	0	0	$2Z_i\tau a$	0
4[a]	$\gamma = 90°$ (1 2 3 4)	90	1l-1r-2r-2l	0	$2.828 \times Z_i\tau$	0	$1.73Z_i a$	0	$1.73Z_r a$
6	$\gamma = 90°$ (1 2 3)	120	1l-3l-2l-2r-1r-3r	0	0	0	$1.73Z_i a$	$2.5Z_i\tau a$	$1.73Z_r a$
8	$\gamma = 90°$ (1 2 3 4)	90	1l-3l-3r-2l-2r-1r-4l-4r	0	0	0	$3.16Z_i a$	0	$3.16Z_r a$
12	$\gamma = 60°$ (1 6 2 5 4 3)	120	1l-6r-5l-2r-3l-4r-6l-1r-2l-5r-4r-3l	0	0	0	0	0	0
16	$\gamma = 135°$ (1 8 3 6 7 2 4 5)	90	1l-7r-3l-8r-4l-6r-2l-5r-8l-2r-6l-1r-5l-3r-7l-4r	0	0	0	0	0	0

a) Crankshaft with four throws.

Table B.4 Unbalanced inertial forces and moments in two-stroke V-engines.

				Two-stroke V-engines					
					Unbalanced forces		Unbalanced moments		
No of cyl	Crank configuration	Angle between cranks	Firing order	$\sum F_{iI}$	$\sum F_{iII}$	$\sum K_R$	$\sum M_{iI}$	$\sum M_{iII}$	$\sum M_R$
4	$\gamma = 90°$	180	1l-1r-2l-2r	0	$2.83Z_i\tau$	0	$Z_i a$	0	$Z_r a$
6	$\gamma = 90°$	120	1l-1r-2l-2r-3l-3r	0	0	0	$1.73Z_i a$	$2.45Z_i \tau a$	$1.73Z_r a$
8	$\gamma = 90°$	90	(1l-4r) (3l-1r) (2l-3r) (4l-2r)	0	0	0	$3.16Z_i a$	0	$1.41Z_r a$
12	$\gamma = 90°$	60	1l-5r-6l-1r-2l-6r-4l-2r-3l-4r-5l-3r	0	0	0	0	$4.9Z_i \tau a$	0
16	$\gamma = 45°$	45	(1l-7r) (8l-1r) (2l-8r) (6l-2r) (4l-6r) (5l-4r)	0	0	0	$0.62Z_i a$	0	$0.45Z_r a$

The cylinders shown in a bracket fire simultaneously.

Appendix C

Design Point Calculations – Reciprocating Engines

It is required to determine the engine performance parameters and cylinder size at the design point of a direct injection compression ignition (CI) (diesel) engine with the specifications listed in Table C.1. Table C.2 shows the assumed data for the calculations.

C.1 Engine Processes

Cycle design-point calculations are based on the dual-combustion cycle shown in Figure C.1.

C.1.1 Induction Process

$$p_1 = p_a = 100 \, kPa$$

From Eq. (6.4),

$$T_1 = \frac{T_a + \Delta T + \gamma_{res} T_{res}}{1 + \gamma_{res}} = \frac{298 + 10 + 0.04 \times 860}{1 + 0.04} = 329.2 \, K$$

C.1.2 Compression Process

$$p_2 = p_1 \varepsilon^{n_1} = 100 \times 17.9^{1.36} = 5056.8 \, kPa$$

$$T_2 = T_1 \varepsilon^{n_1 - 1} = 329.2 \times 17.9^{0.36} = 930 \, K$$

C.1.3 Combustion Process

The combustion equation is given by Eq. (6.7):

$$\frac{\varphi H_l}{M_R (1 + \gamma_{res})} + \frac{U_{R2} + \gamma_{res} U_{P2}}{(1 + \gamma_{res})} + \alpha \overline{R} T_2 = \psi (U_{P4} + \overline{R} T_4) \tag{C.1}$$

Fundamentals of Heat Engines: Reciprocating and Gas Turbine Internal Combustion Engines, First Edition. Jamil Ghojel.
© 2020 John Wiley & Sons Ltd. This Work is a co-publication between John Wiley & Sons Ltd and ASME Press.
Companion website: www.wiley.com/go/JamilGhojel_Fundamentals of Heat Engines

Table C.1 Engine and fuel specifications.

Rated power, $\dot{W}_b$ (kW)	70
Rated speed (rpm)	2800
Compression ratio	17.9
Number of cylinders	4
Relative air-fuel ratio (λ)	1.4
Gravimetric fuel composition	$x_c = 0.87$ $x_H = 0.126$ $x_O = 0.04$
Lower heating value (MJ/kg)	43.2

Table C.2 Assumed data.

Ambient temp, T_a (K)	298
Ambient pressure, p_a (kPa)	100
Density of ambient air, ρ_a (kg/m^3)	1.1692
Temperature rise during induction, ΔT (K)	10
Coefficient of residual gases, (γ_r)	0.04
Temperature of residual gases, T_{res} (K)	860 (assumed initially and then refined by iteration)
Compression exponent (n_1)	1.36
Expansion exponent (n_2)	1.26
Pressure rise ratio (α)	1.5 (max cycle pressure not to exceed 80 bar)
Rounding-off factor (σ)	0.925
Coefficient of molar change (ψ)	1.0455
Volumetric efficiency (η_V)	0.83
Heat utilization coefficient (φ)	0.82

Figure C.1 Dual-combustion cycle with rounding-off.

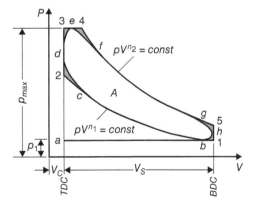

To solve this equation, the quantity of the reactants M_R must be known. We first find the air-fuel ratios on volume and mass bases:

$$\left(\frac{A}{F}\right)_s^v = \frac{1}{0.21}\left(\frac{x_C}{12} + \frac{x_H}{4} - \frac{x_O}{32}\right) = 0.4946 \ kMol \ air/kg \ fuel$$

$$\left(\frac{A}{F}\right)_s^g = \frac{1}{0.23}\left(\frac{8}{3}x_c + 8x_H + x_S - x_O\right) = 14.45 \ kg \ air/kg \ fuel$$

The quantity of the reactive mixture calculated from Eq. (2.31), ignoring the first term, is then

$$M_R = \lambda\left(\frac{A}{F}\right)_s^v = 1.4 \times 0.4946 = 0.6924 \ kmole/kg \ fuel$$

The products of combustion for the given fuel composition are:

$$M_{CO_2} = \frac{x_c}{12} = 0.0725 \ kmole/kg \ fuel$$

$$M_{H_2O} = \frac{x_H}{2} = 0.063 \ kmole/kg \ fuel$$

$$M_{O_2} = 0.21(\lambda - 1)\left(\frac{A}{F}\right)_s^v = 0.0363 \ kmole/kg \ fuel$$

$$M_{N_2} = 0.79\lambda\left(\frac{A}{F}\right)_s^v = 0.5275 \ kmole/kg \ fuel$$

Combustion Eq. (C.1) is solved, assuming a value of T_4 and then finding the internal energies for the reactants and products at points 2 and 4 of the cycle (from the polynomials in Table 2.7 in Chapter 2 or from Table A.2 in Appendix A) and repeating the process until the equation is balanced. In the case of the current task, this process leads to a value of the maximum cycle temperature $T_4 = 2190 \ K$.

The maximum cycle pressure is

$$p_3 = p_4 = \alpha p_2 = 1.5 \times 5056.8 = 7585.2 \ kPa \ (75.85 \ bar)$$

Since $p_3 < 80 \ bar$, the assumption of $\alpha = 1.5$ is valid.

Knowing α and having calculated T_4, the corresponding value of the expansion ratio during the constant-pressure expansion process β can be calculated from Eq. (6.9):

$$\beta = \frac{\psi}{\alpha}\frac{T_4}{T_2} = \frac{1.0439 \times 2190}{1.5 \times 930} = 1.6386$$

C.1.4 Expansion Process

The expansion ratio is calculated from

$$\delta = \frac{\varepsilon}{\beta} = 10.92$$

The pressure and temperature at the end of the expansion process are

$$p_5 = \frac{p_4}{\delta^{n_2}} = \frac{7585.2}{10.92^{1.26}} = 373 \ kPa$$

$$T_5 = \frac{T_4}{\delta^{n_2-1}} = \frac{2190}{10.92^{0.26}} = 1176.2 \ K$$

C.1.5 Performance Parameters

The mean effective pressure of the dual cycle is given by Eq. (6.13):

$$mep_{(CI)} = p_1 \frac{\varepsilon^{n_1}}{(\varepsilon - 1)} \left[\alpha(\beta - 1) + \frac{\alpha\beta}{n_2 - 1} \left(1 - \frac{1}{\delta^{n_2-1}} \right) - \frac{1}{n_1 - 1} \left(1 - \frac{1}{\varepsilon^{n_1-1}} \right) \right]$$

from which

$$mep_{(CI)} = 1059.2 \ kPa$$

From Eq. (8.4),

$$imep_{(CI)} = \sigma mep_{(CI)} = 0.925 \times 1089 = 979.7 \ kPa$$

The brake mean effective pressure is

$$bmep_{(CI)} = imep_{(CI)} - fmep_{(CI)}$$

The frictional mean pressure can be calculated from Eq. (8.36) and data in Table 8.4:

$$fmep = 0.105 + 0.012 C_p \ MP$$

where

$$C_p = SN/30 \ m/s$$

The procedure used here is to assume a value of the mean piston speed C_p (say, $10 \, m/s$) and then adjust the value at the end following the determination of the engine displacement iV_s. The mean piston speed is eventually adjusted to the value of $10.8 \, m/s$. Using this value, the *fmep* is found to be equal to $234.6 \, kPa$. Hence, $bmep_{(CI)} = 979.7 - 234.6 = 745 \, kPa$.

The mechanical efficiency is

$$\eta_m = \frac{bmep}{imep} = \frac{745}{979.7} = 0.76 \ (76\%)$$

The indicated efficiency is calculated using Eq. (8.18):

$$\eta_i = \frac{(A/F)_s^g \ \lambda(imep)}{H_l} \frac{}{\rho_a \eta_v} = \frac{14.45 \times 1.5 \times 979.7}{43.2 \times 1000 \times 1.1692 \times 0.83} = 0.4727 \ (47.27\%)$$

The indicated specific fuel consumption from Eq. (8.16b) is

$$isfc = 3.6 \times 10^6 \times \frac{\rho_a}{(A/F)_s^g \ \lambda(imep)} \frac{\eta_v}{} = \frac{3.6 \times 10^6 \times 1.1692 \times 0.83}{14.45 \times 1.4 \times 979.7} = 0.1763 \ kg/kW.h$$

The brake specific fuel consumption is then

$$bsfc = \frac{isfc}{\eta_m} = \frac{0.1763}{0.76} = 0.232 \ kg/kW.h$$

The brake efficiency is

$$\eta_b = \eta_m \eta_i = 0.76 \times 0.4727 = 0.359 \ (35.9\%)$$

Table C.3 Calculated engine speed characteristics.

N, rpm	1400	1600	1800	2000	2200	2400	2600	2800
W_b kW	41.4	47.5	53.2	58.3	62.7	66.2	68.7	70.0
bsfc, kg/kW. h	0.237	0.229	0.224	0.221	0.220	0.221	0.225	0.232

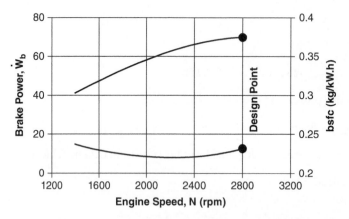

Figure C.2 Estimated speed characteristics of a direct injection CI engine.

The engine swept volume can be calculated from Eq. (8.30):

$$iV_s = \frac{2\dot{W}_b}{N(bmep)} = \frac{120 \times 70}{2800 \times 745} = 4.023 \times 10^{-3} \ m^3 \ (4.023 \ litre)$$

where N is in rev/s.

The swept volume per cylinder is 1.006 l.
The engine stroke can be calculated from $C_p = SN/30$ using the final value of $C_p = 10.8 \ m/s$:

$$S = \frac{30C_p}{N} = 0.1157 \ m \ (115.7 \ mm)$$

The engine bore (cylinder diameter) is then

$$D = \sqrt{\frac{4V_s}{\pi S}} = \sqrt{\frac{4 \times 1.006 \times 10^{-3}}{0.1157\pi}} = 0.1052 \ m \ (105.2 \ mm)$$

The engine power and brake specific fuel consumption for speeds other than the design-point (rated) speed estimated from Eqs. (8.34) and (8.35) together with data from Tables 8.2 and 8.3 are shown in Table C.3 and Figure C.2 (the shaded area in the table denotes the design point).

Appendix D

Equations for the Thermal Efficiency and Specific Work of Theoretical Gas Turbine Cycles

Table D.1 Thermal efficiency and specific output work for the ideal air-standard cycle configurations in Chapter 9.

Cycle type	Cycle configuration	Thermal efficiency (η_{th})	Specific work ($w/c_p T_1$)
Air-standard	Brayton	$\eta_{th} = 1 - \dfrac{1}{r_c^{(\gamma-1)/\gamma}}$	$\dfrac{w}{c_p T_1} = a\left(1 - \dfrac{1}{b}\right) - b + 1$
	Heat exchange	$\eta_{th} = 1 - \dfrac{b}{a}$	$\dfrac{w}{c_p T_1} = a\left(1 - \dfrac{1}{b}\right) - b + 1$
	Reheat	$\eta_{th} = \dfrac{2a - \dfrac{2a}{\sqrt{b}} - b + 1}{2a - \dfrac{a}{\sqrt{b}} - b}$	$\dfrac{w}{c_p T_1} = \left[2a - \dfrac{2a}{\sqrt{b}} - b + 1\right]$
	Intercooling	$\eta_{th} = \dfrac{a\left(1 - \dfrac{1}{b}\right) - 2\sqrt{b} + 2}{a - \sqrt{b}}$	$\dfrac{w}{c_p T_1} = a\left(1 - \dfrac{1}{b}\right) - 2\sqrt{b} + 2$
	Heat exchange and reheat	$\eta_{th} = \dfrac{\left(2a - \dfrac{2a}{\sqrt{b}} - b + 1\right)}{2a\left(1 - \dfrac{1}{\sqrt{b}}\right)}$	$\dfrac{w}{c_p T_1} = \left(2a - \dfrac{2a}{\sqrt{b}} - b + 1\right)$
	Heat exchange and intercooling	$\eta_{th} = \dfrac{a\left(1 - \dfrac{1}{b}\right) - 2\sqrt{b} + 2}{a\left(1 - \dfrac{1}{b}\right)}$	$\dfrac{w}{c_p T_1} = a\left(1 - \dfrac{1}{b}\right) - 2\sqrt{b} + 2$
	Heat exchange, reheat, and intercooling	$\eta_{th} = \dfrac{\left(a - \dfrac{a}{\sqrt{b}} - \sqrt{b} + 1\right)}{a\left(1 - \dfrac{1}{\sqrt{b}}\right)}$	$\dfrac{w}{c_p T_1} = \left(2a - \dfrac{2a}{\sqrt{b}} - 2\sqrt{b} + 2\right)$

Fundamentals of Heat Engines: Reciprocating and Gas Turbine Internal Combustion Engines, First Edition. Jamil Ghojel.
© 2020 John Wiley & Sons Ltd. This Work is a co-publication between John Wiley & Sons Ltd and ASME Press.
Companion website: www.wiley.com/go/JamilGhojel_Fundamentals of Heat Engines

Table D.2 Thermal efficiency and specific output work for the irreversible air-standard cycle configurations in Chapter 10.

Cycle type	Cycle configuration	Thermal efficiency (η_{th})	Specific work ($w/c_p T_1$)
Irreversible air-standard	Simple	$$\eta_{th} = \frac{a\left(1 - \frac{1}{(zb)^{\eta_{pt}}}\right) - b^{1/\eta_{pc}} + 1}{a - b^{1/\eta_{pc}}}$$	$$\frac{w}{c_p T_1} = a\left(1 - \frac{1}{(zb)^{\eta_{pt}}}\right) - b^{1/\eta_{pc}} + 1$$
	Heat exchange	$$\eta_{th} = \frac{a\left(1 - \frac{1}{(zb)^{\eta_{pt}}}\right) - b^{1/\eta_{pc}} + 1}{a - (1-\epsilon)b^{1/\eta_{pc}} - \frac{\epsilon a}{(zb)^{\eta_{pt}}}}$$	$$\frac{w}{c_p T_1} = a\left(1 - \frac{1}{(zb)^{\eta_{pt}}}\right) - b^{1/\eta_{pc}} + 1$$
	Reheat	$$\eta_{th} = \frac{2a\left(1 - \frac{1}{(zb)^{\eta_{pt}/2}}\right) - b^{1/\eta_{pc}} + 1}{2a - b^{1/\eta_{pc}} - \frac{a}{(zb)^{\eta_{pt}/2}}}$$	$$\frac{w}{c_p T_1} = 2a\left(1 - \frac{1}{(zb)^{\eta_{pt}/2}}\right) - b^{1/\eta_{pc}} + 1$$
	Intercooling	$$\eta_{tt} = \frac{a\left(1 - \frac{1}{(zb)^{\eta_{pt}}}\right) - 2b^{1/2\eta_{pc}} + 2}{a - b^{1/2\eta_{pc}}}$$	$$\frac{w}{c_p T_1} = a\left(1 - \frac{1}{(zb)^{\eta_{pt}}}\right) - 2b^{1/2\eta_{pc}} + 2$$
	Heat exchange and reheat	$$\eta_{th} = \frac{2a\left(1 - \frac{1}{(zb)^{\eta_{pt}/2}}\right) - b^{1/\eta_{pc}} + 1}{2a - (1-\epsilon)b^{1/\eta_{pc}} - \frac{a}{(zb)^{\eta_{pt}/2}}(1+\epsilon)}$$	$$\frac{w}{c_p T_1} = 2a\left(1 - \frac{1}{(zb)^{\eta_{pt}/2}}\right) - b^{1/\eta_{pc}} + 1$$
	Heat exchange and intercooling	$$\eta_{th} = \frac{a\left(1 - \frac{1}{(zb)^{\eta_{pt}}}\right) - 2b^{1/2\eta_{pc}} + 2}{a - (1-\epsilon)b^{1/2\eta_{pc}} - \frac{\epsilon a}{(zb)^{\eta_{pt}}}}$$	$$\frac{w}{c_p T_1} = a\left(1 - \frac{1}{(zb)^{\eta_{pt}}}\right) - 2b^{1/2\eta_{pc}} + 2$$
	Heat exchange, reheat, and intercooling	$$\eta_{th} = \frac{2a\left(1 - \frac{1}{(zb)^{\eta_{pt}/2}}\right) - 2b^{1/2\eta_{pc}} + 2}{2a - (1-\epsilon)b^{1/2\eta_{pc}} - \frac{a}{(zb)^{\eta_{pt}/2}}(1+\epsilon)}$$	$$\frac{w}{c_p T_1} = 2a\left(1 - \frac{1}{(zb)^{\eta_{pt}/2}}\right) - 2b^{1/2\eta_{pc}} + 2$$

Nomenclature

$a = T_{max}/T_{min}$, $b = r_c^{(\gamma-1)/\gamma}$

$$z = \left(\frac{1-x}{1+y}\right)^{(\gamma-1)/\gamma}$$

$x = \Delta p_{23}/p_1$, $y = \Delta p_{41}/p_1$

Heat exchanger (HE) effectiveness ϵ:

$$\epsilon = \frac{HE\ exit\ temperature - Compressor\ exit\ temperature}{Turbine\ exit\ temperature - Compressor\ exit\ temperature}$$

η_{pc}, η_{pt}: polytropic efficiencies of the compressor and turbine

r_c: compressor polytropic efficiency

γ: ratio of specific heats of the fluid (usually air)

c_p: specific heat at constant pressure of the fluid

Index

Fundamentals of Heat Engines: Reciprocating and Gas Turbine Internal Combustion Engines, First Edition. Jamil Ghojel.
© 2020 John Wiley & Sons Ltd. This Work is a co-publication between John Wiley & Sons Ltd and ASME Press.
Companion website: www.wiley.com/go/JamilGhojel_Fundamentals of Heat Engines